Toxicological Risks of Selected Flame-Retardant Chemicals

Subcommittee on Flame-Retardant Chemicals

Committee on Toxicology

Board on Environmental Studies and Toxicology

Commission on Life Sciences

National Research Council

NATIONAL ACADEMY PRESS
Washington, D.C.

NATIONAL ACADEMY PRESS 2101 Constitution Ave., N.W. Washington, D.C. 20418

NOTICE: The project that is the subject of this report was approved by the Governing Board of the National Research Council, whose members are drawn from the councils of the National Academy of Sciences, the National Academy of Engineering, and the Institute of Medicine. The members of the committee responsible for the report were chosen for their special competences and with regard for appropriate balance.

This project between the National Academy of Sciences and U.S. Consumer Product Safety Commission (CPSC) was supported by Contract No. CPSC-C-99-1152. This report was prepared by the National Research Council as a result of a direction in the fiscal year 1999 appropriations for the United States CPSC. The views expressed in the report do not necessarily represent the views of CPSC.

International Standard Book Number 0-309-07047-3

Additional copies of this report are available from:

National Academy Press
2101 Constitution Ave., NW
Box 285
Washington, DC 20055

800-624-6242
202-334-3313 (in the Washington metropolitan area)
http://www.nap.edu

Printed in the United States of America

THE NATIONAL ACADEMIES

National Academy of Sciences
National Academy of Engineering
Institute of Medicine
National Research Council

The **National Academy of Sciences** is a private, nonprofit, self-perpetuating society of distinguished scholars engaged in scientific and engineering research, dedicated to the furtherance of science and technology and to their use for the general welfare. Upon the authority of the charter granted to it by the Congress in 1863, the Academy has a mandate that requires it to advise the federal government on scientific and technical matters. Dr. Bruce M. Alberts is president of the National Academy of Sciences.

The **National Academy of Engineering** was established in 1964, under the charter of the National Academy of Sciences, as a parallel organization of outstanding engineers. It is autonomous in its administration and in the selection of its members, sharing with the National Academy of Sciences the responsibility for advising the federal government. The National Academy of Engineering also sponsors engineering programs aimed at meeting national needs, encourages education and research, and recognizes the superior achievements of engineers. Dr. William A. Wulf is president of the National Academy of Engineering.

The **Institute of Medicine** was established in 1970 by the National Academy of Sciences to secure the services of eminent members of appropriate professions in the examination of policy matters pertaining to the health of the public. The Institute acts under the responsibility given to the National Academy of Sciences by its congressional charter to be an adviser to the federal government and, upon its own initiative, to identify issues of medical care, research, and education. Dr. Kenneth I. Shine is president of the Institute of Medicine.

The **National Research Council** was organized by the National Academy of Sciences in 1916 to associate the broad community of science and technology with the Academy's purposes of furthering knowledge and advising the federal government. Functioning in accordance with general policies determined by the Academy, the Council has become the principal operating agency of both the National Academy of Sciences and the National Academy of Engineering in providing services to the government, the public, and the scientific and engineering communities. The Council is administered jointly by both Academies and the Institute of Medicine. Dr. Bruce M. Alberts and Dr. William A. Wulf are chairman and vice chairman, respectively, of the National Research Council.

SUBCOMMITTEE ON FLAME-RETARDANT CHEMICALS

DONALD E. GARDNER (*Chair*), Inhalation Toxicology Associates, Inc., Raleigh, North Carolina
JOSEPH F. BORZELLECA, Virginia Commonwealth University, Richmond, Virginia
DAVID W. GAYLOR, National Center for Toxicological Research, Jefferson, Arkansas
SIDNEY GREEN, Howard University, Washington, D.C.
RICHARD HORROCKS, Bolton Institute, Bolton, U.K.
MICHAEL A. JAYJOCK, Rohm and Haas Company, SpringHouse, Pennsylvania
SAMUEL KACEW, University of Ottawa, Ottawa, Ontario, Canada
JAMES N. MCDOUGAL, Geo-Centers, Inc., Wright-Patterson Air Force Base, Ohio
RICHARD K. MILLER, University of Rochester, Rochester, New York
ROBERT SNYDER, Rutgers University College of Pharmacy, Piscataway, New Jersey
GARY C. STEVENS, University of Surrey, Guildford, Surrey, U.K.
ROBERT G. TARDIFF, The Sapphire Group, Inc., Bethesda, Maryland
MARY E. VORE, University of Kentucky, Lexington, Kentucky

NRC Staff

KULBIR BAKSHI, Project Director
EILEEN ABT, Research Associate
DARRYL ARFSTEN, Research Associate
MICHELLE CATLIN, Post-Doctoral Research Associate
KATHRINE IVERSON, Information Specialist
EVELYN SIMEON, Senior Program Assistant

COMMITTEE ON TOXICOLOGY

BOARD ON ENVIRONMENTAL STUDIES AND TOXICOLOGY

GORDON ORIANS (*Chair*), University of Washington, Seattle, Washington
DONALD MATTISON (*Vice Chair*), March of Dimes, White Plains, New York
DAVID ALLEN, University of Texas, Austin, Texas
INGRID C. BURKE, Colorado State University, Fort Collins, Colorado
WILLIAM L. CHAMEIDES, Georgia Institute of Technology, Atlanta, Georgia
JOHN DOULL, University of Kansas Medical Center, Kansas City, Kansas
CHRISTOPHER B. FIELD, Carnegie Institute of Washington, Stanford, California
JOHN GERHART, University of California, Berkeley, California
J. PAUL GILMAN, Celera Genomics, Rockville, Maryland
BRUCE D. HAMMOCK, University of California, Davis, California
MARK HARWELL, University of Miami, Miami, Florida
ROGENE HENDERSON, Lovelace Respiratory Research Institute, Albuquerque, New Mexico
CAROL HENRY, Chemical Manufacturers Association, Arlington, Virginia
BARBARA HULKA, University of North Carolina, Chapel Hill, North Carolina
JAMES F. KITCHELL, University of Wisconsin, Madison, Wisconsin
DANIEL KREWSKI, University of Ottawa, Ottawa, Ontario
JAMES A. MACMAHON, Utah State University, Logan, Utah
MARIO J. MOLINA, Massachusetts Institute of Technology, Cambridge, Massachusetts
CHARLES O'MELIA, Johns Hopkins University, Baltimore, Maryland
WILLEM F. PASSCHIER, Health Council of the Netherlands
KIRK SMITH, University of California, Berkeley, California
MARGARET STRAND, Oppenheimer Wolff Donnelly & Bayh, LLP, Washington, D.C.
TERRY F. YOSIE, Chemical Manufacturers Association, Arlington, Virginia

Senior Staff

JAMES J. REISA, Director
DAVID J. POLICANSKY, Associate Director and Senior Program Director for Applied Ecology
CAROL A. MACZKA, Senior Program Director for Toxicology and Risk Assessment
RAYMOND A. WASSEL, Senior Program Director for Environmental Sciences and Engineering
KULBIR BAKSHI, Program Director for the Committee on Toxicology
LEE R. PAULSON, Program Director for Resource Management
ROBERTA M. WEDGE, Program Director for Risk Analysis

COMMISSION ON LIFE SCIENCES

OTHER REPORTS OF THE BOARD ON ENVIRONMENTAL STUDIES AND TOXICOLOGY

Strengthening Science at the U.S. Environmental Protection Agency: Research Management and Peer Review Practice (2000)
Scientific Frontiers in Developmental Toxicology and Risk Assessment (2000)
Modeling Mobile-Source Emissions (2000)
Copper in Drinking Water (2000)
Ecological Indicators for the Nation (2000)
Waste Incineration and Public Health (1999)
Hormonally Active Agents in the Environment (1999)
Research Priorities for Airborne Particulate Matter: I. Immediate Priorities and a Long-Range Research Portfolio (1998); II. Evaluating Research Progress and Updating the Portfolio (1999)
Ozone-Forming Potential of Reformulated Gasoline (1999)
Risk-Based Waste Classification in California (1999)
Arsenic in Drinking Water (1999)
Brucellosis in the Greater Yellowstone Area (1998)
The National Research Council's Committee on Toxicology: The First 50 Years (1997)
Toxicologic Assessment of the Army's Zinc Cadmium Sulfide Dispersion Tests (1997)
Carcinogens and Anticarcinogens in the Human Diet (1996)
Upstream: Salmon and Society in the Pacific Northwest (1996)
Science and the Endangered Species Act (1995)
Wetlands: Characteristics and Boundaries (1995)
Biologic Markers (5 reports, 1989-1995)
Review of EPA's Environmental Monitoring and Assessment Program (3 reports, 1994-1995)
Science and Judgment in Risk Assessment (1994)
Ranking Hazardous Waste Sites for Remedial Action (1994)
Pesticides in the Diets of Infants and Children (1993)
Issues in Risk Assessment (1993)
Setting Priorities for Land Conservation (1993)
Protecting Visibility in National Parks and Wilderness Areas (1993)
Dolphins and the Tuna Industry (1992)
Hazardous Materials on the Public Lands (1992)
Science and the National Parks (1992)
Animals as Sentinels of Environmental Health Hazards (1991)
Assessment of the U.S. Outer Continental Shelf Environmental Studies Program, Volumes I-IV (1991-1993)
Human Exposure Assessment for Airborne Pollutants (1991)
Monitoring Human Tissues for Toxic Substances (1991)
Rethinking the Ozone Problem in Urban and Regional Air Pollution (1991)
Decline of the Sea Turtles (1990)

Copies of these reports may be ordered from the National Academy Press
(800) 624-6242
(202) 334-3313
www.nap.edu

Other Reports of the Committee on Toxicology

Review of the U.S. Navy's Exposure Standard for Manufactured Vitreous Fibers (2000)

Submarine Exposure Guidance Levels for Selected Hydrofluorocarbons: HFC-236fa, HFC-23, and HFC-404a (2000)

Health Risk Assessment of Selected Flame-Retardant Chemicals (2000)

Review of the U.S. Army's Health Risk Assessments for Oral Exposure to Six Chemical-Warfare Agents (1999)

Toxicity of Military Smokes and Obscurants, Volume 1(1997), Volume 2 (1999), Volume 3 (1999)

Assessment of Exposure-Response Functions for Rocket-Emission Toxicants (1998)

Review of a Screening Level Risk Assessment for the Naval Air Facility at Atsugi, Japan (Letter Report) (1998)

Toxicity of Alternatives to Chlorofluorocarbons: HFC-134a and HCFC-123 (1996)

Permissible Exposure Levels for Selected Military Fuel Vapors (1996)

Spacecraft Maximum Allowable Concentrations for Selected Airborne Contaminants, Volume 1 (1994), Volume 2 (1996), Volume 3 (1996), Volume 4 (2000)

Preface

THE U.S. Consumer Product Safety Commission (CPSC) is considering a possible flammability standard for upholstered furniture to reduce the incidence of death, injury, and property loss from fires started by small open flames of matches, cigarette lighters, and other sources of ignition of upholstered furniture in U.S. residences. The most likely method of achieving this standard is to treat furniture upholstery with flame-retardant (FR) chemicals. However, concern has been expressed that potential health risks have been inadequately studied for many of the FR chemicals likely to be used. Some of them have been shown to cause a range of toxicity, including cancer, neurological, reproductive, and developmental effects.

In its fiscal 1999 appropriations report, Congress directed CPSC to arrange for an independent study by the National Research Council (NRC) to conduct toxicological assessments for FR chemicals that are likely to be used as FRs for furniture upholstery. The NRC assigned the project to the Committee on Toxicology, which, in turn, established the Subcommittee on Flame-Retardant Chemicals.

This report evaluates toxicological, epidemiological, and exposure data on the 16 specified FR chemicals, and characterizes risks to human health from exposure to furniture upholstery treated with such chemicals. It is hoped that the findings of the report will be useful to CPSC and Congress in making decisions regarding the safe use of FR chemicals in upholstered furniture fabric.

The subcommittee was assisted by many people, including those who presented valuable information and documents during the subcommittee's public sessions at the National Academy of Sciences in Washington, DC, on July 30-31 and October 26-27,1999, and in Irvine, CA, on September 22-24, 1990:

Elizabeth Anderson (Sciences International, Inc.)
Michael Babich (Consumer Product Safety Commission)
Roger Berkley (American Textile Manufacturers Institute's Upholstery Fabric Committee)
Patricia Bittner (Consumer Product Safety Commission)
Joseph Carra (EPA's Office of Pollution Prevention and Toxics)
Marcia Hardy (Albemarle Corporation)
Judith MacGregor (Toxicology Consultant Services)
Ronald Medford, (Consumer Product Safety Commission)
George Miller (National Association of State Fire Marshalls)
Vincent Piccirillo (NPC, Inc.)
Cecilia Powell (Albright & Wilson)
Richard Rose (Great Lakes Chemical Corporation)
The Honorable Roger Wicker (U.S. Representative from Mississippi)
Joseph Ziolkowski (Upholstered Action Furniture Council)

This report has been reviewed by individuals chosen for their diverse perspectives and technical expertise in accordance with procedures for reviewing NRC reports approved by the NRC's Report Review Committee. The purpose of this independent review is to provide candid and critical comments that will assist the NRC in making the published report as sound as possible and to ensure that the report meets institutional standards for objectivity, evidence, and responsiveness to the study charge. The review comments and draft manuscript remain confidential to protect the integrity of the deliberative process. We wish to thank the following individuals, who are neither officials nor employees of the NRC, for their participation in the review of this report: Deepak Bhalla, Wayne State University; Jack Dean, Sanofi-Synthelabo, Inc.; John Doull, University of Kansas; Paul Foster, Chemical Industry Institute of Toxicology; Bruce Fowler, University of Maryland; Robert Goyer, University of Western Ontario; Rogene Henderson, Lovelace Respiratory Research Institute; Loren Koller, Oregon State University; Daniel Krewski, University of Ottawa; Thomas McKone, University of California, Berkeley; Günter Oberdörster, University of Rochester; and Vincent Piccirillo, NPC Incorporated.

The individuals listed above have provided many constructive comments and suggestions. It must be emphasized, however, that responsibility for the final content of this report rests entirely with the authoring committee and the NRC.

The subcommittee gratefully acknowledges the assistance of its consultants, including the staff of Toxicology Excellence for Risk Assessment—Michael Dourson, Kenneth Poirier, Lynne Haber; Syracuse Research Corporation staff—Gary Diamond, Margaret Fransen, Patricia McGinnis, Marc Odin, and Randi Starmer; and Oak Ridge National Laboratory staff—Po-Yung Lu, Sylvia Milanez, Annetta Watson, and Robert Young, for drafting most of the initial health risk assessments, which were reviewed and modified by the subcommittee.

Appreciation is also extended to Edmund Crouch for his assistance with exposure assessments; Judith Fauls for her editorial assistance; and Charles Auer of EPA for providing information on FRs; and Marcia Hardy, Judith MacGregor, and Patricia Adair for providing the subcommittee with numerous current references. Appreciation is also extended to Alan Mann, University of Surrey, U.K.; Stewart Miller and Patricia Martin, Albright & Wilson; William Walt, DaimlerChrysler Corporation; Carl D'Ruiz and James Plautz, Ciba Specialty Chemicals Corporation; Robert J. Fensterheim, Chlorinated Paraffins Industry Association; and Philip Strong, U.S. Borax Incorporated for providing data on FR chemicals.

The subcommittee also thanks the staff of the sponsor of this study, the U.S. Consumer Product Safety Commission, for their assistance and cooperation, including Ronald Medford, assistant executive director for hazard identification and reduction; Marry Ann Danello, associate executive director for health sciences; Lori Saltzman, director, Division of Health Sciences; Patricia Bittner, technical contract officer, Jacqueline Ferrante, Kristina Hatlelid, and Michael Babich, Division of Health Sciences; Dale Ray, Directorate for Economic Analysis (manager, Upholstered Furniture Project); Peter Nerret, contract officer, and Robert Frost, Division of Procurement Services.

The subcommittee was ably assisted by the staff of the Subcommittee on Flame-Retardant Chemicals—especially Kulbir S. Bakshi, project director; Eileen Abt and Darryl Arfsten, research associates; Michelle Catlin, post-doctoral research associate; and Judy Estep, Pamela Friedman, Stephanie Parker, and Evelyn Simeon, administrative staff. These staff members merit special recognition for their thoughtful contributions and extraordinary efforts in producing the report, and for the many extra hours they worked to complete the report. The subcommittee is also thankful to the following NRC officials for their contributions to the project—Warren R. Muir (Executive Director, Commission on Life Sciences), James J. Reisa (Director, Board on Environmental Studies and Toxicology [BEST]), David Policansky (Associate Director, [BEST]), and Carol Maczka (Director, Toxicology and Risk Assessment Program [BEST]).

Finally, we would like to express our thanks and admiration to the members of the subcommittee for their dedicated efforts throughout the development of this report.

Donald E. Gardner
Chair, Subcommittee on Flame-Retardant Chemicals
Bailus Walker, Jr.
Chair, Committee on Toxicology

Contents

Toxicological Risks of Selected Flame-Retardant Chemicals

Executive Summary

IGNITION of upholstered furniture by small open flames from matches, cigarette lighters, and candles is one of the leading causes of residential-fire deaths in the United States. On average, each year since 1990, about 90 deaths (primarily of children), 440 injuries, and property losses amounting to 50 million dollars have resulted from such fires.

The U.S. Consumer Product Safety Commission (CPSC), an independent federal regulatory agency, is required to protect the public from unreasonable risks of injury and death associated with consumer products. In 1993, the National Association of State Fire Marshals petitioned CPSC to issue a performance-based flammability standard for residential upholstered furniture to reduce the risk of residential fires. If such a standard were promulgated, most residential upholstery fabric would be treated with flame-retardant (FR) chemicals.

It is estimated that in the United States, FRs would be applied to as much as 600 million square yards of upholstery fabrics each year. Over the lifetime of the furniture, many consumers might be exposed to FR-treated fabric. Because some FRs are known to have toxic properties and there is a potential for exposure of millions of people to these chemicals, some have expressed concern about the use of these chemicals until they are shown to be safe.

THE CHARGE TO THE SUBCOMMITTEE

In its fiscal year 1999 appropriations report for CPSC, Congress requested an independent study by the National Academy of Sciences' National Research

Council (NRC) concerning health risks posed by exposure to FR chemicals that are likely to be used in residential upholstered furniture to meet a flammability standard that CPSC is considering. CPSC, with the help of the industry, identified the following 16 chemicals (or chemical classes) that became the focus of this NRC study: (1) hexabromocyclododecane, (2) decabromodiphenyl oxide, (3) alumina trihydrate, (4) magnesium hydroxide, (5) zinc borate, (6) calcium and zinc molybdates, (7) antimony trioxide, (8) antimony pentoxide and sodium antimonate, (9) ammonium polyphosphates, (10) phosphonic acid, (3-{[hydroxymethyl]amino}-3-oxopropyl)-dimethyl ester, (11) organic phosphonates, (12) tris(monochloropropyl) phosphate, (13) tris(1,3-dichloropropyl-2) phosphate, (14) aromatic phosphate plasticisers, (15) tetrakis (hydroxymethyl) hydronium salts, and (16) chlorinated paraffins. The NRC assigned the project to the Committee on Toxicology (COT) of the Board on Environmental Studies and Toxicology. COT convened the Subcommittee on Flame-Retardant Chemicals, which prepared this report. Subcommittee members were chosen for their recognized expertise in toxicology, epidemiology, pharmacology, chemistry, exposure assessment, risk assessment, and biostatistics.

The subcommittee was charged to review the toxicological and exposure data on the above 16 FR chemicals to assess potential health risks to consumers and the general population resulting from potential exposure to these chemicals in residential furniture. The subcommittee was also asked to identify data gaps and make recommendations for future research. The subcommittee was not charged or able, with the available data and resources, to evaluate adverse effects from occupational exposures, or the potential ecological effects that might result from the disposal of household furniture. The subcommittee also was not charged to compare FRs for efficacy, or to evaluate the cost, technology, exposure-standard achievability, or the benefits of using FRs on upholstered furniture to reduce fire risk, or to consider the toxicity of combustion products of FR materials from fires involving treated upholstery fabric. Those issues were also beyond the scope of the subcommittee's charge, expertise, and resources.

THE SUBCOMMITTEE'S APPROACH

To help the subcommittee gather information, the NRC commissioned papers by expert consultants on various FR chemicals, reviewed the toxicity assessments by the CPSC staff on 16 chemicals, and held public meetings at which representatives of CPSC, the U.S. Environmental Protection Agency, Congress, the National Association of State Fire Marshals, and industrial groups made presentations on issues of critical importance to the subcommittee's task. Some of the speakers also provided technical documents on FR chemicals.

In characterizing the human health risks from dermal, oral, and inhalation exposures to each of the 16 FR chemicals, the subcommittee generally followed the risk-assessment paradigm first articulated by the NRC in its 1983 report *Risk Assessment in the Federal Government: Managing the Process*. The subcommittee evaluated data relating to key elements of the risk paradigm, as discussed below.

Hazard Identification

In the hazard-identification phase of each assessment, determinations were made as to whether causal relationships exist between the dose of an FR chemical and an adverse health effect. To identify adverse effects associated with an FR chemical, the subcommittee reviewed human (epidemiological studies, clinical observations, and case reports) and laboratory animal data on neurotoxicity, immunotoxicity, reproductive and developmental toxicity, organ toxicity, dermal and pulmonary toxicity, carcinogenicity, and other local and systemic effects. In vitro data were also reviewed to determine the potential for genotoxicity as well as other toxic effects and to understand the mechanisms of toxic action. Toxicokinetic studies were reviewed to understand the absorption, distribution, metabolism, and excretion of the FR chemicals.

Dose-Response Assessment

In the dose-response-assessment phase, the subcommittee reviewed the relationships between increases in the dose of an FR chemical and changes in the magnitude of the incidence or severity of toxic effects. For all types of toxic effects other than cancer, the procedure used to quantify the dose-response relationships involved estimating the highest dose at which no adverse effects were observed. This exposure level is called the no-observed-adverse-effect level (NOAEL). In contrast, the lowest-observed-adverse-effect level (LOAEL) is the lowest dose at which a statistically or biologically significant increase in an adverse effect was reported. The NOAEL is the highest exposure level below the LOAEL at which no statistically or biologically significant increase was observed in the frequency or magnitude of an adverse effect when compared with a control group. The NOAEL or LOAEL (A LOAEL was used only when a NOAEL could not be identified) was then divided by a composite of one or more uncertainty factors (UFs) to obtain a reference dose (RfD), which is an estimate of lifetime daily dose that is believed to have a reasonable certainty of no harm; for inhalation exposures, similarly defined reference concentrations (RfCs) were calculated or mathematically scaled from the RfDs. The UFs were applied to account for interspecies and intraspecies variation, differences in exposure durations and routes, use of a LOAEL when a NOAEL was not identified, adequacy of experiments, and data quality.

For some types of toxic effects, notably most cancers, the subcommittee conservatively assumed that no threshold for a dose-response relationship exists or that, if one does exist, it is very low and cannot be reliably identified. Therefore, the subcommittee's risk-estimation procedure for carcinogens was different from that for noncarcinogens. In the risk-estimation procedure for carcinogens, the relationship between the incidence of cancer and the dose of a chemical reported in an epidemiological study or an experimental animal study was extrapolated linearly to much lower doses at which humans might be exposed. This procedure overestimated conservatively the excess lifetime risk of cancer resulting from lifetime exposure to a chemical at a particular dose rate. This procedure does not provide a "safe" dose with an estimated risk of zero (except at zero dose), although at sufficiently low doses, the estimated risk becomes very low and is regarded to have no public-health significance.

The relationship between average lifetime daily dose and tumor incidence was fitted to a mathematical model to predict the incidence at low doses. Several such models are widely used. The subcommittee used the linearized multistage no-threshold model because it provides a conservative risk estimate (i.e., it tends to overestimate, rather than underestimate the risk). The subcommittee applied a statistical confidence-limit procedure to this risk estimate to generate an upper bound (i.e., cancer potency factor) on cancer risk. Although the actual risk cannot be determined, the subcommittee concludes that the actual risk will not exceed the upper bound. The actual risk is also highly likely to be lower than the upper bound, and it might be zero.

Exposure Assessment

Exposure assessment is the third phase in the risk-assessment process. Because exposures to FRs in treated residential furniture fabrics have not been studied, there are no quantitative measurements of exposures under relevant exposure conditions. The subcommittee assumed that human exposure to FR-treated fabric in homes can occur potentially via skin contact, ingestion (specifically for infants or children who might suck or chew on fabric), inhalation of particles generated during abrasion of surface fibers, and inhalation of vapors off-gassing from treated fabric. In estimating exposures to FRs, the subcommittee evaluated three exposure scenarios that involved different exposure routes: dermal, oral, and inhalation. For dermal exposure, the scenario was that of an adult sitting on FR-treated fabric of a couch for a substantial fraction (25%) of the time, with potential exposure over 25% of the upper torso area; clothing was conservatively assumed to provide no barrier to exposure. For ingestion, the scenario was of an infant or a child repeatedly sucking on FR-treated fabric of a chair or couch.

For inhalation, the scenario was of a person spending time in a closed (but ventilated) room containing FR-treated upholstered furniture that shed FRs as small (respirable) particles, or from which FR chemicals evaporate. In all of those exposure scenarios, the subcommittee intentionally overestimated exposures by using extremely conservative assumptions. Those assumptions are discussed in detail in Chapter 3.

Risk Characterization

In the final phase of the risk-assessment process, the subcommittee integrated data and analyses from the other three phases (hazard identification, dose-response assessment, and exposure assessment) to determine the likelihood that individuals might experience adverse effects from the FR chemical under anticipated conditions of exposure.

To characterize the health risk from exposure to a noncarcinogenic chemical, a hazard-index approach was used to judge whether a particular exposure would be likely to present a noncancer toxicological risk. A hazard index was calculated for each chemical by dividing the estimated human dose by the RfD or RfC.

In the absence of adequate human carcinogenicity data for any FR chemical, the subcommittee's approach for estimating cancer risks from exposure to carcinogenic FRs involved the extrapolation of observations of cancer at relatively high doses in laboratory animals to much lower doses anticipated for humans in residential settings. The upper limit on the cancer potency factor extrapolated from animal experiments was multiplied by the estimated lifetime average dose rate to estimate an upper limit on lifetime cancer risk. The subcommittee did not recommend an acceptable cancer risk level for carcinogenic FR chemicals, because that is a regulatory policy question, not a scientific matter. Some regulatory agencies consider as acceptable excess lifetime cancer risks ranging from 1 in 10,000 (1×10^{-4}) to 1 in 1 million (1×10^{-6}) exposed people.

SUMMARY OF TOXICOLOGICAL RISK ASSESSMENTS

The subcommittee's toxicological risk assessments for each of the 16 FR chemicals are presented in Chapters 4-19 and summarized in Table ES-1. Table ES-1 shows the critical toxicity end points used for derivation of RfDs, RfCs, or cancer potency factors, estimated worst-case human exposure levels, hazards indices for noncancer effects, and upper limits on lifetime excess cancer-risk estimates for carcinogenic chemicals.

In the absence of adequate toxicity data to derive dermal RfDs, the subcommittee characterized potential risks from dermal exposures by using oral RfDs to calculate the hazard indices. In addition, sufficient data to derive inhalation RfCs were available for only two FRs. Inhalation RfCs for other FRs were extrapolated from oral RfDs by using typical body weights and breathing rates.

Table ES-1 shows that for most of the 16 candidate FRs, the hazard indices for noncarcinogenic effects are less than 1 for all three routes of exposure. FRs with hazard indices of less than 1 are not likely to pose noncancer health risks even at the worst-case exposure levels. FR chemicals with hazard indices greater than 1 might possibly pose noncancer health risks. However, the subcommittee does not necessarily expect adverse effects at hazard indices slightly greater than 1, given the highly conservative assumptions it used to estimate risks. Carcinogenic risk assessments performed on the FRs that were found to be or likely to be carcinogenic indicate that some of the estimated excess cancer risks may be greater than 1×10^{-6}. However, the subcommittee believes that actual carcinogenic risk is likely to be much lower because of the extremely conservative (high) exposure estimates.

Several of the 16 chemicals were actually chemical classes rather than single compounds. In some of these cases, one chemical of the class was selected as a surrogate: tetrakis (hydroxymethyl) hydronium chloride for the tetrakis (hydroxymethyl) hydronium salts and their compounds; dimethyl hydrogen phosphite for organic phosphonates; and tricresyl phosphate for aromatic phosphate plasticizers. Surrogates were selected on the basis of representativeness of the class, availability of data, and most potent chemical in the class. Conclusions about the class are based on the properties of the surrogate. The risk from other members of the class might be different from the risk from the surrogate.

The subcommittee's use of several UFs in the derivation of RfDs or RfCs and the intentional overestimation of exposure levels reflects a precautionary approach to the protection of public health. Such an approach is commonly practiced, but the subcommittee is aware that there are potential shortcomings in taking such an approach. Overestimating risks from FRs might result in a net adverse effect on public health if the uses of FRs that could reduce the risks of death and injury from fires were avoided because of minor toxicological risks estimated through such conservative assumptions.

TABLE ES-1 Summary of Health Risk Assessments of 16 Flame-Retardant Chemicals

						Estimated Worst-case Human Exposure Levels			Hazard Index[a] for Non-Cancer Effects			Upper Limits on Lifetime Excess Cancer-Risk Estimate[b]		
Flame-Retardant Chemical	Critical Toxicity End Point for Derivation of RfD or RfC	Dermal RfD (mg/kg-d)	Inhalation RfC (mg/m^3)	Oral RfD (mg/kg-d)	Cancer Potency Factor	Dermal (mg/kg-d)	Inhalation[c] (µg/m^3)	Oral (mg/kg-d)	Dermal[d]	Inhalation[e]	Oral	Dermal	Inhalation	Oral
Hexabromo-cyclododecane	Oral: liver toxicity	N/C	N/C	0.2	N/A	1.3×10^{-6}	0.48 (particles) 3.4 (vapors)	2.6×10^{-2}	6.7×10^{-6}	6.8×10^{-4} (particles) 5.0×10^{-3} (vapors)	0.13	N/A	N/A	N/A
Decabromo-diphenyl oxide	Oral: liver toxicity	N/C	N/C	4.0	9.0×10^{-4} per mg/kg-d (oral) 2.6×10^{-7} per µg/m^3 (inhalation)	1.3×10^{-9}	0.48 (particles) 0.38 (vapors)	2.6×10^{-2}	3.3×10^{-10}	3.4×10^{-5} (particles) 2.7×10^{-5} (vapors)	6.5×10^{-3}	1.2×10^{-12}	1.2×10^{-7} (particles) 9.7×10^{-8} (vapors)	6.7×10^{-7}
Alumina trihydrate	Oral: develop-mental toxicity	N/C	N/C	1.5	N/A	5.9×10^{-2}	0.71 (particles) N/C (vapors)	1.6×10^{-3}	3.9×10^{-2}[f]	1.4×10^{-4}	1.0×10^{-3}	N/A	N/A	N/A
Magnesium hydroxide	Oral: derived from tolerable upper limit for Mg-induced diarrhea in humans	N/C	N/C	12.0	N/A	1.7×10^{-3}	0.38 (particles) N/C (vapors)	2.1×10^{-2}	1.4×10^{-4}	9.1×10^{-6}	1.7×10^{-3}	N/A	N/A	N/A
Zinc borate	Oral: develop-mental toxicity	N/C	N/C	0.6	N/A	6.3×10^{-3}	0.19 (particles) N/C (vapors)	1.7×10^{-4}	1.0×10^{-2}	9.1×10^{-5}	2.8×10^{-4}	N/A	N/A	N/A

TABLE ES-1 *(Continued)*

Flame-Retardant Chemical	Critical Toxicity End Point for Derivation of RfD or RfC	Dermal RfD (mg/kg-d)	Inhalation RfC (mg/m^3)	Oral RfD (mg/kg-d)	Cancer Potency Factor	Estimated Worst-case Human Exposure Levels: Dermal (mg/kg-d)	Estimated Worst-case Human Exposure Levels: Inhalation[c] ($\mu g/m^3$)	Estimated Worst-case Human Exposure Levels: Oral (mg/kg-d)	Hazard Index[a] for Non-Cancer Effects: Dermal[d]	Hazard Index[a] for Non-Cancer Effects: Inhalation[c]	Oral	Upper Limits on Lifetime Excess Cancer-Risk Estimate[b]: Dermal	Upper Limits on Lifetime Excess Cancer-Risk Estimate[b]: Inhalation	Upper Limits on Lifetime Excess Cancer-Risk Estimate[b]: Oral
Calcium and zinc molybdates	Oral: increased uric acid levels; Inhalation: degeneration of respiratory epithelium	N/C	2.0×10^{-3}	6.0×10^{-4}	2.6×10^{-5} per $\mu g/m^3$ (inhalation[f])	6.3×10^{-3}	0.19 (particles) N/C (vapors)	1.7×10^{-4}	10	9.5×10^{-2}	0.28	N/A	5.0×10^{-6}	N/A
Antimony trioxide	Oral: liver toxicity; Inhalation: noncancer pulmonary toxicity; lung tumors	N/C	2.0×10^{-4}	0.2	7.1×10^{-4} per $\mu g/m^3$ (inhalation[f])	2.0×10^{-2}	0.24 (particles) N/C (vapors)	5.2×10^{-4}	0.1	1.2	2.6×10^{-3}	N/A	1.7×10^{-4} (particles)	N/A
Antimony pentoxide and sodium antimonate	Inadequate data for any route	N/C	N/C	N/C	N/A	2.0×10^{-2}	0.24 (particles) N/C (vapors)	5.2×10^{-4}	—[g]	—[g]	—[g]	N/A	N/A	N/A
Ammonium polyphosphates	Oral: calcification of the kidney	N/C	N/C	300	N/A	2.2	0.71 (particles) N/C (vapors)	5.9×10^{-2}	7.3×10^{-3}	6.8×10^{-7}	2.0×10^{-4}	N/A	N/A	N/A

Phosphonic acid	Inadequate data for any route	N/C	N/C	N/C	N/A	2.8×10^{-2}	0.35 (particles) UE (vapors)	7.5×10^{-4}	—[h]	—[h]	—[h]	N/A	N/A	N/A
Organic phosphonates (dimethyl hydrogen phosphite)	Oral: lung hyperplasia and alveolar/ bronchiolar adenomas or carcinomas observed	N/C	N/C	0.12	5.4×10^{-3} per mg/kg-d (oral) 1.5×10^{-6} per μg/m^3 (inhalation[f])	2.2	0.72 (particles) UE (vapors)	5.9×10^{-2}	18.3	1.7×10^{-3} (particles)	0.49	6.1 $\times 10^{-2}$	1.1×10^{-6} (particles) 6.6×10^{-4} (vapors)	9.1×10^{-6}
Tris(mono-chloropropyl) phosphates	Inadequate data for any route	N/C	N/C	N/C	N/A	1.5	0.48 (particles) UE (vapors)	4.0×10^{-2}	—[g]	—[g]	—[g]	N/A	N/A	N/A
Tris(1,3-dichloropropyl-2) phosphate	Oral: testicular atrophy and seminal vesicle effects	N/C	N/C	5.0×10^{-3}	6.0×10^{-2} per mg/kg-d (oral) 1.7×10^{-5} per μg/m^3 (inhalation[f])	2.6×10^{-3}	0.48 (particles) UE (vapors)	4.0×10^{-2}	0.52	2.7×10^{-2} (particles)	8.0	1.6 $\times 10^{-4}$	8.2×10^{-6} (particles) UE (vapors)	6.6×10^{-5}
Aromatic phosphate plasticizers (tricresyl phosphate)	Oral: liver lesions and adrenal gland toxicity	N/C	N/C	7.0×10^{-2}	N/A	3.0×10^{-3}	0.48 (particles) 417 (vapors)	4.0×10^{-2}	4.3×10^{-2}	1.9×10^{-3} (particles) 1.7 (vapors)	0.57	N/A	N/A	N/A
Tetrakis hydroxymethyl phosphonium chloride (THPC)	Oral: liver toxicity	N/C	N/C	3.0×10^{-3}	N/A	—[i]	0.43 (particles) UE (vapors)	9.4×10^{-4}	N/C	4.1×10^{-2} (particles)	0.31	N/A	N/A	N/A

TABLE ES-1 *(Continued)*

Flame-Retardant Chemical	Critical Toxicity End Point for Derivation of RfD or RfC	Dermal RfD (mg/kg-d)	Inhalation RfC (mg/m³)	Oral RfD (mg/kg-d)	Cancer Potency Factor	Estimated Worst-case Human Exposure Levels: Dermal (mg/kg-d)	Inhalation[c] (µg/m³)	Oral (mg/kg-d)	Hazard Index[a] for Non-Cancer Effects: Dermal[d]	Inhalation[e]	Upper Limits on Lifetime Excess Cancer-Risk Estimate[b]: Oral	Dermal	Inhalation	Oral
Chlorinated paraffins	Oral: Liver and kidney toxicity	N/C	N/C	0.3	N/A	0.59	0.28 (particles) N/C (vapors)	1.6×10^{-2}	1.9	2.7×10^{-4}	5.3×10^{-2}	N/A	N/A	N/A

[a]The hazard index is calculated by dividing exposure levels by RfDs or RfCs. A hazard index of <1 indicates that the exposure at the intended levels is not likely to pose noncancer health risks; a hazard index of >1 was considered to possibly pose a concern for noncancer effects.

[b]Lifetime excess cancer risk above the background lifetime cancer incidence was calculated by multiplying the cancer-potency factor by the exposure estimate. For all routes of exposure, lifetime risk calculations were estimated from lifetime average exposures.

[c]Vapor exposure levels were calculated based on the vapor pressure measurements for unreacted starting material. In reality, most or all material becomes bound to upholstery fabric following curing. Vapor exposure levels for chemical entities formed during the curing process were not calculated.

[d]Toxicity information was not available to derive a dermal RfD; the subcommittee used oral RfDs as best estimates for internal dose from dermal exposure.

[e]Toxicity information was not available to derive an inhalation RfC; inhalation RfCs were estimated from oral RfD data using Equation 7 in Chapter 3 to estimate risk.

[f]The cancer-potency factor following inhalation is for exposure to particles and vapors.

[g]There are inadequate toxicity data from any route of exposure to derive RfDs or RfCs for these compounds. However, structurally related compounds were found to be a health concern at the worst-case exposure levels. Therefore, the subcommittee recommends that exposure measurements be made to determine the need for toxicity studies.

[h]There are inadequate toxicity data for phosphonic acid from any route of exposure to derive RfDs or RfCs; this chemical is likely to crosslink with fabric components, precluding dermal exposure to this FR in furniture upholstery.

[i]Not calculated because THPC is likely to polymerize after application to the upholstery fabric.

Abbreviations: N/A: not applicable because the chemical is not carcinogenic by the relevant route; N/C, not calculated because of inadequate data; UE, modeling produces unrealistic short-term exposure estimates for this material; therefore, maximum exposures are unknown.

Despite the lack of a complete database, the subcommittee concludes that the following FRs can be used on residential furniture with minimal risk, even under worst-case assumptions:

- hexabromocyclododecane,
- decabromodiphenyl oxide,
- alumina trihydrate,
- magnesium hydroxide,
- zinc borate,
- ammonium polyphosphates,
- phosphonic acid (3-{[hydroxymethyl]amino}-3-oxopropyl)-dimethyl ester,[1]
- tetrakis hydroxymethyl phosphonium salts (chloride salt)

On the basis of the hazard indices for noncancer effects and/or the potential for cancer, the subcommittee recommends that exposure studies be conducted on the following FRs to determine whether toxicity studies need to be conducted:

- antimony trioxide,
- antimony pentoxide and sodium antimonates,[2]
- calcium and zinc molybdates,
- organic phosphonates (dimethyl hydrogen phosphite),
- tris(monochloropropyl) phosphates,[2]
- tris(1,3-dichloropropyl-2) phosphate,
- aromatic phosphate plasticizers (tricresyl phosphate), and
- chlorinated paraffins.

It is possible that an individual could be exposed by all three routes: oral, dermal, and inhalation. In such cases, the hazard indices or cancer-risk estimates may be summed across the various routes of exposure. This approach is extremely conservative, because it is unlikely that an individual would be

---

[1]There are inadequate toxicity data for phosphonic acid from any route of exposure to derive RfDs or RfCs; this chemical crosslinks with fabric components, precluding human exposure to this FR in furniture upholstery. Therefore, this chemical can be used safely as an FR.

[2]There are inadequate toxicity data from any route of exposure to derive RfDs or RfCs for these compounds. However, structurally related compounds—antimony trioxide and tris(1,3-dichloropropyl-2) phosphate—were found to possibly be health concerns at the worst-case exposure levels. Therefore, the subcommittee recommends that exposure measurements be made to determine the need for toxicity studies.

exposed at the upper limit for one route of exposure and even less likely that the same individual would be exposed at the upper limits for two or more routes. Because one route of exposure typically dominates the risk assessments, summing the hazard indices or cancer-risk estimates does not materially change the conclusions regarding the safety of FRs.

UNCERTAINTIES ASSOCIATED WITH RISK ESTIMATES

The subcommittee recognizes that there are major uncertainties associated with its toxicological risk estimates. The uncertainties stem from the inadequacy or absence of relevant toxicity and exposure data. In the absence of an adequate toxicity database, the subcommittee applied UFs to account for interspecies and intraspecies differences, route-of-exposure differences, less than lifetime exposure studies, absence of a NOAEL, and inadequate or inferior data. Depending on the information available, the subcommittee used several UFs, each ranging from 1 to 10. However, the exact magnitude of each UF that should be used is unknown, and these uncertainties can be reduced only by research to provide the needed information.

The subcommittee identified no quantitative measurements of exposure to FRs under conditions approximating their use in residential furniture upholstery. In the absence of relevant exposure information, the subcommittee made extremely conservative assumptions to overestimate the levels; therefore, there is considerable uncertainty in these exposure estimates. In the absence of any relevant exposure information, the subcommittee is unable to quantify the magnitude of uncertainty associated with these exposure estimates.

The subcommittee could not locate any data on differential toxicity of the FRs in various susceptible human subpopulations or in experimental animals. Therefore, it was unable to develop risk estimates for susceptible subpopulations. The uncertainty regarding their risk can only be reduced by studying the toxicity or toxicokinetics in young or aged animals or in animals with certain pre-existing disease conditions.

In some cases, the actual chemical form of an FR chemical in treated fabric is different from the pure chemical because of chemical reactions (e.g., polymerization and cross-linking) with components of the fabric or reactions with FR-formulation components during the manufacturing or curing processes. If the chemical form of the applied FR changes in the treated fabric, the subcommittee's risk estimates might be inaccurate.

DATA GAPS AND RESEARCH RECOMMENDATIONS

On the basis of its evaluation of toxicity and exposure data on FR chemicals, the subcommittee identified data gaps and research needs. The subcommittee concludes that it is unnecessary to fill all data gaps and recommends that research be done only to reduce key uncertainties for performing toxicological risk assessments. The subcommittee believes that obtaining appropriate exposure measurements for some of the FRs can provide relevant information quickly and inexpensively. If research shows that actual exposures are lower than the subcommittee's conservatively overestimated levels, there may be no need or a reduced need to perform toxicity studies that are much more expensive and time-consuming.

To estimate exposure, the subcommittee recommends that research be done for some FRs to measure (1) extraction of FRs from treated fabric into saline, (2) volatilization of FRs, and (3) potential for FRs to be released from treated fabric during wear that could lead to the generation of airborne particles that contain FRs. The subcommittee recommends that the CPSC collect such information and do its own risk assessments.

Once exposure data are collected, if the CPSC desires to derive RfDs or RfCs with greater confidence for those FRs that have hazard indices of greater than 1, the subcommittee recommends a tiered research approach. In this approach, the first step may include in vitro tests for genotoxicity and other effects, and short-term (e.g., 28-day) toxicity studies via the relevant routes for assessing various toxicity end points (e.g. neurotoxicity, pulmonary toxicity, and reproductive and developmental toxicity). If the results of the research show no basis for concern for toxicity at the subcommittee's worst-case estimated exposure levels, then no further research is recommended, and the FR can be used safely. If the studies indicate a basis for concern about adverse effects for anticipated human exposure conditions, the subcommittee recommends that the chemical be evaluated in a 90-day subchronic study. If the results of the subchronic study do not show a health concern, no further research is recommended. The subcommittee recommends that a chronic toxicity study be done only when the results of short-term and subchronic studies indicate a basis for concern about cancer at expected exposure levels. Because of the extremely conservative assumptions it used in deriving RfDs and RfCs and in estimating exposure levels, the subcommittee does not recommend further research for noncancer effects for those FR chemicals that have hazard indices of less than 1 (see Table ES-1).

1

Introduction

IGNITION of upholstered furniture by small open flames from matches, cigarette lighters, and candles is one of the leading causes of residential-fire deaths in the United States. These fires accounted for about 16% of civilian fire deaths in 1996. On average, each year since 1990, about 90 deaths (primarily of children), 440 injuries, and property losses amounting to 50 million dollars have resulted from fires caused by the ignition of upholstered furniture by small open flames. Certain commercial seating products (such as aircraft and bus seats) are subject to flammability standards and sometimes incorporate FR-treated upholstery cover materials, but there is no federal-government requirement for residential upholstered furniture, and it is generally not treated with FR chemicals. It is estimated that less than 0.2% of all U.S. residential upholstery fabric is treated with flame-retardant (FR) chemicals.

The Consumer Product Safety Act of 1972 created the U.S. Consumer Product Safety Commission (CPSC) as an independent federal regulatory agency whose mission is to protect the public from unreasonable risks of injury and death associated with consumer products. CPSC also administers the Flammable Fabrics Act, under which it regulates flammability hazards and the Federal Hazardous Substances Act (FHSA), which regulates hazardous substances including chemicals. In 1993, the National Association of State Fire Marshals petitioned CPSC to issue a performance-based flammability standard for upholstered furniture to reduce the risk of residential fires. The Commission granted that portion of the petition relating to small open flame ignition risks. The

CPSC staff subsequently developed a small open flame performance standard. If such a standard were promulgated, most residential upholstery fabric would likely be treated with FR chemicals.

It is estimated that FR chemicals would be applied as formulations to as much as 600 million square yards of upholstery fabrics each year (R. Berkley, American Textile Manufacturers Institute, pers. commun., July 30, 1999). If, over the lifetime of the furniture, the chemicals leach out from the formulations, consumers might be exposed to them and possibly suffer adverse health effects. Because there is a potential for exposure of millions of people to FR chemicals, some have recommended banning the use of FR chemicals until they are shown to be safe.

In response to concerns regarding the safety of FR chemicals, Congress, in the fiscal year 1999 appropriations report for CPSC, requested that the National Research Council conduct an independent study of the health risks to consumers posed by exposure to FR chemicals that are likely to be used in residential upholstered furniture to meet a CPSC standard. The National Research Council assigned the project to the Committee on Toxicology (COT) of the Commission on Life Sciences' Board on Environmental Studies and Toxicology. COT convened the Subcommittee on Flame-Retardant Chemicals, which prepared this report. Subcommittee members were chosen for their recognized expertise in toxicology, pharmacology, epidemiology, chemistry, exposure assessment, risk assessment, and biostatistics. Appendix A provides brief biographical sketches of subcommittee members.

THE CHARGE TO THE SUBCOMMITTEE

The charge to the subcommittee was to assess the health risk to consumers and the general population from exposure to 16 FR chemicals that are likely to be applied to fabrics used in residential upholstered furniture. Specifically, the subcommittee was asked to assess the following chemicals or classes:[1] (1) hexabromocyclododecane, (2) decabromodiphenyl oxide, (3) alumina trihydrate, (3) (4) magnesium hydroxide, (5) zinc borate, (6) calcium and zinc molybdates, (7) antimony trioxide, (8) antimony pentoxide and sodium antimo-

---

[1]The list was developed by the Flame Retardant Chemicals Association. The subcommittee arbitrarily chose to discuss risk assessments for the FR chemicals in the following order: (1) brominated compounds, (2) metal compounds, (3) phosphorous compounds, and (4) chlorinated paraffins. It should be noted that chemicals discussed first are not necessarily more toxic or used more often as flame retardants.

nate, (9) ammonium polyphosphates, (10 phosphonic acid, (3-{[hydroxymethyl]amino}-3-oxopropyl)-dimethyl ester, (11) organic phosphonates, (12) tris(monochloropropyl) phosphate, (13) tris(1,3-dichloropropyl-2) phosphate, (14) aromatic phosphates plasticizers, (15) tetrakis(hydroxymethyl) hydronium salts, and (16) chlorinated paraffins. The subcommittee was also asked to identify data gaps and make recommendations for future research.

In addressing its charge, the subcommittee was asked to focus its assessment on the health risks to consumers who might use FR-treated upholstered furniture, including those with an increased susceptibility to FR chemicals. The subcommittee did not consider the toxicity of combustion products of FR materials from fires involving FR-treated upholstered furniture. The subcommittee also did not address ecotoxicological effects, adverse environmental effects of the disposal of household furniture, cost, technology, exposure-standard achievability, or alternative methods. Those issues are beyond the scope of the subcommittee's charge, expertise, and resources. Similarly, the subcommittee did not compare FR chemicals for efficacy or evaluate the benefits of using FR chemicals in upholstered furniture to reduce fire risk; the latter issue has been the subject of recent reports prepared for the U.K. Department of Trade and Industry.

THE SUBCOMMITTEE'S APPROACH TO ITS CHARGE

To gather information, the subcommittee commissioned papers on the various FR chemicals and held public meetings at which representatives of CPSC, the U.S. Environmental Protection Agency, Congress, the National Association of State Fire Marshals, and industrial groups made presentations on issues of critical importance to the subcommittee's task. Some of the speakers provided technical documents on FR chemicals. CPSC staff also provided its toxicity assessments to the NAS.

To characterize the human health risks from dermal, oral, and inhalation exposures to each of the 16 FR chemicals, the subcommittee adopted the NRC risk assessment paradigm (NRC 1983, 1994). Data were evaluated relating to key elements of this risk assessment paradigm-hazard identification, dose-response assessment, exposure assessment, and risk characterization. Specifically for each FR chemical, the subcommittee reviewed data on physical and chemical properties, toxicity, and exposure. Several of the 16 chemicals were actually chemical classes rather than single compounds. In some of these cases, one chemical of the class was selected as a surrogate: tetrakis (hydroxymethyl) hydronium chloride for the tetrakis (hydroxymethyl) hydronium salts and their compounds; dimethyl hydrogen phosphite for organic phosphonates; and tricresyl phosphate for aromatic phosphate plasticizers. Surrogates were selected on the basis of representativeness of the class, availability of data, and most potent chemical in the class. Conclusions about the class are based on the properties of the surrogate. The risk from other members of the class might be different from the risk from the surrogate.

To identify adverse effects associated with an FR, the subcommittee reviewed human and animal data on local and systemic effects, e.g., neurologic, immunologic, reproductive, developmental, and carcinogenic effects. In vitro data were also reviewed to determine the potential for genotoxicity, other toxic effects, and to understand the mechanism of toxic action. Toxicokinetic studies were reviewed to understand the absorption, distribution, metabolism, and excretion of the FRs. Quantitative toxicity assessments were performed to determine the toxic potency for noncancer and cancer effects. The subcommittee used the NOAEL-uncertainty-factor approach to identify reference doses (RfDs) or reference concentrations (RfCs) for noncancer effects. For carcinogens, carcinogenic potency was estimated using the linearized multistage no-threshold model because it provides a conservative risk estimate.

Human exposures to FRs in residential furniture were estimated using extremely conservative assumptions and assuming exposure occurs via skin contact, ingestion (especially for infants or children that might suck or chew on fabric), inhalation of particles generated during abrasion of the surface fibers, and inhalation of vapors off-gassing from treated fabric.

To characterize health risks from exposure to a non-carcinogenic chemical, a hazard index approach was used. The subcommittee's approach for estimating risks from exposure to carcinogenic FRs involved estimating an upper limit on lifetime cancer risk.

To identify data gaps and make recommendations for future research needs, the subcommittee noted deficiencies in the databases on the FR chemicals and, where appropriate, recommended future research that would reduce uncertainty and increase confidence in the risk assessments.

FLAME RETARDANT PROCESSES

Some upholstery cover materials fabrics are inherently resistant to small open flames because of their chemical or physical structure. Others are made flame resistant with flame-retardant additives. FR chemicals are incorporated into fibers, applied to the surface of a textile, or applied to the back of a textile in the form of a polymeric coating. Chemical after-treatments for flame retardancy may include surface or topical treatments, use of coatings, and use of functional finishes that become part of the final fiber structure. Appendix B describes the four basic processes typically used on various types of fabric.

CALIFORNIA, UNITED KINGDOM, AND EUROPEAN EXPERIENCE

California is the only state with mandatory flammability regulations for residential upholstered furniture. A flammability performance standard was developed and is administered by the California Bureau of Home Furnishings and Thermal Insulation. The principal standard, applicable since 1975 to all upholstered furniture sold in the state, is known as Technical Bulletin 117 (TB-117).

Technical Bulletin 117 contains a series of flammability performance tests and minimum requirements for both cigarette ignition and small open flame resistance of furniture component materials. Manufacturers generally rely on FR-treated polyurethane foam or other foam filling materials to meet the flame-prevention requirement for fillings. Cover fabrics do not require FR treatment to comply with TB-117.

As a consequence of UK furniture-fire regulations, which were introduced in 1988, flame retardants are widely used both in upholstery fabrics and in foams and other fillings for furniture to meet performance requirements. Appendix B provides an overview of the types of application methods and level of FRs presently being used, and available information on durability, degradation, and fate of FRs after treatments of furniture upholstery. Flame retardants are incorporated at the fabric-finishing stage before furniture manufacture, so furniture upholsterers have been occupationally exposed to fabrics containing flame retardants and UK consumers who have purchased new furniture or have lived in rented furnished accommodations have been exposed for over a decade to FR-treated fabrics and foams.

Some workers were reported to have experienced dermal sensitization to FR-treated upholstery, but the sensitization was in fact due to fabric finish and fiber factors, not to FRs, and no other adverse effects have been reported. No adverse effects have been reported in the general population exposed to FR-treated furniture or other FR-treated consumer products.

Almost all FR chemicals used in Europe were used commercially at the time the European existing-substances regulations were introduced. These chemicals continue to be used without the requirement of formal risk assessments unless they were identified as high-priority substances that specifically needed such an assessment. The latter is the case for a small number of flame retardants that are undergoing formal risk assessment in Europe, including the polybrominated diphenyl ethers and hexabromocyclododecane.

ORGANIZATION OF THIS REPORT

The remainder of this report is organized into 18 chapters and two appendices. Chapter 2 describes the risk assessment process used by the subcommittee in determining the risk associated with potential exposure to the various FR chemicals. Chapter 3 describes the method the subcommittee used to measure and estimate the intensity, frequency, extent, and duration of human exposure to FR chemicals. Chapters 4-19 provide the subcommittee's review and assessment of health risks posed by exposure to each of the 16 FR chemicals. Data gaps and research needs are provided at the end of these chapters.

REFERENCES

NRC (National Research Council). 1983. Risk Assessment in the Federal Government: Managing the Process. Washington, D.C.: National Academy Press.

NRC (National Research Council). 1994. Science and Judgment in Risk Assessment. Washington, D.C.: National Academy Press.

2

Assessment of Health Risks from the Use of Flame Retardants

THE assessment of health risks from the use of flame retardants (FRs) will be central to the U.S. Consumer Product Safety Commission's (CPSC) determination of whether substances likely to be used as FRs in residential upholstered furniture would present a hazard to consumers. Risk estimates derived from this process are often vital in making informed and balanced decisions to effectively manage risks and to communicate about the significance of known or anticipated risks.

The risk assessment process enables the systematic evaluation of data and the quantitative presentation of complex information. This chapter provides an overview of the risk assessment process and describes procedures applied for interpretation of data on physical and chemical properties, toxicity, and exposure to estimate human health risks from exposures to FRs. Many documents are available to the reader to obtain more detailed information about risk assessment procedures and their applicability; some include: EPA (1986, 1992a,b, 1996, 1999a), NRC (1977, 1980, 1983, 1994), Tardiff and Rodricks (1987), and WHO (1978, 1984, 1986, 1999) . The basic principles guiding the assessment of a substance's toxicity are outlined in the *Guidelines for Carcinogen Risk Assessment* (EPA 1999b); *Chemical Carcinogens: A Review of the Science and Its Associated Principles* (OSTP 1985), *Guidelines for Developmental Toxicity Risk Assessment* (EPA 1991a).

THE SUBCOMMITTEE'S RISK ASSESSMENT PROCESS

Risk assessment is the process of quantitatively determining the likelihood of adverse effects resulting from exposures to FR chemicals. The risk assessment paradigm used in this report and widely acknowledged by the U.S. scientific community to be the current standard used in industrial risk management and public policy settings was first articulated by the U.S. National Research Council in its report, *Risk Assessment in the Federal Government: Managing the Process* (NRC 1983). In a sequel entitled *Science and Judgment in Risk Assessment* (NRC 1994), the National Research Council defined major elements in the estimation of health risks and placed the process within contexts of managing such risks. Among the elements elaborated in detail are (1) the values and limitations of default assumptions and the circumstances under which they should be replaced with empirical findings; (2) the importance of understanding the mode of toxic action of a compound to provide increased accuracy in defining the presence of hazards to human health; (3) the role of variability in human response as a basis for deciding the degree of health protection for susceptible subpopulations; and (4) aggregation of exposures and risks from all sources of exposure so as to provide realistic estimates of risk to specific groups of individuals.

In this report, health risks are characterized for potential effects in the body after repeated and prolonged exposures to FRs. Although acute toxicity information was reviewed, it was not used in developing risk assessments because chronic exposure data are more relevant.

The four basic steps of risk assessment process applied to FRs are hazard identification, dose-response assessment, exposure assessment, and risk characterization. Each is described below.

Hazard Identification

In the hazard identification step, a determination was made of causal relationships that exist between exposure to an FR and its adverse health effects. It involves gathering and critically evaluating toxicity data on the types of health effects that might be produced by an FR and on the conditions of exposure under which adverse effects are produced. Such an evaluation requires the development of the weight of all the evidence related to the toxicity of an FR.

Toxicity data are derived from observations of humans (epidemiological studies, clinical findings, and case reports), from investigations of laboratory animals (most often rodents), and from in vitro studies.

Observations of Humans

Epidemiological studies provide the most relevant kind of information for hazard identification, because they involve observations of humans, not laboratory animals. This obvious and substantial advantage is offset by various difficulties associated with obtaining and interpreting epidemiological findings. Rarely are convincing causal relationships identified with a single or even a few studies. Epidemiologists usually weigh the results from several studies, ideally involving different populations and investigative methods, to determine the degree of consistency between exposures and responses among them, including the dose-response relationships.

Few human data were found for the FRs reviewed in this report.

Studies of Laboratory Animals

When human studies are unavailable or unsuitable, risk assessments for FRs are based on findings in studies on laboratory animals (usually rodents). An advantage of animal studies is that they can be controlled, so establishing causation (assuming that the experiments are well conducted) is not in general difficult. Another advantage is that animals can be used to collect toxicity information on chemicals before being marketed, whereas epidemiological data can be collected only after human exposure has occurred. Indeed, many countries require that some classes of chemicals (e.g., pesticides, food additives, and drugs) be subjected to toxicity testing in animals to demonstrate appropriate safety before the product can be marketed. Other advantages of animal studies include the facts that:

- A quantitative relationship between exposure (or dose) and extent of toxic response can be established.
- Animals and animal tissues (including some that would be inaccessible in humans) can be closely examined by toxicologists and pathologists, so the full range of toxic effects produced by a chemical can be identified and, in some instances, the progression to physiologic impairment can be characterized. The reversibility of adverse effects once exposure ceases can also be studied.

Animal data reviewed included studies of neurotoxicity, immunotoxicity, reproductive and developmental toxicity, organ toxicity, dermal and pulmonary toxicity, and other local and systemic effects. Toxicokinetic studies were also reviewed to understand the absorption, distribution, metabolism, and excretion of the FRs.

In Vitro Studies

The subcommittee also evaluated the result of in vitro studies. These studies are inexpensive and can be done in a relatively short time. The use of in vitro tests is increasing because they reduce the need to use experimental animals. These studies also provide some insight into possible mode of action of a chemical. They also provide information on structure-activity relationships (SAR). The SAR information can be useful when the data on chemical under consideration are inadequate or nonexistent. The results of these studies are important in hazard identification and can provide supporting data when the animal data are inadequate or absent.

Discussion

The subcommittee's hazard-identification step describes the types of toxic responses, if any, that can be caused by the FR under review, the weight of the primary and supporting evidence, the scientific merits of the data, and their value and reliability for estimating human toxicity under defined conditions of exposure. The weight of evidence analyses for the individual FRs takes into account replication, reproducibility, and concordance of results, as well as the degree of correspondence between observations in experimental animals and expected responses in humans for a given form of toxicity. These concepts are described in detail elsewhere (Tardiff and Rodricks 1987).

The product of the hazard identification leads only to a statement about the toxic properties of an FR that may occur in humans. It does not reveal whether the FR poses a risk for specific populations with specific exposure circumstances. That determination requires three additional analytic steps: (1) evaluation of toxic potency of the FRs by examining their dose-response characteristics, (2) characterization of the nature and magnitude of human exposure, and (3) characterization of risk by combining the information on magnitude of exposure with dose-response relationships.

Toxic Potency or Dose-Response Assessment

Toxic potency or dose-response assessment information is used to determine the quantitative relationship between increases in the dose of an FR and changes in magnitude of the incidence and/or severity of adverse health effects.

The slope of the dose-response curve in the low-dose range, combined with the increment in exposure (dose), provides a quantitative estimate of the increase in incidence or severity of some adverse effect within an exposed population group.

Dose-response relationships are often grouped into two classes based on two distinct modes of toxicity: (1) adverse effects expected to have a nonlinear dose-response relationships (sometimes referred to as "biological threshold") and (2) those likely to have a linear (i.e., having no biological threshold) dose-response in the low-dose range. A threshold for a particular toxic effect is defined as the dose or dose rate below which the adverse effect attributable to the specific agent is unlikely to occur (Brown 1987). For the evaluation of FRs, toxic effects other than cancer are considered to have a nonlinear dose-response, whereas carcinogenic responses induced by genotoxic carcinogens are treated as having a linear dose-response. Non-genotoxic carcinogens or tumor promoters may have a nonlinear dose response. For the carcinogenic FRs evaluated in this report, the data are insufficient to determine whether dose responses are linear or nonlinear. The subcommittee assumed that they have linear dose responses because this assumption is conservative or health protective.

For some FRs, more than one dose-response relationship may exist, depending on conditions of exposure and a variety of responses (e.g., cancer, birth defects, kidney damage, etc.). The subcommittee's process of evaluating relevant dose-response relationships takes into account diverse information about the body's ability to generate metabolites that are more toxic than the parent compound, its ability to detoxify potentially toxic compounds or metabolites, and differences between the mechanisms of toxicity in test organisms and humans. Exposure circumstances (e.g., duration, frequency, and route) have considerable impact on toxic potency.

Dose-response assessment includes the process of extrapolating adverse effects observed in experimental animal organism from high to low doses; it also includes extrapolating data from animals to humans. To perform such extrapolations, two fundamentally different approaches are used, one for carcinogenic responses and another for all other forms of toxicity. It should be noted that these procedures used by the subcommittee relate specifically to toxicity resulting from repeated and prolonged exposures, as much as a full lifetime in duration. The subcommittee's approach for determining noncancer and cancer potency of FR chemicals is described below.

Determination of Toxic Potency for Noncancer Effects

For all types of noncancer effects, the subcommittee's procedure used to evaluate the dose-response involves identifying the lowest-observed-adverse-effect level (LOAEL) and the no-observed-adverse-effect level (NOAEL). The LOAEL is the lowest dose at which a statistically or biologically significant increase (when compared with a control group) in an observable adverse (although at times mild, reversible) effect has been reported. The NOAEL is the highest exposure level below the LOAEL at which no such significant increase is observed in the frequency or severity of an adverse effect when compared with a control group.

Conceptually, NOAELs are more conservative than LOAELs as a starting point for extrapolating findings to low doses experienced by humans. The NOAEL-uncertainty-factor (UF) approach was used to identify reference dose (RfD) or reference concentration (RfC). An RfD is defined as an estimate (with uncertainty spanning an order of magnitude or greater) of daily oral or dermal doses that are unlikely to have deleterious effects during a life span. An RfC for an inhalation exposure is similarly defined. Statistical deficiencies and improvements in the NOAEL based uncertainty-factor (UF) approach to derive RfDs and RfCs have been described by several investigators (Crump 1984, Kimmel and Gaylor 1988, Chen and Kodell 1989, Gaylor 1989, Kodell et al. 1991).

Uncertainty Factors

Once a NOAEL is estimated from experimental data or observations of humans, an RfD is obtained by dividing it by one to several uncertainty factors (UFs). The degree of confidence in the data used to derive the NOAEL determines the magnitude of UFs. The major UFs, some or all of which may be applied for derivations of RfDs or RfCs, are described below:

Interspecies Extrapolation

Despite physiological similarities among mammalian species, laboratory animals are not human beings, thereby providing a clear disadvantage in estimating possible adverse human health effects from exposure to FRs. Based on both toxicological principles and empirical observations, reasons exist to support the hypothesis that many forms of biological responses, including toxic responses, can be extrapolated across mammalian species, including humans.

One of the most important reasons for species differences in response to chemical exposures is that toxicity is often a function of the metabolism of a chemical. This is also applicable to FRs. Differences in metabolic handling of a chemical among animal species, or even among strains of the same species, are not uncommon. Such differences can account for toxicity differences. In most cases, because information on a chemical's metabolic profile in humans is lacking (and often unobtainable), identifying the animal species and toxic response most likely to predict the human response accurately is generally not feasible. Therefore, by convention, one assumes that, in the absence of clear evidence that a particular toxic response is not relevant to humans, any compound-related adverse effects found in laboratory animals tested in properly designed studies are potentially predictive of response in at least some humans. This subcommittee agrees with this convention and also adopted that approach.

A UF of 10 is used for FRs when data from laboratory animals are extrapolated to humans; if the toxic potency of a compound is known to be similar in humans and experimental animals, then a factor less than 10 is used.

Intraspecies Extrapolation

Tests conducted in a homogeneous laboratory animal population (or even a small human group) do not account fully for the heterogeneous human population. Individuals vary considerably in their susceptibility to chemical insult due to their genetic make-up, lifestyle factors (e.g., nutrition, smoking), age, hormonal status (e.g., pregnancy), immune system integrity, and pre-existing illness. The subcommittee used a UF of 10 to account for intraspecies differences. If data indicate the existence of a narrow range of susceptibilities, a UF of less than 10 was used. Therefore, an understanding of ranges of metabolic differences is of substantial value in determining the most suitable UF. Even when a NOAEL is based on human data, those data are likely to have been collected in exposure circumstances that differ appreciably from the exposure circumstances of interest. Consequently, there will still be uncertainty in such a NOAEL, and it may also be adjusted with UFs to produce an RfD or RfC.

Extrapolation from Subchronic to Chronic Exposures

In some data sets for FRs, the toxicity data used to derive an RfD or RfC is of a duration of less than a lifetime. The UF for extrapolation from subchronic to chronic exposure ranges from one to 10. Selection of a specific value depends in part on the quality of the data and on expectations for cumulative toxicity and accumulation of the substance. For instance, several short-term studies of varying durations may indicate that toxicity of a FR is or is not cumulative. Likewise, an understanding of the chemistry of a FR may assist in deciding whether extending durations of exposure might yield even lower NOAELs.

Extrapolation from LOAEL to NOAEL

At times, toxicity studies reveal only a LOAEL and no NOAEL. Under these circumstances, the subcommittee used a UF of between one and 10. Selection of a UF is dependent largely on the quality of the entire toxicity data set. A factor of 10 was applied, unless the evidence indicates that lesser values could be used confidently.

Route-to-Route Extrapolation

When the toxicity information on an FR is obtained from one route of administration (e.g., ingestion) yet individuals are exposed by another route (e.g., dermal or inhalation), the resulting differences in toxic potency are taken into account by applying a UF of 10; if evidence indicates little or no such variability, a factor of less than 10 is used. Reliance on toxicokinetic data, elucidating differences in absorption rates and metabolic pathways by different routes of exposure, is useful in selecting an appropriate value of the UF.

Adjustment to Account for Poor Quality of the Database

The quality of toxicity studies vary considerably. The subcommittee considered the quality of data in the derivation of an RfD or RfC. Some deficiencies include the use of small numbers of animals and the lack of replication of results. To account for these deficiencies, the subcommittee used UFs between one and 10 and the actual values used varied with the subcommittee's evaluation of the adequacy and quality of data.

Overall Uncertainty Factor

The RfD is obtained from the adjusted NOAEL by dividing it by an overall UF that is equal to the product of all the UFs discussed in the preceding paragraphs. For example, a composite UF of 100 is applied when the NOAEL is derived from chronic toxicity studies (typically 2-year studies) that are considered to be of high quality and when the purpose is to protect members of the general population who could be exposed daily for a full lifetime (10 to account for interspecies differences and 10 to account for intraspecies differences).

Carcinogenic Potency

Carcinogenic potency of chemicals in humans can be estimated for doses far below those in the range of observations only by the use of appropriate mathematical models that extend dose-response curves (NRC 1983; NRC 1994; EPA 1998). Because such dose-response functions cannot be determined empirically, the actual shapes of such dose-response curves at the lowest dose are unknown, and must currently be hypothesized on the basis of assumptions about biological processes.

For most carcinogens, it is believed that the probability for the occurrence of a cancer increases linearly with dose at sufficiently low doses, so that only certain mathematical models (nonthreshold models) are used to predict carcinogenic responses at low doses. In the case of nonthreshold chemicals, it is assumed that there is no dose (except zero dose) that corresponds to zero risk of injury. In theoretical terms, any dose of a carcinogen results in an incremental increase in the risk of cancer (NRC 1983).

If an FR causes cancer in laboratory animals, the slope of the dose-response curve is used as the unit to describe carcinogenic potency and is called the cancer potency factor (also technically designated as q_1^*) (Crump 1996).

Many models have been developed to assess the effects of low doses of carcinogens. The most frequently used cancer risk dose-response model is the linearized multistage model (Crump 1996), and it was used for FRs that produced tumors in rodents; this model is used because it provides conservative cancer estimates.

A major limitation with low-dose extrapolation models is that they all often fit the data from animal bioassays equally well, and it is not possible to determine their validity based on goodness of fit. Each model may fit experimental data equally well, but they are not all equally plausible biologically. The dose-response curves derived from different models diverge substantially in the dose range of interest (NRC 1983). Therefore, low-dose extrapolation is more than a curve-fitting process, and considerations of biological plausibility of the models must be taken into account before choosing the most suitable model for a particular set of data. Mathematically fitting the multistage model to experimental results in laboratory animals allows estimation of the low-dose slope of a dose-response curve, together with an upper 95% confidence limit (UCL) on that slope. These are often referred to as cancer potency factors. These cancer potency factors are extrapolated to humans, usually by incorporating various other assumptions such as constant exposure for a lifetime of 70 years and a suitable dose metric for the extrapolation. Through the use of the UCL, and by the nature of the animal-to-human extrapolation used in practice, it is possible that the cancer potency factors estimates obtained will overestimate risk. Such potency factors are, therefore, suited more to standard setting than to defining actual risks to a specific population. Quantitative estimate of risk is obtained by multiplying cancer potency factors by lifetime average dose rates. Actual risks are unlikely to be higher than such quantitative estimates, are likely to be lower, and may be zero.

The benchmark dose (BD) offers a variant to estimate the cancer potency of a carcinogenic FR or other compound (EPA 1996, Crump 1984). When applied to a carcinogen data set, the BD process first uses the linearized multi-stage model to define the dose-response curve in the observed response range, and to estimate a dose that is likely to produce an increment in response rate of approximately 10%. However, rather than applying UFs to the dose representing the 10% response rate, a straight line is interpolated to the origin of the dose versus incremental-response curve from the lower confidence limit of the BD at 10% incremental response, and the slope of this line is used as the cancer potency factor. The extrapolation to humans then proceeds as for the linearized multi-stage model already discussed. It is likely that linear extrapolation to zero from the lower confidence limit of the BD at 10% incremental response will be close to cancer risk estimates based on q_1* (Cogliano et al. 1988).

Exposure[1] Assessment for Flame Retardants

This third step in the risk assessment process is used to describe the nature (e.g., distribution of age, sex, and unique conditions such as pregnancy, preexisting illness, and lifestyle) and size of the various populations exposed and the magnitude and duration of their exposures. The assessment might include past, current, and prospective exposures. For FRs, prospective exposures in residential settings are estimated; no attempt has been made to reconstruct previous

---

[1]Exposure is defined as the opportunity for a dose, such as concentrations in food, air, or water, and is generally reported in units such as ppm or ppb or mg/L, etc. Dose is defined as the amount received by the body of the target organism (e.g., humans) or a target organ (e.g., the liver or kidneys); it is generally reported in units of weight of the substance (e.g., mg or µg) per body weight or body weight. Inhalation doses are expressed as concentrations in air either as ppm or as mg/m^3. Doses resulting from skin contact are generally described as concentrations per unit of surface area (e.g., $mg/meter^2$); however, when the toxicity from skin contact is systemic, the dose is also described in the traditional unit of mg/kg-day.

exposures. Human exposure to FRs in residential furniture fabric was assumed to possibly occur via skin contact, ingestion (specifically for infants or children who might chew on the fabric), inhalation of particles generated during abrasion of the surface fibers, and inhalation vapors off-gassing from treated fabric.

An ideal exposure assessment for FRs in furniture fabrics, would include the following steps:

1. determining concentrations of the chemicals at the surface of the fabric, their degree of binding, and their rate of disappearance;
2. estimating the amount of an FR that may come into contact with skin (characterizing the surface area of contact is necessary) and that may be ingested or inhaled, and contact with internal surfaces (gut and living membranes);
3. characterizing human behaviors that directly bear on external dose (e.g., frequency of contact, years of contact, etc.); and
4. establishing or estimating the rate of membrane penetration (absorption), and evaluating the pharmacokinetics of the compound to ascertain doses to target tissues, which might be some distance from the point of entry into the body.

These four steps were simplified by assuming worst-case scenarios. For example, individuals were assumed to be exposed to 100% of the compound of interest for a plausible period of time over plausibly large areas of their body. The approach taken results in upper-bound estimates of risk that have utility in screening compounds—they allow elimination from concern of compounds that are not a risk. In Chapter 3, the details of this process are presented; and for each FR, estimates of doses are described within their respective chapters.

Risk Characterization of Flame Retardants

This final step of a risk assessment process involves integration of data and analyses from the other three steps of risk assessment to determine the likelihood that groups of individuals may not experience any of the various forms of toxicity associated with a chemical under its known or anticipated conditions of exposure. This step includes estimations of risk to individuals and population groups and a full exposition of the uncertainties associated with the conclusions. Scientific knowledge is usually incomplete, so the reliance upon inferences about risk is inevitable. A well-constructed risk assessment relies on inferences that are most strongly supported by general scientific understanding and, to the extent feasible, do not include assumptions derived solely from risk management or policy directives.

Risk Characterization of Noncarcinogenic Effects

For adverse health effects other than cancer, the subcommittee used a "hazard index" approach; it is the ratio of an actual or estimated dose to RfD or RfC. This approach was used by the subcommittee to judge whether a particular exposure is unlikely to present a noncancer risk. For each FR, a hazard index was calculated for noncancer toxicity by dividing the estimated human dose by the estimated RfD or RfC. A hazard index of less than 1 was deemed to provide an adequate margin of safety; a hazard index greater than 1 was considered to possibly pose concern for noncancer effects.

The inability to draw more definitive conclusions for hazard indices greater than 1 arises from four factors:

1. The conservative approach taken to estimate doses—the estimated doses are certainly much higher than may occur in practice, because the methodology used in developing these estimates was deliberately chosen to be ultraconservative.
2. The conservative approach taken to estimate the RfDs—the estimated RfDs are certainly substantially lower than doses that might cause health effects, because the methodology used in developing these estimates was deliberately chosen to be conservative.
3. The interaction between the 2 factors discussed in the previous two paragraphs. For any health effect to occur in any individual, the dose *for that individual* must be higher than the threshold for that *same* individual. The estimates of dose, however, are for the highest exposed individuals, while the RfD is estimated for the most sensitive individuals, and these two will coincide only with low probability.
4. FRs are to be used to prevent or reduce a known risk—that of fires caused by ignition of FR-treated upholstered furniture. Even if FRs were to produce adverse effects, the net effect of using them might nevertheless be a reduction in risks. Determination of the acceptability of the various risks requires consideration of the trade-offs involved. The subcommittee's charge did not include evaluation of trade-offs, and it did not attempt to make such evaluation.

Risk Characterization of Carcinogenic Effects

The approach used by the subcommittee to characterize risks from exposure to carcinogenic FRs involves extrapolation of observations of cancer in animals at relatively high doses to much lower doses anticipated in residential settings. The risks for a specified carcinogen in a defined set of circumstances are estimated by multiplying the cancer potency factor by the various measures of dose. The risk is expressed as a probability, for instance, as the proportion of individuals, among all individuals exposed to a cancer-causing agent, that might develop (or die from) cancer attributable to that agent over a specified time, usually a lifetime of 70 years. A specific example would be a lifetime risk of less than one in a million, meaning that chance exists that less than one excess cancer may occur among a million identically exposed persons. The subcommittee's upper limits of cancer estimates reflect the chance that cancer may occur and not that they must inevitably occur; however, because of limitations in knowledge about the processes of cancer causation, it is also possible that the risk may be zero and that no excess cancers above the background levels would ensue from a specified exposure.

REFERENCES

Brown, C.C. 1987. Approaches to interspecies dose extrapolation. Pp. 237-268 in Toxic Substances and Human Risk, Principles of Data Interpretation. R.G. Tardiff, and J. V. Rodricks, eds. New York: Plenum Press.

Chen, J.J., and R.L. Kodell. 1989. Quantitative risk assessment for teratological effects. J. Am. Stat. Assoc. 84:966-971.

Cogliano, V.J., A.M. Koppikar, J.M. Conis, S.C. Gibson, J.S. Gift, et al. 1988. Methodology for Evaluating Potential Carcinogenicity in Support of Reportable Quantity Adjustments Pursuant to CERCLA Section 102. Office of Health and Environmental Assessment, U.S. Environmental Protection Agency, Washington, DC.

Crump, K.S. 1984. A new method for determining allowable daily intakes. Fundam. Appl. Toxicol. 4(5):854-871.

Crump, K.S. 1996. The linearized multistage model and the future of quantitative risk assessment. Hum. Exp. Toxicol. 15(10):787-798.

EPA (U.S. Environmental Protection Agency). 1986. Guidelines for Carcinogen Risk Assessment. Fed. Regist. 51(September 24):33992-34003.

EPA (U.S. Environmental Protection Agency). 1989. Exposure Factors Handbook. EPA/600/8-89/043. Exposure Assessment Group, Office of Health and Environmental Assessment, U.S. Environmental Protection Agency, Washington, DC.

EPA (U.S. Environmental Protection Agency). 1991a. Guidelines for Developmental Toxicity Risk Assessment. Fed. Regist. 56(Dec. 5):63798-63826. EPA600FR91001. National Service Center for Environmental Publications, Cincinnati, OH.

EPA (U.S. Environmental Protection Agency). 1992a. Guidelines for Exposure Assessment, Fed. Regist. 57(104):22888-22938. May 29, 1992.

EPA (U.S. Environmental Protection Agency). 1992b. Draft Report: A Cross-species Scaling Factor for Carcinogen Risk Assessment Based on Equivalence of mg/kg 3/4/Day. Fed. Regist. 57(109):24152-24173(June 5).

EPA (U.S. Environmental Protection Agency). 1996. Proposed Guidelines for Carcinogen Risk Assessment. EPA/600/P-92/003C. Office of Research and Development, U.S. Environmental Protection Agency, Washington, DC.

EPA (U.S. Environmental Protection Agency). December 1998. Draft Revisions to the Proposed Guidelines for Carcinogen Risk Assessment. Office of Research and Development, U.S. Environmental Protection Agency, Washington, DC.

EPA (U.S. Environmental Protection Agency). 1999a. Guidance for Conducting Health Risk Assessment of Chemical Mixtures - External Scientific Peer Review Draft. NCEA-C-0148. [Online]. Available: http://www.epa.gov/nceawww1/mixtures.htm Office of Research and Development - NCEA, U.S. Environmental Protection Agency, Washington, DC.

EPA (U.S. Environmental Protection Agency). 1999b. Guidelines for Carcinogen Risk Assessment. Review Draft. NCEA-F-0644, Risk Assessment Forum, U.S. EPA, Washington, DC (July 1999). [Online]. Available: http://www.epa.gov/nceawww1/raf/crasab.htm

Gaylor, D.W. 1989. Quantitative risk analysis for quantal reproductive and developmental effects. Environ. Health Perspect. 79:243-246.

Kimmel, C.A., and D.W. Gaylor. 1988. Issues in qualitative and quantitative risk analysis for developmental toxicology. Risk Anal. 8(1):15-20.

Kodell, R.L., R.B. Howe, J.J. Chen, and D.W. Gaylor. 1991. Mathematical modeling of reproductive and developmental toxic effects for quantitative risk assessment. Risk Anal. 11(4):583-590.

NRC (National Research Council). 1977. Chemical contaminants: Safety and risk assessment. Pp. 19-62 in Drinking Water and Health. Washington, DC: National Academy Press.

NRC (National Research Council). 1980. Problems of risk estimation. Pp. 25-65 in Drinking Water and Health, Vol. 3. Washington, DC.: National Academy Press.

NRC (National Research Council). 1983. Risk Assessment in the Federal Government: Managing the Process. Washington, DC: National Academy Press.

NRC (National Research Council). 1994. Science and Judgment in Risk Assessment. Washington, DC: National Academy Press.

OSTP (Office of Science and Technology). 1985. Chemical Carcinogens: A Review of the Science and Its Associated Principles. Fed. Regist. 50:10372-10442.

Tardiff, R.G., and J. V. Rodricks, eds. 1987. Toxic Substances and Human Risk: Principles of Data Interpretation. New York: Plenum Press.

WHO (World Health Organization). 1978. Principles and Methods for Evaluating the Toxicity of Chemicals. Environmental Health Criteria, Vol. 6. Geneva: World Health Organization.

WHO (World Health Organization). 1984. Principles for Evaluating Health Risks to Progeny Associated with Exposure to Chemicals During Pregnancy. Environmental Health Criteria No. 30. Geneva: World Health Organization.

WHO (World Health Organization). 1986. Principles and Methods for the Assessment of Neurotoxicity Associated with Exposure to Chemicals. Environmental Health Criteria No. 60. Geneva: World Health Organization.

WHO (World Health Organization). 1999. Principles for the Assessment of Risks to Human Health from Exposure to Chemicals. Environmental Health Criteria No. 210. Geneva: World Health Organization.

3

Exposure Assessment Methodology

DETERMINATION of whether particular uses of flame-retardant (FR) chemicals in fabrics are safe is complicated by our lack of information in multiple areas. First, all available toxicity studies are on the pure chemicals that are present in fibers before yarn and fabric formation or are applied to the fabrics during processing. The chemicals might be incorporated into backcoatings applied to the fabric, or they might be introduced into the fibers. They might have different physical and chemical forms from those used in the toxicity testing. Thus, the finished or treated fabrics possibly contain derivatives of the applied chemicals as a consequence of cross-linking, polymerization, or oxidation processes having occurred (see Appendix B for a discussion of the finishing and treatment technologies and the processes that occur therein). These derivatives might bear little chemical resemblance to the original species in terms of generic structure, types of functional groups present, solubility, volatility, and general reactivity. Toxicity tests are usually performed by gavage or oral feeding studies using the pure and unreacted chemical mixed with a carrier vehicle or animal food. The chemical is not adsorbed to fabric fibers.

Second, for most of the FRs evaluated, only limited toxicity information is available for the oral route of exposure, and even less is available for the inhalation route. For the dermal route, the minimal toxicity information that is available is not adequate for developing dermal reference doses. In most cases, dermal and inhalation toxicities have to be inferred from the toxicity information available from the oral route.

Third, the toxicity information that is available (for example, reference doses (RfDs) and reference concentrations (RfCs) is expressed in terms of the applied amount—not in terms of the amount that is available inside the body to cause toxicity. The difference in absorption with each route makes the route-to-route extrapolations more uncertain.

Finally, potential exposures to FRs applied to furnishing fabrics within the home have not been studied. Thus, there is little basis for estimates of exposure to such materials. There are few, if any, measurements of exposures under relevant conditions of exposure, and the subcommittee located no quantitative measurements of such exposures.

The volume of literature related to residential pesticide exposures was reviewed but not considered relevant because pesticides are not ordinarily applied to fabrics during the manufacturing process. The effectiveness of pesticides is not dependent on their ability to remain in the fabric during use. The subcommittee does not believe that using the small amount of pesticide data that might relate to fabrics would reduce uncertainty. The subcommittee believes that pesticides are not good surrogates for FR chemicals.

To make progress, therefore, the subcommittee adopted some extremely conservative assumptions (that is, corresponding to high concentration and exposure conditions) about potential exposures. In these estimations, the subcommittee determined whether each FR chemical would pose an acceptable risk even with such assumptions. If the risk was acceptable, then that chemical could be dropped from further consideration. Subsequent iterations of the procedure would then depend on finding more defensible information about the exposure conditions. The subcommittee was unable to find any such information and recommends collection of such information—a process that should be relatively straightforward but is outside the subcommittee's charge.

People can come in contact with furnishing fabrics through direct contact (for example, by sitting on them). However, most of the time the contact is likely to be very small because of the presence of clothing and because the FR is incorporated into the fabric fiber structure or is present in a backcoating formulation added on the reverse face of the fabric. Young children, in particular, might suck on furnishing fabrics; therefore, possible dissolution in saliva and ingestion of the FRs present must be considered unless they can be demonstrated to remain "locked" within the fibers or in the backcoating resin formulation. Finally, as fabrics wear they can shed small fibers, the majority of which are likely to be too large to be inhaled to any substantial degree. However, some inhalation exposure of FR chemical species could result from the generation of particles of respirable size (≤ 10 μm).

Actual exposures to FR chemicals applied to fabrics are likely to be limited by multiple factors. FR chemicals are of little use unless they stay in the backcoating or the fabric throughout its life, so durability is a requirement for their selection. An FR treatment that wears off easily, washes off, or evaporates from the fabric is less suitable for its intended purpose, and might actually fail any prescribed durability test-performance requirements.

The subcommittee chose to evaluate three exposure scenarios that are intended to represent the three routes of exposure (dermal, ingestion, and inhalation). For each of these exposure scenarios, exposures estimates were obtained for the general population[1] with deliberately conservative assumptions, as detailed below.

For dermal exposure, the exposure scenario is that of an adult sitting on a couch for a substantial fraction of his or her time, with potential exposure of the skin directly or indirectly (through clothing) in contact with the couch. For the ingestion exposure scenario, the subcommittee examined the scenario of a child repeatedly sucking on the treated fabric (for example, chair or couch fabric). For inhalation exposure, the scenario is that of a person spending a considerable fraction of his or her time in a room containing FR-treated furniture from which the FR is shed as small (respirable) particles or from which the FR chemical evaporates. In all these scenarios, it is plausible that some quantities of FR can be transferred from furniture to humans. The subcommittee aims to overestimate the quantities by using exaggerated estimates of values for such controlling factors as the time of exposure or the rate of movement of the FR.

DERMAL EXPOSURE SCENARIO

The subcommittee chose to address the problem in two iterations based on ease of determining the information and impact on the final answer. Release from the fabric might result in direct contact of the chemical with the skin surface. Transfer of this chemical through the skin surface would provide an internal dose. In the first iteration, the release rate was the only limiting factor and any chemical on the skin was assumed to completely transfer within the body and chemicals that would not cause health concerns under this scenario were eliminated from further calculations. For the remaining chemicals, an estimate of the transfer across the skin was used to estimate the internal dose and the hazard index.

---

[1]Exposures to workers involved in the manufacture of FR chemicals, manufacture of FR-treated fabric or furniture using FR-treated fabric were not evaluated.

First Iteration

As a first estimate of exposure, it was assumed that the skin and clothing of the person sitting on fabric would present no barrier to movement of a nonionic FR. Further, it was assumed that there would be sufficient water present (for example, from sweat) to allow dissolution of the nonionic FR in that water, with subsequent transfer to the skin and then into the body of the sitting person. With those assumptions, the only limiting factor on the transfer rate of nonionic FRs is assumed to be the limited dissolution rate of the FR on the fabric. All the FR that dissolves is assumed to be immediately absorbed by the sitting body.

It is generally believed that the skin is an extremely good barrier to ionic chemicals (Grasso and Lansdown 1972). Swarbrick et al. (1984) showed that the ionized form of carboxylic acids penetrates the skin four orders of magnitude more slowly than the nonionized form. For ionic FRs, the subcommittee assumed that the permeability of skin to water (10^{-3} cm/hr) would provide a conservative estimate (EPA 1992).

Under the exposure conditions described above, the dose rate for the dermal route for FRs is calculated using Equation 1:

$$D = \frac{S_a A_b \mu_w f_c}{W_a} \quad \textbf{(1)}$$

where the meanings and values assigned to the symbols are as follows:

D = The dose rate of chemical (mass of chemical per unit body weight per unit time). This rate is the desired value calculated using the above formula.

S_a = The area density of the FR (the application rate to the fabric or back-coating—mass per unit surface area). This value is chemical specific and was chosen at the highest value likely to be used. It ranged from 2 to 7.5 mg/cm^2 depending on the treatment type. (The range was chosen from the experience of the UK's textile market in meeting the UK's furniture-fire regulations).

A_b = The area of body in contact with the couch was chosen to be 2,200 cm^2. This value is based on 8,880 cm^2 for the total body surface of the upper extremities (trunk, arms, and neck) of an adult (EPA Exposure Factors Handbook, Table 4-4). A worst-case estimate of body surface repeatedly in contact with furniture for long periods would be about 1/4 of the bare upper torso, or 2,200 cm^2.

μ_w = The fractional rate (per unit time) of FR extraction by water (e.g., sweat) assumed to be present under the given conditions. This rate is chemical specific; it was generally estimated from extraction measurements or laundering tests and ranged from 0.0004 to 0.038 per day.

f_c = The fraction (dimensionless) of time spent on the couch by the adult was assumed to be 1/4, or 6 hr/d (every day). The subcommittee believes, based on measurements of how people spend their time, that 6 hr/d may be considered a reasonable upper bound.

W_a = The adult body weight (mass) was assumed to be 70 kg.

The computed internal dose rate was then divided by the dermal RfD (external dose) to determine a hazard index. In practice, the oral RfD was substituted for the dermal RfD, because the latter was not available. For those FRs considered to be possibly carcinogenic, an overestimate of lifetime risk was obtained by multiplying the dose rate by the carcinogenic potency slope (q_1*) for the FR.

If the hazard index for a particular chemical was less than one with the assumptions just described, the subcommittee considered the exposures via the dermal route to be sufficiently small to merit no further examination. If the hazard index exceeded one in the first iteration, an alternative iteration of the exposure assessment was performed in which some consideration was given to the skin as a barrier to penetration.

Alternative Iteration

For the alternative iteration of the dermal assessment, the exposure assumptions were the same as those in the first iteration, except that the assumption of 100% immediate absorption of all the FR that dissolved was modified. Instead, an estimate of the rate at which the FR could penetrate the skin was made, assuming that the FR dissolved up to its solubility limit in water. That rate of penetration was then factored into the exposure assessment.

The rate of penetration of a chemical through skin may be estimated using the skin permeability coefficient (K_p, with dimensions of velocity)—the total mass penetration rate is the product of water concentration, permeability coefficient, and skin area. Such coefficients have not been measured for the FR chemicals, but they may be estimated from the octanol-water partition coefficient (K_{ow}, dimensionless) and molecular weight (m, mass/unit amount of substance) by using a correlation (Potts and Guy 1992) that may be written in dimensionless form (Equation 2):

$$\log_{10}\left(\frac{K_p}{K_{p0}}\right) = -6.3 + 0.71\log_{10} K_{ow} - 0.0061\frac{m}{m_0} \tag{2}$$

where

K_{p0} = 1 cm/sec is a reference permeability coefficient, and

m_0 = 1 g/mol is a reference molecular weight.

Using the permeability coefficient, the dose rate for the dermal route (alternative iteration) is obtained using Equation 3:

$$D = \frac{C_w K_p A_b f_c}{W_a} \tag{3}$$

where the meanings and values assigned to the symbols are the following:

D = The dose rate of a chemical (mass per unit body weight per unit time). This rate is the desired value calculated using the above formula.

C_w = The water solubility of FR (mass/unit volume), which is different for each chemical.

K_p = The permeability coefficient (length/time), which is different for each FR and calculated from the correlation given above.

A_b = The area of the body in contact with the couch (equal to 2,200 cm^2) as described for the first iteration.

f_c = The fraction (dimensionless) of time spent on the couch by the adult (1/4, or 6 hr/d), as described for the first iteration

W_a = The adult body weight (mass) (70 kg), as described for the first iteration.

Because the exposures of interest were 6 hr per day over a lifetime, this equation uses the historical steady-state-flux relationship instead of an alternative which adjusts for the concentration of chemical in the skin (EPA 1992). Once again, this dose rate was divided by the oral RfD (as the best estimate of the internal dose for comparison with the calculated internal dose from dermal exposures) to obtain a hazard index. If the hazard index was less than one, the subcommittee considered the dermal exposure route to be sufficiently small to merit no further attention. Also, as in the first iteration, for those FRs considered to be possibly carcinogenic, an overestimate of lifetime risk was obtained by multiplying the dose rate by the carcinogenic potency slope (q_1^*) for the FR.

In cases where both iterations were performed for an FR, the lower of the two calculated dose rates was taken as the subcommittee's best available estimate of a conservative dose rate. Each iteration examines the effects of just one mechanism that limits dose rates. The first examines just the dissolution rate of the FR in water, and the second just the barrier presented by the skin to permeation by the FR. In reality, both of these mechanisms, and more besides, act to limit dose rates. It would be possible to include both of these mechanisms in a single model, but to do so would require information beyond that required when modeling each mechanism in assumed isolation. Each of the mechanisms considered introduces a resistance to movement of the FR into the body, and their combination introduces more resistance than either alone. However, the combined effect of both mechanisms cannot reduce the dose rate by more than a factor of two below the lower of the two dose rates estimated in the two iterations–that is, for each mechanism acting in isolation.

There are many uncertainties in the dermal-exposure parameters and the calculations of the hazard indices. The subcommittee believes that the actual exposures are at least 100-fold lower than calculated. Parameters and calculations were deliberately chosen to provide a worst-case estimate. The subcommittee wanted to be sure that chemicals for which no further research was recommended would be safe, so conservative choices were made at every practical juncture.

The most conservative assumptions in the dermal-exposure parameters relate to the body-surface area exposed, fraction of the day on the couch, and the daily exposure for a lifetime. The assumption of 2,200 cm^2 is based on 8,880 cm^2 for the total body surface of the upper extremities (trunk, arms, and neck) of an adult (EPA *Exposure Factors Handbook*, Table 4-4). One-quarter of the body-surface area not covered by clothing and being in contact with a couch is a very high estimate of exposed surface area. According to Table 14-2 of the *Exposure Factors Handbook*, the greatest time spent on any of the activities that might be done on the couch (such as watching TV, reading, or conversing) was 4.4 hr on a weekend for a male 12-17 yr old. For males and females of other ages, exposure time was significantly less. On weekdays, exposure for all groups was less. The assumption that the exposure occurs daily for a lifetime is also very conservative, and it assumes no deviation from the exposure for any reason.

The most conservative assumptions in calculating hazard indices are that the constant release rate of FRs is for the life of the fabric, permeability coefficients are probably overestimated for the high-molecular-weight chemicals, the applied chemicals do not react in the fabric, and there is no attempt to limit the exposures by the actual amount of chemical in the fabric. An assumption that is not conservative is the use of the oral RfD as the internal dose for comparison with the calculated internal dose. The dermal iterative calculations assumed that the lesser of the release rate or the dermal absorption rate was rate limiting, when in actuality both would be rate limiting. Toxicity based on oral studies for chemicals that might have been incompletely absorbed would underestimate risk by the dermal route when the hazard index was calculated.

INHALATION EXPOSURE SCENARIO

Particles

In the inhalation exposure scenario, a person spends some fraction of his or her time in a room containing FR-treated upholstered furniture. Some of the FR might be worn away during everyday use of the upholstery, and some of the particles so eroded might be small enough to be entrained into the air of the room and be inhaled. The concentration of such small particles in the room air will depend on the amount of upholstery in the room, the volume of the room, and how fast air is drawn through the room (the air exchange rate). The average concentration of FR present on the upholstery fabric or as small airborne (respirable) particles is estimated using Equation 4:

$$C_p = \frac{S_a A_c \mu_r}{V_r R_v} \quad \textbf{(4)}$$

where the meanings and values assigned to the symbols are

C_p = The average concentration (mass/unit volume) of FR attached to respirable particles in the room, calculated using the above formula.

S_a = The area density of the FR (application rate to the fabric or back-coating—mass per unit surface area). This value is chemical specific, and the highest value likely to be used was chosen. It ranged from 2 to 7.5 mg/cm^2, depending on the treatment type.

A_c = The area of FR-treated fabric within the room. A suite of furniture using 30 m^2 of fabric was chosen as reasonably large compared with the small room size (see V_r below).

μ_r = The release rate (per unit time) for the FR as respirable particles that are entrained into room air; see the discussion below.

V_r = The room volume, chosen as 30 m^3 (about 12 ft × 11 ft × 8 ft) to represent a fairly small room to contain such a suite of furniture. The ratio of fabric area to room volume drives the FR concentration.

R_v = The air exchange rate (air changes per unit time) within the room, chosen as 0.25 air changes per hour, corresponding to the low end of the range of measured values in houses (EPA 1997, Murray and Burmaster 1995).

The subcommittee is unaware of any measurements of μ_r, the release rate of the FR as respirable particles. It was therefore estimated by considering the possible loss of FR over an average lifetime (about 15 yr) of furniture might amount to 50% in a 25%-fraction of upholstery fabric surfaces (the upper cushion surfaces and the upper backrest of a couch, for example) that might receive heavy wear. Of the worn material, perhaps 1% might be in the form of particles small enough to be considered respirable. Thus, the Release Rate was estimated using Equation 5:

$$\mu_r = \frac{f_l f_w f_r}{T_f} \qquad \textbf{(5)}$$

where the meanings and values assigned to the symbols are the following:

μ_r = The release rate (per unit time) for the FR as respirable particles that are entrained into room air, calculated using this equation.

f_l = The fraction (dimensionless) of FR remaining in worn areas after the lifetime of the upholstery (0.50).

f_w = The fraction (dimensionless) of the upholstery that is relatively heavily worn (0.25).

f_r = The fraction (dimensionless) of particles released by wear that are respirable (0.01).

T_f = A typical lifetime for upholstery (15 yr). The upholstery is assumed to be replaced after this period, so that exposure continues for a lifetime.

With these assumptions, the release rate is approximately 2.3×10^{-7}/d. With the other parameter values discussed above, the resulting estimated FR particle concentration ranges from 0.8 to 3 $\mu g/m^3$. Such concentrations may be compared with typical total indoor air concentrations of respirable particles on the order of 100 $\mu g/m^3$ (EPA 1996).

From the average indoor concentration, a time-averaged exposure concentration for a person using the room was estimated using Equation 6:

$$C_{p,avg} = C_p f_i \qquad \textbf{(6)}$$

where the meanings and values assigned to the symbols are the following:

$C_{p,avg}$ = The time-average exposure concentration (mass/unit volume), calculated using the above equation.
C_p = The average concentration (mass/unit volume) of FR attached to respirable particles in the room, estimated as already described.
f_i = The fraction (dimensionless) of time spent in the room containing the upholstery (0.25).

For each FR, the time-average exposure concentration was divided by the provisional inhalation RfC derived from the chemical's oral RfD in order to calculate a hazard index. When necessary, the estimated RfC was calculated using Equation 7:

$$\mathrm{RfC} = \mathrm{RfD}\frac{W_a}{v_b} \qquad (7)$$

where the meanings and values assigned to the symbols are the following:

RfC = Inhalation reference concentration (mass/unit volume).
RfD = Oral reference dose rate (mass per unit body weight per unit time).
W_a = The adult body weight (mass), assumed to be 70 kg.
v_b = The nominal adult breathing rate (volume/unit time), assumed to be 20 m^3/d.

If the hazard index for a particular chemical was less than one with the assumptions just described, the subcommittee considered that inhalation exposures to FR particles would be sufficiently small to merit no further examination.

For those FRs considered to be possibly carcinogenic, an overestimate of lifetime risk was obtained by multiplying the average dose rate by the carcinogenic potency slope (q_1*) for the FR.

Vapors

In addition to the possibility of release of FR chemicals as particles worn from upholstery fabric, the subcommittee considered the possibility of their release by evaporation. For the ionic chemicals with vapor pressures that are extremely small, such releases are negligible. The vapor pressures of some of the nonionic FR chemicals, however, are not negligible in their as-applied state, but the chemicals might be in considerably altered form (for example, cross-linked or oxidized, chemically bound to the fabric, or physically adsorbed to the fabric). Therefore, the calculations discussed below, using the vapor pressure of the pure and unmodified chemical, must be carefully interpreted in each individual case and might have to be rejected outright as being so far from the truth as to be useless.

The rate of flow of vapor from the room is given by expression 8:

$$V_r R_v \left(C - \lambda C_v\right) \qquad \textbf{(8)}$$

and the rate of emission from the upholstery may be estimated by expression 9:

$$\frac{\xi A_c D_a}{d}\left(\gamma C_v - C\right) \qquad \textbf{(9)}$$

where the meanings and values assigned to the symbols are the following:

C = The equilibrium vapor concentration (mass/unit volume) within the room (calculated as shown below).

C_v = The saturated vapor concentration (mass/unit volume) at room temperature. This varies for each chemical.

λ = The saturation fraction (dimensionless) of the chemical in air entering the room (less than or equal to one). For all the FR chemicals, this value is assumed to be zero, because the air entering the room is not expected to be contaminated with the FR vapor. For water vapor (see below for context), a value of 0.7 is used (i.e., 70% relative humidity).

A_c = The area of FR-treated fabric within the room. A suite of furniture using 30 m^2 of fabric was chosen as reasonably large compared with the small room size discussed below.

V_r = The room volume, chosen as 30 m^3 (about 12 ft × 11 ft × 8 ft) to represent a fairly small room to contain such a suite of furniture. The concern is the ratio of fabric area to room volume.

R_v = The air exchange rate (number of air changes per unit time) within the room, chosen as 0.25 air changes per hour, corresponding to the low end of the range of measured values in houses.

ξ = The fraction (dimensionless) of surface area of the fabric that is not occluded and so might release vapor (the areas beneath cushions would be occluded, for example) (assumed to be 0.75).

D_a = The diffusivity (area/unit time) of the chemical vapor in air. For water (see below), the vapor diffusivity in air is approximately 2.2×10^{-5} m^2/sec. Most organic chemicals have vapor diffusivities in air within a factor of two of 5×10^{-6} m^2/sec, and this value is used in the estimates.

d = A boundary layer thickness (length) corresponding to a relatively undisturbed layer of air around the upholstery (estimated as 0.01 m (1 cm), as explained below).

γ = A dimensionless factor (less than or equal to one) to account for adsorption of the chemical in, or binding of the chemical to, the fabric or other materials incorporated as part of the FR treatment. For these conservative estimates, this factor is assumed to be one.

Equating the rate of flow of vapor from the room and the rate of emission of vapor from the upholstery allows estimation of the equilibrium vapor concentration in room air. It is calculated using Equation 10:

$$C = C_v \frac{\lambda \eta + \gamma}{1 + \eta} \quad \text{where} \quad \eta = \frac{R_v V_r d}{\xi A_c D_a} \qquad \textbf{(10)}$$

where the meanings and values of the symbols are as previously assigned. The time for which this equilibrium vapor concentration could be maintained within the room, before all the FR originally applied to the fabric evaporates may be calculated using Equation 11:

$$t_v = \frac{S_a d}{C_v D_a} \left(\frac{1 + \eta}{\eta(\gamma - \lambda)} \right) \qquad \textbf{(11)}$$

where the meanings and values assigned to the symbols not defined immediately above are the following:

t_v = The time for which the equilibrium vapor concentration could be maintained (computed from the above equation).

S_a = The area density of the FR (application rate to the fabric or back-coating—mass per unit surface area). This value is chemical specific, and the highest value likely to be used was chosen. It ranged from 2 to 7.5 mg/cm^2 depending on the treatment type.

The boundary layer thickness (d) was initially estimated as 1 cm. That value was confirmed approximately by applying the theoretical approach just described to the drying of wet upholstery. With influent air at 70% relative humidity (lamda = 0.7) and with the fabric assumed to be wetted at 1.5 kg/m^2 (150 mg/cm^2), the resultant estimated drying time is about 35 hr if drying is not limited by the air-change rate (as obtained in the equations above by setting $\eta \rightarrow \infty$). That drying time corresponds within a small factor with common experience. With the other parameter values used, estimates of FR concentration are relatively insensitive to the exact value chosen for the boundary-layer thickness or to the diffusivities of the individual FRs—evaporation is principally limited by the room air-exchange rate.

From the equilibrium vapor concentration in room air, the short-term time-average vapor exposure concentration was estimated using Equation 12:

$$C_{s,avg} = Cf_i \quad \textbf{(12)}$$

where the meanings and values assigned to the symbols are the following:

$C_{s,avg}$ = The short-term time-average vapor exposure concentration (mass/unit volume) (calculated from the above equation).

C = The equilibrium vapor concentration in room air (mass/unit volume) (calculated as described earlier).

f_i = The fraction (dimensionless) of time spent in the room containing the upholstery (0.25).

To calculate a hazard index for each FR, the short-term time-average vapor exposure concentration was divided by the RfC or provisional inhalation RfC derived from the oral RfD as calculated in Equation 7. If the hazard index for a particular chemical was less than one with the assumptions just described, the subcommittee considered that exposures via the vapor inhalation route would be sufficiently small to merit no further examination.

For all the FRs examined in this report, the fraction γ used to account for adsorption of the FR to the fabric has been taken to be one, since no better information was available for any of the FRs. In practice, for several of the FRs this assumption leads to results that are plainly implausible, in that the time for which the FR would remain on the fabric would be very limited. Any useful FR obviously has to remain on the fabric for a period of years, so that in such cases it is clear that the assumption $\gamma = 1$ is incorrect—the FR is likely to be strongly bound to the fabric, which would have the effect of making γ much smaller. The subcommittee, nevertheless, lists the results of the calculations and points out the situations where it considers the results implausible. Where appropriate, the results are used to indicate that further information on the evaporation potential of the particular FR would be desirable.

For each FR, the long-term time-average vapor exposure concentration was estimated from the equilibrium vapor concentration in room air using Equation 13:

$$
\begin{aligned}
C_{l,avg} &= Cf_i && \text{for} \quad t_v > T_f \\
C_{l,avg} &= Cf_i \frac{t_v}{T_f} && \text{for} \quad t_v < T_f
\end{aligned}
\qquad \textbf{(13)}
$$

where the meanings and values assigned to the symbols are the following:

$C_{l,avg}$ = The long-term time-average vapor exposure concentration (mass/unit volume) (calculated using the above equation).

C = The equilibrium vapor concentration in room air (mass/unit volume) (calculated as described earlier).

f_i = The fraction (dimensionless) of time spent in the room containing the upholstery (0.25).

t_v = The time for which the equilibrium vapor concentration could be maintained (computed as described earlier).

T_f = A typical lifetime for upholstery (15 yr). The upholstery is assumed to be replaced after this period, so exposure continues for a lifetime.

For those FRs considered to be possibly carcinogenic, an overestimate of lifetime risk was obtained by multiplying the long-term time-average vapor exposure concentration by the inhalation unit risk for the chemical. If an inhalation unit risk was not available, the unit risk was estimated from the oral carcinogenic potency by using Equation 14:

$$
U = q \frac{v_b}{W_a} \qquad \textbf{(14)}
$$

where the meanings and values assigned to the symbols are the following:

U = The estimated unit risk (volume/mass) for the specific FR (calculated using the above equation).

q = The oral carcinogenic potency (time times body weight per unit mass) for the specific FR.

W_a = The adult body weight (mass) (assumed to be 70 kg).

v_b = The nominal adult breathing rate (volume/unit time) (assumed to be 20 m^3/d).

It was pointed out earlier that the estimates for short-term exposure concentrations for the FRs might be substantially erroneous because of lack of information about the binding of the FR to the fabric. For the estimates of lifetime risk presented here, the uncertainty is somewhat smaller, because of the correction introduced to take into account the time for which vapor concentrations could be maintained. The error in short-term estimates comes from an overestimation of the emission rate, but the lifetime risk estimate is independent of the emission rate if the FR completely evaporates within the typical lifetime of the upholstery.

Uncertainty in the Inhalation Exposure Estimates

There are various uncertainties in inhalation exposure estimates. Given the conservative nature of the estimation procedure, the subcommittee believes that actual exposures are likely to be at least 100-fold lower than calculated. The most conservative assumptions relating to the inhalation exposure scenario are that vapor and particles will be released uniformly for the duration of the exposure and the ratio of room volume to fabric surface area. The most conservative assumptions in the calculation of hazard indices are constant release rate for the life of the fabric, complete absorption from the breath, the assumption that applied chemicals did not react in the fabric, no attempt to limit the exposures by the actual amount of chemical in the fabric, and no attempt to limit the vapor concentration by the actual achievable air concentration.

ORAL EXPOSURE SCENARIO

The exposure scenario for oral exposure is a child repeatedly sucking on upholstery fabric. It is assumed that a young child might repeatedly suck on the fabric for some fraction of time, wetting a different area with saliva each time it occurs. Such behavior might continue for a couple of years. The limiting factor is the rate at which the FR dissolves into the saliva. It is assumed that all the dissolved material is ingested by the child. Under those conditions, the dose rate to the child is calculated using Equation 15:

$$D = \frac{S_a A_f \mu_a f_{cc}}{W_c} \tag{15}$$

where the meanings and values assigned to the symbols are the following:

D = The dose rate of chemical (mass per unit body weight per unit time). This rate is the desired value calculated using the formula given.

S_a = The area density of the FR (application rate to the fabric or back-coating—mass per unit surface area). This value is chemical specific, and the highest value likely to be used was chosen. It ranged from 2 to 7.5 mg/cm^2, depending on the treatment type.

A_f = The area of fabric sucked on each occasion. The subcommittee estimated that 50 cm^2 (about 7.75 square inches) would be a suitable value.

μ_a = The fractional rate (per unit time) of FR extraction by saliva under the given conditions. This rate is chemical specific. It was generally estimated from extraction measurements or laundering tests and ranged from 0.0004 to 0.038 per day.

f_{cc} = The fraction (dimensionless) of the time a child sucks FR-treated fabric. The subcommittee considered a suitable estimate to be 1/24 or 1 hr/d.

W_c = The body weight of the child, assumed to be 10 kg, which is close to the average weight of 1-yr-old children.

The calculated dose rate for each FR was then divided by the oral RfD of that FR to determine a hazard index. If the hazard index was less than one, the subcommittee considered the oral exposure route to be sufficiently small for that FR to merit no further attention.

For those FRs considered to be possibly carcinogenic, the lifetime average dose rate was calculated by taking into account the period during which a child might continue sucking behavior. The lifetime average dose rate was calculated using Equation 16:

$$D_{ave} = D\frac{T_s}{T_L} \tag{16}$$

where the meanings and values assigned to the symbols are the following:

D_{ave} = Lifetime average dose rate (mass per unit body weight per unit time) (calculated using the above equation).
D = Average dose rate (mass per unit weight per unit time) during the period of exposure (calculated using the previous formula).
T_s = Length of time of exposure during childhood (assumed to be 2 yr).
T_L = Standard length of lifetime used in the definition of the carcinogenic potency calculations (70 yr).

An overestimate of lifetime risk was obtained by multiplying the lifetime average dose rate by the carcinogenic potency slope (q_1*) for the FR.

Uncertainties in the Oral Exposure Estimate

There are various uncertainties in oral-exposure estimates. The subcommittee believes that actual exposures are likely to be at least 100-fold lower than calculated, given the conservative nature of the estimation procedure. The most conservative assumptions relating to the oral exposure scenario are the surface area sucked, and that this would occur daily for two years. It is hard to imagine a child actually doing this. The most conservative assumptions in the calculation of hazard indices are constant release rate for life of fabric, complete oral absorption, assumption that applied chemicals did not react in the fabric, and no attempt to limit the exposures by the actual amount of chemical in the fabric.

REFERENCES

EPA (U.S. Environmental Protection Agency). 1992. Dermal Exposure Assessment: Principles and Applications. EPA/600/8-91-011B. Office of Health and Environmental Assessment, U.S. Environmental Protection Agency, Washington, DC.

EPA (U.S. Environmental Protection Agency). 1996. Air Quality Criteria for Particulate Matter, Volume 1 of 3. EPA 600/P-95/001aF. National Center for Environmental Assessment, Office of Research and Development, U.S. Environmental Protection Agency, Research Triangle Park, NC.

EPA (U.S. Environmental Protection Agency). 1997. Exposure Factors Handbook, Volume I - General Factors, Update to Exposure Factors Handbook EPA/600/8-89-043. EPA/600/P-95/002Fa. Office of Research and Development, National Center for Environmental Assessment, U.S. Environmental Protection Agency, Washington, DC.

Grasso, P., and A.B.G. Lansdown. 1972. Methods of measuring, and factors affecting percutaneous absorption. J. Soc. Cosmet. Chem. 23:481-521.

Murray, D.M., and D.E. Burmaster. 1995. Residential air exchange rates in the United States: Empirical and estimated parametric distributions by season and climactic region. Risk Anal. 15(4):459-466.

Potts, R.O., and R.H. Guy. 1992. Predicting skin permeability. Pharm. Res. 9(5):663-669.

Swarbrick, J., G. Lee, J. Brom, and N.P. Gensmantel. 1984. Drug permeation through human skin II: Permeability of ionizable compounds. J. Pharm. Sci. 73(10):1352-1355.

4

Hexabromocyclododecane

THIS chapter reviews the physical and chemical properties, toxicokinetics, toxicological, epidemiological, and exposure data on hexabromocyclododecane (HBCD). The subcommittee used that information to characterize the health risk from exposure to HBCD. The subcommittee also identified data gaps and recommended research relevant for determining the health risk from exposure to HBCD.

PHYSICAL AND CHEMICAL PROPERTIES

HBCD is a cyclic aliphatic flame retardant. Its physical and chemical properties are presented in Table 4-1.

OCCURRENCE AND USE

HBCD is a solid, white powder that is used as a flame retardant additive for thermoplastic polymers. Its principal use is in expanded polystyrene foams and other styrene resins. It may also be used in latex binders, unsaturated polyesters, and polyvinyl chloride wire, cable, and textile coatings. When used in textiles, it is applied as a back coating to the fabric, encapsulated in a polymer matrix. Textile applications include residential and commercial furniture, up-

TABLE 4-1 Physical and Chemical Properties of HBCD

Property	Value	Reference
Chemical formula	$C_{12}H_{18}Br_6$	CHEMID 1999
Chemical structure	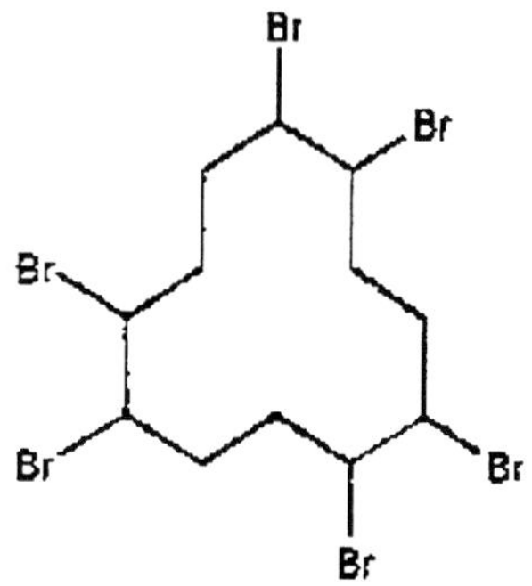	
CAS Registry #	25637-99-4 (mixed isomers)	CHEMID 1999
Molecular weight	641.7	Hatlelid 1999
Melting point	185-195°C	Hatlelid 1999
Vapor pressure	4.7×10^{-7} mm Hg	Stenzel and Nixon 1997
Density	2.24 g/cm^3	Hatlelid 1999
Solubility in water	0.0034 mg/L	Stenzel and Markley 1997
Partition coefficient (Log K_{ow})	5.6	MacGregor and Nixon 1997

holstery seating in transportation, draperies, and wall coverings (FRCA 1998). HBCD is usually applied with antimony trioxide as a back coating in a mass ratio of 2:1 (i.e., about 6-15% HBCD and 4-10% antimony oxide by weight).

TOXICOKINETICS

No human data on the toxicokinetics of HBCD were located for any route. No toxicokinetic studies via the dermal or inhalation exposure routes were reported in experimental animals. However, in a report by Dean and Leong (1977), rats exposed dermally to a high dose of HBCD in saline experienced diarrhea and slight weight loss. This finding indicates that at least some absorption occurs via the dermal route.

In an unpublished study by Vesicol Chemical Corporation (1980), rats administered a single oral dose of 1.93 mg of radiolabeled HBCD eliminated 86% of the dose within 72 hr. (The total dose administered was 7-9 mg/kg body weight.) Absorption from the gastrointestinal tract reportedly occurred rapidly, with a half-life of 2 hr. However, the amount of the absorbed fraction was not reported. HBCD was reported to be rapidly metabolized and eliminated in the feces and urine following absorption, with 70% of the administered radioactivity eliminated in the feces and another 16% eliminated in the urine 72 hr after dosing. A two-compartment model was constructed, with non-adipose tissues in one compartment and adipose tissue in the other. Elimination from the adipose compartment was reported to be slower than elimination from the non-adipose compartment, although elimination half-times were not provided in the review. In another study by Arita et al. (Marcia Hardy, Albermarle Corporation, Pers. Commun., August 3, 1999), HBCD was orally administered to male Wistar rats (number not reported) in olive oil at 500 mg/kg-d for 5 d. HBCD was found to be present only in adipose tissue, and in none of the other organs examined (i.e., spleen, pancreas, liver, kidneys, and heart). HBCD was found to be excreted in the feces, with an average of 32-35% of the cumulative administered dose excreted. No HBCD was found in the urine. Although differences in study design, including the test vehicle and the analytic methods used, may account for some of the difference in the results, both studies by Vesicol Chemical Corporation (1980) and Arita et al. (Maria Hardy, Albermarle Corporation, Pers. Commun., August 3, 1999) suggest that following acute oral doses, HBCD is rapidly absorbed from the gastrointestinal tract, distributed primarily to the body fat, and eliminated rapidly, primarily in the feces.

HAZARD IDENTIFICATION[1]

Dermal Exposure

Irritation

McDonnell (1972) reported no irritant effects in men or women (number not reported) who wore 1-inch squares of Tyvek T-12 fabric treated with 10% HBCD for 6 d on their arms or legs. No details on the method of fabric treatment or description of fabric samples were provided.

[1]In this section, the subcommittee reviewed the data on toxicity of HBCD, including the toxicity assessment prepared by the U.S. Consumer Product Safety Commission (Hatlelid 1999).

Dermal Sensitization

Momma et al. (1993) and Nakamura et al. (1994) reported positive sensitization reactions in guinea pig maximization tests using HBCD induction concentrations of 5% for intradermal injection and 25% for topical application, and HBCD topical challenge concentrations up to 5%. Several studies demonstrated no effects or slight erythema and scaling in rabbits following application of HBCD to the skin for up to 24 hr (Zeller 1962; Dean and Leong 1977; Lewis and Palanker 1978; Crown 1984). A recent guinea pig maximization test conducted by Microbial Associates (1996) found no effects from HBCD. This study used a 5% concentration for the intradermal injection and neat HBCD (moistened with corn oil) for the topical application in both the induction and challenge phases. The previous positive tests (Momma et al. 1993; Nakamura et al. 1994) also used a 5% concentration for the intradermal injection, but used 5% or lower concentrations for the topical applications. The reason for the discrepancy between the Microbial Associates (1996) study and the earlier studies is not apparent. However, the negative results in the Microbial Associates (1996) study, which appears to have been well conducted and used the highest possible concentration for topical induction and challenge, raise questions about the potential of HBCD to produce even a mild sensitization reaction in humans.

Systemic Effects

Several acute toxicity studies in rats and rabbits were conducted via the dermal route of exposure. However, no subchronic or chronic dermal exposure studies were located in the literature. Dean and Leong (1977) shaved skin of two male and two female rabbits and applied HBCD (in 0.9% saline) at a dose of 20 g/kg and occluded the skin for 24 hr. The authors observed diarrhea and slight weight loss in one of two males and one of two female New Zealand white rabbits. No effects were noted in rabbits (number, sex, and strain not reported) after a similar exposure to 8 g/kg of HBCD (Lewis and Palanker 1978).

Other Systemic Effects

No immunological, neurological, reproductive, developmental, or carcinogenic effects were identified following dermal exposure to HBCD.

Inhalation Exposure

Systemic Effects

No subchronic or chronic inhalation exposure studies were located in the literature; however, one acute inhalation exposure study was found. Lewis and Palanker (1978) reported no effects in groups of rats (10 rats/group, 5 males and 5 females) exposed to HBCD at a concentration of 200 g/m^3 for 1 hr. Rats exposed to 200 g/m^3 for 4 hr exhibited only slight dyspnea. The study authors concluded that acute exposures to very high concentrations of HBCD dust were well tolerated by the rats.

Other Systemic Effects

No studies were identified in the literature that investigated the immunological, neurological, reproductive, developmental, or carcinogenic effects of HBCD following inhalation exposures.

Oral Exposure

A summary of toxicity studies from oral exposures to HBCD is presented in Table 4-2.

Systemic Effects

No data on oral exposures to HBCD were located for humans. Several studies reported that a single oral dose of 10 g/kg in rats produced hypoactivity, diarrhea, and matted hair, while a single oral dose of 5 g/kg had no effects and this dose was identified as a no-observed-adverse-effects level (NOAEL) (Dean and Leong 1977; Lewis and Palanker 1978; Nissimov 1984). In mice, a single oral dose of 6.4 g/kg produced apathy, trembling, and death (Schulze 1962).

Zeller and Kirsch (1969) conducted a 28-d feeding study in which groups of 20 rats (10 male and 10 female Sprague-Dawley rats were fed diets containing 0%, 1%, 2.5%, or 5.0% of HBCD for 28 d. Based on food intake and body weight data obtained from the study, the estimated doses in the 0%, 1%, 2.5%, and 5% groups were determined to be 0, 900, 2,400, and 4,700 mg/kg-d in males and 0, 900, 2,300, and 4,900 mg/kg-d in females. No mortality was reported. Rats in the 2.5% and 5% groups were in poor condition after the first

TABLE 4-2 Oral Noncancer Studies of Hexabromocyclododecane

Animal, Species, Sex, Number	Dose	Duration, Route	Effects	NOAEL/LOAEL	Reference
Rat, Sprague-Dawley, M/F, 10/dose	0, 900, 2,350, 4,800 mg/kg-d	28 d, diet	Increased absolute and relative liver weight; thyroid follicular hyperplasia	LOAEL: 900 mg/kg-d	Zeller and Kirsch 1969[b]
Rat, Sprague-Dawley, M/F, 20/dose	0, 100, 200, 450, 925 mg/kg-d	90 d, diet	Dose-related increase in absolute and relative liver weight in both sexes; dose-related increased incidence and severity of liver fatty accumulation; slight decrease in male body weight	NOAEL: 450 mg/kg-d LOAEL: 925 mg/kg-d	Zeller and Kirsch 1970[b]
Rat, Sprague-Dawley, M/F, 6/dose	0, 125, 350, 1,000 mg/kg-d	28 d, gavage in corn oil	Dose-related increase in liver weight	NOAEL: ND LOAEL: ND	Chengelis 1997[b]
Rat, Wistar, F, 13-14/dose	0, 0.01, 0.1, 1% (HDT ≈ 900 mg/kg-d)	Gestation d 0-20, diet	No developmental effects; increase in maternal liver weight	NOAEL: ~500 mg/kg-d	Murai et al. 1985[a]
Rat, Crl:CD®(SD)16 SBR, F, 8/dose	0, 125, 250, 500, 750, 1,000 mg/kg-d	Gestation d 0-20, gavage	No developmental effects; increase in maternal body weight at d 19-20	NOAEL: ~1,000 mg/kg-d	Stump 1999[a]
Mice, NR, M/F, 50/dose	0, 100, 1,000, 10,000 ppm (HDT = ~1,300 mg/kg-d)	18 mo, diet	No effect on growth; no effect on survival; in males, liver hypertrophy, fatty change, vacuolation; altered foci	NOAEL: 100 ppm (13 mg/kg-d) LOAEL: 1,000 ppm (130 mg/kg-d)	Kurokawa et al. (Marcia Hardy, Albermarle Corp., Pers. Commun., Aug. 3, 1999)[c]

F, female; HDT, highest dose tested; LOAEL, lowest-observed-adverse-effect level; M, male; ND, not determined; NOAEL, no-observed-adverse-effect level; NR, not reported.

[a]Published study.

[b]Unpublished Good Laboratory Practice study.

[c]Unpublished non-Good Laboratory Practice study.

2 wk of the study; observations in both dose groups included alopecia and unsteady gait. For males and females in the 5% group, the following variables were reduced: food intake (13-22%), final body weight (14-24%), and body weight gain (37-46%). For all groups, hematology and urinalysis results were similar to those of controls. For males and females in the 1%, 2.5%, and 5% groups, there were statistically significant ($p < 0.001$) dose-related increases in liver weight, liver:body weight ratio, and liver:heart weight ratio. There were no effects on kidney or heart weight. Histopathological examination revealed thyroid follicular hyperplasia in both males and females in the 1%, 2.5%, and 5% groups that increased in severity with dose. Marked hyperplastic thyroid tissue with adenomatous proliferation and epithelial hyperactivity was observed in the 5% group. Females in the 5% group also showed inhibited oogenesis, with reduced numbers of mature and developing follicles present in the ovaries. The testes and epididymides of males in the 5% group were normal. Using this study, a LOAEL of 1% (equivalent to 900 mg/kg-d) for HBCD was identified, based on markedly increased absolute and relative liver weights and thyroid follicular hyperplasia. No NOAEL was identified.

Based on findings in the 28-d study, Zeller and Kirsch (1970) treated groups of 20 male and 20 female Sprague-Dawley rats with 0%, 0.16%, 0.32%, 0.64%, or 1.28% of HBCD in the diet for 90 d. Based on food intake and body weight data obtained from the study, the doses in the 0%, 0.16%, 0.32%, 0.64%, and 1.28% groups were estimated to be 0, 100, 200, 400, and 900 mg/kg-d in males and 0, 100, 200, 500, and 950 mg/kg-d in females. Additional groups of 10 rats of each sex were treated with 0% or 1.28% of HBCD in the diet for 90 d, and observed for an additional 42 d prior to killing.

One male rat in the 1.28% group died on the 43rd day of the study. It is not clear whether this death was related to treatment. No clinical signs were noted in any group. Body weight was slightly, but consistently, reduced (≈4%) throughout the study in the males from the 1.28% group; body weight was not reduced in females. Food intake was not affected in any group. Hematology and urinalysis results were similar to those of controls in all groups. There were statistically significant ($p < 0.05$) differences in liver weight, liver:body weight ratio, and liver:heart weight ratio in males and females in the 0.16%, 0.32%, 0.64%, and 1.28% groups in comparison to controls. Other organ weight changes were not clearly related to treatment. Histopathological examination revealed treatment-related changes in the liver consisting of dose-related increases in the incidence and severity of lipoid phanerosis (fatty accumulation) (see Table 4-3).

In the 1.28% recovery group maintained for 6 wk after the end of the exposure period, food intake and body weight gain were similar to controls. Liver histology was also similar to controls. Liver weight and liver:body weight

TABLE 4-3 Summary of Histological Findings in the Liver of Male and Female Rats Fed Diets Containing up to 1.28% HBCD for 13 Wk (Data from Zeller and Kirsh 1970)

	Male (n = 20)[a]			Female (n = 20)		
	VS	VS	DAD	VS	VS	DAD
0	3	1	1	10	0	5
0.16%	8	0	0	11	0	4
0.32%	8	3	1	9	0	7
0.64%	11[b]	1	2	17[b]	2	6
1.28%	10[b]	9[b]	6[b]	14	2	10

[a]This summary appears to include histological findings in the male from the 1.28% group that died during the study, although this is not stated explicitly in the report.

[b]$p < 0.05$ by Fisher Exact Test conducted by Syracuse Research Corporation.

Abbreviations: DAD, disseminated adipose droplets; MS, moderately slight adipose specks; VS, very slight adipose specks.

ratio remained significantly elevated compared to controls, but the difference was much less than it was at the end of the exposure period.

The 1.28% dose level (approximately 925 mg/kg-d) was chosen as the LOAEL based on increased liver weight (absolute and relative in both sexes) accompanied by an increased incidence of distinctly abnormal fatty accumulation in the liver and, in males, a small reduction in body weight gain. The 0.64% level (approximately 450 mg/kg-d, the mean dose level for males and females) was selected as the NOAEL because the more subtle changes in liver weight and histology were not determined to be clearly adverse.

A recent 28-d study was conducted by Chengelis (1997). Groups of six male and six female Sprague-Dawley rats (about 43-d-old) were administered 0, 125, 350, or 1,000 mg/kg-d of HBCD by gavage in corn oil for 28 consecutive days. Absolute liver weight was statistically increased in high-dose males and in mid- and high-dose females, while relative liver weights were increased in the mid- and high-dose males and in low-, mid-, and high-dose females. These increased liver weights were not, however, accompanied by related histopathological or serum chemistry changes. There was also a slight increase in severity of colloid loss in the thyroid in males in the 1,000-mg/kg-d group, but no other differences in the thyroid in comparison to controls were observed. Both the increases in liver weight and in colloid loss were reduced in additional groups of six rats of each sex that were treated with 1,000 mg/kg-d of HBCD for 28 d, but then were allowed to recover for 2 wk before being killed. The authors concluded that the NOAEL was 1,000 mg/kg-d (Chengelis 1997).

Kurokawa et al. (Marcia Hardy, Albermarle Corporation, Pers. Commun., August 3, 1999) fed groups of 50 male and female mice diets containing 0, 100, 1,000, or 10,000 ppm of HBCD (equivalent to about 0, 13, 130, or 1,300 mg/kg-d) for 18 mo. Survival and growth rates for the treated groups were similar to controls. Liver changes were observed in male but not in female mice in the 1,000- and 10,000-ppm groups, including hypertrophy, vacuolation/fatty change, and altered foci.

Reproductive and Developmental Effects

Two reproductive/developmental studies found in the literature identified no reproductive or developmental effects in treated rats (Murai et al. 1985; Stump 1999). Murai et al. (1985) conducted a reproductive/developmental study in rats. Groups of 20 female Wistar rats were fed a diet containing 0%, 0.01%, 0.1%, or 1% HBCD (determined by study authors to be equivalent to about 0, 5, 50, and 500 mg/kg-d, based on daily food consumption of pregnant rats) on d 0-20 of gestation. Bodies of pregnant rats were observed every day through pregnancy, and body weight and food consumption were measured. Fourteen pregnant rats per group were killed on d 20, and major organs were examined. Six dams from each dose group were delivered naturally, and the growth of each fetus was observed. There was a slight, but significant decrease in maternal food intake and a significant increase in maternal liver weight in the 1% group (magnitude of these changes was not reported). No differences in body weights were observed between the administration and control groups. HBCD was reported as having no effect on the number of implants; the number of resorbed, dead, or live fetuses; the body weight of live fetuses; the incidence of external, visceral, or skeletal anomalies; or delivery, nursing, lactation, or neonatal development. No abnormality in parturition, weaning status, or growth of newborns was observed at the maternal toxic dose of 1%. Based on the available reviews, this study appears to have identified a NOAEL of 1% (500 mg/kg-d) in the diet for developmental effects.

Stump (1999) conducted a developmental toxicity study in rats. Groups of eight female Crl:CD®(SD)IGS BR rats were dosed orally by gavage with 125, 250, 500, 750, or 1,000 mg/kg-d of HBCD dissolved in corn oil. A concurrent control group received corn oil. During gestation, all females were observed twice daily for appearance and behavior, and body weight and food consumption were recorded. On d 20 of gestation, laparohysterectomy was performed, uteri and ovaries were examined, and the numbers of fetuses, early and late resorptions, total implantations, and corpora lutea were recorded. Maternal animals were autopsied and fetuses were examined. Mean maternal body weights were significantly increased in the 250-, 500-, 750-, and 1,000-mg/kg-d dose groups in comparison to controls during gestation d 19-20; however, this increase may be attributed to elevated food consumption observed on d 19-20. No dose-related effects were observed for the mean percentage of viable fetuses per litter or for postimplanation losses. One fetus in the 125-mg/kg-d dose group had multiple external malformations (mandibular micrognathia, microphthalmia [unilateral], and aglossia) that were considered spontaneous in origin. No other external malformations or developmental variations were observed in fetuses. Based on the findings of this study, a NOAEL of 1,000 mg/kg-d was determined for developmental effects.

Cancer

Kurokawa et al. (Marcia Hardy, Albermarle Corporation, Pers. Commun., August 3, 1999), (see Systemic Effects section) fed mice diets containing 0, 100, 1,000, or 10,000 ppm (equivalent to about 0, 13, 130, or 1,300 mg/kg-d) for 18 mo. For the male mice, hypertrophy and vacuolization/fatty changes in the liver were observed in the 1,000- and 10,000-ppm dose groups, and an increase in altered foci was seen in the 1,000-ppm group but not at the 10,000-ppm level. No changes were observed in the female mice. Since no correlation was observed between the dosage and incidence of neoplastic changes in the liver in male mice, the study authors concluded there was no evidence of carcinogenicity.

Other Systemic Effects

No studies were identified that examined immunological or neurological effects following oral exposure to HBCD.

Genotoxicity

The weight of the evidence suggests that HBCD is not genotoxic. Negative results have been reported for HBCD in assays for mutagenicity in yeast and *Salmonella*, and for chromosonal aberrations in human peripheral blood lymphocytes at doses up to the limits of solubility and toxicity (Brusick 1976; Baskin and Phillips 1977; Oesch 1978; Shoichet and Ehrlich 1978; Zeiger et al.

1987; Pinto 1989; Gudi and Schadly 1996). However, a study by Helleday et al. (1999) identified statistically significant increases in recombination frequency in the Sp5 and SPD8 cell lines.

QUANTITATIVE TOXICITY ASSESSMENT[2]

Noncancer

Dermal Assessment

The limited dermal data available suggest that HBCD is at most a mild skin irritant and a mild allergen, if allergenic at all (Momma et al. 1993; Nakamura et al. 1994). Acute exposure to high dermal doses of HBCD apparently produced systemic toxicity (diarrhea and slight weight loss) in rabbits (Dean and Leong 1977), but the systemic effects of long-term dermal exposure have not been published. The lack of information on systemic effects from subchronic or chronic dermal exposure to HBCD precludes the derivation of an RfD based on dermal toxicity data. However, the oral RfD was used in this risk assessment in place of the dermal RfD as the best estimate of internal dose from dermal exposure (derivation of the oral RfD is presented below).

Inhalation RfC

The subcommittee did not identify any inhalation studies of sufficient duration (i.e., subchronic or chronic) for deriving an RfC. The available data are limited to two acute studies (Dean and Leong 1977; Lewis and Palanker 1978). The subcommittee concluded that there are insufficient inhalation toxicity data on HBCD to derive an inhalation RfC.

Oral RfD

On the basis of its review of the oral toxicity data, the subcommittee determined that the 13-wk rat study by Zeller and Kirsch (1970) is appropriate for deriving an RfD for HBCD (see Systemic Effects section for additional de-

---

[2]In this section, the subcommittee reviewed the data on toxicity of HBCD, including the toxicity assessment prepared by the U.S. Consumer Product Safety Commission (Hatlelid 1999).

tails). The 925-mg/kg-d dose level was chosen as the LOAEL based on increased liver weight (absolute and relative in both sexes) accompanied by an increased incidence of distinctly abnormal fatty accumulation in the liver and, in males, a small reduction in body weight gain. The 450-mg/kg-d dose level (the mean value for males and females) was selected as the NOAEL because the more subtle changes in liver weight and histology at this dose level are not considered to be adverse effects. An uncertainty factor of 3,000 was applied (10 for extrapolation from rats to humans, 10 for intraspecies variation, 10 for extrapolation from subchronic to chronic duration, and 3 to account for database deficiencies including lack of a two-generation reproductive study and a developmental toxicity study in a second species). Therefore, based on a NOAEL of 450 mg/kg-d and an uncertainty factor of 3,000, an RfD of 0.2 mg/kg-d was calculated for HBCD (see Table 4-4).

Confidence in the critical study (Zeller and Kirsch 1970 is medium. The study included an adequate number of rats of both sexes and investigated a variety of endpoints, but reporting of experimental methods and results was only marginally adequate. Confidence in the database is low because of the lack of availability of other subchronic/chronic studies. Therefore, confidence in the oral RfD is low.

Cancer

In an 18-mo feeding study of HBCD in mice, Kurokawa et al. (Marcia Hardy, Albermarle Corporation, pers. commun., August 3, 1999) found no

TABLE 4-4 Oral Reference Dose for HBCD

Critical effect	Species	Effect level (mg/kg-d)	Uncertainty factors	RfD (mg/kg-d)	Reference
Increased liver weights, accompanied by abnormal fatty accumulation observed at the LOAEL of 925 mg/kg-d	Male and female rats	NOAEL: 450	UF_A: 10 UF_H: 10 UF_S: 10 UF_D: 3 Total: 3,000	0.2	Zeller and Kirsch (1970)

NOAEL, no-observed-adverse-effect level; RfD, reference dose; UF_A, extrapolation from animals to humans; UF_H, intraspecies variability; UF_S, extrapolation from a study of less-than-lifetime duration; UF_D, inadequate or deficient toxicity database.

evidence of carcinogenicity at dietary concentrations up to 10,000 ppm of HBCD. The available genotoxicity data suggest that HBCD is not genotoxic. The potential carcinogenicity of HBCD in humans cannot be determined based on inadequate data for an assessment of carcinogenicity via the dermal, inhalation, or oral routes.

EXPOSURE ASSESSMENT AND RISK CHARACTERIZATION

Noncancer Assessment

Dermal Exposure

The assessment of noncancer risk for the dermal route of exposure is based on the scenario described in Chapter 3. This exposure scenario assumes that an adult spends 1/4th of his or her time sitting on furniture upholstery backcoated with HBCD and also assumes that 1/4th of the upper torso is in contact with the upholstery, and clothing presents no barrier. Exposure to other chemicals present in the backcoating was not included in this assessment.

First Iteration

As a first estimate of exposure, it was assumed that the skin and clothing of the person sitting on the couch, and the fabric of the couch, would present no barrier to movement of HBCD. In addition, it was assumed that there would be sufficient water present (e.g. from sweat) to allow dissolution of the HBCD in the water, and transfer to the skin and into the body of the sitting individual. The only limiting factor on the transfer rate using these assumptions results from the limited dissolution rate from the fabric—all the HBCD that dissolves is assumed to be absorbed immediately by the sitting individual.

Dermal exposure was estimated using Equation 1 in Chapter 3. For this calculation, the subcommittee estimated an upholstery application rate (S_a) for HBCD of 5 mg/cm^2. The extraction rate (μ_w) by water for HBCD was estimated to be 0.025/d based on extraction data for HBCD in polyester fiber (McIntyre et al. 1995). This release rate was calculated as 0.04/d at 28°C from the fiber, with a correction from fiber to film of a factor of 0.63 ($2d/2\pi R$ for film thickness d, fiber radius R).

Using these values, the estimated dermal absorbed dose rate was determined to be 0.98 mg/kg-d. Although lack of sufficient data precludes deriving a dermal RfD, the oral RfD (0.2 mg/kg-d) is used as the best estimate of internal dose from dermal exposure to calculate the hazard index. The hazard index of 4.9, derived by dividing the dermal absorbed dose rate of 0.98 mg/kg-d by the oral RfD of 0.2 mg/kg-d, indicates that HBCD might pose a noncancer risk by the dermal absorption route when used as an upholstery fabric flame retardant. Therefore an alternative iteration of the exposure assessment was performed.

Alternative Iteration

For the alternative iteration of the dermal assessment, the same exposure assumptions were made as in the first iteration, except that the assumption of immediate absorption of all the HBCD that dissolves was modified. Instead, an estimate of the rate that HBCD could penetrate the skin was determined, assuming that HBCD dissolves up to its solubility limit in water. This rate of penetration was then factored into the exposure assessment.

The rate of penetration of a chemical through skin may be estimated using the skin permeability coefficient (K_p, with dimensions of velocity) — the total mass penetration rate is the product of water concentration, permeability coefficient, and skin area. This coefficient has not been measured for HBCD. However, it was estimated from the octanol-water partition coefficient (K_{ow}, dimensionless) and molecular weight (m, mass/unit amount of substance) using a correlation (Potts and Guy 1992) based on Equation 2 in Chapter 3. The value estimated from this correlation is 4.99×10^{-2} cm/d for HBCD. Using equation (5) in Chapter 3 in conjunction with the permeability coefficient (4.99×10^{-2} cm/d) and the water solubility specific to HBCD (3.40×10^{-3} mg/L), the dose rate of HBCD, using this alternative iteration, was estimated to be 1.33×10^{-6} mg/kg-d. The hazard index was then calculated by dividing the dermal absorbed dose rate (1.33×10^{-6} mg/kg-d) by the oral RfD (0.2 mg/kg-d), as the best estimate for internal dose from dermal exposure. The hazard index of 6.67×10^{-6} indicates that HBCD used as an upholstery fabric flame retardant is not likely to pose a noncancer risk via the dermal exposure route.

Inhalation Exposure

Particles

Inhalation exposure to HBCD in the particulate phase was calculated using the scenario described in Chapter 3. This scenario assumes that a person spends 1/4th of his or her life in a room with low air-change rates (0.25/hr) and with a relatively large amount of fabric upholstery treated with HBCD (30 m^2 in a

30-m^3 room), with this treatment gradually being worn away over 25% of its surface to 50% of its initial quantity over the 15-yr lifetime of the fabric. A small fraction, 1%, of the worn-off HBCD is released into the indoor air as small particles that may be inhaled.

Particle exposure was estimated using Equations 4 through 6 in Chapter 3. The release rate (μ_r) for HBCD from upholstery, 2.3×10^{-7}/d (Equation 5), was used in conjunction with the upholstery application rate (S_a) for HBCD of 5 mg/cm^2 to calculate a room airborne particulate concentration of 1.9 $\mu g/m^3$ (Equation 4). Factoring in the fraction of a day a person spends in the room containing upholstery (0.25), the time-average exposure concentration was determined to be 0.48 $\mu g/m^3$ (Equation 6).

For the purpose of estimating a hazard index for the inhalation of HBCD and in the absence of relevant inhalation exposure data, the subcommittee chose to estimate the inhalation RfC from the oral RfD. The subcommittee, however, recognizes that this is not an ideal approach and also recognizes that the estimated RfC might be considerably different than the actual reference concentration (if inhalation data were available). Extrapolating from one route of exposure (oral) to another (inhalation) requires specific knowledge about the uptake kinetics into the body by each exposure route, including potential binding to cellular sites. The subcommittee believes that its extrapolation of the oral RfD to the inhalation RfC is highly conservative; it assumes that all of the inhaled compound is deposited in the respiratory tract and is completely absorbed into the blood. The NRC Committee on Toxicology (NRC 1985) has used this approach when inhalation exposure data were insufficient to derive inhalation exposure levels. The subcommittee believes that such an approach is justified for conservatively estimating the toxicological risk from exposure to HBCD. This RfC should be used as an interim or provisional level until relevant data become available for the derivation of an inhalation RfC for calculating the hazard index.

Therefore, a provisional RfC of 0.7 mg/m^3 was derived by using the oral RfD of 0.2 mg/kg-d and Equation 7 in Chapter 3. A hazard index of 6.8×10^{-4} was estimated by dividing the exposure concentration (0.48 $\mu g/m^3$) by the provisional inhalation RfC (0.7 mg/m^3). This indicates that under the worst-case exposure scenario, HBCD, used as an upholstery flame retardant, is not likely to pose noncancer risk via inhalation of HBCD in the particulate phase.

Vapors

In addition to the possibility of release of HBCD in particles worn from upholstery fabric, the subcommittee considered the possibility of its release by evaporation. The approach to estimate vapor exposure is described in Chapter 3 and uses a scenario similar to that previously described for exposure to HBCD in the particulate phase.

Using Equations 8 through 10 in conjunction with the saturation vapor concentration (C_v) (0.016 mg/m^3) and the application density (S_a) of 5 mg/cm^2 for HBCD, the equilibrium room-air concentration of HBCD was estimated to be 0.014 mg/m^3. From Equation 11, it was determined that this vapor concentration could be maintained for approximately 1,200 yr. Factoring in the fraction of a day a person spends in the room containing upholstered fabric (0.25), the time-average exposure concentration was determined to be 3.4 μg/m^3. Division of this exposure concentration (3.4 μg/m^3) by the provisional inhalation RfC (0.7 mg/m^3) results in a hazard index of 5×10^{-3}, indicating that under the worst-case scenario, exposure to HBCD, used as an upholstery-fabric flame retardant, is not likely to pose a noncancer risk via the inhalation route, when exposures occur in the vapor phase.

Oral Exposure

The assessment of the noncancer risk for the oral exposure route is based on the scenario described in Chapter 3. This scenario assumes a child is exposed to HBCD through sucking on 50 cm^2 of fabric backcoated with HBCD daily for 2 yr, 1 hr/d. The dose rate to the child was calculated using Equation 15 in Chapter 3.

Using these values, the average oral dose rate was estimated to be 0.026 mg/kg-d, compared with an oral RfD of 0.2 mg/kg-d, giving a hazard index of 0.13. The subcommittee concludes that HBCD used as an upholstery-fabric flame retardant is not likely to pose a noncancer risk by the oral route.

Cancer Assessment

Based on inadequate carcinogenicity data from any route of exposure, the subcommittee concluded that the potential carcinogenicity of HBCD cannot be determined.

RECOMMENDATIONS FROM OTHER ORGANIZATIONS

Occupational exposure limits for HBCD have not been established by the Occupational Safety and Health Administration (OSHA), the American Conference of Government Industrial Hygienists (ACGIH), or the National Institute of Occupational Safety and Health (NIOSH).

DATA GAPS AND RESEARCH NEEDS

There are no subchronic or chronic inhalation or dermal bioassays to evaluate systemic toxicity or carcinogenicity of HBCD. There are no dermal absorption studies. HBCD has a high octanol water partition coefficient, low vapor pressure, and low solubility in water. It is used as a fabric backcoating and is encapsulated in a polymer matrix. Therefore, direct exposure is likely to be minimal. Based on existing review of toxicity and use information, the conservative exposure assumptions, and a hazard index of less than 1 for all exposure routes, the subcommittee concludes that no further research is needed for assessing health risks from HBCD.

REFERENCES

Baskin, A.D., and B.M. Phillips 1977. Mutagenicity of Two Lots of FM-100, Lot 53 and Residue of Lot 3322 in the Absence and Presence of Metabolic Activation. Industrial Bio-Test Laboratories, Northbrook, IL. EPA/OTS Doc. #86-900000267.

Brusick, D. 1976. Mutagenicity Evaluation of 421-32B: Final Report. LBI Project No. 2547. Litton Bionetics Inc., Kensington, Md. Sponsored by Ciba-Geigy Corp., Ardsley, NY. EPA/OTS Doc. #86-900000265.

Chengelis, C.P. 1997. A 28-day Repeated Dose Oral Toxicity Study of HBCD in Rats. Wil Research Laboratories, Inc., Ashland, OH. Laboratory Study No. WIL-186004. Sponsored by Chemical Manufacturers Association, Brominated Flame Retardant Industry Panel, Arlington, VA. Sponsor Project No. BFRIP 2.0-WIL HBCD.

ChemID. 1999. Chemical ID Data Base. National Library of Medicine, National Toxicology Information Program, Bethesda, MD.

Crown, S. 1984. HBCD (FR - 1206) Primary Skin Irritation Study in Rabbits. LSRI Report No. DSB/059/FR. Life Science Research Israel, Ltd., Ness Ziona, Israel. Sponsored by Dead Sea Bromine Works Ltd., Beer Sheva, Israel. EPA/OTS Doc. #86-900000433.

Dean, W.P., and B.K.J. Leong. 1977. Acute toxicity studies in rabbits and rats. International Research and Development Corp. Sponsor: Velsicol Chemical Corporation. Study No. 163-499. EPA/OTS Doc. #86-900000266.

FRCA (Fire Retardant Chemicals Association). 1998. Textile Flame Retardant Applications by Product Classes for 1997 Within and Outside of the United States. Submitted to: U.S. Consumer Product Safety Commission. Prepared by: Fire Retardant Chemicals Association.

Gudi, R., and E.H. Schadly. 1996. Chromosome Aberrations in Human Peripheral

Blood Lymphocytes. Microbiological Associates, Inc., Rockville, Md. Study No. G96AO61.342. Sponsor: Chemical Manufacturers Association, Arlington, VA.

Hatlelid, K. 1999. Toxicity Review of Hexabromocyclododecane. Memorandum, dated March 2, 1999, from Kristina Hatlelid, Toxicologist, Division of Health Sciences to Ronald L. Medford, Assistant Executive Director for Hazard Identification and Reduction. U.S. Consumer Product Safety Commission, Washington, DC.

Helleday, T., K.-L. Tuominen, A. Begman, and D. Jenssen. 1999. Brominated flame retardants induce intragenic recombination in mammalian cells. Mutat. Res. 439:137-147.

Lewis, C.A., and A.L. Palanker. 1978. Technical and Toxicity Data on Brominated Flame Retardants Including Hexabromocyclododecane. Consumer Product Testing Co. Inc., Fairfield, N.J. Experiment Reference No. 78385-1 and 78385-2. Sponsored by Saytech Inc., Sayreville, NJ. EPA/OTS Doc. #FYI-OTS-0794-0947.

MacGregor, J.A., and W.B. Nixon. 1997. Hexabromocyclododecane (HBCD): Determination of n-Octanol/Water Partition Coefficient. Wildlife International LTD. Project No. 439C-104. Sponsor: Chemical Manufacturers Association, Brominated Flame Retardant Industry Panel, Arlington, VA.

McDonnell, M.E. 1972. Studies on Decabromodiphenyl Ether, Hexabromocyclododecane and 4-vinylcyclohexane. Haskell Laboratory Report No. 185-72. Haskell Laboratories. EPA/OTS Doc. #86-900000119S.

McIntyre, J.E., I. Holme, and O.K. Sunmonu. 1995. The desorption of model compounds from poly(ethylene terephthalate) fibre. Colourage 41(13)77-81.

Microbiological Associates. 1996. Maximization Test in Guinea Pigs. Microbiological Associates Study No. M96AO61.1X64. Sponsored by Chemical Manufacturers Association, Arlington, VA.

Momma, J., M. Kaniwa, H. Sekiguchi, K. Ohno, Y. Kawasaki, M. Tsuda, A. Nakamura, and Y. Kurokawa. 1993. Dermatological evaluation of a flame retardant, hexabromocyclododecane (HBCD) on guinea pig by using the primary irritation, sensitization, phototoxicity and photosensitization of skin. [Article in Japanese; English abstract]. Eisei Shikenjo Hokoku 111:18-24.

Murai, T., H. Kawasaki, and S. Kanoh. 1985. Studies on the toxicity of insecticides and food additives in pregnant rats. (7) Fetal toxicity of hexabromocyclododecane. [Article in Japanese]. Pharacometrics 29(6):981-986.

Nakamura, A., J. Momma, H. Sekiguchi, T. Noda, T. Yamano, M.-A. Kaniwa, S. Kojima, M. Tsuda, and Y. Kurokawa. 1994. A new protocol and criteria for quantitative determination of sensitization potencies of chemicals by guinea pig maximization test. Contact Dermatitis 31(2):72-85.

Nissimov, S. 1984. Acute Oral Range Finding Study in the Rat. LSRI Schedule No. DSB/052/HBCD. Life Science Research Israel, Ltd., Ness Ziona, Israel. Sponsored by Dead Sea Bromine Works Ltd., Beer Sheva, Israel. EPA/OTS Doc. #86-900000433.

NRC (National Research Council). 1985. Emergency and Continuous Exposure Limits for Selected Airborne Contaminants, Volume 5. Committee on Toxicology. Board on Toxicology and Environmental Health Hazards, National Research Council. Washington, DC: National Academy Press.

Oesch, F. 1978. Ames Test for Hexabromid S. Pharmakologisches Institute, BASF Institute for Industrial Hygiene and Pharmacology, Federal Republic of Germany. EPA/OTS Doc. #86-900000379.

Pinto, M. 1989. HBCD Blend: Assessment of Mutagenic Potential in Histidine Auxotrophs of Salmonella Typhimurium (The Ames Test). LSRI Project No. DSB/109/HBCD. Life Science Research Israel, Ltd., Ness Ziona, Israel. Sponsored by Bromine Compounds, Ltd., Beer Sheva, Israel. EPA/OTS Doc. #86-900000433.

Potts, R.O., and R.H. Guy. 1992. Predicting skin permeability. Pharm. Res. 9(5):663-669.

Schulze. 1962. Report on the Study of the Acute Oral Toxicity of Hexabromocyclododecane in the Mouse. Original report printed in German, translated into English by B.E. Byrt and reprinted in 1985. BASF, Federal Republic of Germany. EPA/OTS Doc. #86-900000383.

Shoichet, A., and K. Ehrlich. 1978. Mutagenicity Test of GLS-S6-41A. Gulf South Research Institute, New Orleans, LA. EPA/OTS Doc. #FYI-OTS-0794-0947.

Stenzel, J.I., and B.J. Markley. 1997. Hexabromocyclododecane (HBCD): Determination of the Water Solubility. Wildlife International LTD. Project No. 439C-105. Sponsored by Chemical Manufacturers Association, Brominated Flame Retardant Industry Panel, Arlington, VA.

Stenzel, J.I., and W.B. Nixon. 1997. Hexabromocyclododecane (HBCD): Determination of the Vapor Pressure Using a Spinning Rotor Gauge. Wildlife International LTD. Project No. 439C-117. Sponsored by Chemical Manufacturers Association, Brominated Flame Retardant Industry Panel, Arlington, VA.

Stump, D.G. 1999. A Dose Range-Finding Prenatal Developmental Toxicity Study of Hexabromocyclododecane (HBCD) in Rats. Performing laboratory, WIL Research Laboratories, Inc. Laboratory Study No. WIL-186008. Sponsored by Chemical Manufactuers Association, Brominated Flame Retardant Industry Panel, Arlington, VA. CMA Reference No. BFRIP 02 WIL-02 HBCD.

Vesicol Chemical Corporation. 1980. Unpublished laboratory study. Chicago, Il.

Zeiger, E., B. Anderson, S. Haworth, T. Lawlor, K. Mortelmans, and W. Speck. 1987. Salmonella mutagenicity tests: III. Results from the testing of 255 chemicals. Environmental Molecular Mutagenesis 9 (Suppl. 9): 1-10.

Zeller, H. 1962. Report on the Study of the Irritation of Hexabromocyclododecane to the Dorsal Skin of Rabbits After 20-hour Exposure. Original report printed in German, English translation by B.E. Byrt and reprinted in1985. BASF, Federal Republic of Germany. EPA/OTS Doc. #86-900000381.

Zeller, H., and P. Kirsch. 1969. Hexabromocyclododecane: 28-day Feeding Trials With Rats. BASF Institute for Industrial Hygiene and Pharmacology, Federal Republic of Germany. EPA/OTS Doc. #86-900000376.

Zeller, H., and P. Kirsch. 1970. Hexabromocyclododecane: 90-day Feeding Trials With Rats. Pharmakologisches Institute, BASF Institute for Industrial Hygiene and Pharmacology, Federal Republic of Germany. EPA/OTS Doc. #86-900000380.

5

Decabromodiphenyl Oxide

THIS chapter reviews the physical and chemical properties, toxicokinetics, toxicological, epidemiological, and exposure data on decabromodiphenyl oxide (DBDPO). The subcommittee used that information to characterize the health risk from exposure to DBDPO. The subcommittee also identified data gaps and recommended research relevant for determining the health risk from exposure to DBDPO.

PHYSICAL AND CHEMICAL PROPERTIES

DBDPO is a brominated aromatic flame retardant that exists as a white to off-white powder. Its physical and chemical properties are presented in Table 5-1.

OCCURRENCE AND USE

DBDPO is the most widely used of the polybrominated diphenyl ethers. DBDPO is a single well-defined compound, because it has no isomers. Typically, commercial DBDPO has a purity of 97-98%; the major impurities being isomers of nonabromodiphenyl oxide (NBDPO) and octabromodiphenyl oxide (OBDPO) (Marcia Hardy, Albermarle Corporation, Pers. Commun., February 9, 1999).

TABLE 5-1 Physical and Chemical Properties of DBDPO

Property	Value	Reference
Chemical formula	$C_{12}Br_{10}O$	CHEMID 1999
Synonyms	decabromodiphenyl ether	IPCS 1994
Chemical structure	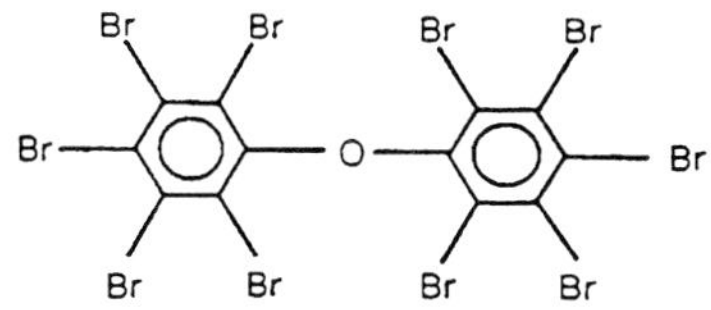	IPCS 1994
CAS registry #	1163-19-5	CHEMID 1999
Molecular weight	959	HSDB 1998
Melting point	295-305°C	HSDB 1998
Vapor pressure	4.63×10^{-6} Pa at 21°C	Stenzel and Nixon 1997
Density	3.0 g/cm^3	HSDB 1998
Solubility in water	<0.1 µg/L in water at 25°C; slightly soluble in acetone, benzene, dichloromethane, *ortho*-xylene, methanol, methyl ethyl ketone, pentane, and toluene	Stenzel and Markley 1997
Partition coefficient (Log K_{ow})	6.26	MacGregor and Nixon 1997
Tradenames	AFR 1021; Berkflam B 10E; BR 55N; Bromkal 82-ODE; Bromkal 83-10DE; Caliban F/R-P39P; EBR 700; FR300; Saytex 102; Saytex 102E; Tardex 100	IPCS 1994

Approximately 12 ,000 tons of DBDPO are used annually worldwide. About two-thirds of the annual use is in high-impact polystyrene applications such as television and radio cabinets. Textile applications, such as those used in polyester fibers and in coatings for residential and commercial furniture, automobile fabrics, wall coverings, draperies, tarpaulins, and carpets, account for an additional 900 tons (IARC 1990). When used in textiles, DBDPO is applied as a back coating to the fabric, encapsulated in a polymer.

DBDPO is commonly used in conjunction with antimony trioxide, which acts as a synergist, contributing to the flame retardancy of the textile. Typically, the flame retardant mixture consists of about 66-75% DBDPO and 25-33% antimony trioxide. The amount of DBDPO added to the mixture varies depending on the desired flame-retardant properties of the fabric.

TOXICOKINETICS

No studies on the absorption of DBDPO were identified following dermal exposure.

Sjodin et al. (1999) compared potential inhalation exposures of three groups of workers (clerks working in front of computer screens full-time, workers at an electronics dismantling plant, and a control group of hospital cleaning workers) to polybrominated diphenyl ethers. The investigators found DBDPO in the blood serum of individuals from all three groups. The median DBDPO concentration in hospital cleaners and computer clerks was <0.7 pmol/g lipid weight (range = <0.3-3.9, respectively). The median DBDPO concentration in electronics dismantlers was 5 pmol/g lipid weight (range <0.3-9.9). Plasma levels of DBDPO were significantly higher in the electronics dismantling workers than in the other two groups, apparently resulting from inhalation of particle-bound DBDPO in the air of the dismantling plant, as high ambient levels of particle-bound DBDPO were recorded at the plant. Electronics dismantling involved grinding plastic goods in a shredder. DBDPO in the blood of cleaning workers and clerks was presumed to be due to exposure of contaminated food, although there was no correlation between plasma levels of DBDPO and fish intake (the only food evaluated in the study).

Pulmonary clearance was evaluated in rats following a single intratracheal injection of 0 or 20 mg DBDPO dust (purity 77.4%) suspended in rat serum (Dow 1976). Based on determinations of total bromine content in the lungs 3, 10, 30, 91, and 365 d following exposure, the half-life for elimination of DBDPO from the lungs was estimated to be 150 d.

Several studies (Norris et al. 1974, 1975; NTP 1986; El Dareer et al. 1987) have demonstrated that DBDPO is very poorly absorbed in rats following oral exposure, generally <1% of the amount ingested. A large percentage of absorbed DBDPO appears to be rapidly metabolized in the liver and excreted in the bile with a half-life of <24 hr (NTP 1986; El Dareer et al. 1987). Analyses of radioactivity from ^{14}C-DBDPO and of bromine from unlabeled compound (composed of 77.4% DBDPO, 21.8% NBDPO, and 0.8% OBDPO) following ingestion in rats indicate that DBDPO did not accumulate in tissues, other than adipose, with only slight increases occurring very slowly over time (Norris et al. 1974, 1975; Great Lakes 1976; IRDC 1976, 1977). Bromine content was not increased in the kidney, muscle, or serum of rats exposed for up to 2 yr in their diets (0.01, 0.1, or 1 mg/kg-d) (purity 77.4%) (Dow 1994).

HAZARD IDENTIFICATION[1]

Dermal Exposure

Irritation/Sensitization

DBDPO does not appear to be a primary irritant based on observations from a skin sensitization study in humans and dermal irritation and acnegenesis studies in animals. A human skin sensitization study was conducted in which 0.03 mL of a 5% suspension of commercial DBDPO in petrolatum (0.02 mg/kg) was applied via patch to the skin of 50 subjects three times per week for 3 wk (Dow 1972a; Norris et al. 1974, 1975). Commercial DBDPO was a mixture that contained 77.4% DBDPO, 21.8% NBDPO, and 0.8% OBDPO. The dermal applications did not result in skin sensitization reactions during the sensitizing period or on challenge 2 wk after the last application. Skin irritation, attributed to the stringency of the test procedure by the investigators, occurred in 9 of the 50 subjects (14/450 total applications; 11 of the reactions were classified as very slight and 3 as mild erythema).

DBDPO caused essentially no dermal response in rabbits when applied as a dry solid (500 mg) to intact shaved skin under occluded conditions for 24 hr, and a slight erythematous and edematous response when similarly applied to abraded skin (Norris et al. 1974, 1975; IRDC 1974). Repeated application of dry solid DBDPO (500 mg) to intact skin of rabbits for 5 d/wk for 2 wk or to abraded skin for 3 d also did not alter their dermal responses.

An acnegenesis study was performed in which 0.1 mL of 0.1%, 1%, 10%, or 100% DBDPO (0.40 mg/kg) in chloroform was rubbed into the external ear canal of four rabbits/dose level once a day, 5 d/wk for 4 wk (Pharmakon 1981). Observations made prior to the initial dose and after 7, 14, 21, and 28 d of dosing showed slight erythema, epidermal sloughing and scaling (effect levels

---

[1]In this section, the subcommittee reviewed data on toxicity of DBDPO, including the toxicity assessment prepared by the U.S. Consumer Product Safety Commission (Bittner 1999).

not specified), but no clear indication of chloracne (a slight response was observed in one animal at the 10% concentration on d 28). Gross necropsy showed no treatment-related systemic effects. Norris et al. (1974, 1975) similarly reported that a 10% chloroform solution of DBDPO caused slight erythema and exfoliation, and no indication of chloracne, when applied to the ear of rabbits for 28 d. Other industry studies also found that 10% DBDPO in chloroform did not induce chloracne in rabbits (Dow 1970, 1971, 1972b, 1972c, 1973).

Ocular exposure to dry solid DBDPO caused transient conjunctival irritation in washed and unwashed rabbit eyes. Instillation of DBDPO (100 mg/eye) into the eye caused very slight conjunctival redness and chemosis and slight or moderate discharge in some rabbits, but the investigators concluded that the effects were not serious enough to be considered primary eye irritation (Norris et al. 1975; IRDC 1974). Pharmakon (1981) similarly reported that DBDPO (Saytex 102) did not cause primary eye irritation when instilled once (100 mg/eye) into the eye of rabbits.

Systemic Effects

DBDPO has low acute toxicity via the dermal route. This conclusion is based on lack of treatment-related effects, including body weight gain and survival, in rabbits following single administrations of 200- or 2,000-mg/kg to clipped intact skin for 24 hr and observed for 14 d (IRDC 1974; Great Lakes 1977). No data were located that investigated the chronic toxicity of DBDPO from dermal exposures.

Other Systemic Effects

No data were identified on immunological, neurological, reproductive, developmental, or carcinogenic effects of DBDPO following dermal exposure.

Inhalation Exposure

Systemic Effects

A higher than normal prevalence of primary hyperthyroidism and a significant reduction in calf sensory and fibula motor nerve velocities were observed in workers exposed to DBDPO during manufacturing (Bahn et al. 1980, as cited in NTP 1986; Bialik 1982, as cited in HSDB 1998). No other effects were observed. It was unclear whether these effects were due to DBDPO, which was not detected in the blood of the workers, or to polybrominated biphenyls (PBBs), which were found in the blood and were previously manufactured at the plant. Surveys of workers during the manufacture and use of DBDPO have determined time-weighted average (TWA) ambient air exposures of 1-4 mg/m^3, with more than 90% of the particles <10 µm in diameter (NTP 1986). However, no definitive conclusion can be drawn on the toxicity of DBDPO in humans because of confounding exposures to PBBs. No other human data are available on the toxicity of DBDPO from inhalation exposure.

No deaths or effects on body weight gain were observed in rats exposed to DBDPO at concentrations of 2,000 or 48,000 mg/m^3 for 1 hr and observed for 14 d (IRDC 1974; Great Lakes 1984). However, dyspnea and ocular porphyrin discharge were observed at both levels, and eye squint was seen at the high concentration only. Effects on pulmonary tissue were evaluated in rats following a single intratracheal injection of 0 or 20 mg DBDPO dust (purity 77.4%) suspended in rat serum (Dow 1976). All rats were observed frequently for changes in appearance and demeanor, body weights were determined weekly for the first month, biweekly up to 5 mo, and monthly through 1 yr. Gross and histopathological examinations of the trachea, lungs, and thoracic lymph nodes were performed on groups of rats that were killed on post-treatment d 10, 30, 416, and 556. Histopathological effects were limited to minimal changes (scattered focal aggregates of alveolar macrophages) consistent with retention of large dust particles that would not normally reach the lungs during inhalation. No toxicity data were identified from chronic inhalation studies.

Other Systemic Effects

No data on immunological, reproductive, developmental or carcinogenic effects of DBDPO were identified following inhalation exposure.

Oral Exposure

A summary of toxicity studies from oral exposures is presented in Table 5-2.

Systemic Effects

No human toxicity data were located from oral exposure to DBDPO. In animals, DBDPO has low acute toxicity following oral exposure in part because of its poor gastrointestinal absorption. No clinical signs of toxicity or death occurred in rats that received single gavage doses as high as 5,000 mg/kg

TABLE 5-2 Oral Noncancer Studies of DBDPO

Species, Strain, Sex, Number	Purity	Dose (mg/kg-d)[a]	Duration	Effects	NOAEL/ LOAEL	References
Rats, Sprague-Dawley, M, 5/dose	77.4%	0, 8, 80, 800	30 d	Thyroid hyperplasia (80, 800); increased liver weight (80); increased liver weight, pathology (800); renal tubular degeneration (800)	NOAEL: 8 mg/kg-d LOAEL: 80 mg/kg-d	Norris et al. 1973, 1974, 1975
Rats, NS, M/F, 10/dose	NS	0, 7.4, 74	3 studies, 28 d	No histology liver or thyroid	NOAEL: 74 mg/kg-d	Great Lakes 1976; IRDC 1976, 1977
Rats, F344, M/F, 10/dose	97-99%	0%, 0.31%, 0.62%, 1.25%, 2.5%, 5.0%	13 wk	No effects on body weight, survival, organ weights, histology	NOAEL: 5.0%	Hazelton 1979a,b; NTP 1986
Mice, B6C3F1, M/F, 10/dose	97-99%	0%, 0.31%, 0.62%, 1.25%, 2.5%, 5.0%	13 wk	No effects on body weight, survival, organ weights, histology	NOAEL: 5.0%	Hazelton 1979a, b; NTP 1986
Rats, Sprague-Dawley, M/F, 25/dose	77.4%	0, 0.01, 0.1, 1	2 yr	No histology	NOAEL: 1 mg/kg-d	Norris et al. 1974, 1975; Dow 1994; Kociba et al. 1975

Rats, F344, M/F, 50/dose	94-97%	M: 0, 1,120, 2,240 F: 0, 1,200, 2,550	103 wk	In males, liver thrombosis (1/50, 0/50, 9/50)/ degeneration (13/50, 19/50, 22/50), splenic fibrosis, lymphoid hyperplasia	NOAEL: 1,120 mg/kg-d LOAEL: 2,240 mg/kg-d	NTP 1986
Mice, B6C3F1, M/F, 50/dose	94-97%	M: 0, 3,200, 6,650 F: 0, 3,760, 7,780	103 wk	In males, thyroid hyperplasia 2/50, 10/50, 19/50; liver centrilobular hypertrophy	LOAEL: 3,200 mg/kg-d	NTP 1986
Rats, Sprague-Dawley, M: 10-15/dose, F: 20-30/dose	77.4%	0, 3, 30, 100	60 d pre-mating, mating, gestation, lactation (~115 d)		NOAEL (reproductive): 100 mg/kg-d	Norris et al. 1975; Dow 1975
Rats, NS, F	77.4%	0, 10, 100, 1,000	Gestation d 6-15 gavage in corn oil	No maternal toxicity; no fetal malformations; subcutaneous edema and delayed skull ossification in pups	NOAEL (maternal): 1,000 mg/kg-d NOAEL (fetal): 100 mg/kg-d LOAEL (fetal): 1,000 mg/kg-d	Norris et al. 1974; 1975; Dow 1985; 1994

F, female; LOAEL, lowest-observed-adverse-effect level; M, male; ND, not determined; NOAEL, no-observed-adverse-effect level; NS, not specified.

[a]Administered in diet, unless otherwise indicated.

(IRDC 1974; Great Lakes 1984; Nissimov 1984). When DBDPO was administered to rats by single gavage doses of up to 2,000 mg/kg, no signs of toxicity were noted up to 14 d after dosing (Norris et al. 1974, 1975). Liver weights increased without enzymatic changes in rats treated with 96 mg/kg-d by gavage for 14 d (Carlson 1980). Rats and mice that ingested DBDPO for 14 d in dietary doses up to 10,850 and 23,100 mg/kg-d, respectively, showed no clinical signs or effects on body weight, survival, or gross pathology (NTP 1986). Liver weight and histology were not evaluated.

Longer-term oral exposure at high doses to DBDPO caused changes in the thyroid gland. Thyroid hyperplasia was induced in male rats fed DBDPO at doses of 80 or 800 mg/kg-d, but not at a dose of 8 mg/kg-d, for 30 d (Norris et al. 1973, 1974, 1975). Incidences of this lesion were not reported, only five rats/dose were tested, and the test material was 77.4% DBDPO, 21.8% NBDPO, and 0.8% OBDPO. The doses in the Norris et al. (1974, 1975) and Sparschu et al. (1971) reports are estimated based on 0.01%, 0.1%, and 1.0% dietary concentrations of DBDPO. The Norris et al. (1974, 1975) study is the same study as reported in Sparschu et al. (1971).

No histological changes in the thyroid were found in three 28-d studies in which groups of 10 male and 10 female rats ingested similar doses of DBDPO (7.4 or 74 mg/kg-d, corresponding to 0.01% or 0.1% in the diet) (Great Lakes 1976; IRDC 1976, 1977) as used in the Norris et al. (1974, 1975) studies. However, these studies may be limited by possible use of different formulations or batch numbers within the same study and/or an absence of statistical analyses. Rats and mice that ingested DBDPO for 13 wk up to 2,500 and 7,000 mg/kg-d, respectively, showed no clinical signs or effects on survival, body weight, food consumption, or gross or microscopic pathology (NTP 1986).

Additional evidence for DBDPO-related thyroid changes is provided by the occurrence of thyroid follicular cell hyperplasia in 2/50, 10/50, and 19/50 male mice that were fed 0, 3200, and 6650 mg/kg-d DBDPO (purity 94-97%) in the diet, respectively, for 103 wk (NTP 1986). Follicular cell hyperplasia was not observed in female mice or male or female rats that were similarly exposed to DBDPO at doses as high as 7,780, 2,240, or 2,550 mg/kg-d, respectively. DBDPO doses in these rat and mouse studies corresponded to dietary concentrations of 25,000 or 50,000 ppm (2.5% or 5.0%). A NOAEL for the thyroid hyperplasia could not be determined because the lesion was induced at the lowest dose tested in the male mice. The LOAEL is therefore 3,200 mg/kg-d. The follicular cell hyperplasia observed in the male mice after 103 wk was not observed in preliminary 13-wk mouse and rat studies using the same maximum dietary test concentrations of DBDPO (Hazleton Labs 1979a, 1979b; NTP 1986). Rats (25/sex/dose) that ingested much lower dietary doses of DBDPO (0.01, 0.1, or 1 mg/kg-d; 77.4% pure containing 21.8% NBDPO and

0.8% OBDPO) for up to 2 yr also had no exposure-related histopathological changes in the thyroid or other tissues (Norris et al. 1974, 1975; Kociba et al. 1975; Dow 1994).

Dietary exposure to DBDPO also caused liver changes in animals. Male rats that were fed 8, 80, or 800 mg/kg-d DBDPO for 30 d had increased liver weight at 80 and 800 mg/kg-d, accompanied by hepatic centrilobular cytoplasmic enlargement and vacuolation at 800 mg/kg-d (Norris et al. 1974, 1975). Incidences of hepatic effects were not found, only five rats/dose were tested, and the test material was 77.4% DBDPO, 21.8% NBDPO, and 0.8% OBDPO. No DBDPO-related changes in liver weight or histology were observed in three 28-d studies in which groups of 10 male and 10 female rats ingested similar dietary doses of DBDPO (7.4 or 74 mg/kg-d) (Great Lakes 1976; IRDC 1976, 1977). However, as stated above, these studies are limited by possible use of different formulations or batch numbers within the same study and/or an absence of statistical analyses.

Centrilobular hypertrophy was observed in male mice fed 3,200 or 6,650 mg/kg-d DBDPO (purity 94-97%) for 103 wk, but not in female mice, or male and female rats that were similarly exposed to doses as high as 7,780, 2,240, or 2,550 mg/kg-d, respectively (NTP 1986). The centrilobular hypertrophy consisted of enlarged hepatocytes with frothy vacuolated cytoplasm. A NOAEL was not established for liver effects because the centrilobular hypertrophy occurred at the lowest dose tested in male mice. The LOAEL is therefore 3,200 mg/kg-d. Incidences of thrombosis and degeneration of the liver were increased in male rats at 2,240 mg/kg-d (LOAEL), but not at 1,120 mg/kg-d (NOAEL); these hepatic effects were not observed in the female rats or mice of either sex. Thrombosis was characterized by a near total occlusion of a major hepatic blood vessel by a dense fibrin coagulum. No liver effects were observed in the preliminary 13-wk rat and mouse studies using the same maximum dietary test concentrations of DBDPO (NTP 1986; Hazleton Labs 1979a, 1979b). Rats (25/sex/dose) that ingested much lower dietary doses of 0.01, 0.1, or 1 mg/kg-d DBDPO (77.4% DBDPO, 21.8% NBDPO and 0.8% OBDPO) for up to 2 yr had no exposure-related histopathological changes in the liver or other tissues (Norris et al. 1974, 1975; Kociba et al. 1975; Dow 1994).

The subchronic and chronic studies summarized above showed few effects on target organs other than the thyroid and liver. The chronic dose that caused liver thrombosis and degeneration in male rats (2,240 mg/kg-d for 103 wk) also induced fibrosis of the spleen and lymphoid hyperplasia (NTP 1986). Body weight gain was decreased by 13% in female rats fed 11,110 mg/kg-d for 14 d (Hazleton Labs 1978) and decreased by 14% in female rats fed 3,650 mg/kg-d for 13 wk (Hazleton Labs 1979a). However, these effects may not be treatment related because they were not observed in the other studies, some of which included animals treated at higher doses for longer durations. Renal tubular hyaline degeneration was found in male rats fed 800 mg/kg-d DBDPO for 30 d (Norris et al. 1975), but this may be indicative of α-2μ globulin nephropathy and therefore not relevant to humans.

Reproductive and Developmental Effects

No data were located on reproductive and developmental toxicity of DBDPO in humans from oral exposure. No signs of toxicity were observed in a single-generation reproductive study in which male and female rats were fed 3, 30, or 100 mg/kg-d DBDPO (77.4% pure containing 21.8% NBDPO and 0.8% OBDPO) for 60-90 d prior to mating, and subsequently throughout mating, gestation, and lactation (Dow 1975; Norris et al. 1975). The NOAEL for reproductive toxicity was 100 mg/kg-d; a LOAEL was not established. Rats that were treated with 10, 100, or 1,000 mg/kg-d DBDPO (77.4% pure containing 21.8% NBDPO and 0.8% OBDPO) by gavage on gestation d 6-15 showed no maternal toxicity or teratogenic effects, although fetotoxicity was observed at the highest dose (Norris et al. 1974, 1975; Dow 1985, 1994). There were significant increases in numbers of litters with subcutaneous edema and delayed ossification of skull bones at 1,000 mg/kg-d (Norris et al. 1974). A significant increase in the percentage of resorptions in all treated groups was also observed, but this effect was not considered to be treatment related because of the lack of a dose-response relationship at higher doses. The NOAEL and LOAEL for fetal effects were 100 and 1,000 mg/kg-d, respectively, and the NOAEL for maternal effects was 1,000 mg/kg-d.

Cancer

Information on the carcinogenicity of DBDPO is available from two chronic feeding studies in rodents (Kociba et al. 1975; NTP 1986). The NTP (1986) bioassay was performed using groups (50/sex/dose) of F344/N rats and B6C3F1 mice that were fed DBDPO (94-97% pure) at dietary concentrations of 0, 25,000, or 50,000 ppm for 103 wk. (This is equivalent to 1120, 1200, and 2240 mg/kg-d in male rats; 1120, 1200, and 2550 mg/kg-d in female rats; 3200, 3760, and 6650 mg/kg-d in male mice; and 3200, 3760, and 7780 mg/kg-d in female mice.) Incidences of liver neoplastic nodules were significantly increased in low- and high-dose male rats (7/50 and 15/49, respectively, compared to 1/50 in controls) and high-dose female rats (9/50 compared to 1/50 and 3/49 in control and low-dose groups, respectively); this lesion appeared to be compound related. Incidence of hepatocellular carcinomas was low in all rat groups and apparently not compound related. There was a positive trend in mononuclear cell leukemia in male rats (30/50 controls, 33/50 low-dose rats, 35/50 high-dose rats), but the increase was marginal and not considered to be biologically significant because of the unusually high incidence in controls. A significant positive trend and marginally greater incidence of acinar cell adenomas in the pancreas of high-dose male rats were also observed, but this lesion was considered to not be compound related (NTP 1986). Hepatocellular adenomas or carcinomas (combined) were significantly increased in low- and high-dose male mice (8/50 controls, 22/50 low-dose mice, 18/50 high-dose mice). The incidence of hepatocellular carcinomas alone was significantly elevated in male mice in the low-dose group, but not in the high-dose group, as compared with controls. Thyroid gland follicular cell adenomas or carcinomas (combined) were marginally, but not significantly increased in male mice (0/50 controls, 4/50 low-dose mice, 3/50 high-dose mice). The possible significance of this finding was strengthened by increased incidences of follicular cell hyperplasia in the male mice (2/50 controls, 10/50 low-dose mice, 19/50 high-dose mice), but was weakened by increased mortality in control animals. There was no evidence of carcinogenicity in the female mice at either dose. The NTP study (1986) concluded that there was "some evidence of carcinogenicity" for male and female rats based on significantly increased incidences of neoplastic nodules of the liver, and "equivocal evidence of carcinogenicity" for male mice based on a significantly increased incidence of hepatocellular tumors in only the low-dose group and non-statistically significant increases in thyroid follicular cell tumors in both dose groups. The conclusion of "some evidence of carcinogenicity" in rats appears to be based on the finding that the only chemical related effect was benign liver neoplasms (NTP 1986). The conclusion of "equivocal evidence of carcinogenicity" in male mice appears to be based on the interpretation that the increases in liver and thyroid tumors are marginal and chemical related.

There was no evidence of carcinogenicity in groups of 25 male and 25 female Sprague-Dawley rats that were fed 0, 0.01, 0.1, or 1 mg/kg-d DBDPO in the diet for up to 2 yr (Kociba et al. 1975). The test material was 77.4% pure containing 21.8% NBDPO and 0.8% OBDPO. It has been suggested that the doses and numbers of animals used in this study were inadequate to determine carcinogenic potential (Ulsamer et al. 1980, as cited in NTP 1986; IARC 1990).

Other Systemic Effects

No data on immunological or neurological effects were located from studies of oral exposures to DBDPO.

Genotoxicity

DBDPO does not appear to be genotoxic. DBDPO did not induce gene mutations in *Salmonella typhimurium* in most Ames assays (Great Lakes 1976; Litton Bionetics 1976; Gulf South Research 1977; Huntingdon Research 1978; Haworth et al. 1983; Nissimov 1984; NTP 1986; Wagner and Klug 1998). DBDPO did not induce mutations in mouse L5178Y lymphoma cells, chromosomal aberrations in mouse bone marrow cells, or sister-chromatid exchanges or cell transformation in Chinese hamster ovary cells (Norris et al. 1975; NTP 1986; McGregor et al. 1988; Myhr et al. 1990; LeBoeuf et al. 1996).

QUANTITATIVE TOXICITY ASSESSMENT

Noncancer

Dermal Assessment

Available data suggest that DBDPO is not irritating to the skin and is not a dermal sensitizer. Systemic effects of short- or long-term dermal exposures to DBDPO have not been adequately studied. There were no treatment-related changes in body weight gain or survival in rabbits following a single application of ≤2,000 mg/kg (IRDC 1974; Great Lakes 1977). There were no treatment-related changes in body weight or gross pathological effects in rabbits treated with 40 mg/kg-d DBDPO (rubbed into external ear canal skin) for 4 wk in an acnegenesis assay (Pharmakon 1981). There is insufficient information on toxicity of DBDPO from subchronic or chronic dermal exposures to estimate the dermal RfD.

Inhalation RfC

The subcommittee identified no inhalation studies of sufficient duration (i.e., subchronic or chronic) for deriving an RfC, since the available data are limited to an acute inhalation study (IRDC 1974; Great Lakes 1984) and an acute intratracheal study (Dow 1976). Therefore, the subcommittee did not derive an inhalation RfC for DBDPO.

Oral RfD

EPA's reference dose (RfD) of 1×10^{-2} mg/kg-d for DBDPO is based on the 1 mg/kg-d NOAEL for histopathology and other toxicity endpoints in rats exposed via diet for 2 yr (Kociba et al. 1975). Doses higher than 1 mg/kg-d were not tested in this study, precluding identification of a LOAEL. The reason the NTP (1986) 2-yr toxicology/carcinogenesis bioassay for DBDPO was not considered in the current RfD summary in IRIS (EPA 1999) is because it was not available at the time of the RfD derivation (1984-1985). The subcommittee believes that it was appropriate to reevaluate the RfD considering the NTP data, because of the higher compound purity (≈99% versus 77.4%), reflecting the actual chemical composition applied as a flame retardant (Marcia Hardy, Albermarle Corporation, Pers. Commun., February 9, 1999); the larger number of animals (50 versus 25 rats/sex/dose); the higher dose levels; and the use of a second species in the NTP (1986) bioassay in comparison to the Kociba et al. (1975) study.

The subcommittee derived an oral RfD for DBDPO by using the chronic NOAEL of 1,120 mg/kg-d, based on liver thrombosis and degeneration observed in rats at the next higher dose (NTP 1986), and a composite uncertainty factor of 300, resulting in an RfD of 4 mg/kg-d (RfD = NOEL ÷ 300) (see Table 5-3). The composite uncertainty factor is compromised of 3 uncertainty factors: 10 for interspecies extrapolation, 10 for intraspecies variability, and 3 for database uncertainties ($10_A \times 10_H \times 3_D = 300$). The RfD is based on a well-designed chronic toxicity study of DBDPO in two species. Data on chronic, developmental, and reproductive toxicity are available from other studies in rats. However, limitations in these studies (particularly compound purity (77.4%), lack of a second species, and use of low dose levels in the chronic study; lack of longer than one-generation testing in the reproductive study) indicate that there is some uncertainty in the DBDPO database. Based on these considerations, an uncertainty factor of 3, instead of 10, for database insufficiency was used.

Confidence in the key study (NTP 1986) is high because it was well conducted and because it used two species, a sufficient number of animals, a

TABLE 5-3 Oral Reference Dose for DBDPO

Critical effect	Species	Effect level (mg/kg-d)	Uncertainty factors	RfD (mg/kg-d)	Reference
Liver thrombosis and degeneration observed at the LOAEL of 2,240 mg/kg-d	Male and female rats	NOAEL: 1,120	UF_A: 10 UF_H: 10 UF_D: 3 Total: 300	4.0	NTP (1986)

NOAEL, no-observed-adverse-effect level; RfD, reference dose; UF_A, extrapolation from animals to humans; UF_H, extrapolation for intraspecies variation; UF_D, inadequate or deficient toxicity database.

dose range adequate to identify a NOAEL and LOAEL for a known sensitive effect, and a high-purity test formulation. However, confidence in the database is low because of limitations in the available developmental, reproductive, and supporting chronic toxicity studies of DBDPO, as well as use of low-purity compound, lack of testing in species other than the rat, lack of multigenerational reproductive tests, and a low range of chronic dose levels. Therefore, confidence in the provisional RfD is medium to low.

Cancer

Oral

There are no epidemiological data available on the carcinogenicity of DBDPO. However, the carcinogenicity of DBDPO has been assessed in two chronic bioassays (Kociba et al. 1975; NTP 1986). No evidence for carcinogenicity was observed in male or female Sprague-Dawley rats fed dose levels of 0, 0.01, 0.1, or 1 mg/kg-d DPDPO in their diet for 2 yr (Kociba et al. 1975). However, this study has several limitations including use of an inadequate number of animals (25/sex/dose), dosing with impure DBDPO (77.4% DBDPO, 21.8% NBDPO, 0.8% OBDPO), and utilization of dose levels that were probably below the maximum tolerated dose (MTD) (see NTP 1986). "Some evidence of carcinogenicity" was reported by NTP (1986) for male and female rats fed DBDPO in their diet at dose levels of up to 50,000 ppm for 2 yr. NTP (1986) reports that there was "equivocal evidence of compound-related carcinogenicity" for mice exposed to DBDPO in their diet for 2 yr at dose levels of 25,000 or 50,000 ppm.

The EPA weight-of-evidence cancer classification for DBDPO in accordance with the currently used 1986 Guidelines for Carcinogenic Risk Assessment (EPA 1986) is Group C, possible human carcinogen (EPA 1999). This is based on no human data and limited evidence of carcinogenicity in animals (NTP 1986), specifically, statistically significant increases in the incidences of "neoplastic nodules" of the liver in male and female rats and hepatocellular adenomas and carcinomas combined in male mice. Under the newer Proposed Guidelines for Carcinogenic Risk Assessment (EPA 1996), which take into account genotoxicity data, the weight-of-evidence for the carcinogenic potential of DBDPO in humans would be termed "suggestive evidence of carcinogenicity, but not sufficient to assess human carcinogenic potential."

The subcommittee has concluded that the weight-of-evidence, based on the data currently available, suggests that DBDPO is a possible carcinogen in rats. Because there is uncertainty concerning the carcinogenicity of this compound, the subcommittee concluded that derivation of a cancer risk estimate was warranted and should be used for assessing the potential carcinogenic risk associated with this compound when used as a flame retardant in residential furniture. The subcommittee believes that derivation of a cancer potency factor ($0.1/LED_{10}$) as opposed to a hazard index is justified in this case because a NOAEL was not detected for liver nodules in rats (NTP 1986).

The subcommittee has not concluded that DBDPO is a carcinogen in humans but believes that a conservative approach is justified at this time in order to be protective of human health. The subcommittee acknowledges that the increased incidence of "neoplastic nodules" of the liver in male and female rats and male mice does not constitute sufficient evidence for the carcinogenicity of DBDPO and is aware that there is controversy over the significance of these lesions in determining cancer risk (Maronpot et al. 1986). However, the finding of marginal increases in follicular cell tumor incidence in conjunction with an increased frequency of hyperplasia in dosed animals as compared with controls is suggestive of a carcinogenic response and adds weight for the use of a conservative approach for evaluating the carcinogenicity of DBDPO. This approach is further justified by other scientists who have concluded that hyperplasia is a stage in the thyroid follicular cell carcinogenic process (Hill et al. 1989, 1998; EPA 1998; Hard 1998).

The cancer potency factor ($0.1/LED_{10}$) was derived for DBDPO using both the censored (for early death) and uncensored neoplastic nodule incidence data for male rats from NTP (1986). Use of the censored data produced a $0.1/LED_{10}$ of 9×10^{-4}/mg/kg-d as compared with 7×10^{-4}/mg/kg-d when the uncensored data were used in the derivation (see Table 5-4). Use of censored data represents a crude attempt to adjust for differential mortality among male rats (low- and high-dose groups). Survival of the low-dose male rats was reduced as compared to controls and to males in the high-dose group; these differences were statistically significant by the end of the study. The number of neoplastic nodules produced might have been greater in the animals of this dose group if a greater number had survived until the 2-yr termination point.

Data for female rats were not used to calculate a cancer slope factor because LED_{10} values derived from female liver neoplastic nodule incidence data were roughly two-fold greater (less protective) than those for male rats. LED_{10}s were also derived using liver adenoma and carcinoma incidence (combined) in male mice, but model fit was "poor" when either censored or uncensored data for these tumor types were used. This is primarily because of the higher incidence of tumors in the low- versus high-dose groups. These values were higher, and thus less protective, than LED_{10} values derived from female rat liver nodule data.

TABLE 5-4 Cancer Assessment for DBDPO Based on Hepatic Neoplastic Nodules Reported for Male Rats in NTP (1986) Study (oral exposure)

	Daily dose level (mg/kg-d)			0.1/LED$_{10}$ derivation		
Nodules	0	1,120	2,240	LED$_{10}$[b] (mg/kg-d)	LED$_{10}$ (mg/kg-d), adjusted[c]	0.1/LED$_{10}$[d] (mg/kg-d)$^{-1}$
Uncensored	1/50	7/50	15/49	516	137	7×10^{-4}
Censored[a]	1/45	7/38	15/45	435	115	9×10^{-4}

LED, lowest effective dose

[a]Animals in all groups that died prior to the occurrence of the first hepatic neoplastic nodule in either treated group (wk 87) were removed from the denominator on the assumption that these animals had insufficient opportunity to develop the tumor.

[b]Calculated using a multistage model fit to the dose-response data and based on extra risk.

[c]Dose adjusted for human equivalency by taking the ratio of human body weight to rat body weight over human to rat body weight to the 0.75 power (EPA 1992). Defaults used: human body weight, 70 kg; rat body weight, 0.35 kg.

[d]Calculated as 0.1/LED$_{10}$, as per the EPA (1996, 1999) proposed cancer guidelines.

The subcommittee has low-to-moderate confidence that the NTP (1986) bioassay results accurately characterize the carcinogenic potential of DBDPO. While the assay was not conducted at the MTD, the dose levels administered were the highest recommended for use in NTP studies. Mortality was significantly elevated among male rats in the low-dose group and in male control mice and is an issue when judging the quality of the study.

The subcommittee places moderate confidence in the derived LED$_{10}$, and subsequently the 0.1/LED$_{10}$. The NOAEL for liver neoplastic nodules in the NTP (1986) bioassay was not determined, which raises the concern that these effects could occur at lower dose levels than the LED$_{10}$.

EXPOSURE ASSESSMENT AND RISK CHARACTERIZATION

Noncancer

Dermal Exposure

The assessment of noncancer risk for the dermal route of exposure is based on the dermal exposure scenario described in Chapter 3. This exposure scenario assumes that an adult spends 1/4th of his or her time sitting on furniture upholstery backcoated with DBDPO and also assumes that 1/4th of the upper torso is in contact with the upholstery and clothing presents no barrier. Exposure to other chemicals present in the backcoating were not included in this assessment.

First Iteration

As a first estimate of exposure, it was assumed that the skin and clothing of the person sitting on the couch, and the fabric of the couch, would present no barrier to movement of DBDPO. In addition, it was assumed that there would be sufficient water present (e.g., from sweat) to allow dissolution of the DBDPO in the water, and transfer to the skin and into the body of the sitting individual. The only limiting factor on the transfer rate using these assumptions results from the limited dissolution rate from the fabric—all the DBDPO that dissolves is assumed to be absorbed immediately by the sitting individual.

Dermal exposure was estimated using Equation 1 in Chapter 3. For this calculation, the subcommittee estimated an upholstery application rate (S_a) for DBDPO of 5 mg/m^2. The extraction rate (μ_w) by water for DBDPO was estimated to be 0.025/d based on extraction data for hexabromocyclododecane in polyester fiber (McIntyre et al. 1995). This release rate was calculated as 0.04/d at 28°C from the fiber, with a correction from fiber to film of a factor of 0.63 ($2nd/2\pi R$ for film thickness d, fiber radius R).

Using these values specific to DBDPO, the estimated dermal absorbed dose rate was determined to be 0.98 mg/kg-d. Although lack of sufficient data precludes deriving a dermal RfD, the oral RfD (4 mg/kg-d) was used to calculate a hazard index. The hazard index of 0.25, derived by dividing the dermal absorbed dose rate of 0.98 mg/kg-d by the oral RfD of 4 mg/kg-d, indicates that DBDPO does not pose a noncancer risk by the dermal absorption route when used as an upholstery fabric flame retardant. Nevertheless, an alternative iteration of the exposure assessment was performed because of concerns about potential cancer risk (see below).

Alternative Iteration

For the alternative iteration of the dermal assessment, the same exposure assumptions were made as in the first iteration, except that the assumption of immediate absorption of all the DBDPO that dissolves was modified. Instead, an estimate of the rate at which DBDPO can penetrate the skin was determined, assuming that DBDPO dissolves up to its solubility limit in water. This rate of penetration was then factored into the exposure assessment.

The rate of penetration of a chemical through skin may be estimated using the skin permeability coefficient (K_p, with dimensions of velocity)—the total mass penetration rate is the product of water concentration, permeability coefficient, and skin area. This coefficient has not been measured for DBDPO. However, it was estimated from the octanol-water partition coefficient (K_{ow}, dimensionless) and molecular weight (m, mass/unit amount of substance) using a correlation (Potts and Guy 1992) based on Equation 2 in Chapter 3. The value estimated from this correlation is 3.21×10^{-4} cm/d for DBDPO.

Using Equation 5 in Chapter 3 in conjunction with the permeability coefficient (3.21×10^{-4} cm/d) and the water solubility specific to DBDPO (<0.1 µg/L), the dose rate, using this alternative iteration, was estimated to be 1.33×10^{-9} mg/kg-d. The hazard index was then recalculated by dividing the dermal absorbed dose rate (1.33×10^{-9} mg/kg-d) by the oral RfD (4 mg/kg-d), as the best estimate for internal dose from dermal exposure. The hazard index of 3.34×10^{-10}, again demonstrates that DBDPO, used as an upholstery fabric flame retardant, is not likely to pose a noncancer risk from dermal exposure.

Inhalation Exposure

Particles

Inhalation exposure estimates for DBDPO were calculated using the exposure scenario described in Chapter 3. This scenario assumes that a person spending a quarter of his or her life in a room with low air-change rates (0.25/hr) and with a relatively large amount of fabric upholstery (30 m^2 in a 30 m^3 room), with the DBDPO treatment gradually being worn away over 25% of its surface to 50% of its initial quantity over the 15 yr lifetime of the fabric. A small fraction, 1%, of the worn-off DBDPO is released into the indoor air as small particles that may be inhaled.

Particle exposure was estimated using Equations 4 through 6 in Chapter 3. The release rate (μ_w) for DBDPO from upholstery, 2.3×10^{-7}/d (Equation 5), was used in conjunction with the upholstery application rate (S_a) for DBDPO of 5 mg/cm^2 to calculate a room airborne particulate concentration of 1.9 mg/m^3 (Equation 4). Factoring in the fraction of a day a person spends in the room containing upholstery (0.25), the time-average exposure concentration was determined to be 0.48 mg/m^3 (Equation 6).

For the purpose of estimating a hazard index for the inhalation of DBDPO and in the absence of relevant inhalation exposure data, the subcommittee chose to estimate the inhalation RfC from the oral RfD. The subcommittee, however, recognizes that this is not an ideal approach and also recognizes that the estimated RfC might be considerably different than the actual reference concentration (if inhalation data were available). Extrapolating from one route of exposure (oral) to another (inhalation) requires specific knowledge about the uptake kinetics into the body by each exposure route, including potential binding to cellular sites. The subcommittee believes that its extrapolation of the oral RfD to the inhalation RfC is highly conservative; it assumes that all of the inhaled compound is deposited in the respiratory tract and is completely absorbed into the blood. The NRC Committee on Toxicology (NRC 1985) has used this approach when inhalation exposure data were insufficient to derive inhalation exposure levels. The subcommittee believes that such an approach is justified for conservatively estimating the toxicological risk from exposure to DBDPO. The provisional RfC should be used as an interim or provisional level until relevant data become available for the derivation of an inhalation RfC for calculating the hazard index.

Based on this, a provisional RfC of 14 mg/m^3 was derived from the oral RfD of 4.0 mg/kg-d and Equation 7 in Chapter 3. A hazard index of 3.4×10^{-5} was estimated by dividing the exposure concentration (0.48 µg/m^3) by the provisional inhalation RfC (14 mg/m^3). This indicates that under the worst case exposure assumptions, DBDPO, used as an upholstered flame retardant, does not pose any noncancer risk via inhalation of DBDPO in the particulate phase.

Vapors

In addition to the possibility of release of DBDPO in particles worn from upholstery fabric, the subcommittee considered the possibility of its release by evaporation. The approach is described in Chapter 3 and uses a scenario similar to that previously described for exposure to DBDPO in the particulate phase.

Using Equations 8 through 10 in conjunction with the saturation vapor concentration (C_v) (1.8×10^{-3} mg/m^3) and the application density (S_a) (5 mg/cm^2) for DBDPO, the equilibrium room-air concentration of DBDPO was estimated to be 1.52×10^{-3} mg/m^3. From Equation 11, it was determined that this vapor concentration could be maintained for approximately 390 yr. Factoring in the fraction of a day a person spends in the room containing upholstered fabric (0.25), the time-average exposure concentration was determined to be 3.8×10^{-4} mg/m^3.

Division of this exposure concentration (3.8×10^{-4} mg/m^3) by the provisional inhalation RfC (14 mg/m^3) results in a hazard index of 2.71×10^{-5}, indicating that under the worst case scenario, exposure to DBDPO, used as an upholstery flame retardant, is not likely to pose a noncancer risk via the inhalation route, when exposure occurs in the vapor phase.

Oral Exposure

The assessment of the noncancer risk for the oral exposure route is based on the scenario described in Chapter 3. This scenario assumes a child is exposed to DBDPO through sucking on 50 cm^2 of fabric back coated with DBDPO daily for 2 yr, 1 hr/d. The dose rate to the child was calculated using Equation 15 in Chapter 3. DBDPO specific parameters used in this calculation included an upholstery application rate (S_a) of 5 mg/m^2 and an extraction rate (μ_w) by saliva of 0.025/d. This extraction rate was based on data for hexabromocyclododecane in polyester fiber (McIntyre et al. 1995) and was calculated as 0.04/d at 28°C from the fiber, with a correction from fiber to film of 0.63 ($2d/2\pi R$ for film thickness d, fiber radius R).

Using these values, the average oral dose rate was estimated to be 2.6×10^{-2} mg/kg-d, compared with an oral RfD of 4 mg/kg-d, giving a hazard index of 6.5×10^{-3}. It was concluded that DBDPO used as an upholstery fabric flame retardant does not pose any noncancer risk via the oral route.

Cancer

Dermal Exposure

Human cancer risk for dermal exposure to DBDPO was calculated by multiplying the lifetime oral cancer potency factor for DBDPO by the lifetime average dermal dose rate. Using the lifetime average dermal dose rate of 1.33×10^{-9} mg/kg-d, obtained in the alternative dermal exposure iteration (see the Noncancer Dermal Exposure section), and multiplying this by the cancer potency estimate of 9×10^{-4} kg-d/mg, a lifetime risk estimate of 1.20×10^{-12} is obtained. This estimate is small enough that the cancer risk through dermal contact with DBDPO used as an upholstery-fabric flame retardant, can be considered negligible.

Inhalation Exposure

For DBDPO, no inhalation cancer unit risk is available. However, an inhalation cancer unit risk of 2.57×10^{-7} per µg/m^3 was estimated from the oral carcinogenic potency using Equation 14 in Chapter 3.

Particles

The average room-air concentration and average exposure concentration to DBDPO were obtained as described in the Noncancer section. Using the estimated unit risk (2.57×10^{-7} per µg/m^3), the lifetime risk estimate from exposure to DBDPO as particles is 1.2×10^{-7}. From this estimate, DBDPO, used as an upholstered flame retardant, poses a negligible cancer risk via inhalation in the particulate phase.

Vapors

The equilibrium concentration of vapor-phase DBDPO in room air was estimated as described in the Noncancer Inhalation Exposure Section. The long-term time-average vapor exposure concentration was estimated from the equilibrium vapor concentration in room air using Equation 13 in Chapter 3.

Using the estimated unit risk of 2.57×10^{-3} per µg/m^3, the lifetime risk estimate for exposure to DBDPO in the vapor phase is 9.74×10^{-8}. This estimate indicates that DBDPO, used as a flame retardant, poses a negligible cancer risk via inhalation in the vapor phase.

Oral Exposure

For DBDPO, the lifetime average dose rate estimate by the oral route was 7.4×10^{-4} mg/kg-d. This dose rate estimate is multiplied by the cancer unit risk of 9.0×10^{-4}/mg/kg-d, giving a lifetime cancer risk estimate of 6.7×10^{-7}. This estimate is small enough that the cancer risk via the oral route can be dismissed as negligible when DBDPO is used as an upholstery fabric flame retardant.

RECOMMENDATIONS FROM OTHER ORGANIZATIONS

OSHA, ACGIH, and NIOSH have not established recommended exposure levels for DBDPO. EPA has derived an RfD for DBDPO (EPA 1999) based on the 1-mg/kg-d NOAEL from the study by Kociba et al.(1975).

DATA GAPS AND RESEARCH NEEDS

There are inadequate subchronic and chronic dermal and inhalation toxicity data to establish either a dermal RfD or an inhalation RfC for DBDPO. In addition, there are no dermal absorption data. DBDPO is nearly insoluble in water and has a relatively low vapor pressure. It is used as a fabric backcoating with antimony trioxide and is encapsulated in a polymer matrix. Based on existing review of toxicity and use information, and the subcommittee's conservative exposure assumptions, the subcommittee concludes that no further research is needed for assessing health risks from DBDPO.

REFERENCES

Bahn, A., O. Bialik, J. Oler, et al. 1980. Health assessment of occupational exposure to polybrominated biphenyl (PBB) and polybrominate biphenyloxide (PBBO). Report. ISS EPA 560/6-80-001. NTIS No. PB81-159675.

Bialik, O. 1982. Endocrine function of workers exposed to PBB and PBBO. Terminal Progress Report. Grant No. I-R01-OH-01034-0163 (as cited in HSDB 1998).

Bittner, P. 1999. Toxicity Review for Decabromodiphenyl Oxide. Memorandum, dated March 3, 1999, from Patricia Bittner, Toxicologist, Division of Health Sciences, to Ronald Medford, Assistant Executive Director for Hazard Identification and Reduction, U.S. Consumer Product Safety Commission, Washington, DC.

Carlson, G.P. 1980. Induction of xenobiotic metabolism in rats by short-term administration of brominated diphenyl ethers. Toxicol. Lett. 5:19-25.

ChemID. 1999. Chemical ID Data Base. National Library of Medicine, National Toxicology Information Program, Bethesda, MD.

Dow (Dow Chemical Co.). 1970. Toxicological Properties and Industrial Handling Hazards of Decabromodiphenyl Ether. Dow Chemical Co., Biomedical Research Laboratory. TSCATS 405861. EPA/OTS Doc. #86-900000208.

Dow (Dow Chemical U.S.A.). 1971. Chloracne Study Conducted on Two Samples of Decabromodiphenyl Oxide and One Sample of Decabromodiphenyl Oxide Mother Liquor. Dow Chemical U.S.A. TSCATS 405864. EPA/OTS Doc. #86-900000211. OTS #522281.

Dow (Dow Chemical U.S.A.). 1972a. Human Skin Sensitization Study on FR 300BA (Decabromodiphenyl Oxide) (IBT #F2366). Dow Chemical Co. TSCATS 405836. EPA/OTS Doc. #86-900000182. OTS #522253.

Dow (Dow Chemical U.S.A.). 1972b. Chloracne Studies Conducted on Three Samples of Decabromodiphenyl Oxide (FR-300-BA). Dow Chemical U.S.A. TSCATS 405853. EPA/OTS Doc. #86-900000200.

Dow (Dow Chemical U.S.A.). 1972c. Chloracne Study Conducted on Decabromodiphenyl Oxide Mother Liquor. Dow Chemical U.S.A. EPA/OTS Doc. #86-900000212.

Dow (Dow Chemical U.S.A.). 1973. Chloracne Studies Conducted on Decabromodiphenyl Oxide. Dow Chemical U.S.A. TSCATS 405850. EPA/OTS Doc. #86-900000197.

Dow (Dow Chemical U.S.A.). 1975. Results of a Reproduction Study in Rats Maintained on Diets Containing Decabromodiphenyl Oxide. Dow Chemical U.S.A. TSCATS 405835. EPA/OTS Doc. #86-900000181. OTS #522252.

Dow (Dow Chemical U.S.A.). 1976. Pulmonary Clearance and Tissue Response Following a Single Intratracheal Injection of Decabromodiphenyl Oxide (DBDPO) Dust in Male Rats. Dow Chemical U.S.A. TSCATS 0522264. EPA/OTS Doc #86-900000194.

Dow (Dow Chemical Co.). 1985. Decabromodiphenyloxide: A Summary of an Oral Teratology Study in Sprague-Dawley Rats. Dow Chemical Co., Midland, MI.

Dow (Dow Chemical Co.). 1994. Results of a Two-Year Dietary Feeding Study With Decabromodiphenyl Oxide (DBDPO) in Rats. TSCATS #452235. EPA/OTS Doc. #FYI-OTS-0794-1103.

El Dareer, S.M., J.R. Kalin, K.F. Tillrey, and D.L. Hill. 1987. Disposition of decabromodiphenyl ether in rats dosed intravenously or by feeding. J. Toxicol. Environ. Health 22(4):405-415.

EPA (U.S. Environmental Protection Agency). 1986. Guidelines for Carcinogen Risk Assessment. Fed. Regist. 51(185):33992-34003.

EPA (U.S. Environmental Protection Agency). 1992. Dermal Exposure Assessment: Principles and Applications. EPA/600/8-91-011B. Office of Health and Environmental Assessment, U.S. Environmental Protection Agency, Washington, D.C.

EPA (U.S. Environmental Protection Agency). 1996. Proposed guidelines for carcinogen risk assessment. Fed. Regist. 61(Apr. 23):17960-18011.

EPA (U.S. Environmental Protection Agency). 1998. Assessment of Thyroid Follicular Cell Tumors. Risk Assessment Forum, U.S. Environmental Protection Agency. Washington, D.C. EPA/630/R-97/002. March 1998.

EPA (U.S. Environmental Protection Agency). 1999. Chemical Substance Query. Integrated Risk Information System (IRIS). [Online]. Available: http://www.epa.gov/ngispgm3/iris/subst/index.html National Center for Environmental Assessment, Cincinnati, OH.

Great Lakes (Great Lakes Chemical Corporation). 1976. Twenty-eight Day Toxicity Study in Rats. Performed by International Research and Development Corp. for Great Lake Chemical Corporation. TSCATS/407321. EPA/OTS Doc. #86-900000329. OTS #523321.

Great Lakes (Great Lakes Chemical Corporation). 1977. Toxicity Data: Decabromodiphenyl Oxide. Great Lakes Chemical Corporation, West Lafayette, IN. EPA/OTS Doc. I.D. 44-8227036.

Great Lakes (Great Lakes Chemical Corporation). 1984. Initial submission: Letter from Great Lakes Chemical to U.S. EPA Re: Tetrabromobisphenol A, Pentabromobenzene, Decabromodiphenyl Ether, and Dibromopropyl Acrylate with attachments dated 11/11/84. OTS-0794-1105.

Gulf South Research (Gulf South Research Institute). 1977. Mutagenicity Testing of

HFO 102. Gulf South Research Institute, New Orleans, LA.
Hard, G.C. 1998. Recent developments in the investigation of thyroid regulation and thyroid carcinogenesis. Environ. Health Perspect. 106(8):427-436.
Haworth, S., T. Lawlor, K. Mortelmans, W. Speck, and E. Zeiger. 1983. Salmonella mutagenicity test results for 250 chemicals. Environ. Mutagen. 5(Suppl. 1):1-142.
Hazleton Labs (Hazleton Laboratories America, Inc.). 1978. Subject: Final Report 14-Day Repeated Dose Feeding Study - Rats, Material: Decabromodiphenyl Oxide. EPA-OTS 86-900000370.
Hazleton Labs (Hazleton Laboratories America, Inc.). 1979a. Final Report: 13-Week Sub-Chronic Feeding Study - Rats, Material: Decabromodiphenyl Oxide (C44287). Hazleton Laboratories America, Inc. EPA/OTS Doc. ID FYI-OTS-0794-1093.
Hazleton Labs (Hazleton Laboratories America, Inc.). 1979b. Final Report: 13-Week Sub-chronic Feeding Study in Mice with Decabromodiphenyl Oxide. Performed by Hazleton Laboratories. Sponsor: Tracor-Jitco, Inc. TSCATS #452234. EPA/OTS Doc. #FYI-OTS-0794-1102.
Hill, R.N., L.S. Erdreich, O.E. Paynter, P.A. Roberts, S.L. Rosenthal, and C.F. Wilkinson. 1989. Thyroid follicular cell carcinogenesis. Fundam. Appl. Toxicol. 12(4):629-697.
Hill, R.N., T.M. Crisp, P.M. Hurley, S.L. Rosenthal, and D.V. Singh. 1998. Risk assessment of thyroid follicular cell tumors. Environ. Health Perspect 106(8):447-457.
HSDB (Hazardous Substance Data Bank). 1998. Hazardous Substance Data Bank. National Library of Medicine, National Toxicology Information Program, Bethesda, MD.
Huntingdon Research (Huntingdon Research Centre). 1978. Ames Metabolic Activation Test to Assess the Potential Mutagenic Effect of Bromkal 82-O-DE. Huntingdon Research Centre. TSCATS 406623. EPA/OTS Doc. #86-900000366. OTS #522929.
IARC (International Agency for Research on Cancer). 1990. Decabromodiphenyl Oxide. Pp. 73-84 in IARC Monographs on the Evaluation of Carcinogenic Risks to Humans: Some Flame Retardants and Textile Chemicals, and Exposures in the Textile Manufacturing Industry, Vol. 48. World Health Organization, International Agency for Research on Cancer. Lyon, France: IARC Press.
IPCS (International Programme on Chemical Safety). 1994. Decabromodiphenyl Ether. Pp. 59-281 in Brominated Diphenyl Ethers, Environmental Health Criteria 162. Geneva: World Health Organization.
IRDC (International Research and Development Corporation). 1974. Compound: Decabromodiphenyl Ether (DE-83), Subject: Acute Toxicity Studies in Rats and Rabbits. Sponsor: Great Lakes Chemical Corporation. EPA/OTS Doc. #86-900000327. OTS #523319.
IRDC (International Research and Development Corporation). 1976. Twenty-eight Day Toxicity Study in Rats. International Research and Development Corporation. Sponsor: Great Lakes Chemical Corporation. TSCATS/407322. EPA/OTS Doc. #86-900000330. OTS #523322.
IRDC (International Research and Development Corporation). 1977. Twenty-eight

Day Toxicity Study in Rats. International Research and Development Corporation. Sponsor: Great Lakes Chemical Corporation. TSCATS/407323. EPA/OTS Doc #86-900000331. OTS #523323.

Kociba, R.J., L.O. Frauson, C.G. Humiston, J.M. Norris, C.E. Wade, R.W. Lisowe, J.F. Quast, G.C. Jersey, and G.L. Jewett. 1975. Results of a two-year dietary feeding study with decabromodiphenyl oxide (DBDPO) in rats. J. Combust. Toxicol. 2:267-285.

LeBoeuf, R.A., G.A. Kerckaert, M.J. Aardema, D.P. Gibson, R. Brauninger, R.J. Isfort. 1996. The pH 6.7 Syrian hamster embryo cell transformation assay for assessing the carcinogenic potential of chemicals. Mutat. Res. 356(1):85-127.

Litton Bionetics (Litton Bionetics, Inc). 1976. Mutagenicity Evaluation of Compound 277-10, Final Report. LBI Project No. 2547. Litton Bionetics, Inc., Kensington, Md. TSCATS 407324. EPA/OTS Doc. #86-90000332. OTS #523324.

MacGregor, J.A., and W.B. Nixon. 1997. Decabromodiphenyl oxide (DBDPO): Determination of n-Octanol/Water Partition Coefficient. Wildlife International Ltd. Project No. 439C-101. Sponsor: Chemical Manufacturers Association's Brominated Flame Retardant Industry Panel. Wildlife International Ltd., Easton, MD.

Maronpot, R.R., C.A. Montgomery, Jr., G.A. Boorman, and E.E. McConnell. 1986. National Toxicology Program nomenclature for hepatoproliferative lesions of rats. Toxicol. Pathol. 14(2):263-273.

McGregor, D.B., A. Brown, P. Cattanach, I. Edwards, D. McBride, C. Riach, W.J. Caspary. 1988. Responses of the L5178Y tk^+/tk^- mouse lymphoma cell forward mutation assay: III. 72 coded chemicals. Environ. Mol. Mutagen.12(1):85-154.

McIntyre, J.E., I. Holme, and O.K. Sunmonu. 1995. The desorption of model compounds from poly(ethylene terephthalate) fibre. Colourage 41(13)77-81.

Myhr, B., D. McGregor, L. Bowers, C. Riach, A.G. Brown, I. Edwards, D. McBride, R. Martin, and W.J. Caspary. 1990. L5178Y mouse lymphoma cell mutation assay results with 41 compounds. Environ. Mol. Mutagen. 16(Suppl. 18):138-167.

Nissimov, S. 1984. Acute Oral Range-Finding Study in the Rat. LSRI Schedule No. DSB/055/DEC. Life Science Research Israel, Ltd. Sponsored by Dead Sea Bromine Co., Ltd., Beer Sheva, Israel.

Norris, J.M., J.W. Ehrmantraut, C.L. Gibbons, R.J. Kociba, B.A. Schwetz, J.Q. Rose, C.G. Humiston, G.L. Jewett, W.B. Crummett, P.J. Gehring, J.B. Tirsell, and J.S. Brosier. 1973. Toxicological and environmental factors involved in the selection of decabromodiphenyl oxide as a fire retardant chemical. Appl. Polym. Symp. 22:195-219.

Norris, J.M., J.W. Ehrmantraut, C.L. Gibbons, R.J. Kociba, B.A. Schwetz, J.Q. Rose, C.G. Humiston, G.L. Jewett, W.B. Crummett, P.J. Gehring, J.B. Tirsell, and J.S. Brosier. 1974. Toxicological and environmental factors involved in the selection of decabromodiphenyl oxide as a fire retardant chemical. JFF/Combust. Toxicol. 1:52-77.

Norris, J.M., R.J. Kociba, B.A. Schwetz, J.Q. Rose, C.G. Humiston, G.L. Jewett, P.J. Gehring, and J.B. Mailhes. 1975. Toxicology of octabromobiphenyl and decabromodiphenyl oxide. Environ. Health Perspect. 11:153-161.

NRC (National Research Council). 1985. Emergency and Continuous Exposure Limits for Selected Airborne Contaminants, Volume 5. Committee on Toxicology. Board on Toxicology and Environmental Health Hazards, National Research Council. Washington, DC: National Academy Press.

NTP (National Toxicology Program). 1986. Toxicology and Carcinogenesis Studies of Decabromodiphenyl Oxide (CAS No. 1163-19-5) in F344/N Rats and $B6C3F_1$ Mice (Feed Studies). NTP TR 309. NIH publication no. 86-2565. National Toxicology Program, U.S. Department of Health and Human Services, Public Health Service, National Institutes of Health.

Pharmakon. 1981. Assay of comedogenicity in the rabbit ear, PH 425-ET-001-81, Saytex 102, Lot# 4-1478J. Pharmakon Laboratories, Waverly, PA. EPA/OTS Doc.# FYI-OTS-0794-1095.

Potts, R.O., and R.H. Guy. 1992. Predicting skin permeability. Pharm. Res. 9(5):663-669.

Sjodin, A., L. Hagmar, E. Klasson-Wehler, K. Kronholm-Diab, E. Jakobsson, and A. Bergman. 1999. Flame retardant exposure: Polybrominated diphenyl ethers in blood from Swedish workers. Environ. Health Perspect. 107(8):643-648.

Sparschu, G.L., R.J. Kociba, and A. Clashman. 1971. Results of 30 Day Rat Dietary Feeding Studies on Octabromodiphenyl SA-1902 and Decabromodiphenyl Oxide SA-1892.1. Dow Chemical Co., Midland, MI. TSCATS/406630. EPA/OTS Doc. # 86-90000373.

Stenzel, J.I., and B.J. Markley. 1997. Decabromodiphenyl oxide (DBDPO): Determination of the Water Solubility. Wildlife International Ltd. Project No. 439C-102. Sponsor: Chemical Manufacturers Association's Brominated Flame Retardant Industry Panel. Wildlife International Ltd., Easton, MD.

Stenzel, J.I., and W.B. Nixon. 1997. Decabromodiphenyl oxide (DBDPO): Determination of the Vapor Pressure Using a Spinning Rotor Gauge. Wildlife International Ltd. Project No. 439C-115. Sponsor: Chemical Manufacturers Association's Brominated Flame Retardant Industry Panel. Wildlife International Ltd., Easton, MD.

Ulsamer, A., R. Osterberg, and J. McLaughlin. 1980. Flame-retardant chemicals in textiles. Clin. Toxicol. 17:101-131.

Wagner, V.O., and M.L. Klug. 1998. Bacterial Reverse Mutation Assay. Performed by MA BioServices, Inc. Study No. G98AV87.503. Sponsor: Chemical Manufacturers Association's Brominated Flame Retardant Industry Panel.

6

Alumina Trihydrate

THERE are limited toxicokinetic and toxicity data available on alumina trihydrate. Therefore, this chapter reviews the physical and chemical properties, toxicokinetics, toxicological, epidemiological and exposure data on alumina trihydrate and a number of chemically related aluminum compounds. The bioavailability of aluminum is dependent upon its form, however, the underlying mechanism of toxicity appears to be similar among the different forms (with the exception of aluminum phosphide for which the toxicity is associated with phosphine gas). The effect of bioavailability of the various forms of aluminum on toxicity is discussed in the Quantitative Risk Assessment and the Exposure Assessment and Risk Characterization sections.

The subcommittee used that information to characterize the health risk from exposure to alumina trihydrate. The subcommittee also identified data gaps and recommended research relevant for determining the health risk from exposure to alumina trihydrate.

PHYSICAL AND CHEMICAL PROPERTIES

The physical and chemical properties of alumina trihydrate are summarized in Table 6-1.

TABLE 6-1 Physical and Chemical Properties for Alumina Trihydrate

Characteristic	Value	Reference
Chemical formula	$Al(OH)_3$	Lide 1991-1992
CAS registry #	21645-51-2	Lide 1991-1992
Synonyms	aluminum hydroxide, aluminum hydrate, hydrated alumina	Budavari et al. 1989
Molecular weight	77.99	Lide 1991-1992
Physical state	White powder	Budavari et al. 1989
Solubility	Insoluble in hot or cold water; soluble in acid and alkali; insoluble in alcohol	Lide 1991-1992
Melting point	300°C	Lide 1991-1992
Density	2.42 at 25°C	Lide 1991-1992

OCCURRENCE AND USE

Alumina trihydrate is used as a flame retardant both within and outside the U.S. in the interiors of automobiles, commercial upholstered furniture, draperies, wall coverings and carpets (R.C. Kidder, Flame Retardant Chemical Association, unpublished material, April 21, 1998). It is also used in detergents, antiperspirants, and cosmetics, and used therapeutically as an antacid (e.g., Maalox) and to control phosphate levels.

TOXICOKINETICS

Absorption

Dermal Exposure

No data were found on the dermal absorption of alumina trihydrate. However, two reports were found on the dermal absorption of aluminum chloride. Dermal application of aqueous aluminum chloride (0.025-0.1 μg/cm^2) to shaved Swiss mice increased urine, serum, and whole brain aluminum concentrations (Anane et al. 1995). Dermal application (0.4 μg/d; 20 d of gestation) of aluminum chloride to pregnant Swiss mice resulted in elevated aluminum concentrations in the serum and organs of the dams and fetuses, and in the amniotic fluid (Anane et al. 1997).

Inhalation Exposure

Workers exposed to aluminum dust or fumes had higher urinary aluminum concentrations at the end of a work shift than a control group (Mussi et al. 1984). Plasma aluminum concentrations, however, were not increased. Serum and urinary aluminum concentrations increased in three individuals not previously exposed to aluminum-containing welding fumes following an 8-hr exposure to those fumes (average exposure of 2.4 mg aluminum/m^3) (Sjögren et al. 1985). Sjögren et al. (1988) reported that workers exposed to aluminum from welding fumes had elevated aluminum concentrations in their urine, and that a 16 to 37-d break from exposure resulted in decreased urinary aluminum concentrations (median levels decreased from 54 μg/g creatinine to 29 μg/g creatinine). Serum and urinary aluminum concentrations were higher in workers exposed to aluminum (25 μg/m^3 respirable particles; 100 μg/m^3 total particles) compared with pre-shift concentrations and concentrations in unexposed controls (Gitelman et al. 1995). The percentage of aluminum absorbed was not determined in those studies. No relevant animal data were identified on absorption of aluminum following inhalation exposure.

Oral Exposure

The bioavailability of orally administered aluminum is related to the form in which it is ingested and the presence of dietary constituents with which the metal can complex. Ligands in food can have a marked effect on absorption of aluminum; they can either enhance uptake by forming absorbable (usually water-soluble) complexes (e.g., with carboxylic acids such as citric acid or lactic acid), or reduce absorption by forming insoluble compounds (e.g., with phosphate or dissolved silicate).

In humans, evidence suggests that the most important compound that aluminum complexes with that increases aluminum uptake is citric acid (or its conjugate base citrate). Citric acid is a constituent of many foods and beverages, and can be present in the gut at high concentrations (Reiber et al. 1995). Concomitant exposure to aluminum-containing antacids and orange juice caused a 10-fold increase in absorption of aluminum as compared to exposure to antacids alone (Fairweather-Tait et al. 1994). Milk had no effect on aluminum absorption in that study. Volunteers (n = 7-10) who ingested antacids containing 976 mg of alumina trihydrate (approximately 14 mg/kg) absorbed 0.004%, 0.03%, or 0.2% of the aluminum when the antacids were suspended in tap water (pH 9.2), orange juice (pH 4.2), or citric acid (pH 2.4), respectively (Weberg and Berstad 1986). Priest et al. (1996, as cited in ATSDR 1999) measured aluminum absorption in two male volunteers following administration of a single dose of Al[26]-labeled aluminum citrate (aqueous solution) or alumina trihydrate (colloidal suspension in water) directly into the stomach; 0.5% of the aluminum in aluminum citrate and 0.01% of the aluminum in alumina trihydrate were absorbed. In that same study (Priest et al. 1996, as cited in ATSDR 1999), 0.14% of the aluminum was absorbed after concomitant exposure to alumina trihydrate and trisodium citrate; that exposure scenario is similar to ingestion of aluminum in orange juice. Urinary and plasma aluminum concentrations were significantly higher in women treated with calcium citrate than when they were not treated with calcium citrate, indicating that dietary factors can affect the uptake of aluminum from normal diets (Nolan et al. 1994, as cited in ATSDR 1999).

Infants are able to absorb orally administered aluminum. Plasma aluminum concentrations increased (from 0.64 μmol/L prior to treatment to 3.48 μmol/L after treatment) in 7 infants treated with aluminum-containing antacids (400-800 μmol aluminum for 2 d) (Chedid et al. 1991).

Individuals with young senile dementia of the Alzheimer's type (Taylor et al. 1992) and individuals with Down's Syndrome (Moore et al. 1997) appear to have increased absorption of aluminum.

Evidence in animals indicates that absorption of aluminum is low following oral exposure, and that the form of aluminum ingested and dietary factors can affect aluminum absorption. Only 0.97% of the dose was absorbed in rats gavaged with $Al[26]Cl_3$ (n = 3/group) (Zafar et al. 1997). Following a single gavage dose of alumina trihydrate, aluminum citrate, aluminum citrate with sodium citrate added, or aluminum maltolate, 0.1%, 0.7%, 5.1%, and 0.1% of the aluminum was absorbed, respectively (Schonholzer et al. 1997). Jouhanneau et al. (1997) measured skeletal retention and urinary excretion of aluminum, as an indication of absorption, in 2-mo-old Wistar rats fed aluminum in the diet. In the absence of citrate, 0.05% of the aluminum dose was found in the urine and in the skeleton. The presence of citrate in the diet increased excretion by two- to five-fold (Jouhanneau et al. 1997). Plasma, bone, kidney, cerebral cortical, and cerebellar aluminum concentrations were not increased (compared to untreated controls) in rats fed alumina trihydrate alone, but were increased in rats fed an equivalent concentration of aluminum complexed with citrate, lactate, malate, or tartrate (Testolin et al. 1996). Domingo et al. (1993) investigated the effect of dietary constituents on the absorption of aluminum from the normal diet. The addition of lactic, tartaric, gluconic, malic, succinic, ascorbic, citric, or oxalic acid to drinking water increased the concentration of aluminum in the bone; all except succinic and ascorbic acid increased aluminum concentrations in the brain. Prolonged fasting increased the absorption of aluminum in Wistar rats (Drueke et al. 1997).

Based on the data discussed above, it was concluded that alumina trihydrate is more poorly absorbed than other aluminum compounds. Some data indicate a direct linear relationship between the dose of soluble aluminum and the plasma aluminum level (Partridge et al. 1992, as cited in ATSDR 1999). However, the data on both solubility and bioavailability are inadequate to reliably extrapolate quantitatively from solubility in water to bioavailability, especially with the effects of dietary constituents.

Distribution and Metabolism

Dermal Exposure

Following dermal absorption, aluminum chloride distributes to the brain in Swiss mice (Anane et al. 1995) and to the fetus in pregnant Swiss mice (Anane et al. 1997).

Inhalation Exposure

Autopsy results of men chronically exposed to aluminum via inhalation indicated that aluminum distributes to the lungs, liver, and spleen (Teraoka 1981). Rabbits exposed to low concentrations of aluminum dust (Al_2O_3; 1/20th of the threshold limit value) had 2.5-times higher concentrations of aluminum in the brain compared to controls. Serum concentrations were only slightly increased and concentrations in other tissues were not elevated (Rollin et al. 1991). Rats exposed via inhalation to aluminum acetylacetonate also demonstrated an accumulation of aluminum in the brain (Zatta et al. 1993).

Oral Exposure

Following gavage in rats, the highest accumulation of aluminum is in the bone, followed by the spleen, kidneys and liver, and brain (Zafar et al. 1997). Testolin et al. (1996) also demonstrated that aluminum distributes to the bone, kidneys, cerebral cortex, and cerebellum.

Other Routes of Exposure

Yokel and McNamara (1989) investigated the distribution and half-life of aluminum in rabbits after a 6-hr intravenous infusion. Aluminum concentrations were increased in the bile, kidneys, liver, lungs, serum, and spleen after 4 hr, but not in the brain. The half-life was tissue-dependent, ranging from 12 hr in the bile to 113 d in the spleen. After intravenous injection of aluminum lactate or aluminum citrate in rats and rabbits, aluminum appeared to freely diffuse into liver, but was lower in the brain than the blood, indicating that there is a partial barrier to aluminum entry into the brain (Yokel et al. 1991). Further research, however, indicates that an active process pumps aluminum out of the brain following administration of aluminum citrate (Yokel et al. 1994; Allen et al. 1995; Ackley and Yokel 1997).

Regardless of the route of exposure, once absorbed and distributed in the body, aluminum can exist in different forms. Low concentrations of aluminum exist as free ions. Aluminum can also complex with organic acids, amino acids, nucleotides, phosphates, and carbohydrates. Aluminum can form reversible and practically irreversible complexes with proteins, polynucleotides, and glycosaminoglycans (Ganrot 1986).

Excretion

Dermal Exposure

Aluminum was detected in the urine of Swiss mice following dermal exposure to aluminum chloride (Anane et al. 1995).

Inhalation Exposure

Following inhalation exposure, absorbed aluminum is primarily excreted via the urine. Excretion half-lives of 7.5 and 8 hr have been reported in workers exposed to aluminum from welding fumes (Sjögren et al. 1985; Pierre et al. 1995). The urinary excretion half-life appears to rise with increasing exposure duration (Sjögren et al. 1985). Urinary aluminum concentrations in workers exposed to aluminum were more than 10 times higher than those of individuals not exposed to aluminum in the workplace, and remained elevated many years after the occupational exposure ceased (Elinder et al. 1991).

Oral Exposure

The majority of ingested aluminum is excreted in the feces without being absorbed systemically (Gorsky et al. 1979; Jouhanneau et al. 1997). Absorbed aluminum is primarily excreted in the urine (Kaehny et al. 1977; Recker et al. 1977; Gorsky et al. 1979; Greger and Baier 1983).

HAZARD IDENTIFICATION[1]

Dermal Exposure

Irritation

Skin rashes in sensitive individuals are the only adverse effects observed in humans dermally exposed to aluminum compounds (ATSDR 1999).

Damage to the skin was observed in mice, rabbits, and pigs following exposure to 10% aluminum chloride and aluminum nitrate for 5 d. No dermal effects were observed in animals exposed to 10% alumina trihydrate, aluminum sulfate, aluminum acetate, or aluminum chlorohydrate (Lansdown 1973, as cited in ATSDR 1999).

Systemic Effects

No studies were identified that report immunological, neurological, reproductive, developmental, carcinogenic, or other systemic effects of aluminum following dermal exposure.

Inhalation Exposure

Systemic Effects

No studies were identified that investigated the effects of alumina trihydrate via inhalation exposure. Pulmonary fibrosis is the most common respiratory effect in workers exposed to finely ground aluminum dust (pyropowder) (Ueda et al. 1958; Edling 1961; Mitchell et al. 1961; McLaughlin et al. 1962). However, that effect appears to be associated with a specific type of oil coating on the aluminum dust (Crombie et al. 1944; Meiklejohn and Posner 1957; Posner and Kennedy 1967). Case reports indicate that inhalation exposure to various

---

[1]In this section, the subcommittee reviewed the toxicity data of alumina trihydrate, including the toxicity assessment prepared by the U.S. Consumer Product Safety Commission (Ferrante 1999).

forms of aluminum leads to pulmonary toxicity (Chen et al. 1978; Miller et al. 1984; Park et al. 1996, as cited in ATSDR 1999; Vandenplas et al. 1998). In a study of 17 occupationally exposed individuals, pulmonary fibrosis was associated with inhalation exposure to aluminum silicate dust (Musk et al. 1980). Avolio et al. (1989) reported interstitial fibrosis following inhalation exposure to aluminum. Those occupational studies are limited by concomitant exposures to other chemicals and cigarettes. However, in one study of non-smoking individuals occupationally exposed to aluminum compounds (14 exposed; 28 controls) there were indications of increased alveolar capillary permeability and activation of alveolar macrophage in bronchoalveolar lavage, but no evidence of restrictive lung disease (Eklund et al. 1989).

Granulomatous reactions (at concentrations of 2.5 and 25 mg/m^3 aluminum chlorohydrate), decreases in body weight (at concentrations of 25 mg/m^3), and increases in lung to body weight ratios (at concentrations of 25 mg/m^3) were seen in rats and guinea pigs exposed to aluminum chlorohydrate for 6 mo (Steinhagen et al. 1978). Exposure of female Wistar rats to aluminum fibers for 86 wk resulted in minimal pulmonary reactions (Pigott et al. 1981).

Neurological Effects

Subclinical neurological effects have been observed in workers chronically exposed to aluminum dust, welding fumes, and McIntyre powder (finely ground aluminum and aluminum oxide) (Hosovski et al. 1990; Rifat et al. 1990; White et al. 1992; Bast-Pettersen et al. 1994; Hänninen et al. 1994; Sjögren et al. 1996; Dick et al. 1997; Sim et al. 1997). Those effects include changes in neurobehavioral test performance (e.g., eye-hand coordination, reaction time, cognitive tests) and increased incidences of subjective symptoms (e.g., incoordination, depression, fatigue).

A role of aluminum has been hypothesized in the etiology of Alzheimer's disease (AD). However, in an unmatched case-control study (198 AD cases; 340 controls made up of 164 individuals with non-AD dementias and 176 individuals with no dementias), no significant association (odds ratio = 0.98) between occupational aluminum exposure and AD was reported (Salib and Hillier 1996).

Cancer

There are a number of epidemiological studies on cancer incidence in workers in aluminum reduction plants (Gibbs and Horowitz 1979; Milham 1979;

Theriault et al. 1981; Rockette and Arena 1983; Gibbs 1985; Armstrong et al. 1986; Spinelli et al. 1991). In a review of many of those studies, Ronneberg and Langmark (1992) concluded that some data were suggestive of an increased risks for specific cancers for workers in aluminum reduction plants. However, those conclusions were limited by inadequate information on smoking and exposure to other carcinogenic compounds, including asbestos and polycylic aromatic hydrocarbons. In a retrospective cohort study that was initiated because of a cluster of pituitary adenoma cases (four cases over 5 yr), there was no indication of an increased risk for pituitary adenoma at an aluminum production factory (Cullen et al. 1996). There was no overall excess risk for cancer and no excess risk for bladder or liver cancer among men or women workers in aluminum foundries and scrap aluminum smelters in Sweden (n = 6,454) (Selden et al. 1997). However, risk estimates for lung cancer in males (standardized incidence ratio [SIR] = 1.49), anorectal cancer (SIR = 2.13), and sinonasal cancer (SIR = 4.70) were increased. Socioeconomic status appeared to underlie the increased risk of lung cancer, except for individuals employed in the sand casting of aluminum for 10 yr or more. Epidemiological studies of workers in aluminum smelters report an increased mortality from malignant lung neoplasm, however, many of the workers had evidence of co-exposure to asbestos, silicates, and metal-rich nonfibrous particles, such as chromium and cobalt (Dufresne et al. 1996), or polycyclic aromatic hydrocarbons (Armstrong et al. 1994).

In the only animal study investigating the carcinogenic potential of inhaled aluminum compounds, there was no evidence of an increased incidence of tumors in the lungs of male or female Wistar rats exposed to aluminum fibers (2.18-2.45 mg aluminum/m^3; 96% aluminum oxide) for 86 wk (Pigott et al. 1981).

Other Systemic Effects

No studies were identified on the immunological, reproductive, or developmental effects following inhalation of aluminum.

Oral Exposure

Systemic Effects

Aluminum compounds have low acute toxicity because of their low solubility. The maximum tolerated daily dose for alumina trihydrate in a healthy, 70-

kg adult is 50 to 128 mg/kg (17.5-45 mg aluminum/kg) (Poisindex 1998). Constipation, diarrhea, distension, and/or obstruction with perforation have been reported in individuals on chronic antacid therapy. However, the role of aluminum in that effect is not known (HSDB 1990). Individuals with chronic renal failure who ingest large amounts of aluminum trihydrate to treat hyperphosphatemia can accumulate aluminum in the body, resulting in hypercalcemia, microcytic anemia, proximal myopathy, osteomalacia, and progressive dialysis encephalopathy (Sideman and Manor 1982; HSDB 1990; Ellenhorn 1997). Osteomalacia has also been observed in healthy children treated with aluminum-containing antacids for colic (Pivnick et al. 1995). Preterm infants are at risk for aluminum toxicity from ingestion of some infant formulas that contain aluminum compounds, and from aluminum-containing parenteral nutrition solutions (Sedman et al. 1985; Koo et al. 1992; Golub and Domingo 1996).

There is an extensive oral toxicity database in animals, but many of the studies are limited by a lack of information on background concentrations of aluminum compounds in the diet. Commercial grain-based feeds for laboratory animals contain high concentrations of aluminum compounds which can contribute substantially to total aluminum exposure. The background aluminum concentrations in feed, therefore, should be considered when assessing the toxicity of aluminum compounds. A summary of the studies is presented in Table 6-2.

Most aluminum compounds have LD_{50}s in the range of 1-4 g aluminum/kg (Poisindex 1998). No significant effects on mortality or body weight were observed in Sprague-Dawley rats fed 989 or 1,070 μg aluminum/g of food (as alumina trihydrate; calculated to be equivalent to approximately 158 mg aluminum/kg-d) for 16 d (background concentrations, 9-26 μg aluminum/g food) (Greger and Donnaubauer 1986). Hicks et al. (1987) reported no significant alterations in hematology, clinical chemistry, histopathology, or organ weights in Sprague-Dawley rats fed 302 mg aluminum/kg-d as alumina trihydrate in the diet for 28 d (background concentration, 66 ppm; reported as 5 mg aluminum/kg-d).

In general, subchronic and chronic studies in mice and rats examining a number of systemic end points do not demonstrate adverse effects following dietary or drinking water exposure to aluminum. Oteiza et al. (1993) fed Swiss-Webster mice 1,000 μg aluminum/g in food (background levels; 3 mg aluminum/g food) as aluminum chloride for 5 or 7 wk. No systemic effects were seen. Oneda et al. (1994) fed male and female B6C3F1 mice 1%, 2.5%, 5%, or 10% aluminum potassium sulfate for 20 mo and reported a decrease in liver weight (5-10%), and an increase in kidney weight (2.5%) and heart weight (5%). Relative organ weight and blood parameters were not affected in

TABLE 6-2 Selected Toxicity Studies of Orally Administered Alumina Compounds[a]

Animal Species and Strain/Aluminum Compound/ Duration of Exposure	Dose (mg Al/kg-d)	Effects	NOAEL (mg Al/kg-d)	LOAEL (mg Al/kg-d)	Reference
Systemic toxicity					
Female Sprague-Dawley rats; aluminum nitrate (drinking water); 100 d	25 (control), 51, 77, 284	None	284	ND	Domingo et al. 1987a
Sprague-Dawley rats; aluminum trihydrate (diet); 28 d	5 (control), 302	None	302	ND	Hicks et al. 1987
Female Sprague-Dawley rats; aluminum nitrate (drinking water) 1 mo	25 (control), 52, 79, 133	Hyperemia in the liver, periportal monocytic infiltrate in liver	79	133	Gomez et al. 1986
Sprague-Dawley rats; aluminum trihydrate (diet); 16 d	158	None	158	ND	Greger and Donnaubauer 1986
Neurotoxicity					
Female Swiss-Webster mice; aluminum lactate (diet); 6 wk	3 (control), 62, 130	Decreased motor activity	62	130	Golub et al. 1989
Female Swiss-Webster mice; aluminum lactate (diet); 90 d	4.9 (control), 195	Decreased motor activity, hindlimb grip strength and startle responsiveness	ND	195	Golub et al. 1992a
Female Swiss-Webster mice; aluminum chloride and 3.5% sodium citrate (diet) 5-7 wk	0.6 (control), 195	Decreased forelimb and hindlimb grip strength	ND	195	Oteiza et al. 1993

TABLE 6-2 *(Continued)*

Animal Species and Strain/Aluminum Compound/ Duration of Exposure	Dose (mg Al/kg-d)	Effects	NOAEL (mg Al/kg-d)	LOAEL (mg Al/kg-d)	Reference
Neurodevelopmental toxicity					
Pregnant Swiss-Webster mice; aluminum lactate (diet); gestational d 0 through weaning	7.5 (control), 155, 310	Decreased forelimb and hindlimb grip strength and increased foot splaying in weanlings	ND	155	Donald et al. 1989
Pregnant Swiss-Webster mice; aluminum lactate (diet); gestational d 0 through weaning	7.5 (control), 155, 310	Decreased forelimb and hindlimb grip strength	ND	155	Golub et al. 1995
Pregnant Swiss-Webster mice; aluminum lactate (diet); gestational d 0 through weaning; offspring exposed for another 150 d	7.5 (control), 155, 310	Decreased forelimb and hindlimb grip strength and decreased air puff startle response	ND	155	Golub et al. 1995
Pregnant Swiss-Webster mice; aluminum lactate (diet); gestational d 1-21	1 (control), 250	Decreased forelimb grip strength	ND	250	Golub et al. 1992b
Pregnant Swiss-Webster mice; aluminum lactate (diet); gestational d 1-21 and lactational d 1-21	1 (control), 250	Increased hindlimb grip strength and tail withdrawal times	ND	250	Golub et al. 1992b
Pregnant Swiss-Webster mice; aluminum lactate (diet); lactational d 1-21	1 (control), 250	Increased negative geotaxis latency	ND	250	Golub et al. 1992b

Pregnant Swiss-Webster mice; aluminum lactate (diet); gestational d 1-21 and lactational d 1-21	1.2 (control), 200	Reduced auditory startle responsiveness	ND	200	Golub et al. 1994
Pregnant Swiss-Webster mice; aluminum lactate (diet); gestational d 1 through postnatal d 52	0.7 (control), 200	Reduced auditory startle responsiveness	ND	200	Golub et al. 1994
Swiss-Webster mice; aluminum lactate (diet); gestational d 0 to postnatal d 21	4.1 (control), 17.5, 28.3	Treated animals lagged behind controls on the Wahlsten neurobehavioral test battery	ND	17.5	Golub et al. 1987
Developmental toxicity					
Pregnant Sprague-Dawley rats; aluminum citrate (gavage); gestational d 6-15	25 (control), 158	Increase in number of fetuses with absent xiphioides	ND	158	Gomez et al. 1991
Pregnant Sprague-Dawley rats; aluminum hydroxide (gavage); gestational d 6-15	25 (control), 158	None (developmental end points examined included number of implantation sites, resorptions, dead and live fetuses, and external malformations)	158	ND	Gomez et al. 1991
Pregnant Sprague-Dawley rats; aluminum nitrate (gavage); gestational d 6-14	25 (control), 38, 51, 77	Decreased fetal weight, decreased tail length, increased rib, skull, and sternebral malformations	ND	38	Paternain et al. 1988
Reproductive toxicity					
Female Sprague-Dawley rats; aluminum nitrate (gavage); 14 d prior to mating and throughout gestation and lactation	13 (control), 38, 52	No effects on fertility were observed	52	ND	Domingo et al. 1987b

TABLE 6-2 *(Continued)*

Animal Species and Strain/Aluminum Compound/ Duration of Exposure	Dose (mg Al/kg-d)	Effects	NOAEL (mg Al/kg-d)	LOAEL (mg Al/kg-d)	Reference
Reproductive toxicity					
Pregnant Sprague-Dawley rats; aluminum hydroxide (gavage); gestational d 6-15	25 (control), 158	None (reproductive end points examined included numbers of litters, corpora lutea per dam, implantation, preimplantation loss, viable implants, nonviable implants, postimplantation loss, and sex ratio)	158	ND	Gomez et al. 1991

ND, not determined

[a]Studies that did not provide information on the concentration of aluminum in the base diet were not included in the table.

Sprague-Dawley rats (10/group) exposed to aluminum nitrate (0, 375, 750, and 1,500 mg/kg-d) for 1 mo (Gomez et al. 1986). Mild histopathological changes occurred in the two highest dose groups. Domingo et al. (1987a) exposed female Sprague-Dawley rats to aluminum nitrate (0, 360, 720, and 3,600 mg/kg-d) for 100 d and no effects were seen on organ weights or histology. An increase in serum glutamic pyruvic transaminase was seen in the 3,600 mg/kg-d group and an increase in blood urea in the 720 mg/kg-d group.

Neurological Effects

A role of aluminum in the etiology of AD has been suggested, but remains controversial (see reviews Savory et al. 1996; Exley 1998; Forbes and Hill 1998; Munoz 1998; Smith and Perry 1998). Because of the conflicting data and the lack of quantitative data for a risk assessment, the literature pertaining to the role of aluminum in AD is not reviewed in this report.

Neurotoxicity and neurobehavioral studies in animals provide strong evidence that the nervous system is the most sensitive target organ for aluminum toxicity. Neurobehavioral effects have been observed in animals exposed as adults, weanlings, during gestation and lactation, and during gestation through adulthood. Although studies lacking information on background aluminum concentrations in the diet provide valuable hazard identification information, NOAELs and LOAELs from these studies cannot be derived with confidence because they are not suitable for dose-response assessment. Therefore, the following discussion of oral toxicity studies focuses on those reports that provide information on the concentrations of aluminum in the control diet. Some studies that used alumina trihydrate but did not report control diet aluminum concentrations are also discussed.

A diminished learning ability was observed in Long-Evans rats exposed to 30 or 100 mg/kg aluminum chloride (6 or 20 mg aluminum/kg-d), 300 mg/kg aluminum trihydrate (104 mg aluminum/kg-d), or 100 mg/kg aluminum trihydrate plus citric acid (35 mg aluminum/kg-d) (Bilkei-Gorzo 1993). Although aluminum content was not measured in the control diet, the aluminum levels were measured in the brains of all animals in that study. Thorne et al. (1987) did not observe neurological effects in rats following oral exposure to aluminum trihydrate (NOAEL = 14 mg aluminum/kg-d) for 60 d during the weaning period. Background aluminum intake was not measured.

Comprehensive neurobehavioral testing of N: NIH Swiss-Webster mice exposed to 195 mg aluminum/kg-d as dietary aluminum lactate for 90 d found reduced motor activity, decreased hindlimb grip strength, decreased startle response, and increased tissue concentrations of aluminum (in brain and liver, but not bone), but no overt clinical signs of neurotoxicity (Golub et al. 1992a). Oteiza et al. (1993) fed adult N: NIH Swiss-Webster mice 195 mg aluminum/kg-d as aluminum chloride in a diet that also contained 3.5% sodium citrate. Neurobehavioral effects similar to those observed by Golub et al. (1992a) were seen, except grip strength was reduced in forelimbs as well as hindlimbs. Aluminum concentrations were elevated in the bone, brain, and liver. The more pronounced effects are most likely due to increased absorption of aluminum in the presence of citrate.

Golub et al. (1987) fed pregnant Swiss-Webster mice diets containing aluminum lactate (100 [control], 500, and 1,000 ppm elemental aluminum; approximately equivalent to 4.1, 17.5, and 28.3 mg/kg-d based on average food intake during gestation and lactation) on gestational d 0 to postnatal d 21. There were no consistent adverse effects on organ weight that were not seen in the pair-fed control group. Performance in the Wahlsten neurobehavioral test battery was affected in the 500 and 1,000 ppm groups on postnatal d 14 and 16 (but not postnatal d 11-13, 15, and 17-18).

Golub et al. (1989) fed female Swiss-Webster mice (n = 16) diets containing 500 or 1,000 ppm aluminum as aluminum lactate (doses estimated as 3 (control), 62, and 150 mg aluminum/kg-d) for 6 wk. During the 5th wk of exposure, motor activity in a 24-hr period was measured using an automated method. No overt signs of neurotoxicity were observed. Total activity was significantly decreased (20%) in the 130-mg aluminum/kg-d group as compared with controls, with vertical movement affected more than horizontal movement. Those animals were less active than controls during the diurnal period of peak activity and had shorter periods of activity (130 versus 200 min), but there were no shifts in the diurnal activity cycle or any prolonged periods of inactivity. Motor activity was not significantly affected in the 62-mg aluminum/kg-d group. A LOAEL of 130 mg aluminum/kg-d and a NOAEL of 62 mg aluminum/kg-d were identified.

Golub et al. (1995) fed pregnant Swiss-Webster mice (group size inadequately reported) diets containing 500 or 1,000 ppm aluminum as aluminum lactate throughout gestation and lactation (control diet contained 25 ppm aluminum). Daily doses of 7.5 (control), 155, and 310 mg aluminum/kg-d were estimated by averaging reported estimated doses at the beginning of pregnancy and during lactation. At weaning, each litter was assigned to continue with a diet similar to the dam or transferred to a control diet, therefore there were subgroups of offspring that received pre- and post-weaning exposure (continuous exposure group) or only pre-weaning exposure (developmental exposure group). A neurobehavioral test battery that assessed strength, responsiveness, and coordination was administered on d 150-170. Additionally, one male and one female from each litter were tested in either a discrimination reversal test or a delayed spatial alternation testing paradigm. In the developmentally exposed animals, there was a significant decrease in forelimb and hindlimb grip strength in the 155- and 310-mg/kg-d exposure groups and air puff startle response was decreased in the 155-mg/kg-d group. In the continuously exposed animals, significant decreases in forelimb and hindlimb grip strength and air puff startle responses were observed in the 155- and 310-mg/kg-d groups. No differences in grip strength or startle responses were observed between the developmental exposure group and the continuous exposure group. No significant alterations in auditory startle response, temperature sensitivity, or negative geotaxis were observed in any of the aluminum-exposed offspring. Mice in the developmental exposure group required fewer operant training sessions to reach criterion than controls. The continuous exposure group reached the criterion in fewer sessions than the developmental exposure group. Aluminum exposure (developmental or continuous) did not markedly affect learning of the spatial alternation task, or performance of the delayed spatial alternation task or the discrimination reversal task. A LOAEL of 155 mg aluminum/kg-d was identified. No NOAEL was identified.

Donald et al. (1989) fed 16 pregnant Swiss-Webster mice diets containing 500 or 1,000 ppm aluminum as aluminum lactate throughout gestation and lactation (control diet contained 25 ppm aluminum). Approximate doses at the beginning of gestation and maximal intake during lactation were averaged to estimate doses of 7.5 (control), 155, and 310 mg aluminum/kg-d. On postnatal d 8-18, neurobehavioral tests were conducted on one male and one female from each litter. The offspring were also tested on postnatal d 25 and 39 using a neurobehavioral test battery. Composite scores on the neurobehavioral maturation tests did not differ significantly between groups, although results on specific days and tests did differ between groups. On postnatal d 9 and 16, pups in the 310-mg/kg-d group had lower test scores than controls, and only 72% of the pups in the 310-mg/kg-d group reached the criterion by postnatal d 18 (as compared to 100% in the control and 155-mg/kg-d groups). On postnatal d 25, significant increases in forelimb grip strength (310 mg aluminum/kg-d) and hindlimb grip strength (155 and 310 mg aluminum/kg-d) were observed. On postnatal d 39, forelimb grip strength in the 155-mg/kg-d group was significantly lower than in control animals, and hindlimb grip strength in both treated groups was similar to controls. Higher latencies for the temperature aversion test were seen in the 310-mg/kg-d group offspring at postnatal d 25 and 39. An increase in foot splay distance was observed in the 155- and 310-mg/kg-d group offspring on postnatal d 21 and in the 155-mg/kg-d group on postnatal d 35. Startle response and negative geotaxis were not consistently affected by aluminum exposure. A LOAEL of 155 mg aluminum/kg-d was identified. A NOAEL was not identified.

Findings in other mouse studies using similar or higher estimated doses of aluminum lactate corroborate the neuromotor alterations summarized above (Golub et al. 1992b). The findings include increased grip strength, increased tail withdrawal time from hot water, and increased negative geotaxis latency in weanling mice following gestation and/or lactation exposure to 250 mg aluminum/kg-d (Golub et al. 1992b). Reduced auditory startle responsiveness was also seen in pups exposed during gestation and lactation, or from gestation continuing into the post-weaning period and tested at 52 d of age (maternal dose was 200 mg aluminum/kg-d) (Golub et al. 1994).

Reproductive and Developmental Effects

Neurodevelopmental effects are discussed in the section on Neurological Effects.

No marked maternal or developmental effects were seen in Swiss mice treated orally with aluminum trihydrate during organogenesis (0, 66.5, 133, or 266 mg/kg-d; gestational d 6-15) (Domingo et al. 1989).

Colomina et al. (1992) fed pregnant Swiss albino mice 57.5 mg aluminum/kg as alumina trihydrate, aluminum lactate, or alumina trihydrate concurrently with lactic acid. Alumina trihydrate alone had no marked effect on maternal body weight or organ weights (uterine, liver, and kidney), or on reproductive end points or skeletal development. Sporadic effects on maternal body weight and relative liver weight were seen with the other treatments. Aluminum lactate decreased fetal body weight per litter and increased the occurrence of cleft palate, dorsal hyperkiphosis, and delayed parietal ossification. Lactic acid alone increased delayed parietal ossification (Colomina et al. 1992).

Colomina et al. (1994) gavaged pregnant Swiss mice with 103.8 mg aluminum/kg as alumina trihydrate in the presence or absence of ascorbic acid on gestational d 6-15. Dams were killed on gestational d 18. No marked effects of aluminum were seen on the number of resorptions per litter, number of dead and live fetuses per litter, percentage of postimplantation loss, sex ratio, or fetal body weight per litter. There were also no apparent malformations or developmental variations based on gross external, visceral, and skeletal parameters (Colomina et al. 1994).

Gomez et al. (1991) gavaged pregnant Sprague-Dawley rats on gestational d 6 to 15 with aluminum (133 mg/kg-d) as alumina trihydrate, aluminum citrate, or alumina trihydrate concurrently with citric acid. Gestational body weight, food consumption, body, and organ (liver, kidney, and brain) weights were measured at the end of the study, and reproductive and developmental endpoints were examined. Neither alumina trihydrate alone nor aluminum citrate had any marked effects on the end points studied. In the presence of citric acid, alumina trihydrate significantly decreased gestational body weight gain during the treatment period (gestational d 6-15), but significantly increased it during the post-treatment period (gestational d 16-20). Combined treatment with alumina trihydrate and citric acid also significantly decreased fetal body weight per litter, increased the incidence of delayed occipital and sternebrae ossification, and increased the absence of xiphoides (Gomez et al. 1991).

Domingo et al. (1987b) gavaged male and female Sprague-Dawley rats with aluminum nitrate at doses of 0, 180, 360, or 720 mg/kg-d for 60 d (males) or 14 d (females) prior to mating, and throughout the mating period, gestation, delivery, and lactation. A decrease in the number of corpora lutea on gestational d 13 in the high-dose group was the only effect seen on fertility measures. Survival of the treated offspring was affected, with significant decreases in the number of living offspring and increases in the number of dead offspring at the two highest doses. Body weight was also decreased in the offspring at all three dose levels. The baseline aluminum concentration was not reported in the study (Domingo et al. 1987b), but were provided at a later date (25 mg aluminum/kg-d) (Domingo et al. 1993; Colomina et al. 1998).

Cranmer et al. (1986) observed an increased incidence of resorptions in mice exposed during gestation to aluminum chloride (100, 150, 200, 300 mg $AlCl_3$/kg-d) via gavage. A decrease in sperm count was observed in rats exposed to aluminum chloride for 6-12-mo, but this study did not assess reproductive function (Krasovskii et al. 1979). Misawa and Shigeta (1993) found that administration of a single dose of aluminum chloride (0, 900, or 1,800 mg/kg, by gavage) on d 15 of gestation resulted in a decrease in body weight and affected the timing of pinna detachment and eye opening in the offspring.

Other Systemic Effects

No studies were identified that investigated the immunological or carcinogenic effects of alumina trihydrate following oral exposure.

Genotoxicity

An increase in chromatid-type aberrations occurred in mice injected intraperitoneally with aluminum chloride (0.01, 0.05, or 0.1 M aluminum chloride), but no apparent dose-response relationship was identified (Manna and Das 1972).

Aluminum chloride caused cross-linking of proteins to DNA in intact Novikoff ascites hepatoma cells, with optimal cross-linking occurring at 0.5 mM (Wedrychowski et al. 1986). Aluminum compounds were negative in Syrian hamster cell transformation experiments (DiPaolo and Casto 1979), in recombination repair assays with *Bacillus subtilis* (Kanematsu et al. 1980), and in the Ames assay with *Salmonella typhimurium* (Marzin and Phi 1985).

QUANTITATIVE TOXICITY ASSESSMENT

Noncancer

Dermal Assessment

There are inadequate dermal toxicity data on aluminum compounds to derive a dermal RfD.

Inhalation RfC

There are inadequate inhalation toxicity data on aluminum compounds to derive an RfC.

Oral RfD

There is an extensive database on the oral toxicity of aluminum in animals. Collectively, the results of the animal studies provide strong evidence that the nervous system is the most sensitive target organ for aluminum toxicity. Golub et al. (1989) identified a NOAEL of 62 mg aluminum/kg-d and a LOAEL of 130 mg aluminum/kg-d based on neurobehavioral effects in adults. That study was not selected as the critical study for the derivation of the RfD because it was only of six weeks duration. Golub et al. (1995) identified a LOAEL of 155 mg aluminum/kg-d based on neurodevelopmental effects following exposure to aluminum lactate throughout pregnancy and lactation, and into adulthood in mice. No NOAEL was identified. The subcommittee selected that study as the critical study for derivation of the oral RfD because exposure occurred from conception until adulthood. The results of the study by Golub et al. (1995) are supported by Donald et al. (1989) who also identified 155 mg aluminum/kg-d as the LOAEL for neurobehavioral effects.

To derive the RfD, the LOAEL of 155 mg aluminum/kg-d was divided by a composite uncertainty factor of 300 (10 for interspecies extrapolations, 10 for intraspecies variability; and 3 for use of a LOAEL rather than a NOAEL; see Table 6-3 for summary) to yield an RfD of 0.5 mg aluminum/kg-d. An UF of 3 for the use of a LOAEL rather than a NOAEL was used rather than the default factor of 10 because the observed effects appear to be marginal in severity. The RfD of 0.5 mg aluminum/kg-d is equivalent to an RfD of 1.5 mg alumina trihydrate/kg-d.

Current estimated dietary intakes of aluminum are 0.10 to 0.12 mg aluminum/kg-d (Pennington and Schoen 1995), which are below the RfD that the subcommittee is recommending.

It should be noted that because of the lack of data available for alumina trihydrate, a critical study (Golub et al. 1995) was selected in which exposure was to aluminum lactate. The form of aluminum can affect the bioavailability of aluminum, but data suggest that alumina trihydrate is less bioavailable, and consequently less toxic, than other aluminum compounds. Therefore, the use of data on aluminum lactate should yield a conservative RfD for alumina trihydrate.

The subcommittee's confidence in this RfD is medium. That confidence rating is based on medium-to-high confidence in the principal studies and medium confidence in the database. The study by Golub et al. (1995) was well conducted. Confidence in the study is low to moderate because of inadequate reporting of the number of offspring tested. The medium confidence in the database is reflective of the number of studies which have assessed the systemic toxicity of aluminum in several species, developmental toxicity studies in two species, and a large number of studies assessing neurotoxicity and neurodevel-

TABLE 6-3 Derivation of Oral Reference Dose for Alumina Trihydrate

Critical effect	Species	Effect level (mg Al/kg-d)	Uncertainty factors	RfD	Reference
Neurodevelopmental effects	Mouse	LOAEL: 155	UF_A: 10 UF_H: 10 UF_L: 3 Total: 300	0.5 mg Al/kg-d (equivalent to 1.5 mg $Al(OH)_3$/kg-d)	Golub et al. 1995

LOAEL, lowest-observed-adverse-effect level; RfD, reference dose; UF_A, uncertainty factor for extrapolation from animals to humans; UF_H, uncertainty factor for intraspecies variability; UF_L, uncertainty factor for the use of a LOAEL rather than a no-observed-adverse-effect level (NOAEL).

opmental toxicity. Although a multigeneration reproductive study was not identified, the available single-generation studies suggest that reproductive toxicity is not a sensitive end point. The database lacks studies that identify a NOAEL for neurodevelopmental effects and a study that adequately assesses potential differences in the toxicity of various aluminum compounds.

Cancer

The potential carcinogenicity of alumina trihydrate cannot be determined based on inadequate data for an assessment of carcinogenicity via the dermal, inhalation, and oral routes.

EXPOSURE ASSESSMENT AND RISK CHARACTERIZATION

Noncancer

Dermal Exposure

The assessment of noncancer risk by the dermal route of exposure is based on the scenario described in Chapter 3. This exposure scenario assumes that an adult spends 1/4th of his or her time sitting on furniture upholstery treated with alumina trihydrate, that 1/4th of the upper torso is in contact with the upholstery, and that clothing presents no barrier. Alumina trihydrate is considered to be ionic, and is essentially not absorbed through the skin. However, to be conservative, the subcommittee assumed that ionized alumina trihydrate permeates the skin at the same rate as water, with a permeability rate of 10^{-3} cm/hr (EPA 1992). Using that permeability rate, the highest expected application rate for alumina trihydrate (7.5 mg/cm^2), and Equation 1 in Chapter 3, the subcommittee calculated a dermal exposure level of 5.9×10^{-2} mg/kg-d. The oral RfD for alumina trihydrate (1.5 mg/kg-d; see Oral RfD in Quantitative Toxicity section) was used as the best estimate of the internal dose for dermal exposure. Dividing the exposure level by the oral RfD yields a hazard index of 3.9×10^{-2}. Thus it was concluded that alumina trihydrate used as a flame retardant in upholstery fabric is not likely to pose a noncancer risk by the dermal route.

Inhalation Exposure

Particles

The assessment of the noncancer risk by the inhalation route of exposure is based on the scenario described Chapter 3. This scenario corresponds to a person spending 1/4th of his or her life in a room with a low air-change rate (0.25/hr) and with a relatively large amount of fabric upholstery treated with alumina trihydrate (30 m^2 in a 30-m^3 room), with this treatment gradually being worn away over 25% of its surface to 50% of its initial quantity over the 15-yr lifetime of the fabric. A small fraction, 1%, of the worn-off alumina trihydrate is released into the indoor air as inhalable particles and may be breathed by the occupant. Equations 4 through 6 in Chapter 3 were used to estimate the average concentration of alumina trihydrate present in the air. The highest expected application rate for alumina trihydrate is about 7.5 mg/cm^2. The estimated release rate for alumina trihydrate is 2.3×10^{-7}/d. Using those values, the estimated time-averaged exposure concentration for alumina trihydrate is 0.71 $\mu g/m^3$.

Although lack of sufficient data precludes deriving an inhalation RfC for alumina trihydrate, the oral RfD (1.5 mg alumina trihydrate/kg-d; see Oral RfD in Quantitative Toxicity Assessment section), which represents a very conservative estimate (see Chapter 4 for the rationale), was used to estimate an RfC of 5.25 mg/m^3.

Division of the exposure concentration (0.71 $\mu g/m^3$) by the estimated RfC (5.25 mg/m^3) results in a hazard index of 1.4×10^{-4}, indicating that under the worst-case exposure scenario, exposure to alumina trihydrate, used as an upholstery fabric flame retardant, is not likely to pose a noncancer risk from exposure to alumina trihydrate particles.

Vapors

In addition to the possibility of release of alumina trihydrate in particles worn from upholstery fabric, the subcommittee considered the possibility of its release by evaporation. However, because of alumina trihydrate's negligible vapor pressure at ambient temperatures, the subcommittee concluded that exposure to alumina trihydrate vapors from its use as an upholstery fabric flame retardant is not likely to pose a noncancer risk.

Oral Exposure

The assessment of the noncancer risk by the oral exposure route is based on the scenario described in Chapter 3. That exposure assumes a child is exposed to alumina trihydrate through sucking on 50 cm^2 of fabric backcoated with alumina trihydrate daily for two yr, one hr/d. The highest expected application rate for alumina trihydrate is about 7.5 mg/cm^2. A fractional rate (per unit time) of alumina trihydrate extraction by saliva is estimated as 0.001/d, based on leaching of antimony from polyvinyl chloride cot mattresses (Jenkins et al. 1998). Using those assumptions and Equation 15 in Chapter 3, the average oral dose rate was estimated to be 0.0016 mg/kg-d. Division of that exposure estimate (0.0016 mg/kg-d) by the oral RfD (1.5 mg/kg-d; see Oral RfD in Quantitative Toxicity Assessment Section) results in a hazard index of 1.0×10^{-3}. Therefore, under the worst-case exposure assumptions, alumina trihydrate, used as a flame retardant in upholstery fabric, is not likely to pose a noncancer risk by the oral exposure route.

Cancer

There are inadequate data to characterize the carcinogenic risk from exposure to alumina trihydrate from any route of exposure.

RECOMMENDATIONS FROM OTHER ORGANIZATIONS

The Agency for Toxic Substances and Disease Registry (ATSDR 1999) has established an intermediate-duration oral minimal risk level (MRL) for aluminum of 2.0 mg aluminum/kg-d.

The Occupational Safety and Health Administration (OSHA) permissible exposure limit (PEL) for aluminum dust is 15 mg/m^3 (for total dust) and 5 mg/m^3 (for respirable dust) (OSHA 1974). The American Conference of Governmental Industrial Hygienists (ACGIH 1999) has set a Threshold Limit Value (TLV) for alumina trihydrate of 10 mg/m^3.

DATA GAPS AND RESEARCH NEEDS

Although there are toxicity data on other aluminum compounds, data on aluminum trihydrate are lacking. In addition, chronic carcinogenic studies following dermal, inhalation, and oral exposure, and reproductive and developmental studies following dermal and inhalation exposure are lacking for any relevant aluminum compound. However, alumina trihydrate is used extensively in antacids (e.g., "Maalox") and cosmetics, and the hazard indices are less than 1 for all routes of exposure using the subcommittee's conservative assumptions. Therefore, the subcommittee concludes that further research is not needed to assess the health risks from alumina trihydrate when used as a flame-retardant chemical in furniture upholstery fabric.

REFERENCES

ACGIH (American Conference of Government Industrial Hygienists). 1999. Threshold Limit Values and Biological Exposure Indices. Cincinnati, OH: American Conference of Government Industrial Hygienists, Inc.

Ackley, D.C., and R.A. Yokel. 1997. Aluminum citrate is transported from brain into blood via the monocarboxylic acid transporter located at the blood-brain barrier. Toxicology 120(2):89-97.

Allen, D.D., C. Orvig, and R.A. Yokel. 1995. Evidence for energy-dependent transport of aluminum out of brain extracellular fluid. Toxicology 98(1-3):31-39.

Anane, R., M. Bonini, J.M. Grafeille, and E.E. Creppy. 1995. Bioaccumulation of water soluble aluminium chloride in the hippocampus after transdermal uptake in mice. Arch. Toxicol. 69(8):568-571.

Anane, R., M. Bonini, and E.E. Creppy. 1997. Transplacental passage of aluminum from pregnant mice to fetus organs after maternal transcutaneous exposure. Hum. Exp. Toxicol. 16(9):501-504.

Armstrong, B.G., C.G. Tremblay, D. Cyr, and G.P. Theriault. 1986. Estimating the relationship between exposure to tar volatiles and the incidence of bladder cancer in aluminum smelter workers. Scand. J. Work Environ. Health 12(5):486-493.

Armstrong, B., C. Tremblay, D. Baris, and G. Theriault. 1994. Lung cancer mortality and polynuclear aromatic hydrocarbons: A case-cohort study of aluminum production workers in Arvida, Quebec, Canada. Am. J. Epidemiol. 139(3):250-262.

ATSDR (Agency for Toxic Substances and Disease Registry). 1999. Toxicological Profile for Aluminum. U.S. Department of Health & Human Services, Public Health Service, Atlanta, GA.

Avolio, G., F. Galietti, M. Iorio, and A. Oliaro. 1989. Aluminum lung as an occupational disease. Case reports. [Article in Italian]. [Abstract in English]. Minerva Med. 80(4):411-414.

Bast-Pettersen, R., P.A. Drablos, L.O. Goffeng, Y. Thomassen, and C.G. Torres. 1994. Neuropsychological deficit among elderly workers in aluminum production. Am. J. Ind. Med. 25(5):649-662.

Bilkei-Gorzo, A. 1993. Neurotoxic effect of enteral aluminum. Food Chem. Toxicol. 31(5):357-361.

Budavari, S., M.J. O'Neil, A. Smith, and P.E. Heckelman. 1989. The Merck Index. 11th Ed. Rahway, NJ: Merck and Co., Inc.

Chedid, F., A. Fudge, J. Teubner, S.L. James, and K. Simmer. 1991. Aluminium absorption in infancy. J. Paediatr. Child Health 27(3):164-166.

Chen, W.J., R.J. Monnat, Jr., M. Chen, and N.K. Mottet. 1978. Aluminum induced pulmonary granulomatosis. Hum. Pathol. 9(6):705-711.

Colomina, M.T., M. Gomez, J.L. Domingo, J.M. Llobet, and J. Corbella. 1992. Concurrent ingestion of lactate and aluminum can result in developmental toxicity in mice. Res. Commun. Chem. Pathol. Pharmacol. 77(1):95-106.

Colomina, M.T., M. Gomez, J.L. Domingo, and J. Corbella. 1994. Lack of maternal and developmental toxicity in mice given high doses of aluminum hydroxide and ascorbic acid during gestation. Pharmacol. Toxicol. 74(4-5):236-239.

Colomina, M.T., J.L. Esparza, J. Corbella, and J.L. Domingo. 1998. The effect of maternal restraint on developmental toxicity of aluminum in mice. Neurotoxicol. Teratol. 20(6):651-656.

Cranmer, J.M., J.D. Wilkins, D.J. Cannon, and L. Smith. 1986. Fetal-placental-maternal uptake of aluminum in mice following gestational exposure: Effect of dose and route of administration. Neurotoxicology 7(2):601-608.

Crombie, D.W., J.L. Blaisdell, and G. MacPherson. 1944. The treatment of silicosis by aluminum powder. Can. Med. Assoc. J. 50:318-328.

Cullen, M.R., H. Checkoway, and B.H. Alexander. 1996. Investigation of a cluster of pituitary adenomas in workers in the aluminum industry. Occup. Environ. Med. 53(11):782-786.

Dick, R.B., E.F. Krieg, Jr., M.A. Sim, B.P. Bernard, and B.T. Taylor. 1997. Evaluation of tremor in aluminum production workers. Neurotoxicol. Teratol. 19(6):447-453.

DiPaolo, J.A., and B.C. Casto. 1979. Quantitative studies of in vitro morphological transformation of Syrian hamster cells by inorganic metal salts. Cancer Res. 39(3):1008-1013.

Domingo, J.L., J.M. Llobet, M. Gomez, J.M. Tomas, and J. Corbella. 1987a. Nutritional and toxicological effects of short-term ingestion of aluminum by the rat. Res. Commun. Chem. Pathol. Pharmacol. 56(3):409-419.

Domingo, J.L., J.L. Paternain, J.M. Llobet, and J. Corbella. 1987b. The effects of aluminum ingestion of reproduction and postnatal survival in rats. Life Sci. 41(9):1127-1131.

Domingo, J.L., M. Gomez, M.A. Bosque, and J. Corbella. 1989. Lack of teratogenicity of aluminum hydroxide in mice. Life Sci. 45(3):243-247.

Domingo, J.L., M. Gomez, D.J. Sanchez, J.M. Llobet, and J. Corbella. 1993. Effect of various dietary constituents on gastrointestinal absorption of aluminum from drinking water and diet. Res. Commun. Chem. Pathol. Pharmacol. 79(3):377-380.

Donald, J.M., M.S. Golub, M.E. Gershwin, and C.L. Keen. 1989. Neurobehavioral effects in offspring of mice given excess aluminum in diet during gestation and lactation. Neurotoxicol. Teratol. 11(4):345-351.

Drueke, T.B., P. Jouhanneau, H. Banide, B. Lacour, F. Yiou, and G. Raisbeck. 1997. Effects of silicon, citrate and the fasting state on the intestinal absorption of aluminium in rats. Clin. Sci. 92(1):63-67.

Dufresne, A., P. Loosereewanich, B. Armstrong, G. Theriault, and R. Begin. 1996. Inorganic particles in the lungs of five aluminum smelter workers with pleuro-pulmonary cancer. Am. Ind. Hyg. Assoc. J. 57(4):370-375.

Edling, N.P.G. 1961. Aluminum pneumoconiosis: A roentgendiagnostic study of five cases. Acta. Radiol. 56:170-178.

Eklund, A., R. Arns, E. Blaschke, J. Hed, S.O. Hjertquist, K. Larsson, H. Lowgren, J. Nystrom, C.M. Skold, and G. Tornling. 1989. Characteristics of alveolar cells and soluble components in bronchoalveolar lavage fluid from nonsmoking aluminium potroom workers. Br. J. Ind. Med. 46(11):782-786.

Elinder, C.G., L. Ahrengart, V. Lidums, E. Pettersson, and B. Sjogren. 1991. Evidence of aluminum accumulation in aluminium welders. Br. J. Ind. Med. 48(11):735-738.

Ellenhorn, M.J. 1997. Ellenhorn's Medical Toxicology: Diagnosis and Treatment of Human Poisoning. 2nd Ed. Baltimore: Williams & Wilkins.

EPA (U.S. Environmental Protection Agency). 1992. Dermal Exposure Assessment: Principles and Applications. EPA/600/8-91-011B. Office of Health and Environmental Assessment, U.S. Environmental Protection Agency, Washington, DC.

Exley, C. 1998. Does antiperspirant use increase the risk of aluminium-related disease, including Alzheimer's disease? Mol. Med. Today 4(3):107-109.

Fairweather—Tait, S., K. Hickson, B. McGaw, and M. Reid. 1994. Orange juice enhances aluminium absorption from antacid preparation. Eur. J. Clin. Nutr. 48(1):71-73.

Ferrante, J. 1999. Toxicity review for alumina trihydrate. Memorandum, dated March 9, 1999 from Jacqueline Ferrante, Pharmacologist, Division of Health Sciences to Ronald Medford, Assistant Executive Director for Hazard Identification and Reduction, U.S. Consumer Product Safety Commission, Washington, DC.

Forbes, W.F., and G.B. Hill. 1998. Is exposure to aluminum a risk factor for the development of Alzheimer disease? -- Yes. Arch. Neurol. 55(5):740-741.

Ganrot, P.O. 1986. Metabolism and possible health effects of aluminum. Environ. Health Perspect. 65:363-441.

Gibbs, G.W. 1985. Mortality of aluminum reduction plant workers, 1950 through 1977. J. Occup. Med. 27(10):761-770.

Gibbs, G.W., and I. Horowitz. 1979. Lung cancer mortality in aluminum reduction plant workers. J. Occup. Med. 21(5):347-353.

Gitelman, H.J., F.R. Alderman, M. Kurs-Lasky, and H.E. Rockette. 1995. Serum and urinary aluminum levels of workers in the aluminum industry. Ann. Occup. Hyg. 39(2):181-191.

Golub, M.S., and J.L. Domingo. 1996. What we know and what we need to know about developmental aluminum toxicity. J. Toxicol. Environ. Health 48(6):585-597.

Golub, M.S., M.E. Gershwin, J.M. Donald, S. Negri, and C.L. Keen. 1987. Maternal and developmental toxicity of chronic aluminum exposure in mice. Fundam. Appl. Toxicol. 8(3):346-357.

Golub, M.S., J.M. Donald, M.E. Gershwin, and C.L. Keen. 1989. Effects of aluminum ingestion on spontaneous motor activity of mice. Neurotoxicol. Teratol. 11(3):231-235.

Golub, M.S., B. Han, C.L. Keen, and M.E. Gershwin. 1992a. Effects of dietary aluminum excess and manganese deficiency on neurobehavioral endpoints in adult mice. Toxicol. Appl. Pharmacol. 112(1):154-160.

Golub, M.S., C.L. Keen, and M.E. Gershwin. 1992b. Neurodevelopmental effect of aluminum in mice: Fostering studies. Neurotoxicol. Teratol. 14(3):177-182.

Golub, M.S., B. Han, C.L. Keen, and M.E. Gershwin. 1994. Auditory startle in Swiss Webster mice fed excess aluminum in diet. Neurotoxicol. Teratol. 16(4):423-425.

Golub, M.S., B. Han, C.L. Keen, M.E. Gershwin, and R.P. Tarara. 1995. Behavioral performance of Swiss Webster mice exposed to excess dietary aluminum during development or during development and as adults. Toxicol. Appl. Pharmacol. 133(1):64-72.

Gomez, M., J.L. Domingo, J.M. Llobet, J.M. Tomas, and J. Corbella. 1986. Short-term oral toxicity study of aluminum in rats. Arch. Farmacol. Toxicol. 12(2-3):145-151.

Gomez, M., J.L. Domingo, and J.M. Llobet. 1991. Developmental toxicity evaluation of oral aluminum in rats: Influence of citrate. Neurotoxicol. Teratol. 13(3):323-328.

Gorsky, J.E., A.A. Dietz, H. Spencer, and D. Osis. 1979. Metabolic balance of aluminum studied in six men. Clin. Chem. 25(10):1739-1743.

Greger, J.L., and M.J. Baier. 1983. Excretion and retention of low or moderate levels of aluminum by human subjects. Food Chem. Toxicol. 21(4):473-477.

Greger, J.L., and S.E. Donnaubauer. 1986. Retention of aluminum in the tissues of rats after the discontinuation of oral exposure to aluminum. Food Chem. Toxicol. 24(12):1331-1334.

Hänninen, H., E. Matikainen, T. Kovala, S. Valkonen, and V. Riihimaki. 1994. Internal load of aluminum and the central nervous system function of aluminum welders. Scand. J. Work Environ. Health 20(4):279-285.

Hicks, J.S., D.S. Hackett, and G.L. Sprague. 1987. Toxicity and aluminium concentration in bone following dietary administration of two sodium aluminium phosphate formulations in rats. Food Chem. Toxicol. 25(7):533-538.

Hosovski, E., Z. Mastelica, D. Sunderic, and D. Radulovic. 1990. Mental abilities of workers exposed to aluminum. Med. Lav. 81(2):119-123.

HSDB (Hazardous Substances Data Bank). 1990. National Library of Medicine. [Online]. Available: http://sis.nlm.nih.gov/cgi-bin/sis/htmlgen?HSDB

Jenkins, R.O., P.J. Craig, W. Goessler, and K.J. Irgolic. 1998. Antimony leaching from cot mattresses and sudden infant death syndrome (SIDS). Hum. Exp. Toxicol. 17(3):138-139.

Jouhanneau, P., G.M. Raisbeck, F. Yiou, B. Lacour, H. Banide, and T.B. Drueke. 1997. Gastrointestinal absorption, tissue retention, and urinary excretion of dietary aluminum in rats determined by using 26A1. Clin. Chem. 43(6 Pt. 1):1023-1028.

Kaehny, W.D., A.P. Hegg, and A.C. Alfrey. 1977. Gastrointestinal absorption of aluminum from aluminum-containing antacids. N. Engl. J. Med. 296(24):1389-1390.

Kanematsu, N., M. Hara, and T. Kada. 1980. rec assay and mutagenicity studies on metal compounds. Mutat. Res. 77(2):109-116.

Koo, W.W., S.K. Krug-Wispe, P. Succop, R. Bendon, and L.A. Kaplan. 1992. Sequential serum aluminum and urine aluminum: Creatinine ratio and tissue aluminum loading in infants with fractures/rickets. Pediatrics 89(5 Pt. 1):877-881.

Krasovskii, G.N., L.Y. Vasukovich, and O.G. Chariev. 1979. Experimental study of biological effects of leads and aluminum following oral administration. Environ. Health Perspect. 30:47-51.

Lansdown, A.B. 1973. Production of epidermal damage in mammalian skins by some simple aluminum compounds. Br. J. Dermatol. 89(1):67-76.

Lide, D.R. 1991-1992. Handbook of Chemistry and Physics, 72nd Ed. Boca Raton, FL: CRC Press.

Manna, G.K., and R.K. Das. 1972. Chromosome aberrations in mice induced by aluminum chloride. Nucleus 15:180-186.

Marzin, D.R., and H.V. Phi. 1985. Study of the mutagenicity of metal derivatives with

Salmonella-typhimurium TA-102. Mutat. Res. 155:49-52.
McLaughlin, A.I.G., G. Kazantzis, E. King, D. Teare, R.J. Porter, and R. Owen. 1962. Pulmonary fibrosis and encephalopathy associated with the inhalation of aluminum dust. Br. J. Ind. Med. 19:253-263.
Meiklejohn, A., and E. Posner. 1957. The effect of the use of calcined alumina in china biscuit placing on the health of the workman. Br. J. Ind. Med. 14:229-231.
Milham, S., Jr. 1979. Mortality in aluminum reduction plant workers. J. Occup. Med. 21(7):475-480.
Miller, R.R., A.M. Churg, M. Hutcheon, and S. Lom. 1984. Pulmonary alveolar proteinosis and aluminum dust exposure. Am. Rev. Respir. Dis. 130(2):312-315.
Misawa, T., and S. Shigeta. 1993. Effects of prenatal aluminum treatment on development and behavior in the rat. J. Toxicol. Sci. 18(1):43-48.
Mitchell, J., G.B. Manning, M. Molyneux, and R.E. Lane. 1961. Pulmonary fibrosis in workers exposed to finely powdered aluminum. Br. J. Ind. Med. 18:10-20.
Moore, P.B., J.A. Edwardson, I.N. Ferrier, G.A. Taylor, D. Lett, S.P. Tyrer, J.P. Day, S.J. King, and J.S. Lilley. 1997. Gastrointestinal absorption of aluminum is increased in Down's syndrome. Biol. Psychiatry 41(4):488-492.
Munoz, D.G. 1998. Is exposure to aluminum a risk factor for the development of Alzheimer disease?—No. Arch. Neurol. 55(5):737-739.
Musk, A.W., H.W. Greville, and A.E. Tribe. 1980. Pulmonary disease from occupational exposure to an artificial aluminum silicate used for cat litter. Br. J. Ind. Med. 37(4):367-372.
Mussi, I., G. Calzaferri, M. Buratti, and L. Alessio. 1984. Behaviour of plasma and urinary aluminum levels in occupationally exposed subjects. Int. Arch. Occup. Environ. Health 54(2):155-161.
Nolan, C.R., J.J. DeGoes, and A.C. Alfrey. 1994. Aluminum and lead absorption from dietary sources in women ingesting calcium citrate. South Med. J. 87(9):894-898.
Oneda, S., T. Takasaki, K. Kuriwaki, Y. Ohi, Y. Umekita, S. Hatanaka, T. Fujiyoshi, A. Yoshida, and H. Yoshida. 1994. Chronic toxicity and tumorigenicity study of aluminum potassium sulfate in B6C3F1 mice. In Vivo 8(3):271-278.
OSHA (Occupational Safety and Health Administration). 1974. Occupational Safety and Health Standards. Code of Federal Regulations. 29 CFR 1910.1000. Occupational Safety and Health Administration. U.S. Department of Labor.
Oteiza, P.I., C.L. Keen, B. Han, and M.S. Golub. 1993. Aluminum accumulation and neurotoxicity in Swiss-Webster mice after long-term dietary exposure to aluminum and citrate. Metabolism 42(10):1296-1300.
Park, H.S., S.T. Uh, and C.S. Park. 1996. Increased neutrophil chemotactic activity is noted in aluminum-induced occupational asthma. Korean J. Intern. Med. 11(1):69-73.
Partridge, N.A., F.E. Regnier, W.M. Reed, J.L. White, and S.L. Hem. 1992. Contribution of soluble aluminum species to absorption of aluminum from the rat gut in situ. Clin. Sco. (Colch.) 83(4):425-430.
Paternain, J.L., J.L. Domingo, J.M. Llobet, and J. Corbella. 1988. Embryotoxic and teratogenic effects of aluminum nitrate in rats upon oral administration. Teratology 38(3):253-257.

Pennington, J.A., and S.A. Schoen. 1995. Estimates of dietary exposure to aluminum. Food Addit. Contam. 12(1):119-128.

Pierre, F., F. Baruthio, F. Diebold, and P. Biette. 1995. Effect of different exposure compounds on urinary kinetics of aluminum and fluoride in industrially exposed workers. Occup. Environ. Med. 52(6):396-403.

Pigott, G.H., B.A. Gaskell, and J. Ishmael. 1981. Effects of long term inhalation of alumina fibres in rats. Br. J. Exp. Pathol. 62(3):323-331.

Pivnick, E.K., N.C. Kerr, R.A. Kaufman, D.P. Jones, and R.W. Chesney. 1995. Rickets secondary to phosphate depletion. A sequela of antacid use in infancy. Clin. Pediatr. 34(2):73-78.

Poisindex (Poisindex® System). 1998. Toxicologic Managements: Aluminum. Micromedex, Inc. Volume 97, expiration date 8/31/98.

Posner, E., and M.C. Kennedy. 1967. A further study of china biscuit placers in Stoke-on-Trent. Br. J. Ind. Med. 24(2):133-142.

Priest, N.D., R.J. Talbot, J.G. Austin, J.P.Day, S.J. King, K. Fifield, and R.G. Cresswell. 1996. The bioavailability of 26Al-labelled aluminum citrate and aluminum hydroxide in volunteers. Biometals 9(3):221-8.

Recker, R.R., A.J. Blotcky, J.A. Leffler, and E.P. Rack. 1977. Evidence of aluminum absorption from the gastrointestinal tract and bone deposition by aluminum carbonate ingestion with normal renal function. J. Lab. Clin. Med. 90(5):810-815.

Reiber, S., W. Kukull, and P. Standish-Lee. 1995. Drinking water aluminium and bioavailability. J. Am. Water Works Assoc. 87(5):86-100.

Rifat, S.L., M.R. Eastwood, D.R. McLachlan, and P.N. Corey. 1990. Effect of exposure of miners to aluminum powder. Lancet 336(8724):1162-1165.

Rockette, H.E., and V.C. Arena. 1983. Mortality studies of aluminum reduction plant workers: Potroom and carbon department. J. Occup. Med. 25(7):549-557.

Rollin, H.B., P. Theodorou, and T.A. Kilroe-Smith. 1991. Deposition of aluminum in tissues of rabbits exposed to inhalation of low concentrations of Al2O3 dust. Br. J. Ind. Med. 48(6):389-91.

Ronneberg, A., and F. Langmark. 1992. Epidemiologic evidence of cancer in aluminum reduction plant workers. Am. J. Ind. Med. 22(4):573-590.

Salib, E., and V. Hillier. 1996. A case-control study of Alzheimer's disease and aluminum occupation. Br. J. Psychiat. 168(2):244-249.

Savory, J., C. Exley, W.F. Forbes, Y. Huang, J.G. Joshi, T. Kruck, D.R. McLachlan, and I. Wakayama. 1996. Can the controversy of the role of aluminum in Alzheimer's disease be resolved? What are the suggested approaches to this controversy and methodological issues to be considered? J. Toxicol. Environ. Health 48(6):615-35.

Schonholzer, K.W., R.A. Sutton, V.R. Walker, V. Sossi, M. Schulzer, C. Orvig, E. Venczel, R.R. Johnson, D. Vetterli, B. Dittrich-Hannen, P. Kubik, and M. Suter. 1997. Intestinal absorption of trace amounts of aluminum in rats studied with 26 aluminum and accelerator mass spectrometry. Clin. Sci. (Colch.) 92(4):379-383.

Sedman, A.B., G.L. Klein, R.J. Merritt, N.L. Miller, K.O. Weber, W.L. Gill, H. Anand, and A.C. Alfrey. 1985. Evidence of aluminum loading in infants receiving intravenous therapy. N. Engl. J. Med. 312(21):1337-1343.

Selden, A.I., H.B. Westberg, and O. Axelson. 1997. Cancer morbidity in workers at aluminum foundries and secondary aluminum smelters. Am. J. Ind. Med. 32(5):467-477.

Sideman, S., and D. Manor. 1982. The dialysis dementia syndrome and aluminum intoxication. Nephron 31(1):1-10.

Sim, M., R. Dick, J. Russo, B. Bernard, P. Grubb, E. Krieg Jr., C. Mueller, and C. McCammon. 1997. Are aluminum potroom workers at increased risk of neurological disorders? Occup. Environ. Med. 54(4):229-235.

Sjögren, B., V. Lidums, M. Hakansson, and L. Hedstrom. 1985. Exposure and urinary excretion of aluminum during welding. Scand. J. Work Environ. Health 11(1):39-43.

Sjögren, B., C.-G. Elinder, V. Lidums, and G. Chang. 1988. Uptake and urinary excretion of aluminum among welders. Int. Arch. Occup. Environ. Health 60(2):77-79.

Sjögren, B., K.G. Ljunggren, O. Almkvist, W. Frech, and H. Basun. 1996. A follow-up study of five cases of aluminosis. Int. Arch. Occup. Environ. Health 68(3):161-164.

Smith, M.A., and G. Perry. 1998. What are the facts and artifacts of the pathogenesis and etiology of Alzheimer disease? J. Chem. Neuroanat. 16(1):35-41.

Spinelli, J.J., P.R. Band, L.M. Svirchev, and R.P. Gallagher. 1991. Mortality and cancer incidence in aluminum reduction plant workers. J. Occup. Med. 33(11):1150-1155.

Steinhagen, W.H., F.L. Cavender, and B.Y. Cockrell. 1978. Six month inhalation exposures of rats and guinea pigs to aluminum chlorhydrate. J. Environ. Pathol. Toxicol. 1(3):267-277.

Taylor, G.A., I.N. Ferrier, I.J. McLoughlin, A.F. Fairbairn, I.G. McKeith, D. Lett, and J.A. Edwardson. 1992. Gastrointestinal absorption of aluminium in Alzheimer's disease: Response to aluminum citrate. Age Ageing 21(2):81-90.

Teraoka, H. 1981. Distribution of 24 elements in the internal organs of normal males and the metallic workers in Japan. Arch. Environ. Health 36(4):155-165.

Testolin, G., D. Erba, S. Ciappellano, and G. Bermano. 1996. Influence of organic acids on aluminium absorption and storage in rat tissues. Food Addit. Contam. 13(1):21-27.

Theriault, G., L. De Guire, and S. Cordier. 1981. Reducing aluminum: An occupation possibly associated with bladder cancer. Can. Med. Assoc. J. 124(4):419-422; 425.

Thorne, B.M., A. Cook, T. Donohoe, S. Lyon, D.M. Medeiros, and C. Moutzoukis. 1987. Aluminum toxicity and behavior in the weanling Long-Evans rat. Bull. Pyschon. Soc. 25(2):129-132.

Ueda, M., Y. Mizoi, Z. Maki, R. Maeda, and R. Takada. 1958. A case of aluminum dust lung: A necropsy report. Kobe J. Med. Sci. 4:91-99.

Vandenplas, O., J.P. Delwiche, M.L. Vanbilsen, J.Joly, and D. Roosels. 1998. Occupational asthma caused by aluminum welding. Eur. Respir. J. 11(5):1182-1184.

Weberg, R., and A. Berstad. 1986. Gastrointestinal absorption of aluminum from single doses of aluminum containing antacids in man. Eur. J. Clin. Invest. 16(5):428-32.

Wedrychowski, A., W.N. Schmidt, and L.S. Hnilica. 1986. The in vivo cross-linking of proteins and DNA by heavy metals. J. Biol. Chem. 261(7):3370-3376.

White, D.M., W.T. Longstreth, Jr., L. Rosenstock, K.H. Claypoole, C.A. Brodken, and B.D. Townes. 1992. Neurologic syndrome in 25 workers from an aluminum smelting plant. Arch. Intern. Med. 152(7):1443-1448.

Yokel, R.A., and P.J. McNamara. 1989. Elevated aluminum persists in serum and tissues of rabbits after a six-hour infusion. Toxicol. Appl. Pharmacol. 99(1):133-138.

Yokel, R.A., V. Lidums, P.J. McNamara, and U. Ungerstedt. 1991. Aluminum distribution into brain and liver of rats and rabbits following intravenous aluminum lactate or citrate: A microdialysis study. Toxicol. Appl. Pharmacol. 107(1):153-163.

Yokel, R.A., D.D. Allen, and J.J. Meyer. 1994. Studies of aluminum neurobehavioral toxicity in the intact mammal. Cell. Mol. Neurobiol. 14(6):791-808.

Zafar, T.A., C.M. Weaver, B.R. Martin, R. Flarend, and D. Elmore. 1997. Aluminum (26Al) metabolism in rats. Proc. Soc. Exp. Biol. Med. 216(1):81-85.

Zatta, P., M. Favarato, and M. Nicolini. 1993. Deposition of aluminum in brain tissues of rats exposed to inhalation of aluminum acetylacetonate. Neuroreport 4(9):1119-1122.

7

Magnesium Hydroxide

THIS chapter reviews the physical and chemical properties, toxicokinetics, toxicological, epidemiological, and exposure data on magnesium hydroxide, $Mg(OH)_2$. The subcommittee used that information to characterize the health risk from exposure to $Mg(OH)_2$. The subcommittee also identified data gaps and recommended research relevant for determining the health risk from exposure to $Mg(OH)_2$.

PHYSICAL AND CHEMICAL PROPERTIES

Magnesium hydroxide, $Mg(OH)_2$, is one of a number of inorganic compounds that has been used as a flame retardant (FR) in furniture upholstery. The physical and chemical properties of $Mg(OH)_2$ are listed in Table 7-1.

OCCURRENCE AND USE

$Mg(OH)_2$ is most commonly used as an antacid and is the active ingredient in the laxative milk of magnesia. It is also used as a residual fuel-oil additive, an alkali drying agent in food, a color-retention agent, an ingredient of tooth-

TABLE 7-1 Physical and Chemical Properties of Magnesium Hydroxide

Property	Value	Reference
Chemical formula	$Mg(OH)_2$	HSDB 1998
CAS registry no.	1309-42-8	HSDB 1998
Synonyms	Caustic magnesite, magnesia magma, magnesium dihydroxide, magnesium hydrate, magnesium hydroxide gel, magnesium oxide, Marinco H, Marinco H 1241, Milk of magnesia, Mint-o-mag, Nemalite	HSDB 1998
Molecular weight	58.34	HSDB 1998
Physical state	White very fine powder	HSDB 1998
Solubility	0.0009 g in 100 cm^3 of H_2O at 18°C 0.004 g in 100 cm^3 of H_2O at 100°C	HSDB 1998
pH	9.5-10.5 (aqueous slurry)	HSDB 1998
Temperature of decomposition	350°C	HSDB 1998
Density	2.36	HSDB 1998

paste and frozen desserts, a clarifier in sugar refining, and a neutralizing agent in the chemical industry (HSDB 1998). $Mg(OH)_2$ is categorized by the U.S. Food and Drug Administration (FDA) as a GRAS (generally recognized as safe) food ingredient and is approved for use as a nutritional supplement and a pH-control agent in foods (FDA 1999).

$Mg(OH)_2$ is used as an FR in commercial furniture applications in the United States and in commercial and residential furniture in the United Kingdom (Fire Retardant Chemicals Association 1998). The stability of $Mg(OH)_2$ at temperatures above 300°C allows it to be incorporated into several polymers (IPCS 1997). Market-volume data published in 1993 suggest increasing the use of $Mg(OH)_2$ as a FR. About 2,000 and 3,000 tons of $Mg(OH)_2$ were marketed as an FR in the United States in 1986 and 1993, respectively (IPCS 1997).

Mg^{2+} is the fourth most abundant cation in the human body (Rude and Singer 1981). The average adult's body contains about 24 g of magnesium (Sutton and Dirks 1986; Elin 1987). The normal range of magnesium in human serum is 1.5-2.5 mEq/L (18.2-30.3 mg/L) and about two-third of magnesium is present as free cation and one-third is bound to plasma proteins (Elin 1987).

TOXICOKINETICS

Absorption

Dermal

No information was found on the dermal absorption of $Mg(OH)_2$ in humans or experimental animals.

Inhalation

Absorption of Mg^{2+} from inhalation exposure has been shown to take place in male Wistar rats after inhalation of $MgSO_4$ /5$Mg(OH)_2$ • $3H_2O$ filaments (Hori et al. 1994). That few filaments were detected in the rat lungs after 1 d of exposure suggested to the authors that the filaments were dissolved and absorbed through the alveolar capillaries. An intratracheal-instillation experiment in hamsters showed that the half-life for pulmonary clearance of short filaments (4.9 mm) was 17.6 min (Oberdörster 1996). However, this experiment could not determine whether fibers were removed from the lungs by alveolar macrophages or were absorbed by alveolar capillaries.

Oral

When administered orally, $Mg(OH)_2$ dissociates in stomach acids to Mg^{2+} cations. About 5-15% of the dissociated Mg^{2+} cations are absorbed (HSDB 1998) through the epithelial lining of the small intestine (Sutton and Dirks 1986; Elin 1987). Absorption of Mg^{2+} can be affected by the presence of food or other substances that readily complex with Mg^{2+} cations. $Mg(OH)_2$ and MgO, which have relatively low solubilities at neutral and alkaline pH, are less completely absorbed than the more water-soluble Mg^{2+}compounds–magnesium chloride ($MgCl_2$), magnesium citrate, and magnesium lactate (Benech et al. 1998; Brunton 1996). Determination of increased plasma or urinary Mg^{2+} cations after oral administration of $Mg(OH)_2$ is not possible, because of rapid homeostasis of exogenous Mg^{2+} in humans (Benech et al. 1998).

A single study in human volunteers measured the oral absorption of Mg^{2+} cations (Benech et al. 1998). In that study, six healthy males were administered a single oral dose of 360 mg of $^{26}Mg^{2+}$ as magnesium lactate or citrate and absorption of Mg^{2+} over 5 d was found to be 34.5% ± 18.8% and 39.8% ±

24.3% (mean ± SD) based on urinary excretion. Absorption of Mg^{2+} was 25.6% ± 34.5%; this estimate was based on fecal excretion.

Distribution

Dermal and Inhalation

No studies were found that investigated distribution of Mg^{2+} cations following dermal or inhalation exposure of humans or experimental animals to $Mg(OH)_2$.

Oral

Twelve volunteers that received a single oral dose of 16 mmol of $MgCl_2$ (389 mg of magnesium) exhibited a rise in serum magnesium during the first 4 hr after ingestion accompanied by an increase in urinary magnesium excretion (White et al. 1992). In another study, 23 women and 12 men (21-50 yr old) received two oral doses of 400 mg of MgO/d (476 mg Mg^{2+}/d) for 60 d and had no change in serum magnesium (Marken et al. 1989).

Metabolism

No studies were found that investigated the biotransformation of $Mg(OH)_2$ in humans or animals after exposure by any route. $Mg(OH)_2$ probably does not undergo biotransformation by the liver, lungs, skin, or intestinal epithelium after oral, dermal, or inhalation exposure.

Excretion

Dermal and Inhalation

No studies were found that investigated excretion of Mg^{2+} cations following dermal or inhalation exposure of humans or experimental animals to $Mg(OH)_2$.

Oral

Following oral exposure, undissolved $Mg(OH)_2$ antacids pass through the intestines and are eliminated in the feces (Brunton 1996). Unabsorbed Mg^{2+} is excreted unchanged in the feces (Brunton 1996). When 19 normal volunteers were given $Mg(OH)_2$ for 4 d in addition to magnesium in their normal diet (total magnesium, 27, 54, or 107-mmol/d), soluble Mg^{2+} output in feces reached a maximum of 14.6 mmol/d mg/kg-d (Fine et al. 1991). The average fecal output of soluble Mg^{2+} in these volunteers was 5.06.

Urinary excretion is the major route of elimination of absorbed Mg^{2+} when kidney function is normal. Urinary excretion accounts for about 80-90% of Mg^{2+} under steady-state conditions (Sutton and Dirks 1986). However, much of the absorbed dose of Mg^{2+} enters Mg^{2+} pools within the body and is not readily eliminated. Only about 7.4% ± 4% of a 50-mg intravenous dose of soluble Mg^{2+} or 2.22% ± 0.43% of an oral dose of 360 mg of soluble Mg^{2+} was excreted in urine over 5 d in six healthy male volunteers. White et al. (1992) measured a 22% increase in urinary magnesium in 12 healthy volunteers (nine men and three women, 23-46 yr old) during the first 8 hr after ingestion of $MgCl_2$ of Mg^{2+} at 5.6 mg/kg-d.

HAZARD IDENTIFICATION[1]

The toxicology database on $Mg(OH)_2$ is sparse. Therefore, the subcommittee decided that this review should also include toxicity data on magnesium salts that are expected to dissociate in stomach acid to Mg^{2+} cations. These include $MgCl_2$, magnesium lactate, and magnesium citrate.

Dermal Exposure

No studies were found on the toxic effects of $Mg(OH)_2$ following dermal exposure.

Inhalation Exposure

Systemic Effects

No human studies were found that investigated the toxic effects of $Mg(OH)_2$ following inhalation exposure. Exposure of male Wistar rats to short (4.9 × 0.31 mm) or long (12 × 0.44 mm) $MgSO_4/5Mg(OH)_2 \bullet 3H_2O$ filaments by

---

[1]In this section, the subcommittee reviewed the toxicity data on $Mg(OH)_2$, including the toxicity assessment prepared by the U.S. Consumer Product Safety Commission (Ferrante 1999).

inhalation, 6 hr/d, 5 d/wk for up to a year was associated with a slight increase in the incidence of pulmonary lesions 1 yr after cessation of exposure (Hori et al. 1994). A year after cessation of exposure, histopathological examination of treated animals revealed a slight increase in segmental calcification of the pulmonary artery and thickening of the lung pleura in rats exposed to either short or long filaments for 4 wk or 1 yr. Differences between exposed and unexposed animals were statistically significant. No significant differences in body, lung, liver, kidney, or spleen weights were detected between animals sacrificed 1 d or 1 yr after a 1-yr exposure to short or long filaments. No significant differences in survival were observed between animals sacrificed 1 d or 1 yr after a 1-yr exposure to short or long filaments.

Other Systemic Effects

No inhalation toxicity studies were found that investigated the immunological, neurological, reproductive, or developmental effects of $Mg(OH)_2$.

Cancer

The incidence of all cancers among 2,391 Norwegian males who worked between 1951 and 1974 in a factory producing magnesium metal was not significantly increased when compared with cancer incidence for the Norwegian national population of the same age (Heldaas et al. 1989). The number of cases of lip, as well as stomach, and lung cancers were significantly increased. Workers in this study were also exposed to magnesium oxide dust, coal dust, chlorine gas, hydrochlorine aerosols, chlorinated aromatics, and sulphur dioxide. Therefore, it is not possible to determine whether exposure to magnesium dust alone is responsible for the observed elevations in cancer incidence.

Exposure of male Wistar rats to short (4.9 × 0.31 mm) or long (12 × 0.44 mm) $MgSO_4/5Mg(OH)_2 \bullet 3H_2O$ filaments by inhalation (6 hr/d, 5 d/ wk for 1 yr) was not associated with an increase in the incidence of any tumor types in animals sacrificed 1 d or 1 yr after cessation of exposure (Hori et al. 1994). One yr after exposure, one pulmonary adenoma was observed in animals that had been exposed to long filaments for 4 wk and none in controls. One yr after exposure, neoplastic lesions were observed in control animals and short- and long-filament treated rats that had been exposed for 1 yr. Two pulmonary adenomas were observed in the exposed animals and one in control animals. No hepatocellular adenomas or carcinomas occurred in controls, one hepatocellular adenoma was found in the long-filament group, and one hepatocellular carcinoma was found in the short-filament group, respectively.

Oral Exposure

Systemic Effects

Human Data

Most available toxicity data on $Mg(OH)_2$ describe effects of acute exposure to $Mg(OH)_2$ or of prolonged exposure to antacid or laxative products containing $Mg(OH)_2$. Magnesium intoxication has been reported in infants (2-42 d old) that received Mg^{2+}-containing oral laxatives at Mg^{2+} doses of 224-917 mg/kg-d for 2-11 d (Alison and Bulugahapitiya 1990, Brand and Greer 1990, Humphrey et al. 1981, Mofenson and Caraccio 1991). Whereas normal serum magnesium ranges from 1.4 to 2.4 mEq/L, these infants had concentrations of 3.5 -11.7 mEq/L. In one case, Mg^{2+} body burden was high enough to cause perforation of the bowel (Mofenson and Caraccio 1991).

In adults, serious toxic effects associated with excess magnesium intake occur at very high intake levels equating to serum concentrations of 4 mEq/L (Mordes and Wacker 1978; Rude and Singer 1981). Toxicity has been limited to persons with intestinal or renal disease (Poisindex 1998). The Hazardous Substance Data Bank (HSDB) entry for $Mg(OH)_2$ states that the probable oral lethal dose of $Mg(OH)_2$ in humans is 5-15 g/kg in a 70-kg person (HSDB 1998). Cardiac arrest has been reported at serum Mg^{2+} concentrations of 15-16 mEq/L (Dreisbach 1977). Respiratory depression, depression of the central nervous system, and coma occur in adult patients with plasma Mg^{2+} concentrations of 10-14 mEq/L (Ferdinandus et al. 1981). Hypotension, nausea, and vomiting occur at plasma concentrations of 3-8 mEq/L.

In its review of clinical studies, the Institute of Medicine (IOM 1997) concluded that Mg^{2+} in the diet is never high enough to cause adverse effects. The IOM set a "tolerable upper intake level" (TUL)[2] for the ingestion of magnesium (Mg^{2+}) supplements of 5 mg/d for anyone over 1 yr old. The TUL was based on the approximate no-observed-adverse-effects level (NOAEL) for osmotic diarrhea in humans reported by Marken et al. (1989), Fine et al. (1991), Ricci et al. (1991), and Bashir et al. (1993).

Gastrointestinal discomfort occurred in six of 21 patients with stable congestive heart failure who received 15.8 mmol $MgCl_2$/d (5.5 mg/kg-d) for 6 wk (Bashir et al. 1993). Five of the six patients reported epigastric burning or distension and two reported diarrhea.

Five of 50 pregnant women developed adverse gastrointestinal effects (nau-

---

[2]A tolerable upper intake level is the maximal total chronic daily intake of a nutrient or food component that is unlikely to pose risks of adverse effects.

sea, soft stool, or diarrhea) given intravenous and oral doses of Mg^{2+} to suppress preterm delivery (Ricci et al. 1991). The women were initially given a 4-g bolus dose of $MgSO_4$ by intravenous injection followed by intravenous infusion of $MgSO_4$ at a dose of 2 g/hr for 12 hr. This approximates to a minimum dose of 353 mg/kg (assuming a female body weight of 55 kg).

Increases in the incidence of adverse neonatal outcomes were observed in women that ingested $MgCl_2$ tablets (7 mg/kg-d over an average of 29 d) as compared with controls given no $MgCl_2$ tablets (Ricci et al. 1991). Eleven of 25 neonates in treated mothers were diagnosed with jaundice as compared with 6 out of 25 infants born to control mothers. One fatal case of respiratory distress syndrome, one case of intraventricular hemorrhage, and 2 cases of necrotizing enterocolitis occurred in women taking $MgCl_2$ tablets as compared with no cases among infants born to control mothers. These increases were judged not to be statistically significant by the study investigators.

Diarrhea occurred in 18 of 50 (36%) healthy adult volunteers who received magnesium (as MgO) doses of 476 mg/d for 60 d (Marken et al.1989). Diarrhea was also observed in 14 healthy male subjects who ingested $Mg(OH)_2$ at a dose rate of 16.7, 33.3, or 67 mg/kg-d for 4 d (Fine et al.1991).

Animal Data

The key oral toxicity studies for $Mg(OH)_2$ are summarized in Table 7-2.

Decreased body weight was found to be the critical effect in B6C3F1 mice fed diets containing 0%, 0.3%, 0.6%, 1.25%, 2.5% or 5% $MgCl_2 \cdot 6H_2O$ for 13 wk (Tanaka et al. 1994). Intake of Mg^{2+} added to the diet was calculated to be 73, 146, 322, 650, or 1,368 mg/kg-d in treated males and 92, 190, 391, 817, and 1,660 mg/kg-d in treated females (the amount of magnesium in the basal diet was not provided). The 5%-treatment group of both sexes showed a significant decrease in weight gain (15% in males and 10% in females). Males in the 2.5 and 5% group exhibited an increased incidence of renal tubular vacuolation. The authors determined that the LOAEL for this study was 650 mg/kg-d.

Decreased body weight and increased renal vacuolation were observed in male, but not female B6C3F1 mice fed a diet that contained 5% $MgCl_2 \cdot 6H_2O$ (Mg^{2+} at 840 mg/kg-d) for 13 wk (Kurata et al. 1989). No treatment-related effects were reported for male and female mice fed a diet containing 0, 0.3, 0.6, 1.25, or 2.5% $MgCl_2 \cdot 6H_2O$ for 13 wk. The NOAEL for Mg^{2+} in this study was determined to be 587 mg/kg-d for females and 420 mg/kg-d for males.

Decreased body weight gain (about 25% at termination of the exposure) and increases in relative brain, heart, and kidney weights compared with controls were observed in female B6C3F1 mice fed diets for 96 wk that contained 2%

TABLE 7-2 Selected Oral Animal Toxicity Data on Magnesium Hydroxide

Species, strain, sex	Dose or concentration, Mg^{2+}	Duration	Effects	NOAEL/LOAEL (Mg^{2+})	Reference
ORAL EXPOSURE					
Mouse, B6C3F1, M/F	M: 73, 146, 322, 650, 1,368 mg/kg-d F: 92, 190, 391, 817, 1,660 mg/kg-d	90 d	Significant decrease in weight gain in high-dose groups for both males and females. Increased incidence of renal tubular vacuolation in male mice in the high- and 650 mg/kg-d dose groups.	LOAEL: 650 mg/kg-d	Tanaka et al. 1994
Mouse, B6C3F1, M/F	0.3, 0.6, 1.25, 2.5%	90 d	Significant decrease in weight gain and increased incidence of renal tubular vacuolation in males in the high-dose group (840 mg/kg-d).	NOAEL: 587 mg/kg-d (F) 420 mg/kg-d (M)	Kurata et al. 1989
Mouse, B6C3F1, M/F	M: 68, 336 mg/kg-d F: 87, 470 mg/kg-d	96 wk	In high-dose females, significant body weight gains, increases in food and water consumption, and increases in relative brain, heart, and kidney weights compared to controls at 96 wk. No treatment-related effects in male mice.	LOAEL: 470 mg/kg-d (F)	Kurata et al. 1989
Rat, Wistar, F	24, 47, or 96 mg/kg-d	Gestation d 6-15	No treatment-related effects reported.	NOAEL: >96 mg/kg-d	Usami et al. 1996

F: female; LOAEL: lowest-observed-adverse-effect level; M: male; ND: not determined; NOAEL: no-observed-adverse-effect level.

$MgCl_2 \cdot 6H_2O$ (470 mg Mg^{2+}/kg-d) (Kurata et al. 1989). No treatment-related effects were observed in male mice fed diets that contained 0.5% or 2% of $MgCl_2 \cdot 6H_2O$ (68, or 336 mg/kg-d) or female mice fed diets that contained 0.5% of $MgCl_2 \cdot 6H_2O$ (87 mg/kg-d) for 96 wk. Histopathological examination after 104 wk of exposure revealed no treatment-related changes. Urinary, hematological, and clinical chemistry parameters and histopathological measures were not affected by treatment, except for a significant increase in serum albumin in high-dose females. Survival rates were comparable between treated and control animals. The LOAEL for this study is 470 mg/kg-d based on the treatment-related effects in high-dose female mice.

Immunological and Neurological Effects

No studies were found that investigated the immunological or neurological effects of $Mg(OH)_2$ from oral exposures.

Reproductive and Developmental Effects

There are no studies in humans that evaluated reproductive or developmental effects associated with the ingestion of $Mg(OH)_2$.

Oral administration of $MgCl_2$ solution caused no toxic signs in pregnant Wistar rats and no increases in the incidences of fetal malformations that were given doses of 0, 200, 400, or 800 mg/kg-d (Mg^{2+} at 0, 24, 47, and 96 mg/kg-d) on d 6 through 15 of pregnancy (Usami et al. 1996). Pregnant dams were killed on d 20 of pregnancy and all fetuses underwent pathological examination for skeletal and visceral malformations. No malformations were observed at any dose tested. The authors concluded that the NOAEL for developmental and maternal toxicity was over 800 mg/kg-d (equivalent to 96 mg Mg^{2+}/kg-d) in this study.

Cancer

The subcommittee found no oral chronic toxicity studies or epidemiological studies that investigated the carcinogenicity of $Mg(OH)_2$ in rodents or humans.

Mice fed 0.5% or 2% of aqueous $MgCl_2$ in their diet for 96 wk (68, or 336 mg/kg-d for males; 87 or 470 mg/kg-d for females) showed no significant change in the incidence of malignant lymphoma and leukemia (Kurata et al. 1989). Dose-related increases in incidence of malignant lymphoma and leukemia occurred in male mice (controls, five of 50; low dose, seven of 50; high dose, eleven of 50), but not in females (controls, nine of 49; low dose, 17 of 50; high dose, 11 of 50). The incidence of hepatocellular carcinomas in male mice was decreased in a dose-related manner (controls, 13 of 50; low dose, six of 50; high dose, four of 50) and the incidence in high-dose males was significantly different from that in controls. Toxicity in female mice (i.e., decreased body weight) suggests that the study was conducted at or near the maximum tolerated dose (MTD) for females.

Several studies in rats have shown that dietary $Mg(OH)_2$ can protect against chemically induced bowel carcinogenesis by suppressing hyperproliferation of the colon epithelium. Dietary levels of 250 ppm $Mg(OH)_2$ inhibited the incidence of colon adenoma and adenocarcinoma in rats given carcinogens methylazoxymethanol acetate (MAM acetate) or 1,2-dimethylhydrazine (Tanaka et al.1989; Morishita et al.1991; Mori et al. 1993). Administration of $Mg(OH)_2$ in the diet and the bowel carcinogen cholic acid reduced cell proliferation in bowel tissue (Wang et al. 1994). Dietary $Mg(OH)_2$ also prevented the expression of *c-myc* gene in colon mucosa cells of MAM acetate-treated rats (Wang et al. 1993).

Genotoxicity

The subcommittee found no mutagenicity data on $Mg(OH)_2$. However, there are studies that have investigated the genotoxicity of other magnesium salts. Most of theses studies report negative genotoxicity findings for these compounds.

$MgCl_2$ was judged to be a nonmutagen in the Ames assay when tested with and without metabolic activation and it did not induce chromosomal aberrations in Chinese hamster fibroblast cells in vitro (Ishidate et al. 1984, as cited in Tanaka et al. 1994). Chromatid gaps, breaks, and exchanges were observed in Chinese hamster lung fibroblasts treated with $MgCl_2$ at concentrations of 8.0 and 12.0 mg/ml but not at or below concentrations of 4 mg/mL (Ashby and Ishidate 1986). Since positive results occurred at only high concentrations, the authors suggest that the clastogenic effects observed may be an artifact induced by hypertonic solutions. $MgCl_2$ did not induce mutations in mouse lymphoma L5178/$TK^{+/-}$ cells at concentrations of 5.7-18.1 mg Mg^{2+}/ml (Amacher and Paillet 1980).

$MgSO_4$ was not mutagenic in *Salmonella typhimurium* (strains TA100, TA1535) and *Escherichia coli* WP2 uvrA at concentrations of 313-5,000 mg/plate (Oguma et al. 1998). $MgSO_4$ was not mutagenic in *Salmonella* strain TA98 tested without metabolic activation and strain TA1537 tested with metabolic activation at a concentration of 156-5000 mg/plate (Oguma et al. 1998).

QUANTITATIVE TOXICITY ASSESSMENT

Noncancer

Dermal Assessment

There are no adequate studies that investigated the subchronic or chronic toxicity of $Mg(OH)_2$ administered by the dermal route of exposure. Therefore, the subcommittee did not estimate a dermal RfD for $Mg(OH)_2$.

Inhalation RfC

One study investigated the chronic toxicity of manufactured mineral filaments that contain $Mg(OH)_2$ (Hori et al. 1994). However, the subcommittee concluded that this study is not useful for deriving an inhalation RfC for $Mg(OH)_2$, because of the threefold difference in solubility between $Mg(OH)_2$ and the composite mineral filaments and the great variation in the size and shape of $Mg(OH)_2$ particles as opposed to the composite filaments used in the study. Therefore, the subcommittee concludes that there are inadequate human or animal data on the inhaled toxicity of $Mg(OH)_2$ to derive an inhalation RfC.

Oral RfD

There are inadequate toxicity data on $Mg(OH)_2$ from oral exposure studies. However, the Institute of Medicine's Committee on Recommended Dietary Allowance (IOM 1997) has derived a tolerable upper limit (TUL) for intake of Mg^{2+} (from nonfood sources) based on the LOAEL for osmotic diarrhea based on the results of several studies in adults (Marken et al. 1989; Fine et al. 1991; Ricci et al. 1991; Bashir et al. 1993). The TUL for Mg^{2+} is 5 mg/kg-d (0.21 mmol/kg-d) for all population groups 1 yr old and older (see Hazard Identification section for more detail). In its calculation, IOM (IOM 1997) used an uncertainty factor of 1.0 reasoning that diarrhea is a mild and reversible toxic effect.

Based on differences in molecular weight, the equivalent TUL for $Mg(OH)_2$ is estimated to be about 2.4 times that of Mg^{2+} or 12 mg/kg-d. The subcommittee believes that this TUL can be safely assumed to also be the oral RfD.

Carcinogenicity

There are insufficient data to assess the carcinogenicity of $Mg(OH)_2$. EPA, the National Toxicology Program (NTP), and the International Agency for Research on Cancer (IARC) have not evaluated the carcinogenicity of $Mg(OH)_2$.

A chronic study in mice exposed to $Mg(OH)_2$ filaments did not find evidence of carcinogenicity. Studies in rats suggest that $Mg(OH)_2$ incorporated into the diet can protect against some chemically induced cancers (Tanaka et al.1989; Morishita et al. 1991; Mori et al. 1993; Wang et al. 1993, 1994). The subcommittee is not aware of any mutagenicity data on $Mg(OH)_2$. However, genotoxicity studies conducted with several magnesium salts have all been negative.

On the basis of the data available, the subcommittee concludes that there are insufficient data on oral carcinogenicity of $Mg(OH)_2$ to determine its carcinogenicity.

EXPOSURE ASSESSMENT AND RISK CHARACTERIZATION

Noncancer

Dermal Exposure

Dermal exposure to $Mg(OH)_2$ was estimated using the dermal exposure scenario described in Chapter 3. This exposure scenario assumes that an adult spends 1/4th of his or her time sitting on furniture upholstery treated with $Mg(OH)_2$ and also assumes that 1/4th of the upper torso is in contact with the upholstery and clothing presents no barrier.

The subcommittee concluded that $Mg(OH)_2$ is an ionic substance and, therefore, is essentially not absorbed through the skin and should not pose a health risk from the dermal route of exposure when used as an FR in furniture upholstery. However, to be conservative, the subcommittee assumed that ionized $Mg(OH)_2$ permeates the skin at the same rate as water, with a permeability rate of 10^{-3} cm/hr (EPA 1992). Using that permeability rate, the highest expected application rate for $Mg(OH)_2$ of 4 mg/cm^2 and Equation 1 in Chapter 3, the subcommittee calculated a worst-case dermal exposure level of 1.7×10^{-3} mg/kg–d. The oral RfD for $Mg(OH)_2$ (12 mg/kg-d; see Oral RfD in Quantitative Toxicity section) was used as the best estimate of the internal dose for dermal exposure. Dividing the exposure level by the oral RfD yields a hazard index of 1.4×10^{-4}. Thus, it was concluded that $Mg(OH)_2$ used as an FR in upholstery fabric is not likely to pose any noncancer risk by the dermal route.

Inhalation Exposure

Particles

The characterization of the noncancer health risk from the inhalation of upholstery particles containing $Mg(OH)_2$ is based on the inhalation exposure scenario described in Chapter 3. In this scenario, a person is exposed to upholstery particles containing $Mg(OH)_2$. It is assumed that particles are generated from wear of the upholstery and 50% of the $Mg(OH)_2$ present in 25% of the treated surface are released as particles over the 15-yr lifetime of the fabric. It is also assumed that only 1% of the worn-off $Mg(OH)_2$ is released into the indoor air as particles that may be inhaled and that a person spends 1/4th of his or her life in a 30-m^3 room that contains 30 m^2 of treated upholstery with an air-change rate of 0.25/hr.

Particle exposure was estimated using Equations 4 and 5 in Chapter 3. The subcommittee estimated an upholstery application rate (S_a) for $Mg(OH)_2$ of 4 mg/cm^2. The release rate (μ_r) for $Mg(OH)_2$ from upholstery fabric was estimated to be 2.3×10^{-7}/d, yielding a room airborne particle concentration (C_p) of 1.5 $\mu g/m^3$. These values were used in Equation 8 in Chapter 3 to yield a short-time-average exposure concentration of 0.38 $\mu g/m^3$. The time-averaged exposure concentration for particles was calculated using Equation 6 in Chapter 3.

There are inadequate human or animal data on the inhaled toxicity of $Mg(OH)_2$ to derive an inhalation RfC. However, for the purpose of estimating a hazard index for characterizing the noncancer risk from the inhalation of $Mg(OH)_2$, a provisional inhalation RfC was derived from the TUL, which in this case is considered to be equivalent to the oral RfD (see Chapter 4 for the rationale). A provisional inhalation RfC of 42 mg/m^3 was derived for $Mg(OH)_2$ using the oral RfD and Equation 7 in chapter 3.

Division of the time-average exposure concentration of 0.38 $\mu g/m^3$ by the provisional inhalation RfC of 42 mg/m^3 results in a hazard index of 9.1×10^{-6}. This ratio suggests that under the worst-case human-exposure assumptions, $Mg(OH)_2$, when used as a FRs in upholstery, is not likely to pose a noncancer risk by the inhalation route of exposure.

Vapors

$Mg(OH)_2$ has negligible vapor pressure at ambient temperatures. Therefore, inhalation of $Mg(OH)_2$ vapor is not anticipated to pose a noncancer toxic risk when incorporated into furniture upholstery.

Oral Exposure

The characterization of noncancer health risk from oral exposure to $Mg(OH)_2$ is based on the oral exposure scenario described in Chapter 3. This scenario assumes a child is exposed to $Mg(OH)_2$ by sucking on 50 cm^2 of fabric treated with $Mg(OH)_2$, 1 hr/d for 2 yr. The subcommittee estimated an upholstery application rate (S_a) for $Mg(OH)_2$ of 4 mg/cm^2 and a fractional rate of $Mg(OH)_2$ extraction (μ_a) by saliva of 0.025/d based on levels reported by Jenkins et al. (1998). Oral exposure was calculated by using Equation 15 in Chapter 3.

The average oral daily dose for $Mg(OH)_2$ was estimated as 0.021 mg/kg-d. Division of the dose estimate by the oral TUL (RfD) for $Mg(OH)_2$ of 12 mg/kg-d results in the hazard index of 1.7×10^{-3}. Therefore, the subcommittee concluded that $Mg(OH)_2$ is not likely to pose a noncancer risk at the worst-case exposure levels from upholstered furniture.

Cancer

The subcommittee concludes that $Mg(OH)_2$ is not likely to be carcinogenic to humans by the oral route. No adequate data are available to assess the carcinogenicity of $Mg(OH)_2$ by the dermal or inhalation or routes of exposure.

RECOMMENDATIONS FROM OTHER ORGANIZATIONS

The current threshold limit value (TLV) for magnesium oxide is 10 mg/m^3 (ACGIH 1999). The subcommittee is not aware of any recommended exposure limits for $Mg(OH)_2$.

DATA GAPS AND RESEARCH NEEDS

There are inadequate subchronic or chronic toxicity data from dermal or inhalation exposure to $Mg(OH)_2$. There are no exposure data from dermal, inhalation, or oral routes of exposure to $Mg(OH)_2$ when used as an FR.

Because the hazard indices for noncancer effects for dermal, inhalation, and oral routes of exposure are less than 1, the subcommittee concludes that no further research is needed for assessing health risks from $Mg(OH)_2$.

REFERENCES

ACGIH (American Conference of Government Industrial Hygienists). 1999. Threshold Limit Values and Biological Exposure Indices. Cincinnati, OH: American Conference of Government Industrial Hygienists, Inc.

Alison, L.H., and D. Bulugahapitiya. 1990. Laxative induced magnesium poisoning in a 6 week old infant. BMJ 300(6717):125.

Amacher, D.E., and S.C. Paillet. 1980. Induction of trifluorothymidine-resistant mutants by metal ions in L5178Y/TK+/- cells. Mutat. Res. 78(3):279-288.

Ashby, J., and M. Ishidate, Jr. 1986. Clastogenicity in vitro of the Na, K, Ca and Mg salts of saccharin; and of magnesium chloride; consideration of significance. Mutat. Res. 163(1):63-73.

Bashir, Y., J.F. Sneddon, H.A. Staunton, G.A. Haywood, I.A. Simpson, W.J. McKenna, and A.J. Camm. 1993. Effects of long-term oral magnesium chloride replacement in congestive heart failure secondary to coronary artery disease. Am. J. Cardiol. 72(15):1156-1162.

Benech, H., A. Pruvost, A. Batel, M. Bourguignon, J.L. Thomas, and J.M. Grognet. 1998. Use of the stable isotopes technique to evaluate the bioavailability of a pharmaceutical form of magnesium in man. Pharm. Res. 15(2):347-351.

Brand, J.M., and F.R. Greer. 1990. Hypermagnesemia and intestinal perforation following antacid administration in a premature infant. Pediatrics 85(1):121-124.

Brunton, L.L. 1996. Agents affecting gastrointestinal water flux and motility; emesis and antiemetics; bile acids and pancreatic enzymes. Pp. 917-936 in Goodman & Gilman's The Pharmacological Basis of Therapeutics, 9th Ed. J.G. Hardman, L.E. Limbird, P.B. Molinoff, R.W. Ruddon, and A.G. Gilman, eds. New York: McGraw-Hill.

Dreisbach, R.H. 1977. Handbook of Poisoning: Diagnosis and Treatment, Ninth Edition. Los Altos, CA: Lange Medical Publications.

Elin, R.J. 1987. Assessment of magnesium status. Clin. Chem. 33(11):1965-1970.

EPA (U.S. Environmental Protection Agency). 1992. Dermal Exposure Assessment: Principles and Applications. EPA/600/8-91-011B. Office of Health and Environmental Assessment, U.S. Environmental Protection Agency, Washington, DC.

FDA (U.S. Food and Drug Administration). 1999. Direct Food Substances Affirmed as Generally Recognized as Safe; Magnesium Hydroxide. Fed. Regist. 50(Apr. 5, 1985):13557, 21 CFR Part 184.1428, as amended at Fed. Regist. 64(Jan. 5, 1999):404-405.

Ferdinandus, J., J.A. Pederson, and R. Whang. 1981. Hypermagnesemia as a cause of refractory hypotension, respiratory depression, and coma. Arch. Intern. Med. 141(5):669-670.

Ferrante, J. 1999. Toxicity review for Magnesium hydroxide. Memorandum, from Jacqueline Ferrante, Pharmacologist, Division of Health Sciences, to Ronald Medford, Assistant Executive Director for Hazard Identification and Reduction. U.S. Consumer Product Safety Commission. Washington, DC.

Fine, K.D., C.A. Santa Ana, and J.S. Fordtran. 1991. Diagnosis of magnesium-induced diarrhea. N. Engl. J. Med. 324(15):1012-1017.

Fire Retardant Chemicals Association. 1998. Textile Flame Retardant Applications by

Product Classes for 1997 Within and Outside of the United States: Magnesium Hydroxide. Fire Retardants Chemicals Association, Lancaster, PA.

Heldaas, S.S., S. Langård, and A. Andersen. 1989. Incidence of cancer in a cohort of magnesium production workers. Br. J. Ind. Med. 46(9):617-623.

Hori, H., T. Kasai, J. Haratake, S. Ishimatsu, T. Oyabu, H. Yamoto, T. Higashi, and I. Tanaka. 1994. Biological effects of inhaled magnesium sulphate whiskers in rats. Occup. Environ. Med. 51(7):492-499.

HSDB (Hazardous Substances Data Bank). 1998. Magnesium hydroxide. [Online]. Available: http://sis.nlm.nih.gov/cgi-bin/sis/search/f?./temp/~AAA99aimz:1 Last updated 3/16/98.

Humphrey, M., S. Kennon, and A.K. Pramanik. 1981. Hypermagnesemia from antacid administration in a newborn infant. J. Pediatr. 98(2):313-314.

IOM (Institute of Medicine). 1997. Phosphorus. Pp. 145-190 in Dietary Reference Intakes for Calcium, Phosphorus, Magnesium, Vitamin D, and Fluoride. Food and Nutrition Board, Institute of Medicine. Washington, DC: National Academy Press. [Online]. Available: http://books.nap.edu/books/0309063507/html/index.html

IPCS (International Programme on Chemical Safety). 1997. Flame Retardants: A General Introduction. Environmental Health Criteria 192. International Programme on Chemical Safety. Geneva: World Health Organization.

Ishidate, M., Jr., T. Sofuni, K. Yoshikawa, M. Hayashi, T. Nohmi, M. Sawada, and A. Matsuoka. 1984. Primary mutagenicity screening of food additives currently used in Japan. Food Chem. Toxicol. 22(8):623-636.

Jenkins, R.O., P.J. Craig, W. Goessler, and K.J. Irgolic. 1998. Antimony leaching from cot mattresses and sudden infant death syndrome (SIDS). Hum. Exp. Toxicol. 17(3):138-139.

Kurata, Y., S. Tamano, M.A. Shibata, A. Hagiwara, S. Fukushima, and N. Ito. 1989. Lack of carcinogenicity of magnesium chloride in a long-term feeding study in B6C3F1 mice. Food Chem. Toxicol. 27(9):559-563.

Marken, P.A., C.W. Weart, D.S. Carson, J.G. Gums, and M.F. Lopes-Virella. 1989. Effects of magnesium oxide on the lipid profile of healthy volunteers. Atherosclerosis 77(1):37-42.

Mofenson, H.C., and T.R. Caraccio. 1991. Magnesium intoxication in a neonate from oral magnesium hydroxide laxative. J. Toxicol. Clin. Toxicol. 29(2):215-222.

Mordes, J.P., and W.E.C. Wacker. 1978. Excess magnesium. Pharmacol. Rev. 29(4):273-300.

Mori, H., Y. Morishita, T. Shinoda, and T. Tanaka. 1993. Preventive effect of magnesium hydroxide on carcinogen-induced large bowel carcinogenesis in rats. Basic Life Sci. 61:111-118.

Morishita, Y., T. Tanaka, T. Kojima, A. Okumura, S. Sugie, and H. Mori. 1991. Effect of magnesium hydroxide on 1,2-dimethylhydrazine-induced intestinal carcinogenesis in rats. J. Toxicol. Pathol. 4:153-157.

Oberdörster, G. 1996. Evaluation and use of animal models to assess mechanisms of fibre carcinogenicity. Pp. 107-125 in Mechanisms of Fibre Carcinogenesis. A.B. Kane, P. Boffetta, R. Saracci, and J.D. Wilbourn, eds. IARC Scientific Publication No. 140.

Oguma, Y., F. Yokota, K. Inoue, and K. Shimamura. 1998. Mutagenicity studies of magnesium sulfate—reverse mutation test with bacteria and chromosomal aberration test with mammalian cells in culture. [Abstract]. J. Toxicol. Sci. 23(Suppl. 1):81-90.

Poisindex®. 1998. Toxicologic management of magnesium. 8/31/98. Micromedex, Inc. [Online]. Avaliable: http://www.micromedex.com/products/pd-poisindex.htm

Ricci, J.M., S. Hariharan, A. Helfgott, K. Reed, and M.J. O'Sullivan. 1991. Oral tocolysis with magnesium chloride: A randomized controlled prospective clinical trial. Am. J. Obstet. Gynecol. 165(3):603-610.

Rude, R.K., and F.R. Singer. 1981. Magnesium deficiency and excess. Ann. Rev. Med. 32:245-259.

Sutton, R.A.L., and J.H. Dirks. 1986. Calcium and magnesium: Renal handling and disorders of metabolism. Pp. 551-618 in The Kidney, 3rd Ed. B.M. Brenner, and F.C. Rector, Jr., eds. Philadelphia: Saunders.

Tanaka, H., A. Hagiwara, Y. Kurata, T. Ogiso, M. Futakuchi, and N. Ito. 1994. Thirteen-week oral toxicity study of magnesium chloride in B6C3F1 mice. Toxicol. Lett. 73(1):25-32.

Tanaka, T., T. Shinoda, N. Yoshimi, K. Niwa, H. Iwata, and H. Mori. 1989. Inhibitory effect of magnesium hydroxide on methylazoxymethanol acetate-induced large bowel carcinogenesis in male F344 rats. Carcinogenesis 10(3):613-616.

Usami, M., K. Sakemi, M. Tsuda, and Y. Ohno. 1996. Teratogenicity study of magnesium chloride hexahydrate in rats. [Abstract]. Eisei Shikenjo Hokoku 114:16-20.

Wang, A., N. Yoshimi, T. Tanaka, and H. Mori. 1993. Inhibitory effects of magnesium hydroxide on c-myc expression and cell proliferation induced by methalazoxymethanol acetate in rat colon. Cancer Lett. 75(2):73-78.

Wang, A., N. Yoshimi, T. Tanaka, and H. Mori. 1994. The inhibitory effect of magnesium hydroxide on the bile acid-induced cell proliferation of colon epithelium in rats with comparison to the action of calcium lactate. Carcinogenesis 15(11):2661-2663.

White, J., L. Massey, S. Kynast Gales, K. Dittus, and K. Campbell. 1992. Blood and urinary magnesium kinetics after oral magnesium supplements. Clin. Therapeut. 14(5):678-687.

8

Zinc Borate

THERE is limited information in the literature on the toxicity of zinc borate. Zinc borate readily breaks down in the stomach to zinc oxide (ZnO) and boric acid (H_3BO_3). Therefore, this chapter reviews the physical and chemical properties, toxicokinetics, toxicological, epidemiological, and exposure data on both those compounds. When data on zinc oxide are lacking, data on other zinc compounds are reviewed. According to the International Programme on Chemical Safety (IPCS), at low concentrations and under the same conditions, an equivalent amount of boron as boric acid or borax has similar chemical and toxicological properties (IPCS 1998). Therefore, data from boric acid and borax are considered in this chapter. Regardless of the zinc or boron compound of exposure, body burdens are measured as the concentration of the element (zinc or boron), and are discussed as such in this review. Doses are given in boron and zinc equivalents for comparison between different zinc and boron compounds.

The subcommittee used the toxicity, toxicokinetic, and exposure data on those compounds to characterize the health risk from exposure to zinc borate. The subcommittee also identified data gaps and recommended research relevant for determining the health risk from exposure to zinc borate.

PHYSICAL AND CHEMICAL PROPERTIES

The physical and chemical properties of zinc borate, zinc oxide, and boric acid are summarized in Table 8-1.

TABLE 8-1 Physical and Chemical Properties of Zinc Borate, Zinc Oxide, and Boric Acid

	Value			
Property	Zinc Borate	Zinc Oxide	Boric Acid	Reference
Chemical formula	$3ZnO{:}2B_2O_3$	ZnO	H_3BO_3	Lide 1991-1992
CAS registry #	138265-88-0	1314-13-2	10043-35-3	Lide 1991-1992
Synonyms	—	Zinc white, pigment white	Orthoboric acid, boracic acid, borofax	Budavari et al. 1989
Molecular weight	383.4	81.4	61.3	Lide 1991-1992
Physical state	Crystal or white powder	Crystal or white powder	Transparent crystals or white granules or powder	Budavari et al. 1989
Solubility	Soluble in cold water; crystal insoluble in HCl; amorphous soluble in HCl	0.00016 g/100 mL water at 29°C; soluble in acid, alkaline, ammonium chloride; insoluble in alcohol	6.35 g/100 mL water at 30°C; 27.6 g/100 mL water at 100°C	Lide 1991-1992
Melting point	980°C	1,975°C	169°C $\pm$ 1	Lide 1991-1992
Octonal/water partition coefficient	ND	ND	0.175	IPCS 1998
Density	Crystal: 4.22 g/cm^3; Powder: 3.64 g/cm^3	5.6 g/cm^3	1.435 g/mL at 15°C	Lide 1991-1992

Abbreviations: HCl, hydrochloric acid; ND, not determined

OCCURRENCE AND USE

Zinc borate is typically composed of 45% ZnO and 34% boric anhydride (B_2O_3), with 20% water of hydration. Zinc borate is used as a flame retardant in conjunction with other chemicals, including antimony trioxide, magnesium hydroxide, alumina trihydrate, and some brominated flame retardants. Zinc borate is used as a flame retardant on commercial furniture, draperies, wall coverings, and carpets (R.C. Kidder, Flame Retardant Chemical Association, unpublished material, April 21, 1998). In addition, zinc borate is used as a fungicide.

Zinc oxide is used as a pigment in paint, cosmetics, and dental and quick drying cements. Therapeutically, zinc oxide is used as an astringent and as a topical protectant.

Boric acid is used in enamels, porcelain, soaps, cosmetics, and as an insecticide. Therapeutically, boric acid is used as an astringent and an antiseptic.

TOXICOKINETICS

Absorption

Zinc Borate

No studies were identified that investigate the toxicokinetics of zinc borate following dermal, inhalation, or oral exposure.

Zinc Oxide

Agren (1991) reported that zinc is present in human interstitial fluid at the site of application following dermal application of zinc oxide in gum resin or hydrocolloids to human forearms. No evidence for absorption into systemic circulation was provided. Zinc readily permeates intact and damaged human skin following dermal application; however, absorption of zinc into systemic circulation was not determined (Hallmans 1977; Agren 1990, as cited in ATSDR 1994; Agren et al. 1991, as cited in ATSDR 1994). Keen and Hurley (1977) determined that when zinc (as zinc chromate) was dissolved in oil and topically applied to rats, absorption of zinc into the bloodstream occurred. No other animal studies were identified that investigated the dermal absorption of zinc.

Workers exposed occupationally (via inhalation) to zinc fumes (zinc compound not specified) had elevated blood zinc concentrations (Hamdi 1969). Exposure of cats to zinc oxide fumes for up to 3.25 hr resulted in increased concentrations of zinc in the pancreas, kidney, and liver (Drinker and Drinker 1928, as cited in ATSDR 1994). In both studies, oral absorption of zinc particles following ciliary clearance and swallowing could account for all, or a significant portion, of the absorbed zinc. In the Drinker and Drinker (1928, as cited in ATSDR 1994) study, the swallowing of zinc particles during grooming activities might also account for the increased tissue zinc levels.

No data are available on the oral absorption of zinc oxide. The estimated rate of oral absorption of zinc (all zinc compounds) in humans is between 8% and 81%, depending on an individual's diet (ATSDR 1994). People who are not deficient in zinc absorb about 20-30%, while zinc-deficient individuals absorb more (ATSDR 1994). Two studies measured the peak blood concentrations of zinc in volunteers following oral ingestion of zinc sulfate; peak blood Zn^{2+} concentrations were reached within 3 hr (Neve et al. 1991; Sturniolo et al. 1991, as cited in ATSDR 1994). The presence of cadmium, mercury, copper, or other trace metals can diminish zinc absorption by inhibiting zinc transport across the intestinal wall (ATSDR 1994).

Zinc absorption in male Wistar rats was approximately 40-48% when diets contained 0.81 mg of radio-labeled zinc (as zinc chloride or zinc carbonate) per kg of body weight (Galvez-Morros et al. 1992). ATSDR (1994) noted that the fraction of ingested zinc absorbed in immature organisms usually exceeds the fraction of ingested zinc absorbed in adult organisms.

Boric Acid

Wester et al. (1998) exposed the back of the hand of volunteers to a 5% aqueous solution of boric acid or borax and measured urinary boron concentrations to determine the extent of absorption, the flux, and the permeability constants (Kp) for intact skin. Following exposure to boric acid, 0.23% of the applied dose was excreted, flux was calculated as 0.01 $\mu g/cm^2/hr$ and Kp was 1.9×10^{-7} cm/hr. Following exposure to borax, 0.21% of the applied dose was excreted, flux was calculated as 0.01 $\mu g/cm^2/hr$, and Kp was 1.8×10^{-7} cm/hr. Draize and Kelly (1959) has also reported low dermal absorption of boric acid, with no increase in urinary boron concentrations following a 4-hr exposure in a volunteer. Blood boron concentrations did not increase in infants after treatment with ointment (3% boric acid), indicating a lack of dermal absorption of boric acid (Friis-Hansen et al. 1982).

No absorption of boric acid (measured as boron in the blood) occurred 1-9 d after a single topical application of boric acid in an anhydrous, water-emulsifying ointment (Stuttgen et al. 1982). However, blood boron concentrations were increased within 2-6 hr after application of the same amount of boric acid in a water-based jelly, indicating that the vehicle in which boric acid is applied to the skin affects absorption.

Boron was detected in the urine of infants who had moderate to marked diaper rash, but not in the urine of infants who had minor or no diaper rash, after application of a commercial talcum powder containing 5% boric acid (Mulinos et al. 1953, as cited in Moore 1997; Vignec and Ellis 1954, as cited in IPCS 1998). In rabbits, intact skin acts as a barrier to dermal absorption of boric acid, whereas absorption was much greater through damaged skin (Draize and Kelly 1959).

In an in vitro absorption assay, 0.05%, 0.5%, or 5% boric acid solution were applied to human skin, and 1.2%, 0.28%, and 0.7%, respectively, of the boric acid was absorbed (Wester et al. 1998). From those data, flux values of 0.25, 0.58, and 14.58 μg/cm^2/hr, and permeability constants (Kp) of 5.0×10^{-4}, 1.2×10^{-4}, and 2.9×10^{-4} cm/hr for the 0.05%, 0.5%, and 5.0% boric acid solutions, respectively, were calculated (Wester et al. 1998).

In contrast to the lack of dermal absorption, boric acid is readily absorbed following inhalation and oral exposure. Kent and McCance (1941, as cited in Moore 1997) demonstrated in two female subjects that at least 90% of ingested boric acid is absorbed and excreted in the urine. More recently, Jansen et al. (1984) demonstrated greater than 90% recovery of administered boron in the urine of six male volunteers following ingestion of boric acid. An occupational study of workers involved in packaging and shipping borax ($Na_2B_4O_7 \cdot 10H_2O$) showed elevated boron levels in the urine after inhalation exposure (Culver et al. 1994).

Metabolism and Distribution

Zinc Borate

Zinc borate is metabolized to zinc oxide and boric acid prior to being absorbed.

Zinc Oxide

No relevant human or animal studies were identified that investigated the distribution of zinc following dermal exposure to zinc compounds.

No inhalation exposure studies were identified that investigated the distribution of zinc compounds in humans. Cats exposed to zinc oxide fumes (12-61 mg Zn^{2+}/kg-d) for 3 hr had increased zinc levels in the pancreas, liver, and kidneys (Drinker and Drinker 1928, as cited in ATSDR 1994), however, oral absorption through swallowing or grooming cannot be ruled out.

There are no data on the metabolism and distribution of zinc oxide following oral exposure. Weigand and Kirchgessner (1992) demonstrated that more zinc is distributed to the kidneys and pancreas than to the liver in rats fed 1.1 mg

Zn^{2+}/kg-d (duration unspecified). Administration of zinc acetate to rats (191 mg Zn^{2+}/kg-d in feed for 3 mo) increased zinc levels in the heart, spleen, kidneys, liver, bone, and blood (Llobet et al. 1988). Mice fed 76.9 mg Zn^{2+}/kg-d as zinc sulfate for 1 mo had increased levels of Zn^{2+} in the kidneys and liver (Schiffer et al. 1991). Newborn, young, and adult mice given a single oral dose of 4.6 mg zinc/kg as zinc chloride generally had the highest level of zinc in the liver, kidneys, lungs, bone, muscle, and carcass 1 d after dosing (He et al. 1991, as cited in ATSDR 1994).

Once in the body, zinc induces and binds to metallothionein (a metal binding protein) (Goyer 1996). Retention of zinc bound to metallothionein in tissues provides a source of zinc for essential cell functions even when zinc intake is deficient. In humans and rats, zinc and metallothionein are distributed throughout the body, with a linear relationship between zinc and metallothionein concentrations in the liver (Heilmaier et al. 1987).

Boric Acid

In humans, boron has been measured in the brain and liver following boric acid poisonings (see review, Moseman 1994). No other data were found on the distribution of boron in humans following exposure to boric acid or borax.

In Fischer rats fed 9,000 ppm boron as boric acid (93-96 mg boric acid/kg-d) in the diet, boron was distributed throughout the soft tissues. Accumulation occurred in the bone, but not in the testis or the brain (Ku et al. 1991).

Regardless of the compound or route of exposure, once in vivo, boron can form weak complexes with hydroxyl, amino, or thiol groups (Moore 1997). No data were found demonstrating that boron interacts with metallothionein.

Excretion

Zinc Borate

No data were identified that investigated the excretion of zinc borate in humans or animals following any route of exposure.

Zinc Oxide

No studies were identified that investigated the excretion of zinc in humans or animals following dermal application of any zinc compounds.

Elevated levels of zinc were found in the urine of workers exposed to zinc oxide fumes containing unknown concentrations of Zn^{2+} (Hamdi 1969). No other studies were identified that investigated the excretion of zinc following inhalation of zinc compounds.

Following oral exposure to zinc compounds, the primary route of zinc excretion in humans and rats is the feces. Zinc can also be excreted in the urine, saliva, hair, and sweat (ATSDR 1994).

Malnutrition or low dietary levels of zinc can promote urinary zinc excretion, possibly as a result of increased levels of tissue breakdown and catabolism (ATSDR 1994).

Boric Acid

As discussed in the section on Absorption, boron has been detected in the urine after exposure to boric acid via the dermal, inhalation, and oral routes. Following ingestion of boric acid by six male volunteers, greater than 90% of an ingested dose was excreted in the urine within 96 hr (Jansen et al. 1984).

HAZARD IDENTIFICATION[1]

Dermal Exposure

Irritation and Sensitization

Zinc Borate

Zinc borate produced only mild conjunctivitis in albino rabbits in the eye irritation test and is not considered to be an irritant or corrosive (U.S. Borax 1996).

Zinc borate was negative in the guinea pig sensitization test (U.S. Borax 1996).

Zinc Oxide

There are two case studies in the literature that suggest that dermal occupational exposure to zinc oxide might cause or contribute to a skin condition

[1]In this section, the subcommittee reviewed toxicity data on zinc borate, including the toxicity assessment prepared by the U.S. Consumer Product Safety Commission (Hatlelid 1999).

referred to as "zinc oxide pox" (itchy papular-pustular eruptions that occur in the pubic region, inner surface of the thigh, and axilla and inner surface of the arms). Turner (1921, as cited in ATSDR 1994) found that 14 out of 17 men developed zinc oxide pox at least once during their employment in the bagging or packaging of zinc oxide. However, that effect has been attributed to poor hygiene among the workers, and not necessarily exposure to zinc oxide. In a similar study, Batchelor et al. (1926, as cited in ATSDR 1994) found that only 1 out of a total of 24 workers with occupational exposure to zinc dusts developed zinc oxide pox.

Agren (1990, as cited in ATSDR 1994) reported that application of patches containing 25% zinc oxide (equivalent to 2.9 mg Zn^{2+}/m^3) to the skin of human volunteers did not produce dermal irritation following 48 hr of exposure.

The dermal irritancy of several zinc compounds has been investigated in mice, rabbits, and guinea pigs (Lansdown 1991). Animals were treated topically once a d for 5 consecutive days with zinc oxide (20% suspension in Tween 80), zinc chloride (1% aqueous solution), zinc sulfate (1% aqueous solution), zinc pyrithione (20% suspension), or zinc undecylenate (20% suspension). In open patch tests, zinc chloride was a strong irritant in all three species, and caused the formation of epidermal hyperplasia and ulceration. All other compounds produced less severe erythema than zinc chloride. None of the compounds caused ulceration or scaling over the 5-d test period. Zinc chloride produced severe dermal irritation in rabbits within 3-5 d of application in occlusive patch tests, zinc acetate produced moderate irritation, and little dermal irritation was caused by the other zinc compounds. Histological examination of skin samples from animals treated with zinc chloride or zinc acetate showed evidence of acanthosis, parakeratosis, hyperkeratosis, and inflammatory changes in the epidermis.

No studies were identified that investigated the ability of zinc oxide to act as a senzitizer.

Boric Acid

Dermatitis has been reported following occupational exposure to borax (Birmingham and Key 1963). Boric acid (5 mL, 10% w/v in water) and borax (10 mL, 5% w/v in water) were found to be moderate and mild irritants, respectively, in guinea pigs. Both were mildly irritating to abraded skin after 24-72 hr in rabbits (Roudabush et al. 1965).

No studies were identified that investigated the ability of boric acid to act as a skin sensitizer.

Systemic Effects

Zinc Borate

The LD_{50} in male and female albino rabbits following dermal exposure to zinc borate is estimated to be >10 g/kg (U.S. Borax 1996).

Zinc Oxide

DuBray (1937, as cited in ATSDR 1994) reported that a worker developed microcytic anemia and had low platelet counts after making zinc chloride solutions. The dose of zinc was not reported. No systemic effects were identified following dermal exposures of animals to zinc oxide.

Boric Acid

No studies were found on the systemic effects of boric acid following dermal absorption.

Other Systemic Effects

No studies were found that investigated the immunological, neurological, reproductive, developmental, or carcinogenic effects of zinc borate, zinc oxide, or boric acid following dermal exposure in humans or experimental animals.

Inhalation Exposure

No data were found on toxic effects of zinc borate following inhalation exposure. Data on the toxic effects of zinc oxide and boric acid following inhalation exposure are discussed below.

Systemic Effects

Zinc Oxide

Although "metal fume fever" has been reported after exposure to zinc oxide, this syndrome is seen following exposure to extremely high concentrations in the occupational setting and is not relevant to exposures from zinc borate-treated upholstered furniture. A number of deaths have been reported in humans following inhalation exposure to airborne mixtures containing very high concentrations of zinc. Ten out of 70 people died within 4 d of exposure to a smoke mixture containing approximately 33,000 mg Zn^{2+}/kg as zinc chloride, in addition to unknown concentrations of hexachloroethane, calcium silicate, and an igniter (Evans 1945, as cited in ATSDR 1994). Milliken et al. (1963, as cited in ATSDR 1994) and Hjortso et al. (1988) reported fatalities after exposure to high, but unknown concentrations of a smoke mixture generated from zinc chloride smoke bombs. Autopsies revealed diffuse microvascular obliteration, widespread occlusion of the pulmonary arteries, and extensive interstitial and intra-alveolar fibrosis of the lungs (Hjortsu et al. 1988). Although zinc oxide is associated with metal fume fever, and a large amount of research has been carried out in that area, those results are not a focus of this report because exposure to such fumes created by welding are not relevant to our exposure scenario.

Nausea has been reported following exposure to high concentrations of zinc oxide in humans (Hammond 1944, as cited in ATSDR 1994; Rohrs 1957, as cited in ATSDR 1994). McCord et al. (1926, as cited in ATSDR 1994) reported that several workers from the galvanized industry had decreased red blood cell counts, but Hamdi (1969) reported that workers exposed to zinc compounds had normal red blood cell counts. Routine blood analysis did not reveal liver disease among 12 workers with 4-21 yr of exposure to zinc oxide (Hamdi 1969).

Pulmonary toxicity and reduced survival were reported in female rodents following exposure to zinc oxide/hexachloroethane smoke (119 mg Zn^{2+}/m^3 for 1 hr/d, 5 d/wk for up to 20 wk) (Marrs et al. 1988). However, the smoke contained a number of other toxic chemicals (e.g., carbon tetrachloride), therefore, the effects can not be attributed to zinc. A single exposure in rats and rabbits to 88-482 mg zinc/m^3 as zinc oxide resulted in pulmonary congestion and leukocytic infiltration (Drinker and Drinker 1928; as cited in ATSDR 1994). Amdur et al. (1982) demonstrated a decreased lung compliance following exposure of guinea pigs to 0.73 mg zinc/m^3 as zinc oxide for 1 hr. Lam et al. (1982) did not see an effect on ventilation, but did see an effect on functional residual capacity after exposure of guinea pigs to 6.3 mg zinc/m^3 as zinc oxide for 3 hr. Deficits in lung function were seen in guinea pigs exposed to zinc oxide dust (3.7-5.6 mc zinc/m^3) 3 hr/d for up to 6 d (Lam et al. 1985, 1988). No effects were seen at a concentration of 2.2 mg zinc/m^3 as zinc oxide (Lam et al. 1988). However, guinea pigs appear to be more sensitive to the pulmonary effects of zinc oxide than humans because of differences in lung structure (Lam et al. 1982).

Boric Acid

In the occupational setting, toxic effects following exposure to boron are generally acute, and include nosebleed, nasal irritation, sore throat, cough, and shortness of breath (IPCS 1998). Garabrant et al. (1984) found an increase in reports of eye irritation, dry mouth, nose or throat irritation, and productive cough in workers in a borax mining and refining plant. Chronic bronchitis, without any abnormal regions on a chest X-ray or impairment of pulmonary function, was also found in the borax plant (Garabrant et al. 1985). In a prospective study of workers exposed to sodium borate dust in a mine and processing plant, Wegman et al. (1994) found increased nasal, eye, and throat irritation, cough, and breathlessness. No long-term effects were found in that study.

Exposure to boric acid 6 hr/d, 5 d/wk had no effect on body weight gain, hematology, blood chemistry, urinalysis, or microscopic analysis in rats (77 mg/m^3 for 24 wk; 175 mg/m^3 for 12 wk, or 470 mg/m^3 for 20 wk) or dogs (57 mg/m^3 for 23 wk) (Wilding et al. 1959, as cited in ATSDR 1992).

Immunological Effects

No studies were identified that investigated immunological effects of boric acid following inhalation exposure. Data from three case reports suggest that inhalation exposure to high concentrations of zinc-containing compounds stimulates changes in immune parameters. Farrell (1987) reported that a worker developed hives and angioedema (suggestive of an immediate or delayed IgE response) following exposure to a low dose of zinc oxide fumes. The symptoms reappeared in a challenge test, suggesting a sensitization to zinc compounds. A correlation between exposure to zinc oxide and the proportion of activated helper-, inducer-, suppressor-, and killer-T-cells was observed among 14 welders approximately 20 hr after exposure to zinc oxide (77-153 mg Zn^{2+}/m^3) (Blanc et al. 1991, as cited in ATSDR 1994). Ameille et al. (1992) reported elevated levels of lymphocytes in the bronchoalveolar lavage fluid of a smelter worker exposed to unknown concentrations of zinc fumes. Cytokine responses have been observed in bronchoalveolar lavage after inhalation of zinc oxide fumes from welding (Blanc et al. 1993; Kuschner et al. 1995, 1997).

Marrs et al. (1988) did not observe abnormalities in the lymph nodes, thymus, or spleen tissue of female rats, mice, or guinea pigs killed 18 mo after a 20-wk exposure to zinc oxide/hexachloroethane smoke at concentrations as high as 119.3 or 121.7 mg Zn^{2+}/m^3 for 1 hr/d, 5 d/wk.

Reproductive and Developmental Effects

Zinc Oxide

No studies were identified in humans that investigated reproductive or developmental effects after inhalation exposure to zinc compounds. Pathological examination 17 mo after exposure of rats, guinea pigs, and mice to zinc oxide/hexachloroethane smoke (1 hr/d, 5 d/wk for 20 wk) did not demonstrate any treatment-related abnormalities in mammary glands, ovaries, or fallopian tubes (Marrs et al. 1988).

Boric Acid

Reproductive effects of boron compounds have been investigated after occupational exposure. The number of live births in employees at a boron mine was not statistically different than the number of live births in the general U.S. population (Whorton et al. 1994). However, that study has been criticized because of its limitations and a lack of sensitivity (Moore 1997). A study (Tarasenko et al. 1972), summarized by Moore (1997), found a decrease in sexual activity in 28 workers exposed to very high concentrations of boron (10 mg/m^3). Examination of the semen from six of the workers demonstrated a reduction in semen volume, a decrease in the number of spermatozoa, and decreased sperm motility.

Cancer

No studies were identified that investigated the carcinogenicity of boric acid.

Excess lung cancer mortality was detected among persons living in an abandoned zinc/lead mining area in the midwestern United States as compared with state and national age- and sex-specific lung cancer rates (Neuberger and Hollowell 1982). However, potential confounding factors (such as smoking) were not controlled for.

Logue et al. (1982) investigated mortality among a cohort of 4,802 male workers from two zinc and seven copper refining plants. Overall mortality and death from specific cancers were not elevated for the whole cohort. Cancer mortality rates, however, were not computed separately for the 978 zinc refinery workers.

Marrs et al. (1988) found that female mice, but not female rats or guinea pigs, exposed to 1.3, 12.8, or 121.7 mg zinc /m^3 as zinc oxide/hexachloroethane smoke (1 hr/d, 5 d/wk for 20 wk) had a statistically significant trend in the prevalence of alveologenic carcinoma, with the frequency of this tumor reaching significance in the high-dose group at 13 mo postexposure. However, the effects cannot be attributed to zinc oxide because the smoke contained a number of other chemicals, including carbon tretrachloride, which is known to be an animal carcinogen.

Other Systemic Effects

No studies were found that investigated the neurological effects of zinc oxide and boric acid following inhalation exposure.

Oral Exposure

Systemic Effects

Zinc Borate

U.S. Borax (1996) lists an acute oral LD_{50} for zinc borate in male rats of >10 g/kg.

Zinc Oxide

A summary of oral toxicity studies on zinc compounds is presented in Table 8-2. When considering the oral toxicity of zinc compounds, it is important to note that zinc is an essential nutrient. NRC (1989) established recommended dietary allowances for zinc of 15 mg/d for males and 12 mg/d for females. However, chronic supplementation of more than 15 mg/d is not recommended without medical supervision because zinc can aggravate copper deficiency in individuals who are already marginally copper deficient (NRC 1989).

No human studies report death following ingestion of high zinc concentrations and toxicity normally occurs only after ingestion of more than 2 grams of zinc (Prasad 1976, as cited in ATSDR 1994; NRC 1989).

Vomiting, abdominal cramps, and diarrhea have been observed following ingestion of high levels of zinc sulfate. A wk after an English school girl ingested 440 mg zinc sulfate-d (2.6 mg zinc/kg-d) in capsules (a medically prescribed treatment for acne), she was admitted to the hospital, diagnosed with anemia, and had indications of gastrointestinal bleeding (Moore 1978). Gastrointestinal

TABLE 8-2 Selected Animal Toxicity Data for Boron, Zinc, and Zinc Borate

Species, Strain, Sex	Dose	Duration, Route	Effects	NOAEL/LOAEL	Reference
BORON					
Rat, Sprague-Dawley, F (pregnant)	Boric acid: 3.3, 6.6, 9.6, 13.3, 25 mg/kg-d	Gestation d 0-20, diet	No maternal deaths or overt signs of toxicity. Significant reduction in fetal body weight observed in dams treated with 13.3 and 25 mg/kg-d. Significant increasing trend in percentage of fetuses with skeletal malformations/litter in two highest dose groups (short XII- or wavy ribs).	NOAEL (developmental toxicity): 9.6 mg boron/kg-d LOAEL (developmental toxicity): 13.3 mg boron/kg-d	Price et al. 1996a
Rabbit, New Zealand, F (pregnant)	Boric acid: 10.9, 21.9, 43.7 mg/kg-d	Gestation d 6-19, gavage	Prenatal survival decreased and number of malformations (mainly cardiovascular) was increased in high dose group.	NOAEL (developmental toxicity): 21.9 mg/kg-d LOAEL (developmental toxicity): 43.7 mg boron/kg-d	Price et al. 1996b
Rat, Sprague-Dawley, F (pregnant)	Boric acid: 3.3, 6.5, 9.8, 12.9, 25.3 mg/kg-d	Gestation d 0-20, diet	Incidence of skeletal malformations was increased in the low- and high-dose groups on postnatal d 21, along with clinical signs of maternal toxicity. The incidence of short rib was increased at postnatal d 21. There was no association between treatment and wavy rib at postnatal d 21.	ND	Heindel et al. 1992
Mouse, CD-1, F (pregnant)	Boric acid: 43, 79, 175 mg/kg-d	Gestation d 0-17, diet	Dose-dependent decrease in fetal body weight at the two highest doses; increased skeletal malformations were also seen. Maternal toxicity was also seen at those doses.	NOAEL (developmental toxicity): 43 mg boron/kg-d LOAEL (developmental toxicity): 79 mg boron/kg-d	Heindel et al. 1992

Mouse, CD-1 and field, M	Boric acid: 1,000, 4,500, 9,000 mg/kg-d	27 wk, diet	Testicular atrophy, decreased spermatogenesis, in 4,500- and 9,000-mg/kg-d dose groups in CD-1 mice. Same effects observed at the high dose in wild field mice.	NOAEL (testicular effects): 4,500 mg/kg-d in field mice NOAEL for testicular effects not established for CD-1 mice	Fail et al. 1989, as cited in Moore 1997
Mouse, B6C3F1, M/F	Boric acid: 1,200, 2,500 5,000, 10,000, 20,000 ppm	91 d, diet	Decreased weight gain in both sexes at the three highest doses. Increased mortality in the highest dose group. Extramedullary hematopoiesis of the spleen occurred in all groups except 20,000 ppm group. Stomach hyperkeratosis and/or acanthosis in both sexes at 20,000 ppm. Testicular degeneration and atrophy of seminiferous tubules in three highest dose groups.	NOAEL (weight loss): 2,500 ppm boric acid in feed LOAEL (weight loss): 5,000 ppm boric acid in feed	NTP 1987
Mouse, B6C3F1, M/F	Boric acid: 2,500, 5,000 ppm	2 yr, diet	Survival decreased in males of both dose groups. Decreased weight gain in high dose group for both sexes. Increased incidence of lung hemorrhage in high-dose females. Dose-related increase in incidence of testicular atrophy and interstitial cell hyperplasia in male mice.	LOAEL (systemic effects): 2,500 ppm boric acid in feed (400-500 mg/kg-d)	NTP 1987
Rat, Sprague-Dawley, M/F	Boric acid: 117, 350, 1,170 ppm	2 yr, diet	Decreased food consumption and suppressed growth, rough fur coat, swollen paws, other gross signs of toxicity in all animals in high-dose group. Increased incidence of testicular and seminiferous atrophy in males in high-dose group.	NOAEL (systemic effects): 350 ppm LOAEL (systemic toxicity): 1,170 ppm	Weir and Fisher 1972
Dog, beagle, M/F	Boric acid: 58, 117, 350, 1,170 ppm	38 wk, diet	Seminiferous tubular atrophy and decreased spermatogenesis observed in 1,170 dose group at 38 wk. See note in text about expert committee findings, i.e., Moore (1997).	NOAEL (systemic effects): 1,170 ppm (8.8 mg/kg-d)	Weir and Fisher 1972

TABLE 8-2 *(Continued)*

Species, Strain, Sex	Dose	Duration, Route	Effects	NOAEL/LOAEL	Reference
BORON *(Continued)*					
Rat, NS, M/F	Boron: 500 mg/kg-d	21-27 d, diet	Reduced weight gain; effect on nucleotide levels in numerous tissues.	ND	Dani et al. 1971
ZINC COMPOUNDS					
Healthy adult human volunteers, F	50 mg/kg-d as zinc gluconate	10 wk, oral (capsules)	Serum zinc significantly increased. Significant decrease in erythrocyte superoxide dismutase activity, serum ferritin, and hematocrit values.	ND	Yadrick et al. 1989
Guinea pig, Dunkin-Hartley, F	Zinc oxide smoke[a] (20% Zn by weight): 1.3 ± 0.4, 12.8 ± 2.6, 119.3 ± 31.7 mg/m^3	1 hr/d, 5 d/wk, inhalation (100 exposures; high dose guinea pigs received 15 exposures)	No increase in tumor frequency. High death rate in high-dose group.	ND	Marrs et al. 1988
Rat, Wistar, F	Zinc oxide smoke[a] (20% Zn by weight): 1.3 ± 0.4, 12.8 ± 2.6, 21.7 ± 17.8 mg/m^3	1 hr/d, 5 d/wk, inhalation (100 exposures)	No increase in tumor frequency.	ND	Marrs et al. 1988
Mice, Porton, F	Zinc oxide smoke[a] (20% Zn by weight): 1.3 ± 0.4, 12.8 ± 2.6, 21.7 ± 17.8 mg/m^3	1 hr/d, 5 d/wk, inhalation (100 exposures)	Treatment-related, fatty changes in the liver in the middle- and high-dose groups. Statistically significant trend in the prevalence of alveologenic carcinoma, with the frequency of this tumor reaching significance in the high-dose group at 13 mo postexposure.	LOAEL (lung tumors): 21.7 ± 17.8 mg/m^3 NOAEL (lung tumors): 12.8 ± 2.6 mg/m^3	Marrs et al. 1988

Healthy adult human volunteers, M	4.3 mg/kg-d as zinc gluconate	6 wk, oral	Impaired mitogenic response elicited from peripheral blood lymphocytes. Impaired chemotactic and phagocytic responses of polymorphonuclear leukocytes.	LOAEL (immune effects): 4.3 mg/kg-d	Chandra 1984
Healthy adult human volunteers, M	50 mg/kg-d as zinc gluconate	6 wk, oral (capsules)	Serum zinc significantly increased at wk 2. Significant decrease in erythrocyte superoxide dismutase activity (15%).	LOAEL (enzyme inhibition): 50 mg/kg-d	Fischer et al. 1984
Rat, Sprague-Dawley, M/F	160, 320, 640 mg/kg-d as zinc acetate	90 d, drinking water	No effects observed on hematocrit or hemoglobin levels, serum glucose	ND	Llobet et al. 1988
Mink, F (pregnant)	20.8 mg/kg-d as zinc sulfate	Diet	No effects on gestation length or litter size	ND	Bleavins et al. 1983
Mice, NS, M/F	10, 104, 1,110 mg/kg-d as zinc sulfate	90 d, diet	Decreased leukocyte count in male mice in the high-dose group. Significantly lower hematocrit values and hemoglobin concentrations in the two highest-dose groups. Regressive kidney lesions in female mice of the highest dose group.	NOAEL (hematocrit values): 10 mg/kg-d LOAEL (hematocrit values): 104 mg/kg-d	Maita et al. 1981
Rats, NS, M/F	100, 200 mg/kg-d as zinc oxide	21 d before to mating through gestation, diet	4-29% increase in resorptions during gestation in 200 mg/kg-d dose group.	LOAEL (developmental toxicity): 200 mg/kg-d NOAEL (developmental toxicity): 100 mg/kg-d	Schlicker and Cox 1968

F, female; LOAEL, lowest-observed-adverse-effect level; M, male; ND, not determined; NOAEL, no-observed-adverse-effect level; NS, not specified.

[a] Authors note that zinc oxide smoke also contained animal carcinogens carbon tetrachloride and hexachloroethane.

upset (abdominal cramps, vomiting, nausea) was reported in 26 out of 47 healthy volunteers following ingestion of zinc sulfate tablets (150 mg Zn^{2+} in three divided doses/d; 2 mg zinc/kg-d) for 6 wk (Samman and Roberts 1987).

Yadrick et al. (1989) investigated the effects of oral zinc supplements on copper and iron balance in a 10-wk clinical study in 18 healthy women. Women were given supplements (as capsules) of 50 mg zinc/d as zinc gluconate. Erythrocyte superoxide dismutase (ESOD) activity declined over the 10-wk period and after 10 wk were significantly decreased ($p<0.05$) compared to ESOD activity during the pretreatment period. Serum ferritin and hematocrit values were also significantly lower than pre-treatment values after 10 wk. Serum zinc was significantly increased. Ceruloplasmin levels were not altered.

Fischer et al. (1984) reported that zinc supplementation in healthy adult male volunteers (50 mg zinc/d as zinc gluconate) for 6 wk significantly decreased ESOD activity (15%). There were no differences in serum copper levels or ceruloplasmin activity in the exposed group as compared to the controls. Serum zinc levels were significantly increased in the exposed group after 2 wk.

No oral LD_{50} value has been reported for zinc oxide, but LD_{50} values for other zinc compounds range from 237 to 623 mg zinc/kg-d in rats, and 86 to 390 mg zinc/kg-d in mice. Ferrets that ingested 390 mg zinc/kg-d as zinc oxide for 2 wk experienced intestinal hemorrhages and a 75% reduction in food intake (Straube et al. 1980). Mice fed a diet containing 1,110 mg zinc/kg-d developed ulcers in the forestomach. No gastrointestinal effects were observed in rats fed 565 mg zinc/kg-d (Maita et al. 1981).

Llobet et al. (1988) fed zinc acetate dihydrate to 40 Sprague-Dawley rats in drinking water at levels of 0, 160, 320, or 640 mg/kg-d for 3 mo. No significant differences were observed in hematocrit and hemoglobin levels between dose groups. The concentrations of glucose and enzymes in plasma were within the normal range in treated and untreated animals. Zaporowska and Wasilewski (1992, as cited in ATSDR 1994) identified a LOAEL of 12 mg zinc/kg-d as zinc chloride in a 4-wk drinking water study in 2-mo-old rats, based on decreased hemoglobin (85% of control values). Maita et al. (1981) fed mice (12/sex/group) zinc sulfate at 0, 300, 3,000, and 30,000 ppm (equivalent to 0, 10, 104, 1,110 mg zinc/kg-d) in the diet for 13 wk. Significantly lower hematocrit and hemoglobin concentrations were measured in the 3,000- and 30,000-ppm groups compared to controls, however, no dose-dependent relationship was observed. Leukocyte counts in male mice exposed to 30,000 ppm were also decreased.

Several animal studies have demonstrated renal effects in animals exposed to zinc oxide, zinc sulfate, and zinc acetate. Zinc sulfate caused an increase in absolute and relative kidney weights, and kidney lesions in female mice fed

1,110 mg zinc/kg-d, but not 565 mg zinc/kg-d, as zinc sulfate in the diet for 13 wk (Maita et al. 1981). Rats exposed to 191 mg zinc/kg-d as zinc acetate for 3 mo had epithelial cell damage in the glomerulus and proximal convoluted tubules, and increased plasma creatinine and urea levels (Llobet et al. 1988). A NOAEL 95 mg zinc/kg-d was identified.

Boric Acid

The free ion of boric acid, boron, is an essential nutrient for plants, and there is some evidence supporting essentiality in animals, including humans (see reviews, Woods 1994; Nielsen 1996). The essentiality of boron in humans is under consideration by the Institute of Medicine; however, no dietary intakes are currently recommended.

Information on the acute toxicity of boron compounds, including boric acid, in humans comes from severe poisonings, often related to old medical treatments or accidental exposures. Following ingestion of large amounts of boric acid, gastrointestinal symptoms occur first (nausea, vomiting, and diarrhea), followed by erythema, exfoliation, and desquamation of the skin (Moore 1997). No consistent lethal dose of boric acid has been reported in adults, but lethal doses in infants and children of 2-3 g and 5-6 g, respectively, have been reported (Moore 1997). Death occurred following ingestion of a large amount of boric acid (2.5% solution), which was accidently used instead of water to prepare an infant's formula (Wong et al. 1964, as cited in Moore 1997). A review of recent reports of boric acid poisoning at a number of poison centers indicate that acute boric acid ingestion generally produces minimal or no toxicity, with no cases of severe toxicity out of 784 poisonings. See Table 8-2 for a summary of oral toxicity studies of boric acid.

Weir and Fisher (1972) determined LD_{50} values for boric acid of 3.45, 4.08, and 3.16 g/kg (equivalent to 0.60, 0.71, and 0.55 g boron/kg) in male and female Sprague-Dawley rats, and male Long-Evans rats, respectively. The corresponding LD_{50} values for borax were 4.5, 5.0, and 6.1 g/kg (equivalent to 0.51, 0.56, and 0.69 g boron/kg). Administration of boric acid in the diet reduced weight gain and affected nucleotide levels in numerous tissues in rats treated with 10,000 ppm (500 mg boron/kg-d) for 21-27 d (Dani et al. 1971).

Changes in spermatogenesis were seen after a 14-d treatment with approximately 1 g/kg-d boron as boric acid (Silaev et al. 1977). Dose- and duration-dependent changes were seen in the activities of enzymes found in spermatogenic cells in male Sprague-Dawley rats treated with 60 or 125-131 mg boron/kg-d for 30 or 60 d (Lee et al. 1978).

In F-344 rats fed boric acid (0 or 60.9 mg boron/kg-d), inhibition of spermiation, significant loss of spermatocytes and spermatids from all tubules, and decreased testosterone concentrations occurred (Treinen and Chapin 1991). Krasovskii et al. (1976) dosed white random-bred rats (number of animals not given) with boric acid (0, 0.015, 0.05, and 0.3 mg boron/kg body weight) for 6 mo. Statistically significant decreases in mobility time, acid resistance, and osmotic resistance were seen at 0.3 mg boron/kg body weight. Mobility time and acid resistance were also decreased at 0.05 mg boron/kg body weight. The authors identified a NOAEL of 0.015 mg boron/kg and a LOAEL of 0.05 mg boron/kg.

The effects of a 14-d exposure to 0, 600, 1,200, 2,400, 4,900, or 9,800 ppm boric acid, and 0, 6,200, 12,500, 25,000, 50,000, or 100,000 ppm in the diet of B6C3F$_1$ mice (five/sex/group) were investigated in two studies conducted under the National Toxicology Program (NTP) (NTP 1987). No dietary intake levels were provided for those studies. Mortality and hyperplasia and/or dysplasia of the forestomach occurred at doses of 25,000 ppm or above.

B6C3F$_1$ mice (10/sex/group) were fed 0, 1,200, 2,500, 5,000, 10,000, or 20,000 ppm boric acid for 13 wk (NTP 1987). Unreliable feed consumption data precluded the estimation of the corresponding doses. Mortality was seen in the two highest concentration groups. The animals in the highest concentration group appeared nervous and dehydrated, and decreased weight gain occurred at the three highest doses. Mild-to-medium extramedullary hematopoiesis of the spleen occurred in all dose groups except the highest; stomach hyperkeratosis and/or acanthosis occurred in the highest dose group. Testicular degeneration or atrophy of the seminiferous tubules occurred at 5,000, 10,000, and 20,000 ppm boric acid.

Weir and Fisher (1972) fed Sprague-Dawley rats (10/sex/group) 0, 52.5, 175, 525, 1,750, and 5,250 ppm boron as boric acid or borax in the diet (0, 2.6, 8.8, 26.3, 87.5, or 262.5 mg boron/kg-d). Hematological, biochemical, urinary, necropsy, and histological analyses were performed. Overt toxicity was seen in rats in the two highest dose groups, including rapid respiration, inflamed eyes, swollen paws, and desquamated skin on paws and tails. All animals in the highest-dose group, and one animal in the 2.6- and 87.5-mg boron/kg-d group died. Doses above 26.3 mg boron/kg-d decreased growth and food consumption. Body weights and some organ weights were altered at 26.3 mg boron/kg-d in females, and were consistently altered at the higher dose levels. Upon necropsy, congestion of liver and kidneys, bright red lungs, small gonads, and a thickened pancreas were seen in several animals. There was an increase in ovary weight at 2.6 mg boron/kg-d as borax, whereas 262.5 mg boron/kg-d, given as borax or boric acid, caused a decrease in ovary weight. At 262.5 mg boron/kg-d, both compounds decreased testis/brain weight ratios. Complete atrophy of the testis occurred in one male at 2.6 mg boron/kg-d, in four males at 26.3 mg boron/kg-d as borax, and in all males dosed with 262.5 mg boron/kg-d as either compound. A LOAEL of 2.6 mg boron/kg-d was identified; no NOAEL was identified.

Weir and Fisher (1972) fed beagle dogs (five/sex/group) 17.5, 175, or 1,750 ppm boron as boric acid or borax for 90 d and assessed most of the same end points as in the rat study. No overt toxicity was seen except for one dog in the high-dose group which died of diarrhea. In the high-dose group, two male and three female dogs had decreased packed cell volume and hemoglobin values. There was a decrease in spleen/body weight ratio in one dog at the lowest dose (borax), and decreased testes/body weight ratio in the 175-ppm group (boric acid). Both compounds significantly decreased thyroid/body and testes/body weight ratios in the 1,750 ppm group. No hematological, biochemical, or urinary effects were seen. No effects were seen upon necropsy. Upon histological examination, the animals in the highest-borax-dose group had severe testicular atrophy, degeneration of the spermatogenic epithelium, red blood cell destruction, slightly greater proprortion of solid epithelial nests and minute follicles, and an increased width of the zona reticularis in the adrenal gland. The highest dose of boric acid also increased the width of the zona reticularis and decreased the width of the zona fasciculata in the adrenal gland, caused lymphoid infiltration of the thyroid (two females), and caused thyroid atrophy (one female) (Weir and Fischer 1972).

Weir and Fisher (1972) fed Sprague-Dawley rats diets containing 0 (70/sex), 117 (35/sex), 350 (35/sex), or 1,170 (35/sex) ppm boron in the diet as boric acid or borax (corresponding to 5.9, 17.5, and 58.5 mg boron/kg-d) for 2 yr. Hematological, biochemical, urinary and histopathological examinations, and a necropsy, were conducted as in the 90-d study (see above) on a subset of the animals. Animals treated with the highest dose of boric acid or borax had decreased food consumption, suppressed growth, overt evidence of toxicity, and some hematological disturbances. No histological effects were seen in the 117 ppm and 350 ppm groups. Atrophic testes and seminiferous epithelium, and decreased tubular size were seen in male rats dosed with 1,170 ppm boron as boric acid or borax. Testes weights and testes/body weight ratios were decreased at that dose. Brain/body and thyroid/body weight ratios were increased in the high-dose group.

Beagle dogs (four/sex/group) were fed 58, 117, and 350 ppm boron equivalents as boric acid or borax (Weir and Fisher 1972). There were no apparent effects on appearance, behavior, appetite, body weight, or food consumption. Organ weights and organ weight to body weight ratios were not changed from control levels, and the hematological and biochemical values in the treated animals were within the normal range of values. In a subsequent study, four dogs were exposed to 1,170 ppm boron equivalents as boric acid or borax for

38 wk. In that study, seminiferous tubular atrophy and decreased spermatogenesis were observed. A NOAEL of 8.8 mg boron/kg-d was identified. However, an expert committee (Moore 1997) that had access to the data from that study, noted a high incidence of effects in the control dogs (three out of four), including seminiferous tubule degeneration, decreased spermatogenesis, and atrophy. Therefore, the usefulness of that study (Weir and Fisher 1972) for quantitative risk assessment is questionable.

Research by Fail et al. (1989, as cited in ATSDR 1994) in CD-1 mice demonstrated testicular atrophy and decreased spermatogenesis after a 27-wk exposure to 4,500 or 9,000 mg boric acid/kg, but not 1,000 mg boric acid/kg. Fertility was also decreased in the middle-dose group, and the high-dose group was infertile. Fail et al. (1989, as cited in ATSDR 1994) also treated wild deer mice with the same concentrations of boric acid as the CD-1 mice; infertility was seen at the highest-dose group, along with decreased testis and accessory sex organ weights. LOAELs of 9,000 mg boric acid/kg and 4,500 mg boric acid/kg were identified in field mice and CD-1 mice, respectively. The NOAEL in field mice was 4,500 mg boric acid/kg; no NOAEL was established in the CD-1 mice.

Linder et al. (1990) investigated the dose- and time-response relationship for male reproductive effects in rats. Rats were dosed with a single dose and killed at various times following dosing (2, 14, 28, or 57 d), or dosed with a single dose at different concentrations (44, 87, 175, or 350 mg boron/kg) and killed after 14 d. Effects on epididymal sperm, testicular and seminiferous tubule spermatids, and spermiation were seen 14 d after exposure to 175 or 350 mg boron/kg. Those effects reversed when the exposure had been stopped for 57 d. Other studies in F-344 rats indicate that some reproductive effects of boron, such as inhibition of spermiation, are reversible, while others, such as focal atrophy, were not reversed when the exposure had been stopped for 32 wk (Ku et al. 1993).

In a 2-yr chronic bioassay by NTP (1987), B6C3F$_1$ mice (50/sex/group) were fed 0, 2,500, or 5,000 ppm boric acid (400-500 or 1,100-1,200 mg boric acid/kg). Survival was decreased in the low- and high-dose males and body weight gain was decreased in the high-dose males and females. No chemically related clinical symptoms were observed. All organs and tissues were examined by necropsy for gross lesions and histology was conducted. There were slight increases in non-neoplastic lesions in both sexes. The increases were not dose-dependent and were not found in both sexes, therefore it was concluded that they were not chemically related. There was a statistically significant increase in lung hemorrhage (control, 2/50; low dose, 5/50; and high dose, 12/50) in the high-dose females. Testicular atrophy (control 3/49; low dose, 6/50; high dose, 27/47; at 0, 2,500, and 5,000 ppm boric acid) and interstitial cell hyperplasia

(0/49; 0/50; 7/47) occurred in male mice. The LOAEL for systemic effects in that study is 2,500 ppm. No NOAEL was identified.

Reproductive and Developmental Effects

Zinc Borate

No studies on the reproductive or developmental effects of zinc borate following oral exposure were identified.

Zinc Oxide

In humans, no developmental effects were reported in newborns following ingestion of zinc compounds by mothers during the last two trimesters of pregnancy (Kynast and Saling 1986; Mahomed et al. 1989, as cited in ATSDR 1994; Simmer et al. 1991).

Schlicker and Cox (1968) administered 200 mg zinc/kg-d (as zinc oxide) in the diet of rats for 21 d prior to mating and during gestation. In the treated group, the incidence of resorptions was 4.3% compared with 0% in controls. When the dose was reduced to 100 mg zinc/kg-d, 21 d prior to mating, no fetal resorptions, malformations, or growth reduction were reported.

Administration of 200 mg zinc/kg-d (as zinc oxide) to dams throughout gestation resulted in decreased growth and decreased tissue levels of copper and iron in fetal rats (Cox et al. 1969; Schlicker and Cox 1968). No measurable effect on gestational length or litter size was reported when female mink were fed a time-weighted average dose of 20.8 mg zinc/kg-d as zinc sulfate for 25 wk (Bleavins et al. 1983). Mice fed 1,110 mg zinc/kg-d for 13 wk did not have any histological alterations in the testes or ovaries (Maita et al. 1981).

Boric Acid

In a three-generational study, Weir and Fisher (1972) fed boric acid and borax (117, 350, and 1,170 ppm boron; calculated as 5.9, 17.5, and 58.5 mg boron/kg-d) to Sprague-Dawley rats (8 males and 16 females/group). Animals were dosed for 14 wk prior to mating. Doses of 117 and 350 ppm had no effect on litter size, weight of progeny, or general appearance, but the overall fertility index was increased in those two groups as compared with controls. No pregnancies occurred in animals in the highest-dose group. Decreased fertility, decreased live births, and decreased offspring survival (no offspring survived to weaning) occurred when treated females were mated with control males. The LOAEL from that study is 1,170 ppm (58.5 mg boron/kg-d) and the NOAEL is 350 ppm (17.5 mg boron/kg-d).

NTP has conducted multigenerational continuous-breeding studies in Swiss CD-1 mice (NTP 1990; Fail et al. 1991). Male and female mice (F_0) were fed boric acid (control, 1,000, 4,500, or 9,000 mg/kg feed; equivalent to 0, 19.2, 104.7, and 222.1 mg boron/kg-d for males and 0, 31.9, 148.1, and 290.5 mg boron/kg-d for females) for 27 wk. Animals (F_0) in the mid-dose group had decreased numbers of litters/pair, decreased live pups per litter, and decreased pup weight; animals in the high-dose group were infertile. Cross-mating experiments indicated that the decreased fertility was associated with adverse reproductive effects on the male. Treated animals (F_0) had decreased sperm motility at all dose levels. Significant increases in uterine weight, and kidney plus adrenal weight were seen in the female offspring (F_1) of the low-dose animals, and a decrease in birth weight (3.3%) was seen in the F_2 offspring of this dose group. A LOAEL of 19.2 mg boron/kg-d was identified in this study based on effects in the low-dose males. No NOAEL could be identified.

Heindel et al. (1992) investigated developmental effects in Sprague-Dawley rats with boron as boric acid in feed (0, 13.6, 28.5, or 57.7 mg boron/kg-d) on gestational d 0-20. Another treatment group received 94.2 mg boron/kg-d on gestational d 6-15. Maternal effects (increased liver and kidney weights relative to body weights) were seen at and above 28.5 mg boron/kg-d. Fetal body weight was decreased in a dose-dependent manner, with significant decreases seen at all doses. Significant fetal malformations, including effects on the eyes, central nervous system, cardiovascular system, and axial skeleton, occurred at and above 28.5 mg/kg-d. The percentage of malformations per litter was reduced relative to controls in the 13.6 and 28.5 mg boron/kg-d dose groups. However, skeletal effects were significantly increased at 94.2 mg boron/kg-d. Treatment for part of gestation with 94.2 mg boron/kg-d resulted in increased prenatal mortality. The lowest dose in that study, 13.6 mg boron/kg-d, is a LOAEL in the study.

Price et al. (1996a) conducted a follow-up study to the study by Heindel et al. (1992). The study consisted of one phase to determine a NOAEL and a second phase to investigate the reversibility of the effects of boric acid on fetal body weight. Both phases consisted of two replicate studies. In each replicate of the phase 1 study, pregnant Sprague-Dawley rats (14-17/group) were fed diets containing boric acid (0, 3.3, 6.3, 9.6, 13.3, or 25 mg boron/kg-d) on gestational d 0-20. There were no maternal deaths or overt signs of maternal toxicity from the treatments. A slight increase in relative maternal right kidney weight was seen, but only on gestational d 20. Boric acid did not affect maternal weight gain, liver weight, or percentage of dams delivering pups, the number of ovarian corpora lutea-dam, implantation sites/litter, live litter size, percentage of resorptions, or late fetal deaths. On gestational d 20, significant reductions in fetal body weight were seen at the two highest dose levels (6% at 13.3 mg boron/kg-d and 12% at 25 mg boron/kg-d). A low incidence of external malformations, and visceral malformations and variations occurred on gestational d 20. A significant increasing trend in the percentage of fetuses with skeletal malformations per litter was seen at the two highest dose levels, but there were no significant differences in the overall incidence of skeletal variations. However, at the two highest doses, there was a significant increase in short rib XIII and wavy ribs. In phase 2 of the study, dams and offspring were kept alive until postnatal d 21. There was no difference in body weight between boron exposed and control animals on postnatal d 21. Overall incidence of skeletal malformations was increased in the low- and high-dose groups on postnatal d 21. A dose-dependent increase in the incidence of short rib occurred at postnatal d 21, but the authors concluded that this effect was treatment related only at the highest dose. There was no association between treatment and wavy rib at postnatal d 21. Based on phase 1 effects, Price et al. (1996a) identified a LOAEL of 13.3 mg boron/kg-d and a NOAEL of 9.6 mg boron/kg-d.

Developmental toxicity studies were also conducted in mice (Heindel et al. 1992). Mice fed boric acid in their diet (0, 43, 79, or 175 mg boron/kg-d) had a dose-dependent decrease in fetal body weight, with statistically significant decreases at the two highest doses. Increased skeletal malformations also occurred at those doses. However, maternal toxicity was also seen at those doses. A LOAEL of 79 mg boron/kg-d and a NOAEL of 43 mg boron/kg-d were identified for developmental effects in mice.

Price et al. (1996b) gavaged New Zealand white rabbits with 0, 10.9, 21.9, or 43.7 mg boron/kg-d. Prenatal survival decreased and the number of malformations increased at 43.7 mg boron/kg-d. Cardiovascular defects contributed the most to the malformations. The LOAEL was 43.7 mg boron/kg-d and the NOAEL was 21.9 mg/kg-d.

Developmental and reproductive end points are the most sensitive effects for boron compounds following oral exposure. The lowest NOAEL identified was 8.8 mg boron/kg-d in the dog study by Weir and Fisher (1972). However, in the report by Moore (1997), questions were raised about that study due to a high level of abnormalities in the control group. The number of animals was also quite small in that study. Therefore, the study by Price et al. (1996a), which provided the next lowest values, with a LOAEL of 13.3 mg boron/kg-d and a NOAEL of 9.6 mg boron/kg-d, based on developmental effects, is the critical study for the reproductive and developmental effects of boron.

Immunological Effects

No studies were identified that investigated the immunological effects of zinc borate or boric acid following oral exposure.

Eleven healthy adult men who ingested 4.3 mg zinc/kg-d for 6 wk experienced impaired mitogenic response elicited from peripheral blood lymphocytes and impaired chemotactic and phagocytic responses of polymorphonuclear leukocytes. No effects were observed on total number of lymphocytes or relative number of T cells or B cells (Chandra 1984). Zinc deficiency is generally associated with adverse effects on the immune system in humans and animals (see Prasad et al. 1978; Thurnham 1997; Shankar and Prasad 1998 for reviews).

Neurological Effects

No studies were identified that investigated the neurological effects of zinc borate or boric acid following oral exposure.

Murphy (1970) reported that a 16-yr-old boy developed symptoms of lethargy, light-headedness, staggering, and difficulty in writing clearly following ingestion of about 86 mg zinc/kg-d (as metallic zinc) over a 2-d period in an effort to promote the healing of a wound.

Limited data were located on neurological effects of zinc compounds in animals. Rats dosed with 487 mg zinc/kg-d as zinc oxide for 10 d (Kozik et al. 1980; as cited in ATSDR 1994) experienced minor neuronal degeneration and proliferation of oligodendroglia. Kozick et al. (1981; as cited in ATSDR 1994) reported that rats receiving 472 mg zinc/kg-d as zinc oxide for 10 d had increased levels of secretory material in the neurosecretory nuclei of the hypothalamus.

Cancer

Zinc Borate

No studies were located that investigated the carcinogenicity of zinc borate following oral exposure.

Zinc Oxide

Two epidemiological studies report conflicting results on the association between high zinc soil levels and cancer. In a survey of cancer registry data

(1954-1978) in Shipham, Somerset (Great Britain), an area with a high soil zinc-to-copper ratio (≈17:1), the gastric cancer incidence rate was significantly lower than the regional rate (Philipp et al. 1982). In contrast, Stocks and Davies (1964, as cited in ATSDR 1994) found an association between an excess rate of gastric cancer in people of North Wales and high zinc-to-copper ratios (≈30:1) in the soil of household gardens.

Walters and Roe (1965) reported no increased incidence of tumors following exposure of mice to 951 mg zinc/kg-d as zinc sulfate in drinking water for 1 yr compared to controls. However, the report of that study lacked important details and the study had several limitations, including a high mortality rate in control mice.

Boric Acid

Weir and Fisher (1972) conducted a 2-yr study in which they fed Sprague-Dawley rats (35/sex/group) 0, 117, 350, or 1,170 ppm of boron (0, 67, 200, and 669 mg boron/kg-d) as boric acid or borax. That study, however, was not designed as a cancer bioassay (see Systemic Effects section for study details). A limited number of tissues were examined histopathologically, and the report failed to even mention tumor findings.

Following dose-finding studies (see Systemic Effects section), the NTP conducted a chronic cancer bioassay (NTP 1987). $B6C3F_1$ mice (50/sex/group) were fed diets containing 0, 2,500, or 5,000 ppm boric acid (0, 400-500, or 1,100-1,200 mg/kg-d as measured by NTP). In males dosed with 400-500 mg/kg-d, there was an increase (not statistically significant) in hepatocellular carcinomas (5/50, 12/50, and 8/49 at 0, 400-500, and 1,100-1,200 mg/kg-d, respectively) and hepatocellular adenomas or carcinomas (14/50, 19/50, and 15/49 at 0, 400-500, and 1,100-1,200 mg/kg-d, respectively). Since the increases were not statistically significant and they did not occur in a dose-dependent manner, and since there is variability in historical controls for that tumor type, the NTP did not consider those tumors to be related to boric acid. NTP concluded that there was no evidence of carcinogenicity for boric acid (NTP 1987).

Genotoxicity

Zinc Borate

Zinc borate was not mutagenic in the Ames *Salmonella* mutagenicity bioassay, with or without metabolic activation (U.S. Borax 1996).

Zinc Oxide

Bauchinger et al. (1976) found an increased incidence of chromosomal aberrations in 24 workers in a zinc smelting plant compared with controls. However, those workers also had increased blood levels of lead and cadmium; authors attributed the increase in the incidence of chromosome aberrations to cadmium exposure.

A number of in vivo studies have reported that zinc salts are clastogenic when administered by various routes of exposure (Table 8-3). Many of those studies were carried out with high exposures (>10 mg Zn/kg/d) or with intraperitoneal injection. Inhalation exposure of mice to zinc oxide induced chromosomal aberrations in bone marrow cells, but no exposure concentrations were given (Voroshilin et al. 1978; as cited in ATSDR 1994).

Zinc sulfate and zinc chloride have been found to be negative for mutagenic activity in in vitro bacterial and mammalian systems (ATSDR 1994).

TABLE 8-3 Genotoxicity of Zinc Compounds In Vivo

Species	Zinc Compound	Genotoxicity End Point	Results	Reference
Mouse	Zinc salts	Chromosomal aberrations	Positive	Vilkina et al. 1978, as cited in ATSDR 1994
Mouse	Zinc chloride	Micronucleus test	Positive	Gocke et al. 1981
Mouse	Zinc chloride	Chromosomal aberrations	Positive[a]	Deknudt and Gerber 1979
Mouse	Zinc chloride	Chromosomal aberrations	Positive	Gupta et al. 1991
Rat	Zinc chlorate	Chromosomal aberrations	Positive	Kowalska-Wochna et al. 1988, as cited in ATSDR 1994
Rat	Zinc chlorate	Sister chromatid exchange	Positive	Kowalska-Wochna et al. 1988, as cited in ATSDR 1994
Drosophila	Zinc chloride	Sex-linked recessive lethal mutations	Negative	Gocke et al. 1981

[a]Zinc chloride was genotoxic only when animals were fed a low calcium diet.

Boric Acid

The weight of evidence indicates that boron compounds are not genotoxic (Table 8-4).

TABLE 8-4 Genotoxicity of Boron Compounds

Test System	Boron Compound	Genotoxicity End Point	Result	Reference
Escherichia coli Blr/Scl-4	Boric acid	Gene mutations	Positive	Demerec et al. 1951
Salmonella typhimuirum	Boric acid, borax	Gene mutations	Negative	NTP 1987
Salmonella typhimuirum	Boric acid	Gene mutations	Negative	Stewart 1991
Human fibroblasts	Borax	Gene mutations	Negative	Landolph 1985
V79 Chinese hamster ovary cells	Borax	Gene mutations	Weakly positive	Landolph 1985
C3H/10T1/2	Borax	Gene mutations	Negative	Landolph 1985
Chinese hamster ovary cells	Boric acid, borax	Sister-chromatid exchange and chromosomal aberrations	Negative	NTP 1987
Mouse lymphoma cells	Boric acid	Gene mutations	Negative	Rudd 1991, as cited in Moore 1997
Mouse lymphoma cells	Boric acid, borax	Gene mutations	Negative	NTP 1987
F-344 rat	Boric acid	Hepatocyte unscheduled DNA synthesis	Negative	Bakke 1991, as cited in Moore 1997
Swiss-Webster mice	Boric acid	Micronucleus test	Negative	O'Loughlin 1991, as cited in Moore 1997

QUANTITATIVE TOXICITY ASSESSMENT

Noncancer

Dermal Assessment

There are inadequate dermal toxicity data on zinc borate, zinc oxide, or boric acid to derive a risk value for dermal exposure.

Inhalation RfC

There are inadequate inhalation toxicity data on zinc borate, zinc oxide, or boric acid to derive an inhalation RfC.

Oral RfD

There are inadequate oral toxicity data on zinc borate to derive an oral RfD. However, zinc borate readily breaks down in the stomach to zinc oxide and boric acid. The subcommittee used the available dose-response data for both compounds to derive their RfDs and selected the more conservative RfD of the two values to characterize the health risk of zinc borate.

Zinc Oxide

The EPA has established an oral RfD of 0.3 mg/kg-d for zinc (EPA 1999). That RfD is based on a LOAEL of 1.0 mg zinc/kg-d which was identified by Yadrick et al. (1989) who reported a decrease in erythrocyte superoxide dismutase activity in adult women following 10 wk of exposure to zinc supplements. The change in the enzyme activity reflects an alteration in copper levels. That RfD is supported by data from several other clinical studies demonstrating the effect of zinc on copper balance (Fischer et al. 1984; Prasad et al. 1978). The LOAEL was divided by an uncertainty factor of 3 (less than the default factor of 10 because the effects are not severe and zinc is an essential micronutrient) to yield an RfD of 0.3 mg zinc/kg-d.

Boric Acid

Developmental toxicity appears to be the most sensitive endpoint for boric acid. Price et al. (1996a) identified the lowest LOAEL and the highest NOAEL (see Hazard Identification section). A LOAEL of 13.3 mg boron/kg-d, based on decreased fetal body weight and skeletal malformations in Sprague-Dawley rats, and a NOAEL of 9.6 mg boron/kg-d were identified. Heindel et al. (1992) identified a LOAEL of 13.6 boron/kg-d based on developmental effects.

Allen et al. (1996) calculated a benchmark dose for the developmental toxicity data associated with boric acid exposure in rats and determined that fetal weight, averaged within litters, is the most sensitive, relevant end points to use. Combining the data from Price et al. (1996a) and Heindel et al. (1992), a benchmark dose (BMD_5) of 59 mg boric acid/kg-d (10.3 mg boron/kg-d) was calculated for a 5% decrease in fetal weight. An expert scientific panel (consisting of 28 individuals from academia, industry, and government) concluded that the benchmark dose estimated by Allen et al. (1996) was appropriate (Moore 1997). The subcommittee agrees that a benchmark dose (BMD_5) of 10.3 mg boron/kg-d is appropriate for use in this risk assessment.

The default uncertainty factor for the derivation of the RfD for boric acid would be 100, composed of a factor of 10 for interspecies extrapolation and 10 for intraspecies variability (further broken down to a factor of 4 for toxicokinetic considerations and 2.5 for toxicodynamic considerations). It has been suggested that, in the case of boron compounds, the database is sufficient to warrant adjusting down the UF based on similar absorption, distribution, and elimination of boron in rats and humans, and lack of metabolism in either rats or humans. The International Programme on Chemical Safety (IPCS) (1998) used a total uncertainty factor of 25 when calculating their tolerable intake level for boron. An ad hoc panel, which included representatives from the EPA and Health Canada, that reviewed the data on the elimination of boron has suggested that a total UF of 60 be used (Dourson et al. 1998). The subcommittee, however, decided that an uncertainty factor of 30 should be used. Division of the BMD_5 of 10.3 mg boron/kg-d by the UF of 30 yields an RfD of 0.3 mg boron/kg-d for boric acid. The derivation of the oral RfD for boric acid is summarized in Table 8-5. Confidence in the critical studies for boric acid (Heindel et al. 1992; Price et al. 1996a) is high because they are well-conducted studies. There is a large body of literature indicating developmental and reproductive effects within the same order of magnitude, therefore, the confidence in the overall database is high. Therefore, the subcommittee has high confidence in the oral RfD for boric acid high.

Zinc Borate

In order to derive an oral RfD for zinc borate from the RfDs for zinc compounds and boric acid, the relative contributions of zinc and boron to zinc borate were determined. Boron comprises approximately 11.3% (w/w) of zinc

TABLE 8-5 Oral Reference Dose for Boric Acid

Critical effect	Species	Effect level (mg boron/kg-d)	Uncertainty factors	RfD	References
Fetal weight averaged within litters	Sprague-Dawley rats	BMD_5: 10.3	UF_A: 3 UF_H: 10 Total: 30	0.3 mg boron/kg-d	Heindel et al. 1992; Price et al. 1996a[a]

BMD_5, benchmark dose for a 5% effect; RfD, reference dose; UF_A, extrapolation from animals to humans; UF_H, intraspecies variability.

[a]The BMD_5 was calculated using the data from both of these studies (Allen et al. 1996).

borate ($3ZnO{:}2B_2O_3$) (Lide 1991-1992). The RfD for zinc borate, based on the RfD for boron (0.34 mg boron/kg-d), would be approximately 3.0 mg zinc borate/kg-d. Zinc comprises approximately 51.2% (w/w) of zinc borate ($3ZnO{:}2B_2O_3$). The RfD for zinc borate, based on the RfD for zinc (0.3 mg zinc/kg-d), would be approximately 0.6 mg zinc borate/kg-d.

The oral RfD for zinc borate based on zinc is more conservative than the RfD for zinc borate based on boron. Therefore, the overall oral RfD for zinc borate is 0.6 mg zinc borate/kg-d, based on the RfD for zinc.

The subcommittee also considered the possibility of additive or synergistic effects between zinc and boron when assessing the risks associated with exposure to zinc borate. The main effects of boron are reproductive and developmental effects. Zinc has been shown to have reproductive effects in animals, but only at extremely high doses (≥ 200 mg Zn/kg-d). Effects other than reproductive and developmental have only been seen following exposure to boric acid at very high doses (≥ 1,000 ppm-d). The subcommittee concluded that additive or synergistic effects are not expected because of the lack of target overlap.

The subcommittee's confidence in the zinc RfD is medium. The lack of data on the parent compound makes the overall confidence in the oral RfD (0.6 mg zinc borate/kg-d) low.

Cancer

The potential carcinogenicity of zinc borate, zinc oxide, and boric acid cannot be determined because of inadequate carcinogenicity data from any route of exposure.

EXPOSURE ASSESSMENT AND RISK CHARACTERIZATION

Noncancer

Dermal Exposure

The assessment of noncancer risk by the dermal route of exposure is based on the scenario described in Chapter 3. This exposure scenario assumes that an adult spends 1/4th of his or her time sitting on furniture upholstery treated with zinc borate, that 1/4th of the upper torso is in contact with the upholstery, and that clothing presents no barrier. Zinc borate is considered to be ionic, and is essentially not absorbed through the skin. However, to be conservative, the subcommittee assumed that ionized zinc borate permeates the skin at the same rate as water, with a permeability rate of 10^{-3} cm/hr (EPA 1992). Using that permeability rate, the highest expected application rate for zinc borate (2 mg/cm^2), and Equation 1 in Chapter 3, the subcommittee calculated a dermal exposure level of 6.3×10^{-3} mg/kg-d. The oral RfD for zinc borate (0.6 mg/kg-d; see Oral RfD in Quantitative Toxicity section) was used as the best estimate of the internal dose for dermal exposure. Dividing the exposure level by the oral RfD yields a hazard index of 1.0×10^{-2}. Thus it was concluded that zinc borate used as a flame retardant in upholstery fabric is not likely to pose a noncancer risk by the dermal route.

Inhalation Exposure

Particulates

The assessment of the noncancer risk by the inhalation route of exposure is based on the scenario described Chapter 3. This scenario corresponds to a person spending 1/4th of his or her life in a room with low air-change rate (0.25/hr) and with a relatively large amount of fabric upholstery treated with zinc borate (30 m^2 in a 30-m^3 room), with this treatment gradually being worn away over 25% of its surface to 50% of its initial quantity over the 15-yr life-time of the fabric. A small fraction, 1%, of the worn-off zinc borate is released into the indoor air as inhalable particles and may be breathed by the occupant. Equations 4 through 6 in Chapter 3 were used to estimate the average concentration of zinc borate present in the air. The highest expected application rate for zinc borate is about 2 mg/cm^2. The estimated release rate for zinc borate is

2.3×10^{-7}/d. Using those values, the estimated time-averaged exposure concentration for zinc borate is 0.19 μg/m^3.

Although lack of sufficient data precludes deriving an inhalation RfC for zinc borate, the oral RfD (0.6 mg zinc borate/kg-d; see Oral RfD in Quantitative Toxicity section), which represents a conservative estimate, was used to estimate an RfC of 2.1 mg/m^3 (see Chapter 4 for the rationale).

Division of the exposure concentration (0.19 μg/m^3) by the estimated RfC (2.1 mg/m^3) results in a hazard index of 9.1×10^{-5}. Therefore, the subcommittee concluded that, under the worst-case exposure scenario, exposure to zinc borate particles from its use as an upholstery fabric flame retardant is not likely to pose a noncancer risk.

Vapor

In addition to the possibility of release of zinc borate in particles worn from upholstery fabric, the subcommittee considered the possibility of its release by evaporation. However, because of zinc borate's negligible vapor pressure at ambient temperatures, the subcommittee concluded that exposure to zinc borate vapors from its use as an upholstery fabric flame retardant is not likely to pose a noncancer risk.

Oral Exposure

The assessment of the noncancer risk by the oral exposure route is based on the scenario described in Chapter 3. The exposure assumes a child is exposed to zinc borate through sucking on 50 cm^2 of fabric, backcoated with zinc borate, daily for two yr, one hr/d. The highest application rate for zinc borate is 2 mg/m^2. A fractional rate (per unit time) of zinc borate extraction by saliva is estimated as 0.001/d, based on leaching of antimony from polyvinyl chloride cot mattresses (Jenkins et al. 1998). Using those assumptions in Equation 15 in Chapter 3, the average oral dose rate was estimated to be 0.00017 mg/kg-d. Division of that exposure estimate (0.00017 mg/kg-d) by the oral RfD (0.6 mg/kg-d; see Oral RfD in Quantitative Toxicity Assessment Section) results in a hazard index of 2.8×10^{-4}. Therefore, under the worst-case exposure assumptions, zinc borate, used as a flame retardant in furniture-upholstery fabric, is not likely to pose a noncancer risk by the oral route of exposure.

Cancer

There are inadequate data to characterize the carcinogenic risk from exposure to zinc borate, zinc oxide, or boric acid from any route of exposure.

RECOMMENDATIONS FROM OTHER ORGANIZATIONS

The U.S. Environmental Protection Agency (EPA), as detailed in the Integrated Risk Information System (IRIS), has established an oral RfD for boron of 0.09 mg/kg-d (EPA 1999). The risk assessment for boron has not been updated since 1989. However, that RfD is currently under review and a revised RfD is expected in the yr 2000 (Fed. Regist. 63(December 10, 1998):68353-68364).

The International Programme on Chemical Safety (IPCS) has an Environmental Health Criteria document (IPCS 1998) on boron in which a tolerable intake for boron is set at 0.4 mg boron/kg-d based on recent reproductive and developmental data.

The Agency for Toxic Substances and Disease Registry (ATSDR) has published toxicological profiles for zinc and boron. An oral minimal risk level (MRL) of 0.3 mg Zn/kg-d for zinc was based on hematological effects (ATSDR 1994). ATSDR lists an oral MRL for boron of 0.01 mg boron/kg-d based on developmental effects (ATSDR 1992).

The National Research Council (NRC) has established a recommended dietary allowance (RDA) for zinc of 12-15 mg/d (0.17-0.21 mg/kg-d of zinc for a 70-kg person (NRC 1980).

The Occupational Safety and Health Administration (OSHA) and the American Conference of Government Industrial Hygienists (ACGIH) considered the toxicity from zinc borate and its components in the workplace to be due to "Particulate Not Otherwise Classified" or "Nuisance Dust." Therefore, the OSHA permissible exposure level is 15 mg/m^3 for total dust and 5 mg/m^3 for respirable dust, and the ACGIH Threshold Limit Value for zinc oxide dust is 10 mg/m^3 (ACGIH 1999).

DATA GAPS AND RESEARCH NEEDS

There are little toxicity data available for zinc borate. Once in the body, zinc borate readily breaks down to zinc oxide and boric acid. There are no chronic studies investigating the carcinogenicity of zinc oxide and boric acid. There are no studies that measured exposure levels from the use of zinc borate as a flame retardant in upholstery furniture fabric. However, there are extensive databases on the toxicity of zinc oxide and boric acid, and the hazard indices for zinc borate, based on those data, are less than one for all three routes of exposure, using the subcommittee's conservative assumptions. Therefore, the subcommittee concluded that no further research is needed to assess the health risks from the use of zinc borate as a flame retardant.

REFERENCES

ACGIH (American Conference of Government Industrial Hygienists). 1999. Threshold Limit Values and Biological Exposure Indices. Cincinnati, OH: American Conference of Government Industrial Hygienists, Inc.

Agren, M.S. 1990. Percutaneous absorption of zinc from zinc oxide applied topically to intact skin in man. Dermatologica 180:36-39.

Agren, M.S. 1991. Influence of two vehicles for zinc oxide on zinc absorption through intact skin and wounds. Acta Derm. Venereol. 71(2):153-156.

Agren, M.S., M. Krusell, and L. Franzen. 1991. Release and absorption of zinc from zinc oxide and zinc sulfate in open wounds. Acta Derm. Venereol. 71(4):330-333.

Allen, B.C., P.L. Strong, C.J. Price, S.A. Hubbard, and G.P. Daston. 1996. Benchmark dose analysis of developmental toxicity in rats exposed to boric acid. Fundam. Appl. Toxicol. 32(2):194-204.

Amdur, M.O., J.F. McCarthy, and M.W. Gill. 1982. Respiratory response of guinea pigs to zinc oxide fume. Am. Ind. Hyg. Assoc. J. 43(12):887-889.

Ameille, J., J.M. Brechot, P. Brochard, F. Capron, and M.F. Dore. 1992. Occupational hypersensitivity pneumonitis in a smelter exposed to zinc fumes. Chest 101(3):862-863.

ATSDR (Agency for Toxic Substances and Disease Registry). 1992. Toxicological Profile for Boron. U.S. Department of Health and Human Services, Public Health Service, Agency for Toxic Substances and Disease Registry, Atlanta, GA. 86 pp.

ATSDR (Agency for Toxic Substances and Disease Registry). 1994. Toxicological Profile for Zinc (Update). U.S. Department of Health and Human Services, Public Health Service, Agency for Toxic Substances and Disease Registry, Atlanta, GA. TP-93/15. 230 pp.

Bakke, J.P. 1991. Evaluation of the potential of boric acid to induce unscheduled DNA synthesis in the in vitro hepatocyte DNA repair assay using the male F-344 Rat. SRI International (Study No. 2389-V500-91 (EPA MRID 420389-03).

Batchelor, R.P., J.W. Fehnel, R.M. Thompson, et al. 1926. A clinical and laboratory investigation of the effect of metallic zinc, of zinc oxide, and of zinc sulphide upon the health of workmen. J. Ind. Hyg. 8:322-363.

Bauchinger, M., E. Schmid, H.J. Einbrodt, and J. Dresp. 1976. Chromosome aberrations in lymphocytes after occupational exposure to lead and cadmium. Mutat. Res. 40(1):57-62.

Birmingham, D.J., and M.M. Key. 1963. Preliminary survey: U.S. Borax plant, California. Cincinnati, Ohio, Occupational Health Research and Training Facility, Division of Occupational Health.

Blanc, P., H. Wong, M.S. Bernstein, and H.A. Boushey. 1991. An experimental human model of metal fume fever. Ann. Intern. Med. 114(11):930-936.

Blanc, P.D., H.A. Boushey, H. Wong, S.F. Wintermeyer, and M.S. Bernstein. 1993. Cytokines in metal fume fever. Am. Rev. Respir. Dis. 147(1):134-138.

Bleavins, M.R., R.J. Aulerich, J.R. Hochstein, T.C. Hornshaw, and A.C. Napolitano. 1983. Effects of excessive dietary zinc on the intrauterine and postnatal development of mink. J. Nutr. 113(11):2360-2367.

Budavari, S., M.J. O'Neil, A. Smith, and P.E. Heckelman eds. 1989. The Merck Index. 11th Edition. Rahway, N.J.: Merck and Co., Inc.

Chandra, R.K. 1984. Excessive intake of zinc impairs immune responses. JAMA 252(11):1443-1446.

Cox, D.H., S.A. Schlicker, and R.C. Chu. 1969. Excess dietary zinc for the maternal rats, and zinc, iron, copper, calcium, and magnesium content and enzyme activity in maternal and fetal tissues. J. Nutr. 98(4):459-466.

Culver, B.D., P.T. Shen, T.H. Taylor, A. Lee-Feldstein, H. Anton-Culver, and P.L. Strong. 1994. The relationship of blood- and urine-boron to boron exposure in borax-workers and the usefulness of urine-boron as an exposure marker. Environ. Health Perspect. 102(Suppl. 7):133-137.

Dani, H.M., H.S. Saini, I.S. Allag, B. Singh, and K. Sareen. 1971. Effect of boron toxicity on protein and nucleic acid contents of rat tissue. Res. Bull. Panjab. Univ. Sci. 22(1/2):229-235.

Deknudt, G., and G.B. Gerber. 1979. Chromosomal aberrations in bone-marrow cells of mice given a normal or a calcium-deficient diet supplemented with various heavy metals. Mutat. Res. 68(2):163-168.

Demerec, M., G. Bertani, and J. Flint. 1951. A survey of chemicals for mutageic action on *E. coli*. Am. Nat. 85:119-136.

Dourson, M., A. Maier, B. Meek, A. Renwick, E. Ohanian, and K. Poirier. 1998. Boron tolerable intake: Re-evaluation of toxicokinetics for data-derived uncertainty factors. Biol. Trace Elem. Res. 66(1-3):453-463.

Draize, J.H., and E.A. Kelly. 1959. The urinary excretion of boric acid preperations following oral administration and topical applications to intact and damaged skin of rabbits. Toxicol. Appl. Pharmacol. 1:267-276.

Drinker, K., and P. Drinker. 1928. Metal fume fever: V. Results of the inhalation by animals of zinc and magnesium oxide fumes. J. Ind. Hyg. 10:56-70.

DuBray, E.S. 1937. Chronic zinc intoxication. JAMA 108:383-385.

EPA (U.S. Environmental Protection Agency). 1992. Dermal Exposure Assessment: Principles and Applications. EPA/600/8-91-011B. Office of Health and Environmental Assessment, U.S. Environmental Protection Agency, Washington, D.C.

EPA (U.S. Environmental Protection Agency). 1999. Integrated Risk Information System. Boron. Last updated 1995. [Online]. Available: http://www.epa.gov/iris/subst/0410.htm

Evans, E.H. 1945. Casualties following exposure to zinc chloride smoke. Lancet ii:368-370.

Fail, P.A., J.D. George, H.R. Sauls, S.W. Dennis, and J.C. Seely. 1989. Effect of boric acid on reproduction and fertility of rodents. Adv. Contracept. Delivery Syst. 5:186-196.

Fail, P.A., J.D. George, J.C. Seely, T.B. Grizzle, and J.J. Heindel. 1991. Reproductive toxicity of boric acid in Swiss (CD-1) mice: Assessment using the continuous breeding protocol. Fundam. Appl. Toxicol. 17(2):225-239.

Farrell, F.J. 1987. Angioedema and urticaria as acute and late phase reactions to zinc fume exposure, with associated metal fume fever-like symptoms. Am. J. Ind. Med. 12(3):331-337.

Fischer, P.W., A. Giroux, and M.R. L'Abbe. 1984. Effect of zinc supplementation on copper status in adult man. Am. J. Clin. Nutr. 40(4):743-746.

Friis-Hansen, B., B. Aggerbeck, and J.A. Jansen. 1982. Unaffected blood boron levels in newborn infants treated with a boric acid ointment. Food Chem. Toxicol. 20(4):451-454.

Galvez-Morros, M., O. Garcia-Martinez, A.J.A. Wright, and S. Southon. 1992. Bioavailability in the rat of zinc and iron from the basic salts $Zn_5(OH)_8Cl_2.H_2O$, $Fe(OH)SO_4$, and $Fe_4(OH)_{11}NO_3.2H_2O$. Food Chem. 43:377-381.

Garabrant, D.H., L. Bernstein, J.M. Peters, and T.J. Smith. 1984. Respiratory and eye irritation from boron oxide and boric acid dusts. J. Occup. Med. 26(8):584-586.

Garabrant, D.H., L. Bernstein, J.M. Peters, T.J. Smith, and W.E. Wright. 1985. Respiratory effects of borax dust. Br. J. Ind. Med. 42(12):831-837.

Gocke, E., M.T. King, K. Eckhardt, and D. Wild. 1981. Mutagenicity of cosmetics ingredients licensed by the European Communities. Mutat. Res. 90(2):91-109.

Goyer, R.A. 1996. Toxic Effects of Metals. Pp. 691-736 in Casarett and Doull's Toxicology: The Basic Science of Poisons, C.D. Klaassen, M.O. Amdur, and J. Doull, eds. New York: Macmillan Publishing Company.

Gupta, T., G. Talukder, and A. Sharma. 1991. Cytotoxicity of zinc chloride in mice in vivo. Biol. Trace Elem. Res. 30(2):95-101.

Hallmans, G. 1977. Treatment of burns with zinc-tape. A study of local absorption of zinc in humans. Scand. J. Plast. Reconstr. Surg. 11(2):155-161.

Hamdi, E.A. 1969. Chronic exposure to zinc of furnace operators in a brass foundry. Br. J. Ind. Med. 26(2):126-134.

Hammond, J.W. 1944. Metal fume fever in crushed stone industry. J. Ind. Hyg. Toxicol. 26:117.

Hatlelid, K. 1999. Toxicity review for zinc borate. Memorandum, dated March 2, 1999, from Kristina Hatlelid, Toxicologist, Division of Health Sciences, to Ronald Medford, Assistant Executive Director for Hazard Identification and Reduction. U.S. Consumer Product Safety Commission, Washington, DC.

He, L.S., X.S. Yan, and D.C. Wu. 1991. Age-dependent variation of zinc-65 metabolism in LACA mice. Int. J. Radiat. Biol. 60(6):907-916.

Heilmaier, H.E., G.A. Drasch, E. Kretschmer, and K.H. Summer. 1987. Metallothionein, cadmium, copper and zinc levels of human and rat tissues. Toxicol. Lett. 38(3):205-211.

Heindel, J.J., C.J. Price, E.A. Field, M.C. Marr, C.B. Myers, R.E. Morrissey, and B.A. Schwetz. 1992. Developmental toxicity of boric acid in mice and rats. Fundam. Appl. Toxicol. 18(2):266-277.

Hjortso, E., J. Qvist, M.I. Bud, J.L. Thomsen, J.B. Andersen, F. Wiberg-Jorgensen, N.K. Jensen, R. Jones, L.M. Reid, and W.M. Zapol. 1988. ARDS after accidental inhalation of zinc chloride smoke. Intensive Care Med. 14(1):17-24.

IPCS (International Programme on Chemical Safety). 1998. Environmental Health Criteria 204: Boron. Geneva: World Health Organization.

Jansen, J.A., J.S. Schou, and A. Aggerbeck. 1984. Gastro-intestinal absorption and in vitro release of boric acid from water-emulsifying ointments. Food Chem. Toxicol. 22(1):49-53.

Jenkins, R.O., P.J. Craig, W. Goessler, and K.J. Irgolic. 1998. Antimony leaching from cot mattresses and sudden infant death syndrome (SIDS). Hum. Exp. Toxicol. 17(3):138-139.

Keen, C.L., and L.S. Hurley. 1977. Zinc absorption through skin: correction of zinc deficiency in the rat. Am. J. Clin. Nutr. 30(4):528-530.

Kent, N.L., and R.A. McCance. 1941. The absorption and excretion of minor elements in man: silver, gold, lithium, boron, vanadium. Biochem. J. 35:837-844.

Kowalska-Wochna, E. J. Moniuszko-Jakoniuk, E. Kulikowska, et al. 1988. The effect of orally applied aqueous solutions of lead and zinc on chromosome aberrations and induction of sister chromatid exchanges in the rat (Rattus sp.) Genetica Polonica 29(2):181-189.

Kozik, M.B., L. Maziarz, and A. Godlewski. 1980. Morphological and histochemical changes occurring in the brain of rats fed large doses of zinc oxide. Folia Histochem. Cytochem. 18:201-206.

Kozik, M.B., G. Gramza, and M. Pietrzak. 1981. Neurosecretion of the hypothalamo-hypophyseal system after intragastric administration of zinc oxide. Folia Histochem. Cytochem. (Krakow). 19(2):115-122.

Krasovskii, G.N., S.P. Varshavskaya, and A.I. Borisov. 1976. Toxic and gonadotropic effects of cadmium and boron relative to standards for these substances in drinking water. Environ. Health Perpect. 13:69-75.

Ku, W.W., R.E. Chapin, R.F. Moseman, R.E. Brink, K.D. Pierce, and K.Y. Adams. 1991. Tissue disposition of boron in Fischer rats. Toxicol. Appl. Pharmacol. 111(1):145-151.

Ku, W.W., R.E. Chapin, R.N. Wine, and B.C. Gladen. 1993. Testicular toxicity of boric acid (BA): Relationship of dose to lesion development and recovery in the F344 rat. Reprod. Toxicol. 7(4):305-319.

Kuschner, W.G., A. D'Alessandro, S.F. Wintermeyer, H. Wong, H.A. Boushey, and P.D. Blanc. 1995. Pulmonary responses to purified zinc oxide fume. J. Investig. Med. 43(4):371-378.

Kuschner, W.G., A. D'Alessandro, H. Wong, and P.D. Blanc. 1997. Early pulmonary cytokine responses to zinc oxide fume inhalation. Environ. Res. 75(1):7-11.

Kynast, G., and E. Saling. 1986. Effect of oral zinc application during pregnancy. Gynecol. Obstet. Invest. 21(3):117-123.

Lam, H.F., R. Peisch, and M.O. Amdur. 1982. Changes in lung volumes and diffusing capacity in guinea pigs exposed to a combination of sulfur dioxide and submicron zinc oxide mixed in a humidified furnace. Toxicol. Appl. Pharmacol. 66(3):427-433.

Lam, H.F., M.W. Conner, A.E. Rogers, S. Fitzgerald, and M.O. Amdur. 1985. Functional and morphologic changes in the lungs of guinea pigs exposed to freshly generated ultrafine zinc oxide. Toxicol. Appl. Pharmacol. 78(1):29-38.

Lam, H.F., L.C. Chen, D. Ainsworth, S. Peoples, and M.O. Amdur. 1988. Pulmonary function of guinea pigs exposed to freshly generated ultrafine zinc oxide with and without spike concentrations. Am. Ind. Hyg. Assoc. J. 49(7):333-341.

Landolph, J.R. 1985. Cytotoxicity and negligible genotoxicity of borax and borax ores to cultured mammalian cells. Am. J. Ind. Med. 7(1):31-43.

Lansdown, A.B. 1991. Interspecies variations in response to topical application of selected zinc compounds. Food Chem. Toxicol. 29(1):57-64.

Lee, I.P., R.J. Sherrins, and R.L. Dixon. 1978. Evidence for induction of germinal aplasia in male rats by environmental exposure to boron. Toxicol. Appl. Pharmacol. 45(2):577-590.

Lide, D.R., ed. 1991-1992. Handbook of Chemistry and Physics. 72nd Ed. Boca Raton, FL: CRC Press.

Linder, R.E., L.F. Strader, and G.L. Rehnberg. 1990. Effect of acute exposure to boric acid on the male reproductive system of the rat. J. Toxicol. Environ. Health 31(2):133-146.

Llobet, J.M., J.L. Domingo, M.T. Colomina, E. Mayayo, and J. Corbella. 1988. Subchronic oral toxicity of zinc in rats. Bull. Environ. Contam. Toxicol. 41:36-43.

Logue, J.N., M.D. Koontz, and M.A.W. Hattwick. 1982. A historical prospective mortality study of workers in copper and zinc refineries. J. Occup. Med. 24(5):398-408.

Mahomed, K., D.K. James, J. Golding, et al. 1989. Zinc supplementation during pregnancy: A double blind randomized controlled trial. Br. Med. J. 299:826-833.

Maita, K., M. Hirano, and K. Mitsumori. 1981. Subacute toxicity studies with zinc sulfate in mice and rats. J. Pesticide Sci. 6:327-336.

Marrs, T.C., H.F. Colgrave, J.A. Edington, R.F. Brown, and N.L. Cross. 1988. The repeated dose toxicity of a zinc oxide/hexachloroethane smoke. Arch. Toxicol. 62(2-3):123-132.

McCord, C.P., A. Friedlander, W.E. Brown, et al. 1926. An occupational disease among zinc workers. Arch. Intern. Med. 37:641-659.

Milliken, J.A., D. Waugh, and M.E. Kadish. 1963. Acute interstitial pulmonary fibrosis caused by a smoke bomb. Can. Med. Assoc. J. 88:36-39.

Moore, J.A. 1997. An assessment of boric acid and borax using the IEHR evaluative process for assessing human developmental and reproductive toxicity of agents. Rep. Toxicol. 11(1):123-160.

Moore, R. 1978. Bleeding gastric erosion after oral zinc sulfate. Br. Med. J. 1(6115):754.

Moseman, R.F. 1994. Chemical disposition of boron in animals and humans. Environ. Health Perspect. 102(Suppl. 7):113-117.

Mulinos, M., C. Conant, and E. Hauser. 1953. The toxicity of boric acid and the clinical implications of barated baby powders. Bull. N.Y. Med. Coll. 16:92-101.

Murphy, J.V. 1970. Intoxication following ingestion of elemental zinc. JAMA 212(12):2119-2120.

Neuberger, J.S., and J.G. Hollowell. 1982. Lung cancer excess in an abandoned lead-zinc mining and smelting area. Sci. Total Environ. 25(3):287-294.

Neve, J., M. Hanocq, A. Peretz, F. Abi Khalil, F. Pelen, J.P. Famaey, and J. Fontaine. 1991. Pharmacokinetic study of orally administered zinc in humans: Evidence for an enteral recirculation. Eur. J. Drug Metab. Pharmacokinet. 16(4):315-323.

Nielsen, F.H. 1996. How should dietary guidance be given for mineral elements with beneficial actions or suspected of being essential? J. Nutr. 126(Suppl. 9):2377S-2385S.

NRC (National Research Council). 1980. Recommended Dietary Allowances, 9th Revised Ed. Washington, D.C.: National Academy Press.

NRC (National Research Council). 1989. Recommended Dietary Allowances, 10th Ed. Washington, D.C.: National Academy Press.

NTP (National Toxicology Program). 1987. Toxicology and Carcinogenesis Studies of Boric Acid (CAS No. 10043-35-3) in $B6C3F_1$ Mice (Feed Studies). NTP TR 324, NIH Publication No. 88-2580. U.S. Department of Health and Human Services, Public Health Service, National Institutes of Health.

NTP (National Toxicology Program). 1990. Final Report on the Reproductive Toxicity of Boric Acid (CAS No. 10043-35-3) in CD-1 Swiss Mice, Volume 1. NTP-90-105. National Toxicology Program, Research Triangle Park, N.C. NTIS # PB90253808.

O'Loughlin, K.G. 1991. Bonde marrow erythrocyte micronucleus assay of boric acid in Swiss-Webster mice. SRI International (Study No. 2389-C400-91 (EPA MRID 420389-04)).

Philipp, R., A.O. Hughes, and M.C. Robertson. 1982. Stomach cancer and soil metal content. Br. J. Cancer 45(3):482.

Prasad, A.S. 1976. Deficiency of zinc in man and its toxicity, pp1-20. In A.S. Prasad [ed.] Trace elements in human health and disease. Academic Press, New York.

Prasad, A.S., G.J. Brewer, E.B. Schoomaker, and P. Rabbani. 1978. Hypocupremia induced by zinc therapy in adults. JAMA 240(20):2166-2168.

Price, C.J., P.L. Strong, M.C. Marr, C.B. Myers, and F.J. Murray. 1996a. Developmental toxicity NOAEL and postnatal recovery in rats fed boric acid during gestation. Fundam. Appl. Toxicol. 32(2):179-193.

Price, C.J., M.C. Marr, C.B. Myers, J.C. Seely, J.J. Heindel, and B.A. Schwetz. 1996b. The developmental toxicity of boric acid in rabbits. Fundam. Appl. Toxicol. 34(2):176-187.

Rohrs, L.C. 1957. Metal-fume fever from inhaling zinc oxide. Arch. Ind. Health 16:42-7.

Roudabush, R.L., C.J. Terhaar, D.W. Fassett, and S.P. Dziuba. 1965. Comparative acute effects of some chemicals on the skin of rabbits and guinea pigs. Toxicol. Appl. Pharmacol. 7(4):559-565.

Rudd, C.J. 1991. Mouse lymphoma cell mutagenesis assay ($tk^{+/-}/tk^{-/-}$) of boric acid. SRI International (Study No. 2389-G300-91 (EPA MRID 420389-02).

Samman, S., and D.C. Roberts. 1987. The effect of zinc supplements on plasma zinc and copper levels and the reported symptoms in health volunteers. Med. J. Aust. 146(5):246-249.

Schiffer, R.B., F.W. Sunderman, Jr., R.B. Braggs, J.A. Moynihan. 1991. The effects of exposure to dietary nickel and zinc upon humoral and cellular immunity in SJL mice. J. Neuroimmunol. 34(2-3):229-239.

Schlicker, S.A., and D.H. Cox. 1968. Maternal dietary zinc, and development and zinc, iron, and copper content of the rat fetus. J. Nutr. 95(2):287-94.
Shankar, A.H., and A.S. Prasad. 1998. Zinc and immune function: The biological basis of altered resistance to infection. Am. J. Clin. Nutr. 68(Suppl. 2):447S-463S.
Silaev, A.A., A.A. Kasparov, V.V. Lorolev, and V. V. Nevstrueva. 1977. Electron-microscopic investigation of the effect of boric acid on the seminiferous tubules of albino rats. [Article in Russian]. [Abstract in English]. Bull. Exp. Biol. Med. (USSR). 83:588-591.
Simmer, K., L. Lort-Phillips, C. James, and R.P. Thompson. 1991. A double-blind trial of zinc supplementation in pregnancy. Eur. J. Clin. Nutr. 45(3):139-144.
Stewart, K.R. 1991. *Salmonella*/microsome plate incorporation assay of boric acid. SRI International (Study No. 2389-A200-91 (EPA MRID 4203-8901)).
Stocks, P., and R.I. Davies. 1964. Zinc and copper content of soils associated with the incidence of cancer of the stomach and other organs. Br. J. Cancer 18:14-24.
Straube, E.F., N.H. Schuster, and A.J. Sinclair. 1980. Zinc toxicity in the ferret. J. Comp. Pathol. 90(3):355-361.
Sturniolo, G.C., M.C. Montino, L. Rossetto, A. Martin, R. D'Inca, A. D'Odorico, R. Naccarato. 1991. Inhibition of gastric acid secretion reduces zinc absorption in man. J. Am. Coll. Nutr. 10(4):372-375.
Stuttgen, G., T. Siebel, and B. Aggerbeck. 1982. Absorption of boric acid through human skin depending on the type of vehicle. Arch. Dermatol. Res. 272(1-2):21-29.
Tarasenko, N.I., A.A. Kasparov, and O.M. Strongina. 1972. Effect of boric acid on the sexual function in males. [Article in Russian]. Gig. Tr. Prof. Zabol. 16(11):13-16.
Thurnham, D.I. 1997. Micronutrients and immune function: Some recent developments. J. Clin. Pathol. 50(11):887-891.
Treinen, K.A., and R.E. Chapin. 1991. Development of testicular lesions in F344 rats after treatment with boric acid. Toxicol. Appl. Pharmacol. 107(2):325-335.
Turner, J.A. 1921. An occupational dermatoconiosis among zinc oxide workers. Public Health Rep. 36:2727-2732.
U.S. Borax. 1996. Material Saftey Data Sheet: Firebrake ZB. U.S. Borax, Valencia, CA.
Vignec, A.J., and R. Ellis. 1954. Inabsorbability of boric acid in infant powder. Am. J. Dis. Child. 88:72-80.
Vilkina, G.A., M.D. Polmerantseva, and L.K. Ramaiia. 1978. Absense of a mutagenic effect of cadmium and zinc salts in mouse somatic and sex cells [article in Russian]. Genetika (Moscow) 14(12):2212-2214.
Voroshilin, S.I., E.G. Plotko, T.V. Fink, and V.I. Nikiforova. 1978. Cytogenic effect of wolfram, zinc, cadmium, and cobalt compounds on human and animal somatic cells [article in Russian]. Tsitol. Genet. 12(3):241-243.
Walters, M., and F.J. Roe. 1965. A study of the effects of zinc and tin administered orally to mice over a prolonged period. Food Cosmet. Toxicol. 3(2):271-276.
Wegman, D.H., E.A. Eisen, X. Hu, S.R. Woskie, R.G. Smith, and D.H. Garabrant. 1994. Acute and chronic respiratory effects of sodium borate particulate exposures. Environ. Health Perspect. 102(Suppl. 7):119-128.
Weigand, E., and M. Kirchgessner. 1992. Absorption, endogenous excretion, and balance of zinc in growing rats on diets with various sugars replacing starch. Biol. Trace Elem. Res. 34(1):67-77.

Weir, R.J., Jr., and R.S. Fisher. 1972. Toxicologic studies on borax and boric acid. Toxicol. Appl. Pharmacol. 23(3):351-364.

Wester, R.C., X. Hui, T. Hartway, H.I. Mailbach, K. Bell, M.J. Schell, D.J. Northington, P. Strong, and B.D. Culver. 1998. In vivo percutaneous absorption of boric acid, borax, and disodium octaborate tetrahydrate in humans compared to in vitro absorption in human skin from infinite and finite doses. Toxicol. Sci. 45(1):42-51.

Whorton, M.D., J.L. Haas, L. Trent, and O. Wong. 1994. Reproductive effects of sodium borates on male employees: Birth rate assessment. Occup. Environ. Med. 51(11):761-767.

Wilding, J.L., W.J. Smith, P. Yevich, et al. 1959. The toxicity of boron oxide. Am. Ind. Hyg. Assoc. J. 20:284-289.

Wong, L.C., M.D. Heimbach, D.R. Truscott, and B.D. Buncan. 1964. Boric acid poisoning: Report of 11 cases. Can. Med. Assoc. J. 90:1018-1023.

Woods, W.G. 1994. An introduction to boron: History, sources, uses, and chemistry. Environ. Health Perpect. 102(Suppl. 7):5-11.

Yadrick, M.K., M.A. Kenney, and E.A. Winterfeldt. 1989. Iron, copper, and zinc status: Response to supplementation with zinc or zinc and iron in adult females. Am. J. Clin. Nutr. 49:145-150.

Zaporowska, H., and W. Wasilewski. 1992. Combined effect of vanadium and zinc on certain selected haematological indices in rats. Comp. Biochem. Physiol. C 103(1):143-147.

9

Calcium and Zinc Molybdates

CALCIUM and zinc molybdates readily dissociate in the body into molybdenum compounds and calcium and zinc ions. Because little data exist on calcium and zinc molydates specifically, this chapter reviews the physical and chemical properties, toxicokinetics, toxicological, epidemiological, and exposure data on molybdenum compounds and zinc. The subcommittee used that information to characterize the health risk from exposure to calcium and zinc molybdates. The subcommittee also identified data gaps and recommended research relevant for determining the health risk from exposure to calcium and zinc molybdates.

PHYSICAL AND CHEMICAL PROPERTIES

The physical and chemical properties of calcium and zinc molybdates are presented in Table 9-1.

OCCURRENCE AND USE

Calcium and zinc molybdates are used as flame retardants in cellulosic materials and other polymers. Textile applications for calcium and zinc molybdates include furniture, draperies, upholstery seating in transportation vehicles, wall coverings, and carpets (FRCA 1998). Calcium and zinc molybdates are

TABLE 9-1 Physical and Chemical Properties of Calcium and Zinc Molybdates

Property	Value	Reference
Calcium molybdate		
Chemical formula	$CaMoO_4$	Budavari et al. 1989
CAS registry #	7789-82-4	Powmet 1999
Synonym	Powellite	Powmet 1999
Molecular weight	200.01	Budavari et al. 1989
Physical state	Solid	Powmet 1999
Melting point	965°C	Powmet 1999
Solubility	0.005 g/100 mL in H_2O at 25°C	Tsigdinos and Moore 1981
Density	4.38-4.53 g/cm^3	Powmet 1999
Zinc molybdate		
Chemical formula	$ZnMoO_4$	Tsigdinos and Moore 1981
CAS registry no.	13767-32-3	Powmet 1999
Synonyms	zinc molybdenum oxide, molybdic acid, zinc salt, Kemguard	Powmet 1999
Molecular weight	225.31	Powmet 1999
Physical state	Solid	Powmet 1999
Solubility	0.5 g/100 mL in H_2O at 25°C	Powmet 1999
Melting point	1,020°C	Tsigdinos and Moore 1981

formed when calcium oxide (CaO) or zinc oxide (ZnO) is complexed with molybdenum trioxide (MoO_3). These molybdates readily dissociate in the body, resulting in molybdemum (Mo) in various valence states, along with zinc and calcium ions (Stokinger 1981). Calcium molybdate can occur naturally as the ore Powellite (Stokinger 1981).

Molybdenum (Mo) exists in six valence states. The most important valence states in biological systems are Mo^{3+}, Mo^{4+}, Mo^{5+}, and Mo^{6+} (Lener and Bibr 1984). In general, higher oxidation states lead to oxygen binding while lower oxidation states favor sulfur or nitrogen binding (EPA 1979). The principal dissolved Mo species in the natural environment is molybdate (EPA 1979). The recommended daily intake of Mo is 75-250 µg/d (NRC 1989). Mo is important biologically to humans, as it is an essential trace element in the Mo-flavoprotein enzyme xanthine oxidase (XO), where it functions as an electron transport agent. XO permits the oxidation of hypoxanthine and xanthine to uric acid. It is also a component of several other metalloenzymes, including aldehyde oxidase and sulfite oxidase (Tsongas et al. 1980).

Zinc is an essential nutrient with a recommended daily allowance of 15 mg/d for males and 12 mg/d for females (NRC 1989). The toxicity of zinc in humans is considered to be quite low; toxicity normally occurs following ingestion of >2 g of zinc (Prasad 1976, as cited in ATSDR 1994).

Calcium is an essential nutrient, with a recommended daily allowance of 800-1,200 mg/d depending upon a person's age (NRC 1989). No adverse effects from consumption of levels of calcium up to 2,500 mg/d in healthy adults have been reported. High calcium intakes may induce constipation, cause increased risks of urinary stone formation in males, and may inhibit the intestinal absorption of iron, zinc, and other essential minerals. Ingestion of very large quantities of calcium may result in hypercalciuria, hypercalcemia, and deterioration of renal function in both sexes (NRC 1989). Additional toxicity data on calcium are not included in this document, because the recommended daily allowance is considerably higher than exposure estimates in flame retardants applied to upholstery fabric.

TOXICOKINETICS

Molybdenum Compounds

There are no toxicokinetic data on Mo compounds following dermal exposure. Only limited data were located regarding the absorption, distribution, metabolism, and excretion by humans of inhaled or ingested Mo compounds. A case study of four humans injected intravenously with ^{99}Mo showed that 5-d cumulative urinary excretion ranged from 16.6% to 27.2% of the dose (50-100 µCi), with the primary excretory pathway being the kidney. Fecal excretion was found to be 6.8% in one patient and less than 1% in another after 10 d (Rosoff and Spencer 1964).

Studies by Fairhall et al. (1945) showed that Mo is rapidly absorbed and eliminated by the kidneys of experimental animals following oral exposure. Six guinea pigs were administered 50 mg of Mo orally, as molybdenum trioxide in a 10% gum arabic solution, and were observed for 4, 16, and 48 hr. The highest concentrations of Mo were excreted in the urine, while much smaller quantities were found in the feces. Rats dosed orally with molybdenum trioxide were found to have Mo levels distributed uniformly in the critical organs within 4 hr, while higher levels of Mo were found in the blood and bile. Two rabbits, administered 100 mg of molybdenum trioxide each via a stomach tube, also demonstrated similar rapid absorption of Mo, with rapidly rising blood levels. Mo was found to be rapidly eliminated from the kidneys, with urinary levels returning to baseline values within 72 hr. Fecal elimination, which comprised about 50% of the urinary levels, also occurred within 72 hr. The authors noted that significant quantities of Mo were stored in the bone.

Arrington and Davis (1955) studied the toxicokinetics of Mo^{99} (in the form of molybdenum trioxide) in Long-Evans rats (18/dose) that were consuming both normal and high-calcium diets. Rats were dosed once orally with 15 μC Mo^{99} at least 5 wk after initiation of the high-calcium diets. No significant differences in the absorption, retention, or excretion of Mo^{99} were observed between the groups of rats consuming the normal versus high-calcium diets. Mo^{99} was eliminated primarily by the kidney and excreted rapidly. Within 6 hr of oral administration of Mo^{99}, about 25% of the dose was excreted in the urine. At 12 hr, 50% of the dose was present in the urine. Mo^{99} was distributed primarily in the kidneys and blood, with smaller amounts in the bone, liver, and muscle.

Mo tissue levels were determined in guinea pigs following 25 d of inhalation exposure to calcium molybdate (121.5 mg/m^3) or molybdenum trioxide (157 mg/m^3) (Fairhall et al. 1945). The highest concentrations of Mo were found in lung, kidneys, spleen, and bone. Analysis 2 wk after termination of exposure to calcium molybdate showed approximately 50-75% of the Mo remained in the tissues. In contrast, 2 wk following exposure to molybdenum trioxide, 25-50% of Mo remained in the tissues.

Zinc

Absorption

Agren (1991) (as cited in ATSDR 1994) reported that zinc was present in human interstitial fluid (site of application) following dermal application of zinc oxide (dissolved in gum resin or hydrocolloids) to human forearms. No evidence for absorption into systemic circulation was provided. Agren (1990) (as cited in ATSDR 1994) and Hallmans (1977) (as cited in ATSDR 1994) determined that zinc readily permeates intact and damaged human skin following dermal application. However, penetration of zinc into systemic circulation was not determined.

Keen and Hurley (1977) (as cited in ATSDR 1994) determined that when zinc (as zinc chromate) was dissolved in oil and topically applied to rats, absorption of zinc in the bloodstream occurred. No other animal studies were identified regarding dermal absorption of zinc.

Data suggest that zinc is absorbed into systemic circulation via the lungs following inhalation exposures. Hamdi (1969) (as cited in ATSDR 1994) found that zinc blood levels were elevated in workers occupationally exposed to zinc fumes. Drinker and Drinker (1928) (as cited in ATSDR 1994) determined that inhalation exposure of cats to zinc oxide fumes for up to 3.25 hr resulted in increased levels of zinc in the pancreas, kidney, and liver. In both studies, oral absorption of zinc particles following ciliary clearance and swallowing could account for all, or a significant portion, of the absorbed zinc. In the Drinker and Drinker (1928) study (as cited in ATSDR 1994), the swallowing of zinc particles during grooming activities may have also accounted for the increased tissue zinc levels.

The estimated rate of oral absorption of zinc in humans is between 8% and 81%, depending on an individual's diet (ATSDR 1994). People who are not zinc-deficient will absorb about 20-30%, while individuals who are zinc-deficient absorb more (ATSDR 1994). Two studies measured the peak blood concentrations of zinc in volunteers following oral ingestion of zinc sulfate and determined that peak blood Zn^{2+} concentrations were reached within 3 hr (Neve et al. 1991, as cited in ATSDR 1994; Sturniolo et al. 1991, as cited in ATSDR 1994). The presence of cadmium, mercury, copper, or other trace metals can diminish zinc absorption by inhibiting zinc transport across the intestinal wall (ATSDR 1994).

Zinc absorption in male Wistar rats was approximately 40-48% when diets contained 0.81 mg of radio-labeled zinc/kg as zinc chloride or zinc carbonate (Galvez-Morros et al. 1992). ATSDR (1994) noted that fractional absorption of zinc in immature organisms usually exceeds the fractional absorption of zinc in adults.

Distribution

No relevant human or animal studies were located that investigated the distribution of zinc following dermal exposure to zinc compounds.

No inhalation studies were identified that investigated the distribution of zinc in humans. Cats exposed to zinc oxide (12-61 mg Zn^{2+}/kg-d) for 3 hr, had increased zinc levels in the pancreas, liver, and kidneys, suggesting that absorption of zinc had taken place in the lungs (Drinker and Drinker 1928, as cited in

ATSDR 1994). Oral absorption through swallowing or grooming, however, cannot be ruled out. Absorption rates were not estimated in this study.

The distribution of zinc to tissues has not been measured in humans following ingestion. However, there are a number of studies in rodents that have investigated the distribution of zinc following oral exposure to zinc compounds. Weigand and Kirchgessner (1992) (as cited in ATSDR 1994) determined that rats fed 1.1 mg Zn^{2+}/kg-d for an unspecified amount of time, had greater amounts of zinc distributed primarily to the kidneys and pancreas than to the liver. Administration of zinc acetate to rats (191 mg Zn^{2+}/kg-d in food for 3 mo) increased zinc levels in the heart, spleen, kidneys, liver, bone, and blood (Llobet et al. 1988). Mice fed either 76.9 mg Zn^{2+}/kg-d as zinc sulfate (Schiffer et al. 1991, as cited in ATSDR 1994) or 38 mg Zn^{2+}/kg-d as zinc nitrate (Cooke et al. 1990, as cited in ATSDR 1994) for 1 mo had increased levels of Zn^{2+} in the kidneys and liver. Newborn, young, or adult mice that received a single oral dose of 4.6 mg Zn^{2+}/kg as zinc chloride generally had the highest level of zinc in the liver, kidneys, lungs, bone, and muscle 1 d after dosing (He et al. 1991, as cited in ATSDR 1994).

Metabolism

Although zinc is not metabolized in the body, it can bind to many molecules in the body. For instance, zinc induces and binds to metallothionein (a metal binding protein) in vivo. Metallothionein therefore acts as a protective mechanism against zinc toxicity (Goyer 1996). Indirect evidence suggests that zinc also complexes with reduced glutathione in the liver in rats following intraperitoneal injection (Alexander et al. 1981, as cited in ATSDR 1994).

Excretion

No studies were located that investigated the excretion of zinc in humans or animals following dermal application of zinc compounds.

Following inhalation exposures, elevated levels of zinc were found in the urine of workers exposed to zinc oxide fumes containing unknown levels of Zn^{2+} (Hamdi 1969, as cited in ATSDR 1994). No other studies were identified that investigated the excretion of zinc following inhalation of zinc compounds.

Following oral exposure, the primary route of zinc excretion in humans and rats is the feces. Zinc can also be excreted in the urine, saliva, hair, and sweat (ATSDR 1994). Malnutrition or low dietary levels of zinc may promote increased levels of urinary zinc excretion and are thought to result from increased levels of tissue breakdown and catabolism (ATSDR 1994).

HAZARD IDENTIFICATION[1]

Dermal Exposure

Irritation

Molybdenum Compounds

No human data on the effects of dermal exposure to Mo compounds were identified. No irritation effects were seen when Mo compounds were applied to intact or abraded skin of rabbits (Stokinger 1981).

Zinc

There are two case studies in the scientific literature that suggest that occupational dermal exposure to zinc at high levels may cause or contribute to a skin condition referred to as "zinc oxide pox" which is described as itchy papular-pustular eruptions that occur in the pubic region, inner surface of the thigh, axilla, and inner surface of the arms. Turner (1921) (as cited in ATSDR 1994) found that 14 out of 17 men developed zinc oxide pox at least once during their employment in the bagging or packaging of zinc oxide. The incidence of zinc oxide pox in the study by Turner (1921) (as cited in ATSDR 1994) has been attributed to poor hygiene among the workers, and not necessarily zinc oxide exposure. In a similar study, Batchelor et al. (1926) (as cited in ATSDR 1994) found that only 1 of a total of 24 workers occupationally exposed to zinc dusts developed zinc oxide pox.

Agren (1990) (as cited in ATSDR 1994) reported that application of patches containing 25% zinc oxide (dose = 2.9 mg Zn^{2+}/m^3) to the skin of human volunteers did not produce dermal irritation following 48 hr of exposure.

The dermal irritancy of several zinc compounds in aqueous solution or suspension has been investigated in mice, rabbits, and guinea pigs (Lansdown 1991, as cited in ATSDR 1994). In this study, animals were treated topically

---

[1]In this section, the subcommittee reviewed toxicity data on calcium and zinc molybdates, including the toxicity assessment prepared by the U.S. Consumer Product Safety Commission (Hatlelid 1999).

once a day for 5 consecutive days with one of the following zinc compounds (w/v): zinc oxide (20% suspension in Tween 80), zinc chloride (1% aqueous solution), zinc sulfate (1% aqueous solution), zinc pyrithione (20% suspension), and zinc undecylenate (20% suspension). In open patch tests, zinc chloride was a potent irritant in all three species and caused the formation of epidermal hyperplasia and ulceration. All other compounds produced less severe erythema than zinc chloride, and none of the compounds caused ulceration or scaling over the 5-d test period. These compounds were also tested in a second group of rabbits using occlusive bandages at the test site. Occlusive patch testing with zinc chloride produced severe dermal irritation in rabbits within 3-5 d of application. Occlusive patch testing of zinc acetate produced moderate irritation. Occlusive patch testing with zinc oxide, sulfate, pyrithione, or undecylenate produced little dermal irritation. Histological examination of skin samples from animals treated with zinc chloride or zinc acetate showed evidence of acanthosis, parakeratosis, hyperkeratosis, and inflammatory changes in the epidermis and in the more superficial aspects of the dermis.

Systemic Effects

Molybdenum Compounds

No data were identified on systemic effects of Mo compounds following dermal exposure.

Zinc

DuBray (1937) (as cited in ATSDR 1994) reported that a worker developed microcytic anemia and had low platelet counts after being exposed to zinc chloride solutions. The concentration of zinc was not reported. No systemic effects following dermal exposures of animals to zinc were identified.

Inhalation Exposure

Systemic Effects

Molybdenum Compounds

Twenty-five workers in a Mo roasting plant (where Mo sulfide is converted to molybdenum oxides) in Colorado were estimated to be exposed to soluble

Mo compounds (primarily molybdenum trioxide) at a concentration of 9.5 mg/m^3 for an 8-hr time-weighted average exposure. Exposure concentrations were estimated from respirable dust samples taken at the plant. Workers showed large increases in serum ceruloplasmin (50.5 mg/dL versus 30.5 mg/dL in controls) (Walravens et al. 1979). Controls consisted of 24 students and research personnel at the University of Colorado Medical Center. Workers were employed at the plant from 0.5 to 20 yr, with the average employment being 4 yr. Plasma and urine Mo levels were elevated in the workers compared with controls (0.9-36.5 μg/dL vs. 0-3.4 μg/dL in plasma and 120-11,000 μg/L vs. 4-347 μg/L in urine). No other adverse health effects were reported. The authors hypothesized that elevated serum ceruloplasmin levels stemmed from mobilization of tissue copper reserves within the hepatocyte, with subsequent ceruloplasmin synthesis and release to prevent intracellular copper toxicity.

A study conducted on 73 workers from a Russian copper-molybdenum processing plant found increased levels of uric acid in the blood (Akopyan 1964). Additional details on the exposure levels of these workers were not provided.

Review articles (Stokinger 1981; ACGIH 1991) reported elevated serum uric acid levels and signs of gout, including pain and deformities of joints in workers and inhabitants in Mo-rich areas in Armenia. Lener and Bibr (1984), in a review article, reported an increased incidence of nonspecific symptoms, including weakness, fatigue, headache, anorexia, and joint and muscle pains among mining and metallurgy workers exposed to 60-600 mg/m^3 Mo. No other reports of effects from industrial Mo exposure were cited.

Mogilevskaya (1967) reported that 3 of 19 workers exposed to Mo compounds (Mo and molybdenum trioxide) at two industrial facilities showed early signs of pneumoconiosis on X-ray examination. These three individuals had worked at the facilities for 4-7 yr and their exposures, although variable, were reported to range from 1-19 mg/m^3.

Fairhall et al. (1945) observed that 5/24 guinea pigs died following exposure via inhalation to 195 mg/m^3 $CaMoO_4$ (125 mg Mo/m^3) dust for 1 hr/d, 5 d/wk for 5 wk, but no other signs of toxicity were observed. Guinea pigs (51 animals, sex not reported), who were exposed to 250 mg MoO_3/m^3 (164 mg Mo/m^3) using the same exposure regime, experienced severe eye and nasal irritation, loss of appetite and weight, diarrhea, muscular incoordination, and loss of hair. Following the 10th exposure, 26/51 animals died.

Two 13-wk studies were conducted by NTP (1997) in which F-344/N rats and B6C3F$_1$ mice (10/sex/group) were exposed to molybdenum trioxide for 6.5 hr/d, 5 d/wk at concentrations of 0, 1, 3, 10, 30, or 100 mg/m^3. All rats and mice survived to the end of the study. Significant increases in liver copper concentrations were observed in female mice exposed to 30 mg/m^3 and in male and female mice exposed to 100 mg/m^3 (males: 11.51 μg/g in the 100-mg/m^3 exposure group versus 8.19 μg/g in controls; females: 6.51 and 6.98 μ/g in the 30-

and 100-mg/m^3 dose groups, respectively, versus 5.68 µ/g in controls). The increased copper concentrations were not regarded as being an adverse effect relevant for deriving a LOAEL and a NOAEL. No other clinical findings were observed in either rats or mice. Additionally, no significant differences in absolute or relative organ weights, sperm counts, or motility were noted in rats or mice.

In the same NTP study (1997), rats (F344/N) and mice (B6C3F$_1$) (50/sex/dose) exposed for 6 hr/d, 5 d/wk at concentrations of 0, 10, 30, or 100 mg/m^3 molybdenum trioxide for 2 yr experienced a significant exposure-dependent increase in blood Mo concentrations. Male and female rats exposed to 30 or 100 mg/m^3 experienced significantly increased incidences of chronic alveolar inflammation. Incidences of hyaline degeneration in the nasal respiratory epithelium in male rats exposed to 30 or 100 mg/m^3 and in all exposed groups of females rats were significantly greater than those of the control groups. Incidences of hyaline degeneration in the nasal olfactory epithelium of all exposed groups of females were also statistically significant. For male mice, the incidences of histiocyte cellular infiltration in all exposed groups were significant. Incidences of hyaline degeneration of the respiratory epithelium of the nasal cavity in female mice at 100 mg/m^3 were significantly greater than those in the controls (NTP 1997). Based on the 2-yr NTP study, the LOAEL is 10 mg/m^3 for increased incidences of hyaline degeneration in the nasal respiratory epithelium and nasal olfactory epithelium in female rats.

Zinc

There are a number of case reports of deaths in humans following high inhalation exposures to airborne mixtures containing zinc. Ten of 70 persons died within four d following intense exposure to a smoke mixture containing approximately 33,000 mg Zn^{2+}/kg as zinc chloride along with other compounds (Evans 1945, as cited in ATSDR 1994). These mixtures were thought to include unknown concentrations of hexachloroethane, calcium silicate, and an igniter. Milliken et al. (1963) (as cited in ATSDR 1994) describes the case of a fireman who died following exposure to a high but unknown concentration of a smoke mixture generated from a zinc chloride smoke bomb. Two soldiers developed severe respiratory distress syndrome and died 25-32 d following exposure to a high concentration of zinc chloride smoke mixture generated from a zinc chloride smoke bomb (Hjortso et al. 1988). No exposure levels were reported in this study. Autopsies performed on the soldiers revealed diffuse microvascular obliteration, widespread occlusion of the pulmonary arteries, and extensive interstitial and intra-alveolar fibrosis of the lungs. Nausea has been reported among persons following intense inhalation exposures to zinc chloride and zinc oxide (Hammond 1944, as cited in ATSDR 1994; Evans 1945, as cited in ATSDR 1994; Rohrs 1957, as cited in ATSDR 1994; Johnson and Stonehill 1961, as cited in ATSDR 1994; Schenker et al. 1981, as cited in ATSDR 1994).

Routine blood chemistries and examinations revealed no liver disease among 12 workers involved in the manufacture of brass alloys, with 4-21 yr of exposure to zinc oxide (Hamdi 1969, as cited in ATSDR 1994). McCord et al. (1926) (as cited in ATSDR 1994) reported that several workers from the galvanized industry had decreased red blood cell counts. Workers investigated by Hamdi (1969) (as cited in ATSDR 1994) had normal red blood cell counts.

Various adverse pulmonary effects and reduced survival rates were reported in female rodents following exposure to zinc oxide/hexachloroethane smoke (119 mg Zn^{2+}/m^3 for 1 hr/d, 5 d/wk for up to 20 wk) (Marrs et al. 1988). The authors noted that the zinc oxide/hexachloroethane smoke contained a number of toxic chemicals including carbon tetrachloride. Therefore, it is not certain whether the toxic effects observed in this study can be solely attributed to the inhalation of zinc particles.

Immunological Effects

Molybdenum Compounds

No data were identified on the immunological effects of Mo compounds following inhalation exposure.

Zinc

There are three case reports in the literature that found that the inhalation of high concentrations of zinc-containing compounds appeared to stimulate changes in the immune system. Farrell (1987) reported a case study of a worker who developed hives and angioedema (suggestive of an immediate or delayed IgE response) following exposure to a low dose of zinc fumes. The signs and symptoms of toxicity were repeated in a challenge test, suggesting that the patient had developed sensitization to zinc compounds. A correlation between exposure to zinc oxide and the proportion of activated T-cells, T-helper cells, T-inducer cells, T-suppressor cells, and activated killer T-cells, was observed among 14 welders approximately 20 hr following exposure to zinc oxide (Blanc et al. 1991, as cited in ATSDR 1994). Zinc oxide exposure levels were estimated to be approximately 77-153 mg Zn^{2+}/m^3. Ameille et al. (1992) (as cited in ATSDR 1994) reported on a case of a smelter worker who had elevated levels of lymphocytes in the bronchoalveolar lavage fluid following exposure to unknown concentrations of zinc fumes.

Marrs et al. (1988) did not observe any abnormalities in the lymph nodes, thymus, or spleen tissue of female rats, mice, or guinea pigs killed at 18 mo following a 20-wk exposure to zinc oxide/hexachloroethane smoke at concentrations as high as 119.3 or 121.7 mg Zn^{2+}/m^3 for 1 hr/d, 5 d/wk.

Reproductive and Developmental Effects

Molybdenum Compounds

No data were identified on reproductive or developmental effects of Mo compounds following inhalation exposure.

Zinc

No studies in humans were identified regarding reproductive or developmental effects following inhalation exposures to zinc. Pathological examination 17 mo after exposure of rats, guinea pigs, and mice to zinc oxide/hexachloroethane smoke (1 hr/d, 5 d/wk for 20 wk) at concentrations as high as 119.3 or 121.7 mg zinc/m^3 produced no treatment-related abnormalities in the mammary glands, ovaries, or fallopian tubes (Marrs et al. 1988).

Carcinogenicity

Molybdenum Compounds

Information on the carcinogenicity of calcium and zinc molybdates was not found. However, data on molybdenum trioxide are available from two chronic inhalation studies conducted in F-344/N rats and $B6C3F_1$ mice (50/sex/group) that were exposed at concentrations of 0, 10, 30, or 100 mg/m^3 of molybdenum trioxide for 6 hr/d, 5 d/wk for 103 wk (NTP 1997). Incidences of alveolar/bronchiolar adenoma or carcinoma (combined) were increased in low-, mid-, and high-dose males (1/50, 1/50, 4/50, respectively, compared to 0/50 in controls). In the larynx, incidences of squamous metaplasia of the epithelium lining the base of the epiglottis in all exposed groups of male and female rats were significantly greater than those in the control groups and rose with increasing exposure concentration (11/50, 16/50, 39/50 males in low-, mid- and high-dose groups, respectively, vs. 0/50 in controls; 18/50, 29/50, 49/50 females in low-, mid-, and high-dose groups, respectively, vs. 0/50 in controls). The incidences of alveolar/bronchiolar carcinoma were significantly greater in all exposed groups of male mice (16/50, 14/49, and 10/50) than in the control group (2/50). In addition, the incidences of alveolar/bronchiolar adenoma or carcinoma (combined), in male mice, exposed at concentrations of 10 or 30 mg/m^3 (27/50 or 21/49), were also significantly greater than the control group (11/50), while the incidence in the 100 mg/m^3 dose group (18/50) was not. In female mice, the incidence of alveolar/bronchiolar adenoma or carcinoma (combined) was significantly greater in the 100 mg/m^3 dose group (15/49) than controls (3/50), but incidences in the 10 and 30 mg/m^3 dose groups were not significant (6/50 and 8/49). Incidences of metaplasia of the alveolar epithelium of minimal severity in the centriacinar region of the lung were significantly increased in all exposed groups of mice. The incidences of squamous metaplasia of the epithelium lining the base of the epiglottis were significantly increased in all exposed groups of males and females. In both male and female mice, the incidences of hyperplasia of the laryngeal epithelium at level II of the larynx rose with increasing exposure concentration, but were statistically significant only in the highest dose group. Based on these results, NTP reported that there was some evidence of carcinogenicity for male and female mice, but equivocal evidence of carcinogenicity in male rats exposed for 2 yr.

Zinc

Excess lung cancer mortality was detected among persons living in an abandoned zinc/lead mining area in the midwestern United States as compared with state and national age- and sex-specific lung cancer rates (Neuberger and Hollowell 1982). However, it was determined that the excess mortality was not related to environmental exposure to zinc or lead. Confounding exposure factors such as smoking and occupation, that might have accounted for the observed elevation in lung cancer mortality, were not addressed.

A second study by Logue et al. (1982) investigated mortality among a cohort of 4,802 male workers from two zinc refining plants and seven copper refining plants. Overall mortality and death from specific cancers was not elevated for the whole cohort. Cancer mortality rates were not computed separately for the 978 zinc refinery workers. Therefore, it cannot be determined from this study whether exposure to zinc increases cancer mortality in male zinc refinery workers.

Marrs et al. (1988) found that female mice exposed to mean concentrations of zinc of 1.3, 12.8, or 121.7 as zinc oxide/hexachloroethane smoke (1 hr/d, 5 d/wk for 20 wk) had a statistically significant trend in the prevalence of alveologenic carcinoma, with the frequency of this tumor reaching significance in the high-dose group at 13 mo postexposure. No increase in tumor frequency occurred in female rats or guinea pigs exposed to similar concentrations by an identical dosing regimen. The authors noted that the zinc oxide/hexachloroethane smoke contained a number of other chemicals including carbon tetrachloride which is an animal carcinogen. Therefore, carcinogenic effects cannot be solely attributed to zinc oxide.

Oral Exposure

Systemic Effects

Molybdenum Compounds

A summary of the toxicity studies following oral exposures to Mo compounds is presented in Table 9-2.

Plants in the village of the exposed population was 30 and 190 times higher, respectively, than that of the control villages. The estimated average intake levels of Mo and copper in the exposed population were 0.14-0.21 and 0.07-0.14 mg/kg-d, respectively, for a 70-kg adult; in the control population the levels were 0.01-0.03 and 0.14-0.21 mg/kg-d for Mo and copper, respectively. Three hundred villagers (84 adults) from the exposed population and 100 from the control group (78 adults) underwent medical examinations. Thirty-one percent of the exposed adult population and 17.9% of the controls demonstrated gout-like symptoms that were characterized by pain, swelling, inflammation, and deformities of the joints. All individuals had increased uric acid content of the blood. In a sub study, 52 adults from the exposed population and five of the controls underwent more detailed clinical examinations, including measurement of copper, Mo, uric acid, and xanthine oxidase concentrations in blood and Mo, copper, and uric acid concentrations in urine. Their average uric acid content was 6.2 mg in comparison to 3.8 mg in controls. Serum Mo and serum xanthine oxidase levels were positively correlated with serum uric acid levels. Increasing urinary excretion of copper was positively correlated with increasing serum levels of Mo. Based on the study findings, an intake level of 0.14 mg/kg-d was designated as the lowest observed adverse effect level (LOAEL) because of concern regarding increased serum uric acid levels (EPA 1999).

TABLE 9-2 Oral Toxicity Molybdenum Compounds

Species, Strain, Sex, Number	Dose (mg/kg-d)	Duration	Effects	NOAEL/LOAEL (mg/kg-d)	Reference
Humans, M/F, 300 exposed, 100 control	Mo: 0.14-0.21 Cu: 0.07-0.14	NS	Increased serum uric acid levels	LOAEL: 0.14	Kovalsky et al. 1961
Humans, M, 4	Mo: up to 1,540	4 d	Increased urinary excretion of copper; no effect on uric acid levels	ND	Deosthale and Gopalan 1974
Rat, NS, 10/dose	$CaMoO_4$	4 mo	LD_{50}: 101 mg/kg-d	ND	Fairhall et al. 1945
	MoO_3		LD_{50}: 125 mg/kg-d		
	$(NH_4)_2MoO_4$		LD_{50}: 333 mg/kg-d		
Rat, Long-Evans, M/F, 4 or 8/dose	Mo: <0.1 or 8 Cu: 2 Mo: <0.1, 2, 8, 14 Cu: 0.5	Weaning to 11 wk	Growth retardation; depigmentation; male infertility (8, 14 mg/kg-d Mo)	LOAEL: ~2	Jeter and Davis 1954
Rat, Long-Evans, F, 4 or 8/dose	Mo: <0.1 or 8 Cu: 2 Mo: <0.1, 2, 8, 14 Cu: 0.5	Gestation and weaning	Decreased newborn weights	ND	Jeter and Davis 1954

$CaMoO_4$, calcium molybdate; Cu, copper; F, female; LD_{50}, lethal dose to 50% of test animals; M, male; Mo, molybdenum; MoO_3, molybdenum trioxide; $(NH_4)_2MoO_4$, ammonium molybdate; ND, not determined; NS, not specified.

The effects of ingestion of Mo in drinking water were investigated in residents of two Colorado cities exposed to low and high levels of Mo over a 2-yr period (EPA 1979, as reported in IRIS). The low-Mo group consisted of 42 individuals from Denver, Colorado (Mo drinking water concentrations ranged from 2 to 50 μg/L). The high-Mo group consisted of 13 college students from Golden, Colorado (Mo drinking water concentrations were greater than or equal to 200 μg/L). Urinary Mo and copper levels, and serum levels of ceruloplasmin and uric acid were compared between the two exposure groups. Subjects in the low-exposure group exhibited no adverse effects. Subjects in the high-exposure group had higher mean urinary Mo (187 vs. 87 μg-d), higher mean serum ceruloplasmin (40.31 vs. 30.41 mg/mL), and lower mean serum uric acid, levels (4.35 vs. 5.34 mg/100 mL) in comparison to the low-exposure group. The authors calculated NOAELs for both the low- and high-exposure groups (incorporating in the dietary Mo) of 4 μg/kg-d and 8 μg/kg-d, respectively, assuming a 2-L/d water consumption rate and a 70 kg body weight.

Deosthale and Gopalan (1974) investigated the effects of dietary Mo on uric acid and copper excretion in four adult males fed diets based on sorghum varieties containing widely varying amounts of Mo for 10 d. The urinary excretion of uric acid was unaltered at Mo intake levels up to 1,540 μg/d (about 0.022 mg/kg-d). Urinary excretion of copper increased in relationship to dietary Mo intake, where intake levels of 0.002 or 0.026 mg/kg-d resulted in urinary excretion of copper at 24 or 77 μg/d, respectively.

Fairhall et al. (1945), in a series of studies in groups of 10 rats, determined that the oral LD_{50} for daily administration of $CaMoO_4$ in food for approximately 4 mo was 101 mg/kg-d. Rats exhibited loss of appetite, weight loss, listlessness, and rough coat. The LD_{50} for MoO_3 and $(NH_4)_2MoO_4$ following similar dosing regimes were 125 and 333 mg/kg-d, respectively. Based on this data it appears that calcium trioxide is a reasonable surrogate for calcium molybdate.

Fairhall et al. (1945) injected male guinea pigs (six or eight/group) intraperitoneally with Mo compounds in 2 mL of isotonic chloride solution and observed their mortality at 4 d, 4 wk, and 4 mo. Mortality of guinea pigs dosed with calcium molybdate, in amounts of 0.1 g Mo, reported mortality ratios of 0/6, 0/6, and 1/6 at 4 d, 4 wk, and 4 mo. In contrast, in animals dosed with Mo trioxide (0.1 g Mo), the mortality ratios were 6/8, 6/8, and 6/8 at 4 d, 4 wk, and 4 mo, respectively. No additional data were provided.

Groups of guinea pigs (eight/dose) received oral doses of either molybdenum trioxide or calcium molybdate (dissolved in a 10% gum arabic solution) for approximately 95 d (Fairhall et al. 1945). Mortality rates for animals dosed with 25, 100, and 200 mg Mo/d were 1/8, 2/8, and 2/8, respectively, for calcium molybdate and 6/8, 8/8, and 8/8, respectively, for molybdate trioxide. Molybdenum trioxide appeared to be more toxic than calcium molybdate. The mortality rate of animals dosed with calcium molybdate did not rise above 25% for any dose group. No additional data on the nonlethal effects of these compounds were provided.

Jeter and Davis (1954) reported on the effects of Mo in the diets of male and female Long-Evans rats from weaning up to 11 wk. Rats (four or eight/sex/ group) were fed combinations of Mo (as NaMoO4· $2H_2O$) and copper (as $CuSo_4 \cdot 5H_20$) in their diets, ad libitum, with concentrations consisting of either <1 or 80 ppm Mo and 20 ppm copper or <1, 20, 80, or 140 ppm Mo and 5 ppm copper. (These doses were equivalent to about <0.1 or 8 mg/kg-d Mo and 2 mg/kg-d copper or <0.1, 2, 8, or 14 mg/kg-d Mo and 0.5 mg/kg-d copper, as reported in EPA 1999.) The growth rates of male rats fed 5 ppm of copper and 20, 80, or 140 ppm of Mo were significantly retarded. Growth rates were significantly retarded in female rats fed 5 ppm of copper and either 80 or 140 ppm of Mo. No effects on growth were reported in rats when the copper content in diets was 20 ppm. Achromotrichia (depigmentation of the hair) and alopecia were observed in some rats fed diets containing 80 or 140 ppm of Mo. Depigmentation was occasionally observed in rats receiving approximately 20 ppm Mo. Male infertility was observed in 75% of male rats fed 80 or 140 ppm of Mo. Examination of the testes revealed degeneration of the seminiferous tubules. The 2-mg/kg-d dose represents a LOAEL in this study based on retarded growth levels in male rats and depigmentation.

Zinc

Zinc is an essential nutrient. The NRC (1989) has established a recommended dietary allowance for zinc of 15 mg/d for males and 12 mg/d for females (NRC 1989). However, chronic ingestion of more than 15 mg/d in addition to dietary intake is not recommended without medical supervision because of the potential of aggravating copper levels in persons who are already marginally copper deficient (NRC 1989). The toxicity of zinc in humans is considered quite low, with toxicity normally occurring following ingestion of more than 2 g of zinc (Prasad 1976, as cited in ATSDR 1994; NRC 1989).

No human studies reported death following oral intake of high doses of zinc. Several LD_{50} values have been reported for rats and mice exposed to various zinc compounds. These LD_{50} values include 237, 293, 528, and 623 mg Zn/kg-d for rats and 86, 204, 605, and 390 mg Zn/kg-d for mice, following oral dosing with zinc acetate, zinc nitrate, zinc chloride, and zinc sulfate, respectively.

Ingestion of zinc-containing compounds has resulted in a variety of gastrointestinal, hematological, and renal effects in humans and animals. Vomiting, abdominal cramps, and diarrhea, in several cases with blood, have been observed following ingestion of zinc sulfate. In one case report, an English school girl who ingested 440 mg zinc sulfate/d (2.6 mg Zn^{2+}/kg-d) in capsules (a medically prescribed treatment for acne) reported epigastric discomfort (Moore 1978, as cited in ATSDR 1994). A week later she was admitted to the hospital. She was diagnosed with anemia and passed melanic stools, indicative of gastrointestinal bleeding. Gastrointestinal upset (abdominal cramps, vomiting, nausea) was reported in 26 of 47 healthy volunteers following ingestion of zinc sulfate tablets (150 mg Zn^{2+} in three divided doses/d, 2 mg Zn^{2+}/kg-d) for 6 wk (Samman and Roberts 1987, as cited in ATSDR 1994).

Gastrointestinal effects have also been observed in animals. Ferrets that ingested 390 mg Zn^{2+}/kg-d as zinc oxide for 2 wk experienced intestinal hemorrhages and a 75% reduction in food intake (Straube et al. 1980, as cited in ATSDR 1994). Mice fed a diet containing 1,110 mg/kg-d developed ulcers in the forestomach. No gastrointestinal effects were observed in rats fed 565 mg Zn^{2+}/kg-d (Maita et al. 1981).

Treatment-related changes in hematological parameters have been observed in humans and animals following oral dosing with zinc. Yadrick et al. (1989) conducted a 10-wk clinical study in 18 healthy women in which they investigated the effects of oral zinc supplements on copper and iron balance. Women were given supplements (as capsules) of 50 mg Zn^{2+}/d as zinc gluconate. Erythrocyte superoxide dismutase (ESOD) activity levels declined over the 10-wk supplementation period and at 10 wk were significantly different ($p < 0.05$) from values during the pretreatment period. Serum ferritin and hematocrit values were also significantly lower than pre-treatment values at 10 wk. Serum zinc was significantly increased. Ceruloplasmin levels were not altered. Fischer et al. (1984) reported on the effects of zinc supplementation in healthy adult male volunteers administered 50 mg Zn^{2+}/d as zinc gluconate for 6 wk. Volunteers had a statistically significant decrease in ESOD activity (15%) following the 6-wk exposure. There were no differences in serum copper levels or ceruloplasmin activity in the exposed group in comparison to the controls. Serum zinc levels were significantly increased in the exposed group after 2 wk.

Decreased hemoglobin, hematocrit, erythrocyte, and/or leukocyte levels have been observed in animals dosed orally with zinc compounds. Zaporowska and Wasilewski (1992) (as cited in ATSDR 1994) reported that the LOAEL in rats for decreased hemoglobin (85% of control values) was 12 mg Zn^{2+}/kg-d as zinc chloride in a 4-wk drinking water study with 2-mo old rats. Maita et al. (1981) fed mice (12/sex/group) a diet containing zinc sulfate at 0, 300, 3,000, or 30,000 ppm (equivalent to 0, 10, 104, 1,110 mg Zn^{2+}/kg-d) for 13 wk, observed significantly lower values in hematocrit and hemoglobin concentrations in the 3,000- and 30,000-ppm groups in comparison to controls; however, no dose response relationship was observed. The leukocyte count in male mice exposed to 30,000 ppm was decreased as well.

No human studies identified renal effects following oral exposures to zinc compounds. However, several animal studies have demonstrated adverse renal effects in animals exposed to zinc oxide, zinc sulfate, and zinc acetate. Zinc sulfate caused an increase in the absolute and relative kidney weights and regressive kidney lesions (not specified) in female mice that consumed 1,110 mg Zn^{2+}/kg-d as zinc sulfate in the diet for 13 wk, but no effects occurred in rats that consumed 565 mg Zn^{2+}/kg-d under similar conditions (Maita et al. 1981). In rats exposed to 191 mg Zn^{2+}/kg-d as zinc acetate for 3 mo, epithelial cell damage in the glomerulus and proximal convoluted tubules and increased plasma creatinine and urea levels were observed (Llobet et al. 1988). The NOAEL for the effects on creatinine and urea was 95 mg Zn^{2+}/kg-d. It is unclear whether the microscopic changes were observed at lower doses.

Immunological Effects

Molybdenum compounds

No human or animal data on immunological effects following oral exposures to Mo compounds were identified.

Zinc

Eleven healthy adult men who ingested 4.3 mg Zn^{2+}/kg-d for 6 wk experienced impaired mitogenic response elicited from peripheral blood lymphocytes and impaired chemotactic and phagocytic responses of polymorphonuclear leukocytes (Chandra 1984, as cited in ATSDR 1994). No effects were observed on total numbers of lymphocytes or relative numbers of T-cells or B-cells. No studies examining the immunological effects in animals following oral dosing with zinc were identified.

Neurological Effects

Molybdenum compounds

No data were identified on neurological effects of Mo compounds following oral exposure.

Zinc

Murphy (1970) (as cited in ATSDR 1994) reported on a 16-yr-old boy who ingested about 86 mg Zn^{2+}/kg-d (as metallic zinc) over a 2-d period in an effort to promote the healing of a wound. The boy developed signs and symptoms of lethargy, light-headedness, staggering, and difficulty in writing clearly.

Very limited data were located on neurological effects in animals. Rats dosed with 487 mg Zn^{2+}/kg-d as zinc oxide for 10 d (Kozik et al. 1980, as cited in ATSDR 1994) experienced minor neuron degeneration and proliferation of oligodendroglia. Kozik et al. (1981) (as cited in ATSDR 1994) reported that rats receiving 472 mg Zn^{2+}/kg-d for 10 d had increased levels of secretory material in the neurosecretory nuclei of the hypothalamus.

Reproductive and Developmental Effects

Molybdenum Compounds

Long-Evans rats (four or eight/sex/dose), fed varying quantities of Mo (as NaMoO4· $2H_2O$) and copper (as $CuSo_4{\cdot}5H_20$) in their diets (ad libitum) were mated with respective animals receiving the same doses (Jeter and Davis 1954). Concentrations of these compounds in diets consisted of either <1 or 80 ppm Mo and 20 ppm copper, respectively or <1, 20, 80, or 140 ppm Mo and 5 ppm copper. No marked effects on female fertility or gestation were observed. However, there was evidence of decreased lactation as observed by the low weaning weights of the litters, particularly for the pups of mothers fed 80 or 100 ppm Mo in their diets. No other effects on the offspring were reported. Mature virgin female rats fed 700 ppm Mo for 10 d showed irregular estrous cycles (Jeter and Davis 1954).

Zinc

No developmental effects were reported among newborns exposed to zinc compounds (0.06-0.3 mg zinc/kg-d) in utero during the second and third trimesters (Kynast and Saling 1986, as cited in ATSDR 1994; Mahomed et al. 1989, as cited in ATSDR 1994; Simmer et al. 1991, as cited in ATSDR 1994).

Bleavins et al. (1983) (as cited in ATSDR 1994) reported no measurable effect on gestational length or litter size when female mink ingested an average dose of 20.8 mg Zn^{2+}/kg-d as zinc sulfate. Maita et al. (1981) fed mice 1,110 mg Zn^{2+}/kg-d for 13 wk and found no effects on the testes or ovaries.

Schlicker and Cox (1968) administered 200 mg Zn^{2+}/kg-d (as zinc oxide) in the diet of rats for 21 d prior to mating and through gestation. During gestation, the 200-mg/kg-d dose group experienced 4-29% resorptions compared with 0% in controls. When the dose was reduced to 100 mg Zn^{2+}/kg-d (21 d prior to mating), no fetal resorptions, malformations, or growth reduction were reported.

Administration of 200 mg Zn^{2+}/kg-d to dams throughout gestation resulted in decreased growth and tissue levels of copper and iron in fetal rats (Cox et al. 1969; Schlicker and Cox 1968). During gestational d 1-18, maternal zinc levels increased in the 100- and 200-mg/kg-d dose groups. However, zinc tissue levels in the 22-d-old fetuses were not elevated in dams dosed with 100 mg/kg-d, suggesting that the placenta was able to act as a barrier to zinc at the lower dietary level. In contrast, Ketcheson et al. (1969) (as cited in ATSDR 1994) reported that newborn and 14-d old rats, from mothers who had consumed 100 mg/kg-d throughout gestation, had elevated levels of total zinc and decreased levels of iron.

Cancer

Molybdenum Compounds

No data were identified on carcinogenic effects of Mo compounds following oral exposure.

Zinc

Limited human and animal data on the carcinogenicity of zinc following oral exposures exist. Two epidemiological studies reported conflicting results on the association between high zinc soil levels and cancer. In a survey of cancer registry data (1954-1978) in Shipham, Somerset (Great Britain), an area with a high soil zinc-to-copper ratio (≈17:1), Philipp et al. (1982) (as cited in ATSDR 1994) found that the gastric cancer incidence rate was significantly lower than the regional rate. In contrast, Stocks and Davies (1964) (as cited in ATSDR 1994) found an association between an excess rate of gastric cancer in people of North Wales and high zinc-to-copper ratios (≈30:1) in the soil of household gardens. It is possible that other factors, not considered by Stocks and Davies (1964) (as cited in ATSDR 1994), may have accounted for the observed association.

Walters and Roe (1965) (as cited in ATSDR 1994) reported that the incidence of tumors following exposure of mice to 951 mg Zn^{2+}/kg-d as zinc sulfate in drinking water for 1 yr was not increased as compared to controls. However, this study lacked important details and had several limitations, including high numbers of deaths in control mice.

Genotoxicity

Molybdenum Compounds

Calcium and zinc molybdates do not appear to be genotoxic based on limited data available. Molybdenum trioxide was reported to be negative in the *Bacillus subtilis* rec assay (Kada et al. 1980, as cited in NTP 1997), and not mutagenic in any of five strains of *Salmonella typhimurium tested*, with or without S9 metabolic activation enzymes (Zeiger et al. 1992, as reported in NTP 1997). Molybdenum trioxide did not induce sister-chromatid exchanges or chromosomal aberrations in cultured Chinese hamster ovary cells.

Zinc

Bauchinger et al. (1976) found that the incidence of chromosomal aberrations was increased in 24 workers in a zinc smelting plant. However, these workers also had increased blood levels of lead and cadmium, and the authors attributed the increase in the incidence of chromosomal aberrations to cadmium exposure.

A number of in vivo studies have reported that zinc salts were clastogenic when administered by various routes of exposure (see Table 9-3a). Kowalska-Wochna et al. (1988) (as cited in ATSDR 1994) found an increase in the incidence of chromosomal aberrations in rats exposed to 14.8 mg Zn^{2+}/kg-d in their drinking water. An increase in the incidence of chromosomal aberrations was observed in mice given intraperitoneal injections of 3.6 mg Zn^{2+}/kg-d (Gupta et al. 1991) and mice exposed to zinc oxide by inhalation (Voroshilin et al. 1978, as cited in ATSDR 1994). An increased incidence of sister-chromatid exchange was observed in bone marrow cells of rats exposed to 17.5 mg Zn^{2+}/kg-d in drinking water (Kowalska-Wochna et al. 1988, as cited in ATSDR 1994). Zinc was negative for genotoxicity when tested in murine somatic or germ cells (Vilkina et al. 1978, as cited in ATSDR 1994).

Zinc has been found to be negative for mutagenic activity in bacterial systems in vitro (see Table 9-3b). In vitro genotoxicity testing of zinc using mouse lymphocytes has resulted in inconsistent findings for mutagenicity. Limited in vitro testing suggests that zinc may have clastogenic activity in human lympho-

TABLE 9-3a Genotoxicity of Zinc In Vivo

Species	Genotoxicity End Point	Results	Reference
Mouse	Dominant lethal mutation	Negative	Vilkina et al. 1978, as cited in ATSDR 1994
Mouse	Micronucleus induction	Positive	Gocke et al. 1981
Mouse	Chromosomal aberrations	Positive	Vilkina et al. 1978, as cited in ATSDR 1994
Mouse	Chromosomal aberrations	Positive	Deknudt and Gerber 1979
Mouse	Chromosomal aberrations	Positive	Gupta et al. 1991
Rat	Chromosomal aberrations	Positive	Kowalska-Wochna 1988, as cited in ATSDR 1994
Rat	Sister chromatid exchange	Positive	Kowalska-Wochna 1988, as cited in ATSDR 1994
Drosophila	Sex-linked recessive lethal mutation	Negative	Gocke et al. 1981

cytes (Deknudt and Deminatti 1978) and Chinese hamster ovary cells (Thompson et al. 1989, as cited in ATSDR 1994).

QUANTITATIVE TOXICITY ASSESSMENT

Noncancer

Dermal Assessment

There are inadequate dermal toxicity data to perform a dermal RfD. The limited data available suggest that calcium/zinc molybdates are not skin irritants. No irritation effects were observed when Mo compounds were applied to the intact or abraded skin of rabbits. Systemic effects following long-term dermal exposure have not been reported.

Inhalation RfC

The 2-yr NTP study (NTP 1997) was used for the derivation of the RfC for calcium and zinc molybdates. This study identified a LOAEL of 10 mg MoO_3/m^3 based on increased incidences of hyaline degeneration in the nasal respiratory epithelium and nasal olfactory epithelium in female rats. No NOAEL was identified in this study.

A composite uncertainty factor of 3,000 was applied to the LOAEL that consists of a factor of 3 to account for database deficiencies including lack of

TABLE 9-3b Genotoxicity of Zinc in Vitro

Species	Genotoxicity End Point	Results: With Metabolic Activation	Results: Without Metabolic Activation	Reference
EUKARYOTIC				
Mouse lymphoma cells	Mutation	NA	Negative	Amacher and Paillet 1980
Mouse lymphoma cells	Mutation	Positive	Positive	Thompson et al. 1989, as cited in ATSDR 1994
Human lymphocytes	Chromosomal aberrations	NA	Positive	Deknudt and Deminatti 1978
Chinese hamster ovary cells	Chromosomal aberrations	Positive	Positive	Thompson et al. 1989, as cited in ATSDR 1994
PROKARYOTIC				
Salmonella typhimurium (TA102)	Mutation	NA	Negative	Marzin and Phi 1985
Salmonella typhimurium (TA98, TA102, TA1535, TA1537)	Mutation	Negative	Negative	Wong 1988
Salmonella typhimurium (TA98, TA100, TA1537, TA 1538)	Mutation	Negative	Negative	Thompson et al. 1989, as cited in ATSDR 1994
Eschericia coli	Mutation	NA	Negative	Nishioka 1975
Eschericia coli	Mutation	NA	Negative	Venitt and Levy 1974

NA, not applicable

a reproductive/developmental study, a factor of 10 for extrapolation from rats to humans, a factor of 10 for intraspecies differences, and a factor of 10 for extrapolation from a LOAEL to a NOAEL in a long-term study. Therefore, the RfC was determined to be 0.003 mg MoO_3/m^3. Since Mo comprises approximately 67%, 48%, and 43% by weight of MoO_3, $CaMoO_4$, and $ZnMoO_4$, respectively, the RfC for MoO_3 was multiplied by $(0.67/0.48)^{-1}$ (for $CaMoO_4$) or $(0.67/0.43)^{-1}$ (for $ZnMoO_4$) to yield RfCs for both calcium and zinc molybdates of 0.002 mg/m^3 (see Table 9-4). Since there are few other supporting data from inhalation studies examining similar effects, confidence in the database and in the inhalation RfC are medium.

TABLE 9-4 Inhalation Reference Concentration for Calcium and Zinc Molybdates

Critical effect	Species	Effect level (mg/m^3)	Uncertainty factors	RfC[b] (mg/m^3-d)	Reference
Increased incidences of hyaline degeneration in nasal respiratory epithelium and olfactory epithelium	Female rats	LOAEL[a]: 10	UF_A: 10 UF_H: 10 UF_L: 10 UF_D: 3 Total: 3,000	0.002	NTP (1997)

LOAEL, lowest-observed-adverse-effect level; RfC, reference concentration; UF_A, extrapolation from animals to humans; UF_H, intraspecies variability; UF_L, extrapolation from a LOAEL to a NOAEL; UF_D, inadequate or deficient toxicity database

[a]LOAEL based on exposure to molybdenum trioxide.

[b]RfC calculated based on percent by weight of molybdenum in the ratio of molybdenum trioxide to calcium molybdates or zinc molybdates (0.67/0.43 or 0.67/0.48, respectively).

Oral RfD

The Kovalsky et al. (1961) study was used for derivation of the oral RfD for calcium and zinc molybdates. In this epidemiological study, increases in uric acid levels and copper excretion, and elevated serum ceruloplasmin, were observed and a LOAEL of 0.14 mg/g-d Mo was identified. No NOAEL was identified in this study for Mo. The RfD was derived with consideration for the estimated safe and adequate daily intake (ESAADI) of 15-40 µg/d for infants, 25-150 µg/d for children, and 75-250 µg/d for adolescents and adults (NRC 1989). Data supporting selection of this key study are provided by several other epidemiological studies (Deosthale and Gopalan 1974; EPA 1979) that found an association between elevated dietary exposure to Mo and increased serum ceruloplasmin and urinary excretion of copper. The LOAEL for Mo is further supported by animal data demonstrating that the toxicological effects of Mo are more pronounced when dietary copper levels are low.

The subcommittee, in calculating the oral RfD, applied an uncertainty factor of 100 to the LOAEL consisting of 10 to account for intraspecies differences and 10 to extrapolate from a LOAEL to a NOAEL in a long-term study (EPA 1999). Based on this calculation, the RfD was determined to be 1×10^{-3} mg Mo/kg-d. Since Mo comprises approximately 43% by weight of zinc molybdate ($ZnMoO_4$) or 48% by weight of calcium molybdate ($CaMoO_4$), the RfD for Mo

TABLE 9-5 Oral Reference Dose for Calcium and Zinc Molybdates

Critical effect	Species	Effect level (mg/kg-d)	Uncertainty factors	RfD[b] (mg/kg-d)	Reference
Increased uric acid levels	Human	LOAEL[a]: 0.14	UF_H: 10 UF_L: 10 Total: 100	0.0006	Kovalsky et al. (1961)

LOAEL, lowest-observed-adverse-effect level; RfD, reference dose; UF_H, intraspecies variability; UF_L, NOAEL for critical effect not determined

[a]LOAEL based on exposure to molybdenum.

[b]RfD calculated based on percent by weight of molybdenum in calcium molybdates and zinc molybdates (43% and 48%, respectively).

was multiplied by 0.43^{-1} ($ZnMoO_4$) or 0.48^{-1} ($CaMoO_4$) to yield RfDs for zinc or calcium molybdates of 0.0006 mg/kg-d (see Table 9-5).

The subcommittee selected an RfD for zinc based on a LOAEL of 1.0 mg Zn^{2+}/kg-d for decreased erythrocyte superoxide dismutase (ESOD) activity in human adult females after 10 wk of exposure to zinc supplements (Yadrick et al. 1989). The change in enzyme activity reflects an alteration in copper levels. This study (Yadrick et al 1989) is supported by data from several other clinical studies demonstrating the effect of zinc on copper balance (Prasad et al. 1978; Fischer et al. 1984). The subcommittee, in calculating an RfD, applied an uncertainty factor of 3, based on the minimal LOAEL of 1.0 mg Zn^{2+}/kg-d from a moderate-duration study of the most sensitive humans and with consideration that zinc is an essential dietary nutrient. Using this uncertainty factor, the RfD was determined to be 0.3 mg Zn^{2+}/kg-d. Because zinc comprises 29% by weight of $ZnMoO_4$, the RfD for zinc was multiplied by 0.29^{-1} to derive an RfD for zinc molybdate of approximately 0.10 mg/kg-d.

The subcommittee selected an RfD for calcium and zinc molybdates of 0.0006 mg/kg-d, based on the toxic effects from exposure to Mo. The subcommittee believes that an RfD based on the toxicity of Mo rather than on zinc, provides a greater margin of safety, because the RfD for Mo is considerably lower than that of zinc (0.001 vs. 0.3 mg/kg-d).

Confidence in the key study is medium. The study examined only gross physical effects and certain hematological parameters associated with gout (Kovalsky et al. 1961). A detailed analysis of blood chemistry and individual dietary habits was not conducted. Confidence in the database is medium because of the lack of other studies examining a broader range of hematological and clinical chemistry parameters. Therefore, confidence in the oral RfD is medium.

Cancer

Dermal

No studies were identified regarding the carcinogenicity of calcium or zinc molybdates or other Mo compounds following dermal exposure in humans or in experimental animals. Therefore, the subcommittee concluded that the carcinogenicity of calcium and zinc molybdates cannot be determined based on available data.

Inhalation

Inhaled molybdenum trioxide was carcinogenic in male and female mice based on a single NTP study (NTP 1997). There was equivocal evidence for its carcinogenicity for male rats. Available data suggests that these compounds are not carcinogenic.

Based on the data currently available, the subcommittee concluded that the weight of evidence suggests that calcium and zinc molybdates may be carcinogenic to humans. Therefore, the subcommittee derived a cancer slope factor for characterizing the carcinogenic risk from exposure to these chemicals.

The cancer slope factor was derived using the multistage model (EPA 1996). Modeling was conducted using the adenoma/carcinoma incidence data (combined) in female mice (3/50, 6/50, 8/49, and 15/49 for the 0-, 10-, 30- and 100-mg/m^3 exposure groups, respectively) (NTP 1997) (see Table 9-6). The female mice data were used instead of the male mice data because female mice were more sensitive. Exposure concentrations were normalized for continuous exposure and were converted to human equivalent concentrations (HEC) using the regional deposited dose ratios (RDDR) based on the aerodynamic particle size generated in the NTP (1997) study. Based on linear extrapolation, the unit risk of lung cancer is less than $2.6 \times 10^{-5}/\mu g/m^3$.

Oral

No studies were identified regarding the carcinogenicity of calcium or zinc molybdates or other Mo compounds following oral exposure in humans or experimental animals. Therefore, the subcommittee concluded that there are insufficient data to determine its carcinogenicity.

TABLE 9-6 Calculation of LED_{10} and $0.1/LED_{10}$ for Molybdenum Trioxide Using Incidence of Lung Carcinoma and Adenoma in Female Mice (NTP 1997)

Assay concentration (mg/m^3)[a]	Assay concentration (mg/m^3), duration adjusted[b]	HEC (mg/m^3)[c]	Tumor response (lung adenomas and carcinomas)	LED_{10} (mg/m^3)[d]	$0.1/LED_{10}$ (per mg/m^3)
0	0	0	3	3.8	0.026
10	1.75	1.8	6		
30	5.25	5.4	8		
100	17.5	17.7	15		

HEC, human equivalent concentrations; LED_{10}, the lower 95% confidence bound on the effective dose that causes a 10% tumor response in animals; MoO_3, molybdenum trioxide.

[a]Female mice were exposed to air concentrations of particulate MoO_3 for 6 hr/d, 5 d/wk for 103 wk.

[b]MoO_3 concentrations were normalized for continuous chronic exposure by multiplying by 6/24 hr and 5/7 d (EPA 1994).

[c]Normalized assay concentrations were converted to HEC using the regional deposited dose ratios for the mouse pulmonary region as recommended by EPA (1994). Particulate mass median aerodynamic diameter (MMAD) and geometric standard deviations listed for each concentration by NTP (1997) were used (μm): 10 mg/m^3: MMAD = 1.3, $\sigma_g = 1.8$; 30 mg/m^3: MMAD = 1.4, $\sigma_g = 1.8$; 100 mg/m^3: MMAD = 1.5, $\sigma_g = 1.8$.

[d]LED_{10} calculated using the multistage model.

EXPOSURE ASSESSMENT AND RISK CHARACTERIZATION

Noncancer

Dermal Exposure

The assessment of noncancer risk for the dermal exposure route is based on the dermal exposure scenario described in Chapter 3. This exposure scenario assumes that an adult spends 1/4th of his or her time sitting on furniture upholstery treated with calcium or zinc molybdates and also assumes 1/4th of the upper torso is in contact with the upholstery and clothing presents no barrier. Calcium and zinc molybdates are considered to be ionic, and are essentially not absorbed through the skin. However, to be conservative, the subcommittee assumed that ionized calcium and zinc molybdates permeate the skin at the same rate as water, with a permeability rate of 10^{-3} cm/hr (EPA 1992). Using that permeability rate, the highest expected application rate for calcium and zinc molybdates (2 mg/cm^3), and Equation 1 in Chapter 3, the subcommittee calculated a dermal exposure level of 6.3×10^{-3} mg/kg-d. The oral RfD for calcium and zinc molybdates (6.0×10^{-4}; see Oral RfD in Quantitative Toxicity section) was used as the best estimate of the internal dose for dermal exposure. Dividing the exposure level by the oral RfD yields a hazard index of 10. Therefore it was concluded that calcium and zinc molybdates used as flame retardants in upholstery fabric may pose a non-cancer risk by the dermal route at the specified concentration and under the given worst-case exposure scenario.

Inhalation Exposure

Particles

Inhalation exposure estimates for calcium and zinc molybdates were calculated using the exposure scenario described in Chapter 3. This scenario assumes that a person spends a quarter of his or her life in a room with low air-change rates (0.25/hr) and with a relatively large amount of fabric upholstery treated with calcium or zinc molybdate (30 m^2 in a 30 m^3 room), with this calcium or zinc molybdate treatment gradually being worn away over 25% of its surface to 50% of its initial quantity over the 15-yr lifetime of the fabric. A small fraction, 1%, of the worn-off calcium or zinc molybdate, is released into the indoor air as inhalable particles, and may be breathed by the occupant.

Particle exposure was estimated using Equations 4 through 6 in Chapter 3. The release rate (μ_r) for calcium and zinc molybdates for upholstery, 2.3×10^{-7}/d (Equation 5), was used in conjunction with the upholstery application rate (S_a) for calcium and zinc molybdates of 2 mg/cm^2 to calculate a room airborne particulate concentration of 0.76 µg/m^3 (Equation 4). Factoring in the fraction of a day a person spends in the room containing upholstery (0.25), the time-average exposure concentration was determined to be 0.19 µg/m^3 (Equation 6).

The inhalation RfC for calcium or zinc molybdate is 2×10^{-3} mg/m^3 (see Inhalation RfC section). A hazard index was calculated as the ratio of the time-averaged exposure concentration to this estimated RfC, yielding a value of 0.095. This indicates that calcium or zinc molybdates, used as upholstery flame retardants, are not likely to pose any noncancer risk via inhalation in the particulate phase.

Vapors

Calcium and zinc molybdates have negligible vapor pressures at ambient temperatures. Therefore calcium or zinc molybdates used as upholstery-fabric flame retardants are not likely to pose any noncancer risks, when exposure occurs in the vapor phase.

Oral Exposure

The assessment of the noncancer risk for the oral exposure route is based on the scenario described in Chapter 3. This scenario assumes a child is exposed to calcium and zinc molybdates through sucking on 50 cm^2 of fabric daily for two yr, 1 hr/d. The dose rate to the child was calculated using Equation 15 in Chapter 3. Parameters specific to calcium and zinc molybdates that were used in this calculation included an upholstery application rate (S_a) of 2 mg/cm^2 and an extraction rate (μ_a) by saliva of 0.0004/d. This extraction rate was based on data from US Borax on zinc and boron extraction from polymer films (PVC and paint film) (Borax 1996).

Using these values, the average oral dose rate was estimated to be 1.7×10^{-4} mg/kg-d. The oral dose rate (1.7×10^{-4} mg/kg-d) was divided by the oral RfD of 0.0006 mg/kg-d, giving a hazard index of 0.28. The subcommittee concluded that calcium or zinc molybdate used as an upholstery fabric flame retardant is not likely to pose any noncancer risk by the oral route.

Cancer

Dermal Exposure

Based on inadequate data on the carcinogencity of calcium and zinc molybdates via the dermal route, the subcommittee concludes that there are insufficient data to assess its carcinogencity.

Inhalation Exposure

Particles

The average room-air concentration and average exposure concentration to calcium or zinc molybdate were obtained as described in the Noncancer section. Using the inhalation unit risk of $2.6 \times 10^{-5}/\mu g/m^3$, the lifetime risk estimate from exposure to calcium or zinc molybdate in the particulate phase is 5.0×10^{-6} (see Table 9-6).

Vapors

Calcium and zinc molybdates have negligible vapor pressures at ambient temperatures. Therefore, calcium or zinc molybdate used as an upholstery-fabric flame retardant is not likely to pose any cancer risk via inhalation in the vapor phase.

Oral Exposure

Based on inadequate data on the carcinogencity of calcium and zinc molybdates via the oral route, the subcommittee concludes that there are insufficient data to assess its carcinogencity.

RECOMMENDATIONS FROM OTHER ORGANIZATIONS

The OSHA permissible exposure limit (PEL) recommended for this compound is 5 mg/m^3 for soluble compounds. The TLV-TWA (Threshold Limit Value–time weighted average) established by the American Conference of Governmental Industrial Hygienists (ACGIH) is also 5 mg/m^3 (ACGIH 1991) for soluble Mo compounds. Additionally, several other countries have adopted a permissible exposure level of 5 mg/m^3 for soluble Mo including Australia, Federal Republic of Germany, Sweden, and the United Kingdom. The EPA, as detailed in IRIS, has established an oral RfD of 5×10^{-3} mg/kg-d for Mo and an oral RfD for zinc of 3×10^{-1} mg/kg-d. The National Research Council has established RDAs for Mo of 75-250 µg/d (1.07-3.57 µg/kg-d for a 70-kg person) and for zinc of 12-15 mg/d (0.17-0.21 mg/kg-d of zinc for a 70-kg person), respectively.

DATA GAPS AND RESEARCH NEEDS

There is a substantial amount of data available on zinc, calcium, and molybdates. For instance, the oral RfD, inhalation RfC, and cancer potency factor determined by the subcommittee for calcium and zinc molybdates are based on molybdenum. Because of the calculated inhalation lifetime cancer risk for calcium and zinc molybdates, the subcommittee believes that the potential of these chemicals to be released as particles from fabric needs to be investigated. Because of a dermal hazard index greater than 1, the dermal absorption of calcium and zinc molybdates from treated fabric should be investigated.

REFERENCES

ACGIH (American Conference of Government Industrial Hygienists). 1991. Molybdenum and Compounds. Pp. 1051-1054 in Documentation of the Threshold Limit Values and Biological Exposure Indices, 6th Ed. Cincinnati, OH: American Conference of Government Industrial Hygienists, Inc.

Agren, M.S. 1990. Percutaneous absorption of zinc from zinc oxide applied topically to intact skin in man. Dermatologica 180(1):36-39.

Agren, M.S., M. Kusell, and L. Frazen. 1991. Release and absorption of zinc from zinc oxide and zinc sulfate in open wounds. Acta Derm. Venereol. 71:330-333.

Akopyan, O.A. 1964. Some biochemical shift in workers having contact with molybdenum-containing dust. [Abstract]. Pp. 65-67 in Materials from the 2nd Summary Scientific Conference on Labor Hygiene and Work-Related Pathology of the Institute of Labor Hygiene and Work-Related Illnesses.

Alexander, J., J. Aaseth, and T. Refsvik. 1981. Excretion of zinc in rat bile - a role of glutathione. Acta Pharmacol. Toxicol. (Copenh.) 49(3):190-194.

Amacher, D.E., and S.C. Paillet. 1980. Induction of trifluorothymidine-resistant mutants by metal ions in L5178Y/TK$^{+/-}$ cells. Mutat. Res. 78(3):279-288.

Ameille, J., J.M. Brechot, P. Brochard, F. Capron, and M.F. Dore. 1992. Occupational hypersensitivity pneumonitis in a smelter exposed to zinc fumes. Chest 101(3):862-863.

Arrington, L.R., and G.K. Davis. 1955. Metabolism of phosphorus^{32} and molybdenum^{99} in rats receiving high calcium diets. J. Nutr. 55(2):185-192.

ATSDR (Agency for Toxic Substances and Disease Registry). 1994. Toxicological Profile for Zinc (Update). U.S. Department of Health and Human Services, Public Health Service, Agency for Toxic Substances and Disease Registry, Atlanta, GA. TP-93/15. 230 pp.

Batchelor, R.P., J.W. Fehnel, R.M. Thompson, et al. 1926. A clinical and laboratory investigation of the effect of metallic zinc, of zinc oxide, and of zinc sulphide upon the health of workmen. J. Ind. Hyg. 8:322-363.

Bauchinger, M., E. Schmid, H.J. Einbrodt, and J. Dresp. 1976. Chromosome aberrations in lymphocytes after occupational exposure to lead and cadmium. Mutat. Res. 40(1):57-62.

Blanc, P., H. Wong, M.S. Bernstein, and H.A. Boushey. 1991. An experimental human model of metal fume fever. Ann. Intern. Med. 114(11):930-936.

Bleavins, M.R., R.J. Aulerich, J.R. Hochstein, T.C. Hornshaw, and A.C. Napolitano. 1983. Effects of excessive dietary zinc on the intrauterine and postnatal development of mink. J. Nutr. 113(11):2360-2367.

Borax (U.S. Borax, Inc.). 1996. Material Safety Data Sheet: Firebrake ZB. U.S. Borax, Inc., Valencia, CA.

Budavari, S., M.J. O'Neil, A. Smith., P.E. Heckelman. 1989. The Merck Index. 11th Ed. Rahway, NJ.

Chandra, R.K. 1984. Excessive intake of zinc impairs immune responses. JAMA 252(11):1443-1446.

Cooke, J.A., S.M. Andrews, and M.S. Johnson. 1990. The accumulation of lead, zinc, cadmium, and fluoride in the wood mouse (Apodemus sylvaticus L.). Water Air Soil Pollut. 51:55-63.

Cox, D.H., S.A. Schlicker, and R.C. Chu. 1969. Excess dietary zinc for the maternal rats, and zinc, iron, copper, calcium, and magnesium content and enzyme activity in maternal and fetal tissues. J. Nutr. 98(4):459-466.

Deknudt, G., and G.B. Gerber. 1979. Chromosomal aberrations in bone-marrow cells of mice given a normal or a calcium-deficient diet supplemented with various heavy metals. Mutat. Res. 68(2):163-168.

Deknudt, G., and M. Deminatti. 1978. Chromosome studies in human lymphocytes after in vitro exposure to metal salts. Toxicology 10(1):67-75.

Deosthale, Y.G., and C. Gopalan. 1974. The effect of molybdenum levels in sorghum (*Sorghum vulgare* Pers.) on uric acid and copper excretion in man. Br. J. Nutr. 31(3):351-355.

Drinker, K., and P. Drinker. 1928. Metal fume fever: V. Results of the inhalation by animals of zinc and magnesium oxide fumes. J. Ind. Hyg. 10:56-70.

DuBray, E.S. 1937. Chronic zinc intoxication. JAMA 108:383-385.

EPA (U.S. Environmental Protection Agency). 1979. Human Health Effects of Molybdenum in Drinking Water. EPA-600A-79-006. Health Effects Research Laboratory, Office of Research and Development, U.S. Environmental Protection Agency, Cincinnati, OH.

EPA (U.S. Environmental Protection Agency). 1992. Dermal Exposure Assessment: Principles and Applications. EPA/600/8-91-011B. Office of Health and Environmental Assessment, U.S. Environmental Protection Agency, Washington, DC.

EPA (U.S. Environmental Protection Agency). 1994. Methods for Derivation of Inhalation Reference Concentrations and Application of Inhalation Dosimetry. EPA/600/8-90/066F. Environmental Criteria and Assessment Office, Office of Health and Environmental Assessment, Office of Research and Development, U.S. Environmental Protection Agency, Research Triangle Park, NC.

EPA (U.S. Environmental Protection Agency). 1996. Proposed Guidelines for Carcinogen Risk Assessment. Fed. Regist. 61(79):17959-18011.

EPA (U.S. Environmental Protection Agency). 1999. Integrated Risk Information System. Office of Research and Development, National Center for Environmental Assessment, Cincinnati, OH. [Online]. Available: http://www.epa.gov/ngispgm3/iris/

Evans, E.H. 1945. Casualties following exposure to zinc chloride smoke. Lancet ii:368-370.

Fairhall, L.T., R.C. Dunn, N.E. Sharpless, and E.A. Pritchard. 1945. Pp. 1-36; 40-41 in Public Health Bulletin No. 293, The Toxicity of Molybdenum. Washington, DC: U.S. Government Printing Office.

Farrell, F.J. 1987. Angioedema and urticaria as acute and late phase reactions to zinc fume exposure, with associated metal fume fever-like symptoms. Am. J. Ind. Med. 12(3):331-337.

Fischer, P.W., A. Giroux, and M.R. L'Abbe. 1984. Effect of zinc supplementation on copper status in adult man. Am. J. Clin. Nutr. 40(4):743-746.

FRCA (Fire Retardant Chemicals Association). 1998. Textile Flame Retardant Applications by Product Classes for 1997 Within and Outside of the United States. Submitted to: U.S. Consumer Product Safety Commission, July 1998. Lancaster, PA: Fire Retardant Chemicals Association.

Galvez-Morros, M., O. Garcia-Martinez, A.J.A. Wright, and S. Southon. 1992. Bioavailability in the rat of zinc and iron from the basic salts $Zn_5(OH)_8Cl_2.H_2O$, $Fe(OH)SO_4$, and $Fe_4(OH)_{11}NO_3.2H_2O$. Food Chem. 43:377-381.

Gocke, E., M.T. King, K. Eckhardt, and D. Wild. 1981. Mutagenicity of cosmetics ingredients licensed by the European Communities. Mutat. Res. 90(2):91-109.

Goyer, R.A. 1996. Toxic Effects of Metals. Pp. 691-736 in Casarett and Doull's Toxicology: The Basic Science of Poisons, C.D. Klaassen, M.O. Amdur, and J. Doull, eds. New York: Macmillan Publishing Company.

Gupta, T., G. Talukder, and A. Sharma. 1991. Cytotoxicity of zinc chloride in mice in vivo. Biol. Trace Elem. Res. 30(2):95-101.

Hallmans, G. 1977. Treatment of burns with zinc-tape. A study of local absorption of zinc in humans. Scand. J. Plast. Reconstr. Surg. 11(2):155-161.

Hamdi, E.A. 1969. Chronic exposure to zinc of furnace operators in a brass foundry. Br. J. Ind. Med. 26(2):126-134.

Hammond, J.W. 1944. Metal fume fever in crushed stone industry. J. Ind. Hyg. Toxicol. 26:117.

Hatlelid, K. 1999. Toxicity Review for Calcium Molybdate and Zinc Molybdate. Memorandum, dated March 2, 1999, from Kristina Hatlelid, Toxicologist, Division of Health Sciences, to Ronald Medford, Assistant Executive Director for Hazard Identification and Reduction, U.S. Consumer Product Safety Commission, Washington, DC.

He, L.S., X.S. Yan, and D.C. Wu. 1991. Age-dependent variation of zinc-65 metabolism in LACA mice. Int. J. Radiat. Biol. 60(6):907-916.

Hjortso, E., J. Qvist, M.I. Bud, J.L. Thomsen, J.B. Andersen, F. Wiberg-Jorgensen, N.K. Jensen, R. Jones, L.M. Reid, and W.M. Zapol. 1988. ARDS after accidental inhalation of zinc chloride smoke. Intensive Care Med. 14(1):17-24.

Jeter, M.A., and G.K. Davis. 1954. The effect of dietary molybdenum upon growth, hemoglobin, reproduction and lactation of rats. J. Nutr. 54:215-220.

Johnson, F.A., and R.B. Stonehill. 1961. Chemical pneumonitis from inhalation of zinc chloride. Dis. Chest 40:619-623.

Kada, T., K. Hirano, and Y. Shirasu. 1980. Screening of environmental chemical mutagens by the rec-assay system with *Bacillus subtilis*. Pp. 149-173 in Chemical Mutagens: Principles and Methods for Their Detection, Vol. 6. F.J. de Serres and A. Hollaender, eds., New York: Plenum Press.

Keen, C.L., and L.S. Hurley. 1977. Zinc absorption through skin: correction of zinc deficiency in the rat. Am. J. Clin. Nutr. 30(4):528-530.

Ketcheson, M.R., G.P. Barron, and D.H. Cox. 1969. Relationship of maternal dietary zinc during gestation and lactation to development and zinc, iron and copper content of the postnatal rat. J. Nutr. 98(3):303-311.

Kovalsky, V.V., G.A. Yarovaya, and D.M. Shmavonyan. 1961. Changes of purine metabolism in humans and animals in molybdenum-rich biogeochemical provinces. [Article in Russian]. Zh. Obshch. Biol. 22(3):179-191.

Kowlaska-Wochna, E. J. Moniuszko-Jakoniuk, E. Kulikowska, et al. 1988. The effect of orally applied aqueous solutions of lead and zinc on chromosome aberrations and induction of sister chromatid exchanges in the rat (*Rattus sp.*) Genetica Polonica 29(2):181-189.

Kozik, M.B., L. Maziarz, and A. Godlewski. 1980. Morphological and histochemical changes occurring in the brain of rats fed large doses of zinc oxide. Folia Histochem. Cytochem. 18:201-206.

Kozik, M.B., G. Gramza, and M. Pietrzak. 1981. Neurosecretion of the hypothalamo-hypophyseal system after intragastric adminstration of zinc oxide. Folia Histochem. Cytochem. 10:115-122.

Kynast, G., and E. Saling. 1986. Effect of oral zinc application during pregnancy. Gynecol. Obstet. Invest. 21(3):117-123.

Lansdown, A.B. 1991. Interspecies variations in response to topical application of selected zinc compounds. Food Chem. Toxicol. 29(1):57-64.

Lener, J., and B. Bibr. 1984. Effects of molybdenum on the organism (A review). J. Hyg. Epidemiol. Microbiol. Immunol. 29(4):405-419.

Llobet, J.M., J.L. Domingo, M.T. Colomina, E. Mayayo, and J. Corbella. 1988. Sub-chronic oral toxicity of zinc in rats. Bull. Environ. Contam. Toxicol. 41:36-43.

Logue, J.N., M.D. Koontz, and M.A.W. Hattwick. 1982. A historical prospective mortality study of workers in copper and zinc refineries. J. Occup. Med. 24(5):398-408.

Mahomed K., D.K. James, J. Golding, and R. McCabe. 1989. Zinc supplementation during pregnancy: A double blind randomised controlled trial. BMJ 299(6703):826-833.

Maita, K., M. Hirano, K. Mitsumori, K. Takahashi, and Y. Shirasu. 1981. Subacute toxicity studies with zinc sulfate in mice and rats. J. Pesticide Sci. 6:327-336.

Marrs, T.C., H.F. Colgrave, J.A. Edington, R.F. Brown, and N.L. Cross. 1988. The repeated dose toxicity of a zinc oxide/hexachloroethane smoke. Arch. Toxicol. 62(2-3):123-132.

Marzin, D.R., and H.V. Phi. 1985. Study of the mutagenicity of metal derivatives with Salmonella typhimurium TA102. Mutat. Res. 155(1-2):49-51.

McCord, C.P., A. Friedlander, W.E. Brown, et al. 1926. An occupational disease among zinc workers. Arch. Intern. Med. 37:641-659.

Milliken, J.A., D. Waugh, and M.E. Kadish. 1963. Acute interstitial pulmonary fibrosis caused by a smoke bomb. Can. Med. Assoc. J. 88:36-39.

Mogilevskaya, O.Y. 1967. Experimental studies on the effect on the organism of rare, dispersed and other metals and their compounds used in industry: Molybdenum. Pp. 12-28 in Toxicology of the Rare Metals. [Translated from Russian] Z.I. Izrael'son, ed. Israel Program for Scientific Translations, Jerusalem.

Moore. R. 1978. Bleeding gastric erosion after oral zinc sulfate. BMJ 1(6115):754.

Murphy. J.V. 1970. Intoxication following ingestion of elemental zinc. JAMA 212(12):2119-2120.

Neuberger, J.S., and J.G. Hollowell. 1982. Lung cancer excess in an abandoned lead-zinc mining and smelting area. Sci. Total Environ. 25(3):287-294.

Neve, J., M. Hanocq, A. Peretz, F. Abi Khalil, F. Pelen, J.P. Famaey, and J. Fontaine. 1991. Pharmacokinetic study of orally administered zinc in humans: Evidence for an enteral recirculation. Eur. J. Drug Metab. Pharmacokinet. 16(4):315-323.

Nishioka, H. 1975. Mutagenic activities of metal compounds in bacteria. Mutat. Res. 31(3):185-189.

NRC (National Research Council). 1989. Recommended Dietary Allowances, 10th Ed., Washington, DC: National Academy Press.

NTP (National Toxicology Program). 1997. Toxicology and Carcinogenesis Studies of Molybdenum Trioxide (CAS No. 1313-27-5) in F344/N Rats and $B6C3F_1$ Mice. (Inhalation Studies). National Toxicology Program, Research Triangle Park, NC.

Philipp, R., A.O. Hughes, and M.C. Robertson. 1982. Stomach cancer and soil metal content. Br. J. Cancer 45(3):482.

Powmet (Powmet, Inc.). 1999. Material Safety Data Sheet, Calcium Molybdate. Powmet, Inc., Rockford, IL.

Prasad, A.S. 1976. Deficiency of zinc in man and its toxicity. In Trace Elements in Health and Disease. Vol. 1, Zinc and Copper, A.S. Prasad, ed. New York: Academic Press.

Prasad, A.S., G.J. Brewer, E.B. Schoomaker, and P. Rabbani. 1978. Hypocupremia induced by zinc therapy in adults. JAMA 240(20):2166-2168.

Rohrs, L.C. 1957. Metal-fume fever from inhaling zinc oxide. Arch. Ind. Health 16:42-7.

Rosoff, B., and H. Spencer. 1964. Fate of molybdenum-99 in man. Nature 202(4930):410-411.

Samman, S., and D.C. Roberts. 1987. The effect of zinc supplements on plasma zinc and copper levels and the reported symptoms in health volunteers. Med. J. Aust. 146(5):246-249.

Schenker, M.B., F.E. Speizer, and J.O. Taylor. 1981. Acute upper respiratory symptoms resulting from exposure to zinc chloride aerosol. Environ. Res. 25(2):317-324.

Schiffer, R.B., F.W. Sunderman, Jr., R.B. Braggs, J.A. Moynihan. 1991. The effects of exposure to dietary nickel and zinc upon humoral and cellular immunity in SJL mice. J. Neuroimmunol. 34(2-3):229-239.

Schlicker, S.A., and D.H. Cox. 1968. Maternal dietary zinc, and development and zinc, iron, and copper content of the rat fetus. J. Nutr. 95(2):287-94.

Simmer, K., L. Lort-Phillips, C. James, and R.P. Thompson. 1991. A double-blind trial of zinc supplementation in pregnancy. Eur. J. Clin. Nutr. 45(3):139-144.

Stocks, P., and R.I. Davies. 1964. Zinc and copper content of soils associated with the incidence of cancer of the stomach and other organs. Br. J. Cancer 18:14-24.

Stokinger, H.E. 1981. The metals. Pp. 1806-1820 in Patty's Industrial Hygiene and Toxicology. G.D. Clayton, and F.E. Clayton, eds. New York: John Wiley and Sons.

Straube E.F., N.H. Schuster, and A.J. Sinclair. 1980. Zinc toxicity in the ferret. J. Comp. Pathol. 90(3):355-361.

Sturniolo, G.C., M.C. Montino, L. Rossetto, A. Martin, R. D'Inca, A. D'Odorico, R. Naccarato. 1991. Inhibition of gastric acid secretion reduces zinc absorption in man. J. Am. Coll. Nutr. 10(4):372-375.

Thompson, E.D., J.A. McDermott, T.B. Zerkle, J.A. Skare, B.L. Evans, and D.B. Cody. 1989. Genotoxicity of zinc in 4 short-term mutagenicity assays. Mutat. Res. 223(3):267-272.

Tsigdinos, G.A., and F.W. Moore. 1981. Simple molybdates of the elements. Pp 9-11 in Molybdenum Chemicals. Ann Arbor, MI: Climax Molybdenum Company of Michigan.

Tsongas, T.A., R.R. Meglen, P.A. Walravens, and W.R. Chappell. 1980. Molybdenum in the diet: An estimate of average daily intake in the United States. Am. J. Clin. Nutr. 33(5):1103-1107.

Turner, J.A. 1921. An occupational dermatoconiosis among zinc oxide workers. Public Health Rep. 36:2727-2732.

Venitt, S., and L.S. Levy. 1974. Mutagenicity of chromates in bacteria and its relevance to chromate carcinogenesis. Nature 250(5466):493-495.

Vilkina, G.A., M.D. Polmerantseva, and L.K. Ramaiia. 1978. Absense of a mutagenic effect of cadmium and zinc salts in mouse somatic and sex cells. [Article in Russian]. Genetika (Moscow). 14(12):2212-2214.

Voroshilin, S.I., E.G. Plotko, T.V. Fink, and V.I. Nikiforova. 1978. Cytogenic effect of wolfram, zinc, cadmium, and cobalt compounds on human and animal somatic cells. [Article in Russian]. Tsitol. Genet. 12(3):241-243.

Walravens, P.A., R. Moure-Eraso, C.C. Solomons, W.R. Chappell, and G. Bentley. 1979. Biochemical abnormalities in workers exposed to molybdenum dust. Arch. Environ. Health 34(5):302-308.

Walters, M., and F.J. Roe. 1965. A study of the effects of zinc and tin administered orally to mice over a prolonged period. Food Cosmet. Toxicol. 3(2):271-276.

Weigand, E., and M. Kirchgessner. 1992. Absorption, endogenous excretion, and balance of zinc in growing rats on diets with various sugars replacing starch. Biol. Trace Elem. Res. 34(1):67-77.

Wong, P.K. 1988. Mutagenicity of heavy metals. Bull. Environ. Contam. Toxicol. 40(4):597-603.

Yadrick, M.K., M.A. Kenney, and E.A. Winterfeldt. 1989. Iron, copper, and zinc status: Response to supplementation with zinc or zinc and iron in adult females. Am. J. Clin. Nutr. 49:145-150.

Zaporowska, H., and W. Wasilewski. 1992. Combined effect of vanadium and zinc on certain selected haematological indices in rats. Comp. Biochem. Physiol. C 103(1):143-147.

Zeiger, E., B. Anderson, S. Haworth, T. Lawlor, and K. Mortelmans. 1992. Salmonella mutagenicity tests: V. Results from the testing of 311 chemicals. Environ. Mol. Mutagen. 19(Suppl 21):2-141.

10

Antimony Trioxide

THIS chapter reviews the physical and chemical properties, toxicokinetics, toxicological, epidemiological, and exposure data on antimony trioxide. The subcommittee used that information to characterize the health risk from exposure to antimony trioxide. The subcommittee also identified data gaps and recommended research relevant for determining the health risk from exposure to antimony trioxide.

PHYSICAL AND CHEMICAL PROPERTIES

The physical and chemical properties of antimony trioxide are summarized in Table10-1.

OCCURRENCE AND USE

Antimony trioxide is formed by reacting antimony trichloride ($SbCl_3$) with water. It is used in combination with some brominated flame retardants, and might also be used in conjunction with zinc borate, both within and outside the United States on commercial furniture, draperies, wall coverings, and carpets (R.C. Kidder, Flame Retardant Chemical Association, unpublished material, April 21, 1998). It is also used in enamels, glasses, rubber, plastics, adhesives, textiles, paper, and as a paint pigment (Budavari et al. 1989).

TABLE 10-1 Physical and Chemical Properties of Antimony Trioxide

Characteristic	Value	Reference
Chemical formula	Sb_2O_3	Lide 1991-1992
CAS registry #	1309-64-4	Lide 1991-1992
Synonyms	diantimony trioxide, flowers of antimony	Budavari et al. 1989
Molecular weight	291.5 g/mol	Lide 1991-1992
Physical state	Crystal	Budavari et al. 1989
Solubility	Very slightly soluble in cold water; slightly soluble in hot water; soluble in KOH, HCl, tartaric acid, and acetic acid	Lide 1991-1992
Vapor pressure	1×10^{-10} mm Hg	Calculated from Lide 1991-1992
Melting point	656 °C	Lide 1991-1992
Boiling point	1,550 °C	Lide 1991-1992
Density	5.2 g/cm^3 (senarmontite), 5.7 g/cm^3 (valentinite)	Lide 1991-1992

TOXICOKINETICS

Absorption

Systemic toxicity and death occurred in rabbits following dermal application of 8g/kg antimony trioxide (Myers et al. 1978), and application of an unspecified dose of antimony trioxide in a paste of "artificial acidic or alkalinic sweat" (Fleming 1938). Both studies indicate that antimony trioxide is absorbed dermally in rabbits.

Elevated blood and urine antimony levels were reported in workers occupationally exposed to antimony, suggesting that antimony trioxide is absorbed following inhalation exposure (Cooper et al. 1968; Lüdersdorf et al. 1987; Kim et al. 1997). However, no quantitative correlation was found between the air concentrations of antimony and the antimony concentration measured in urine (Kim et al. 1997).

Few quantitative data were found regarding the absorption of antimony trioxide following oral exposure. The International Commission on Radiological Protection (ICRP 1981) has recommended that a 1% absorption rate of antimony compounds (including antimony trioxide) be assumed when estimating exposure from the gastrointestinal (GI) tract. That recommendation is based on studies of various organic and inorganic antimony compounds. Toxicity is greater following exposure to 7.9 mg antimony trioxide/kg-d in 5% citric acid than to 101 mg antimony trioxide/kg-d in water, suggesting that solubility can affect antimony absorption (Fleming 1938).

Distribution

No studies were identified on the tissue distribution of antimony trioxide following dermal exposure.

Retired workers occupationally exposed by the inhalation route to antimony were reported to have elevated concentrations of antimony in their lung tissue as compared to non-occupationally exposed individuals (Gerhardsson et al. 1982). Following intratracheal instillation of a single dose of 1.52 mg antimony trioxide/kg in Syrian golden hamsters, the highest concentrations of antimony were measured in the lungs and liver, with lower concentrations present in the kidney, stomach, and trachea (Leffler et al. 1984).

No information was found on the tissue distribution of antimony in humans following oral exposure. In rats, high concentrations of antimony were measured in the thyroid and GI contents following chronic ingestion of 2% antimony trioxide in the feed (Gross et al. 1955a). Detectable levels were also found in the spleen, kidney, heart, bone, muscle, lungs, liver, and GI tissue. Following continuous treatment of rats for 40 d. Antimony was concentrated in the thyroid, with much lower levels found in the other tissues 40 d after cessation of chronic ingestion of 2% antimony trioxide in the feed (Gross et al. 1955a).

Metabolism and Excretion

No data were identified on the metabolism or excretion of antimony trioxide following dermal exposure.

Intraperitoneal injection of rats with 800 μg antimony chloride/kg did not result in detectable levels of any organic form of antimony in the bile or urine, indicating that antimony is not methylated in vivo. Antimony can form a complex with glutathione in vivo (Bailly et al. 1991).

McCallum (1963) reported elevated antimony concentrations in the urine of workers occupationally exposed via inhalation to antimony, indicating that excretion by this pathway occurs in humans.

Gerhardsson et al. (1982) measured post-mortem antimony levels in the lung tissue of workers occupationally exposed via inhalation to antimony. Levels were found to be elevated compared to a control population, even after workers had been retired for up to 20 yr, indicating a long half-life for lung clearance of antimony in humans.

Toxicokinetic studies in adult male Syrian golden hamsters given a single, intratracheal instillation of antimony trioxide (1.52 mg/kg body weight) indicate that 20% of the instilled antimony was cleared from the lung in the first 20 hr (Leffler et al. 1984). Biological half-times of about 40 hr for the initial phase and 20-40 d for the second phase were calculated for lung tissue (Leffler et al. 1984). In rats exposed to 119 mg antimony trioxide dust/m^3 for 80 hr, the majority of urinary excretion occurred within the first 3 d after exposure (Gross et al. 1955a).

Following a single oral dose (200 mg antimony trioxide) of antimony trioxide to rats, 3% of the administered dose was recovered in the urine within 8 d. Only 0.15% was recovered 1 d after treatment, and 3% was recovered between d 2 and 5 post-treatment (Gross et al. 1955a). Following chronic exposure (2% antimony trioxide in the diet; 8 mo), approximately 99% of fecal excretion and the majority of urinary excretion occurred within 7 d after exposure ceased (Gross et al. 1955a). The large amount of antimony excreted in the feces soon after exposure suggests that a substantial portion of the compound is excreted without being absorbed systemically. That is consistent with the low absorption rate (1%) cited by the ICRP (ICRP 1981) (see Absorption section).

HAZARD IDENTIFICATION[1]

Dermal Exposure

Irritation

Dermatitis was reported in workers occupationally exposed to 0.4-70.7 mg antimony/m^3 (Renes 1953; McCallum 1963; Potkonjak and Pavlovich 1983; White et al. 1993). Although antimony trioxide in the work environment was believed to be responsible for the dermatitis, quantitative data on dermal exposure were not available, and the workers were also exposed to other elements such as arsenic. Therefore, the causative agent for the observed dermatitis could not be positively determined.

---

[1]In this section, the subcommittee reviewed toxicity data on antimony trioxide, including the toxicity assessment prepared by the U.S. Consumer Product Safety Commission (Hatlelid 1999).

In a controlled human study (Industrial Bio-Test Laboratories, Inc. 1973), 52 subjects received a series of nine dermal applications of antimony trioxide over a 3-wk period. The antimony trioxide was applied for 24 hr; the dose was not reported. Two wk after the series of applications, a single dose of antimony trioxide was applied. After each application, skin reactions were evaluated. No skin reactions were observed over the course of the study, suggesting that antimony trioxide is neither a skin irritant nor a sensitizer.

Dermal exposure to antimony trioxide generally did not cause dermatitis in tested animals. Only mild skin irritation was observed even after repeated or prolonged exposure to large quantities of antimony trioxide (2-25 g antimony trioxide/kg) in rabbits (Gross et al. 1955a; Ebbens 1972). Skin edema was reported in one study in which antimony trioxide was applied to rabbits in corn oil (8 g antimony trioxide/kg for 24 hr) (Myers et al. 1978). However, that study is limited in that there was no solvent control group, and data on severity and number of animals responding was lacking. In a study by Haskell Laboratory (Haskell Laboratory 1970a), a suspension of 12, 31, or 61 mg antimony trioxide/kg in a fat/acetone/dioxane mixture was applied to intact shaved skin (all doses) or abraded skin (31-mg/kg group only) of 10 albino guinea pigs. The exposure duration was not reported. Irritation was not seen in any of the treated animals. In another study by Haskell Laboratory (1970b), 24 or 49 mg/kg antimony trioxide (suspended in a similar mixture as above) was applied to the intact shaved skin of guinea pigs. One day after the treatment, mild erythema was observed in 2/10 and 5/10 animals treated with 24 mg antimony trioxide/kg and 49 mg antimony trioxide/kg, respectively. All of the responses had disappeared 2 d after the initial dosing.

Sensitization

As mentioned under the Irritation section, no skin reactions were observed in the controlled human study conducted by Industrial Bio-Test Laboratories, Inc. (1973), indicating that antimony trioxide is not a skin sensitizer.

Haskell Laboratory (1970a, b) treated groups of five guinea pigs with nine dermal applications of 31 mg antimony trioxide/kg (25%) or 49 mg antimony trioxide/kg (50%) in a fat/acetone/dioxane mixture on shaved and abraded skin, or four intradermal injections of 1 mg antimony trioxide in either acetone-dimethyl phthalate or propylene glycol solutions, over the course of 3 wk. After a 2-wk rest period, each group of animals received challenge applications of the suspensions on both intact and abraded skin. Sensitization was not observed in any of the test animals.

Systemic Effects

Death occurred in one out of four rabbits following a single dermal exposure to 8 g/kg antimony trioxide (Myers et al. 1978), and in one out of four rabbits exposed to 2 g/kg antimony trioxide (Ebbens 1972). Systemic toxicity and death occurred in three out of eight rabbits, but not in rats, following short-term exposure (20-21 d) to an unspecified dose of antimony trioxide (Fleming 1938). Gross pathologies were seen in the liver, lung, stomach, and kidney.

Other Systemic Effects

No studies were identified that investigated the immunological, neurological, reproductive, developmental, or carcinogenic effects of antimony trioxide following dermal exposure to antimony trioxide.

Inhalation Exposure

Systemic Effects

In humans, the lungs are the primary targets following inhalation exposure to antimony trioxide. Several studies of antimony smelter workers show that workers developed pneumoconiosis, chronic cough, and upper airway inflammation following chronic exposure to antimony trioxide (McCallum 1963, 1967; Cooper et al. 1968; Potkonjak and Pavlovich 1983). In addition, one study reported systemic effects following inhalation exposure in smelter workers, including weight loss, nausea, vomiting, nerve tenderness, and tingling (Renes 1953). In those studies, however, a causal role for antimony trioxide in the observed human health effects could not be confirmed because of the lack of individual exposure data for the workers and exposure to other compounds, including arsenic, lead, and alkali, that could be confounders.

The lungs are also the primary target tissues in animals following inhalation exposure (see Table 10-2). All experimental inhalation studies were conducted using whole-body exposure. Details of particle size and purity are provided in footnotes. Guinea pigs exposed to antimony trioxide[2] (average concentration: 45.4 mg antimony trioxide/m^3, 2-3 hr/d, 6 mo) developed pneumonitis, liver and spleen effects, and decreased white blood cell counts (Dernehl 1945). Similarly, pneumonia was seen following exposure of rats (100-125 mg antimony trioxide/m^3, 100 hr/mo, 14.5 mo) and rabbits (89 mg antimony

[2]Assumed particle size <1 μm and 99.8% pure based on production technique.

TABLE 10-2 Toxic Effects of Antimony Trioxide Following Inhalation Exposure

Species, Strain, Number	Exposure[a] (mg/m^3); Duration	Duration-adjusted Concentrations (mg/m^3)[b]	Effects	NOAEL/LOAEL (mg/m^3)	RDDR	HEC NOAEL/ LOAEL	Reference
Rat, NS, 50	100-125; 100 hr/mo for 14.5 mo	13.9-17.4	Diffuse, interstitial fibrosis; 18% died of pneumonia	LOAEL: 13.9	ND	ND	Gross et al. 1955b
Rabbit, NS, 20	89; 100 hr/mo for 10 mo	12.4	85% died of pneumonia	LOAEL: 12.4	ND	ND	Gross et al. 1955b
Guinea pig, NS, 24	45.4 (average); 2-3 hr/d, 7 d/wk for 6 mo	3.8-5.7	Pneumonitis, liver and spleen effects, decreased white blood cell counts	LOAEL: 3.8	ND	ND	Dernehl 1945
Rat, Wistar, 90/sex	45.5; 7 hr/d, 5 d/wk for 52 wk; 20-wk observation	9.4	Interstitial fibrosis, alveolar-wall cell hypertrophy and hyperplasia, and cuboidal and columnar cell metaplasia of the lungs	LOAEL: 9.4	ND	LOAEL: 5	Groth et al. 1986
Rat, Fischer 344, 55/sex	0.25, 1.08, 4.92, or 23.46; 6 hr/d, 5 d/wk for 13 wk; 27-wk observation	0.04, 0.19, 0.88, 4.19	6% decrease in body weight	NOAEL: 0.88 (M) LOAEL: 4.19 (M)	0.324 0.585	NOAEL: 0.006 LOAEL: 0.51	Newton et al. 1994
			Increased lung weight	NOAEL: 0.19 LOAEL: 0.88	ND	ND	Newton et al. 1994

TABLE 10-2 *(Continued)*

Species, Strain, Number	Exposure[a] (mg/m^3); Duration	Duration-adjusted Concentrations (mg/m^3)[b]	Effects	NOAEL/LOAEL (mg/m^3)	RDDR	HEC NOAEL/ LOAEL	Reference
			Chronic interstitial inflammation, granulomatous inflammation, increased alveolar macrophages	NOAEL: 0.19 (F) 0.88 (M) LOAEL: 0.88 (F) 4.19 (M)	ND	ND	Newton et al. 1994
Rat, Fischer 344, 65/sex)	0.06, 0.51 or 4.5; 6 hr/d, 5 d/wk for 1 yr; 1-yr observation	0.01, 0.09, 0.80	Increased alveolar macrophage	LOAEL: 0.01	0.435	LOAEL: 0.004	Newton et al. 1994
			Interstitial inflammation and granulomatous inflammation	NOAEL: 0.09 LOAEL: 0.80	0.435	NOAEL: 0.039 LOAEL: 0.348	Newton et al. 1994
				BMC (F): 0.16	0.46	BMC(F): 0.074	Newton et al. 1994
Rat, Wistar, 50F	1.9, 5.0; 6 hr/d, 5 d/wk for 1 yr	0.3, 0.9	Discoloration and increased alveolar macrophages	LOAEL: 0.3	ND	ND	Watt 1983

			Interstitial inflammation and granulomatous inflammation	NOAEL: 0.9 LOAEL: 0.3	ND	ND	Watt 1983
Swine, Sinclair S-1, 3F	1.9, 5.0; 6 hr/d, 5 d/wk for 1 yr	0.3, 0.9	Minimal lung fibrosis	ND	ND	ND	Watt 1983
Rat, NS, 6-7	0.027, 0.082, 0.27; 24 hr/d for 21 d of gestation	0.027, 0.082, 0.27	Pre-implantation loss, fetal growth retardation, and pre- and post-implantation embryo death	NOAEL: 0.027 LOAEL: 0.082	ND	ND	Grin et al. 1987

BMC, benchmark concentration; F, female; HEC, human equivalent concentration; LOAEL, lowest-observed-adverse-effect level; M, male; ND, not determined; NOAEL, no-observed-adverse-effect level; RDDR, regional deposited dose ratio.

[a]All inhalation exposures were whole body.

[b]Concentrations are adjusted for continuous exposure.

trioxide/m^3, 100 hr/mo, 10 mo) to antimony trioxide[3] (Gross et al. 1955b). Interstitial fibrosis, hypertrophy, and hyperplasia were seen in male and female Wistar rats (90/sex-group) exposed to antimony trioxide[4] (45.5 mg antimony trioxide/m^3, 7 hr/d, 5 d/wk for 1 yr, followed by a 20-wk observation period) (Groth et al. 1986); those effects were more pronounced in the females.

Watt (1983) investigated the effects of exposure to antimony trioxide[5] (1.6 or 4.2 mg Sb/m^3, equivalent to 1.9 or 5.0 mg antimony trioxide/m^3, 6 hr/d, 5 d/wk for 1 yr) in female CDF Fischer rats (148 animals divided into three dose groups) and Sinclair S-1 miniature swine (eight animals divided into three dose groups). In rats, blood urea nitrogen (BUN) was consistently elevated at the high concentration, but was statistically significant only after 6 mo of exposure. No other changes in hematology, serum biochemistry or histology were reported. A concentration-related increase in lung weight was also observed in the rats. Swine were examined immediately after the treatment, at which time there was minimal fibrosis and no other statistically significant effects were observed.

Newton et al. (1994) conducted a preliminary, subchronic study in which male and female F-344 rats (55/sex/group) were exposed to antimony trioxide[6] (concentrations of 0.25, 1.08, 4.92, and 23.46 mg antimony trioxide/m^3) 6 hr/d, 5 d/wk for 13 wk followed by a 27-wk observation period. A decrease in body weight was seen in the males at the highest concentration tested and an increase in absolute lung weight was seen at the two highest exposure concentrations. Minimal-to-moderate microscopic pathologies were seen in the highest exposure group.

The Newton et al. (1994) pilot study was followed by a 1-yr chronic study

---

[3]Average particle size by electron micrograph = 0.6 μm, few particles up to 1 μm; calculated average particle size by weight = 0.4 μm; purity not reported.

[4]It was noted by the authors that they had difficulty generating the target concentration of 50 mg antimony trioxide/m^3; the mean daily time-weighted average (TWA) = 45.5 mg/m^3; the range was 0-191.1 mg antimony trioxide/m^3; particle size: median circular area equivalent diameter = 0.347 μm, mass median diameter (MMD) = 1.23 μm, mass median aerodynamic diameter (MMAD) = 2.80 μm, analyzed using a scanning electron microscope and image analyzer; purity: 80% antimony by proton-induced X-ray emission, 0.04 mg arsenic/g and 2.3 mg lead/g.

[5]Particle size averaged 0.44 μm (geometric standard deviation [SD] = 2.23) and 0.40 μm (geometric SD = 2.13) for low and high concentration, respectively; purity = 99.4% antimony, 0.02% arsenic, 0.2% lead. MMAD, a critical parameter in inhalation studies, could not be directly derived from the data in this study.

[6]Test material not milled; particle size: count median diameter = 0.485-0.62 μm, MMD = 1.49-2.50 μm, MMAD = 3.05-5.7 μm; purity = 99.68 ± 0.10%, contaminants not reported.

(Bio/dynamics 1990, as cited in EPA 1999). In that study, F-344 rats (65/sex/exposure level) were exposed to antimony trioxide[7] (measured concentrations were 0, 0.06, 0.51, or 4.5 mg antimony trioxide/m^3) 6 hr/d, 5 d/wk for 1 yr, followed by a 1-yr observation period. Five rats/sex/group were killed after 6 and 12 mo of exposure, and at 6 mo postexposure. All survivors were killed 12 mo after the end of the exposure period. Animal body weights were monitored. Complete gross and histopathological examinations were performed on all animals. The sections of the lungs examined included both the right lobes and the major bronchi. The only exposure-related changes occurred in the lungs and included chronic interstitial inflammation, granulomatous inflammation, and increased alveolar macrophages. Pinpoint black foci, thought to be aggregates of macrophages laden with antimony trioxide, were seen in the lungs of exposed animals, most frequently during the post-exposure observation period. In a subsequent analysis performed by the EPA (1999), it was noted that in the low- and mid-exposure groups, there was no indication that the particle-laden macrophages were anything but part of a normal, compensatory response. However, the clearance half-time of the high-exposure group was more than three times that of the mid-exposure group, indicating that clearance mechanisms were severely compromised. In some instances, clearance of particles is slowed by high lung burdens of inert particles, which leads to high lung particle burdens for extended periods (months to years), and pathologies that cannot be directly attributed to the toxicity of the chemical (Witschi and Last 1996). However, in this study (Newton et al. 1994), the decreased clearance appeared to be due to the inherent toxicity of antimony trioxide, rather than a particle overload phenomenon. Newton et al. (1994) reported a 50% increase in the clearance time of antimony trioxide at a dust volume of 270 nanoliter (nL), but benign dust particles have to be at about 1,000 nL to have that effect on clearance (Muhle et al. 1990, as cited in EPA 1999). However, some scientists believe that particle overload could account for the increased clearance time rather than inherent toxicity of antimony trioxide. Despite those observations, the subcommittee considered this study to be appropriate for calculation of an RfC. Based on additional statistical analysis (EPA 1999) of the male and female rats that died spontaneously or were killed at 18 and 24 mo, a LOAEL for interstitial inflammation and granulomatous inflammation of 4.5 mg antimony trioxide/m^3 and a NOAEL of 0.51 mg antimony trioxide/m^3 (NOAEL[HEC] of 0.042 mg antimony trioxide/m^3) were identified from this study.

---

[7]MMAD = 3.7 μm, geometric S.D. = 1.7; purity = 99.68 ± 0.10%, contaminants not reported.

Reproductive and Developmental Effects

Reproductive and developmental effects following inhalation exposure to antimony have been reported in one human study. Based on an English abstract of a study by Belyaeva (1967), women working in an antimony plant had a greater incidence of gynecological problems (not detailed), early interruption of pregnancy, and spontaneous late abortions compared to women working under similar conditions who were not exposed to antimony.

Belyaeva (1967) also reported a reduction in the number of offspring and a disruption of ovulation in rats exposed to 250 mg/m^3 antimony trioxide for 2 mo (particle size and purity not specified).

In a study by Grin et al. (1987) that was translated for the subcommittee, pregnant rats (six to seven/group) were exposed to antimony trioxide (0.027, 0.082, and 0.27 mg antimony trioxide/m^3, 24 hr/d; particle size and purity not reported) throughout gestation (21 d). Changes in clinical parameters at the highest exposure concentration tested included a very large increase in the amount of hemoglobin, blood leukocytes, serum lipids, and total protein in blood. The subcommittee noted that the effects on the hemoglobin and protein levels in the blood might indicate that the dams were sick, and therefore the maternal effects might have impacted the fetal effects. In the fetuses, gross macroscopic changes were seen at the two highest exposure concentrations tested, with increased bleeding in fetal brain membranes and liver, an increase in the size of the kidney cavity and the cerebral ventricles, and isolated cases of ossification at the highest exposure concentration tested. Some of the fetal effects in this study are listed in Table 10-3. Based on these data, 0.082 mg antimony trioxide/m^3 can be considered a LOAEL and 0.027 mg antimony

TABLE 10-3 Results of a Reproductive Toxicity Study on Antimony Trioxide (Grin et al. 1987)

Concentration (mg/m^3)[a]	Number of dead and resorbed/ female	Preimplantation loss (%)	Postimplantation loss (%)	Total death rate (%)
0	0.57 ± 0.20	8.6 ± 1.45	5.90 ± 2.10	14.03 ± 1.92
0.027	0.5 ± 0.22	10.93 ± 2.49	6.25 ± 2.80	16.31 ± 2.03
0.082	0.67 ± 0.33	14.63 ± 1.56	6.08 ± 2.82	19.91 ± 2.31
0.27[b]	1.20 ± 0.48[c]	13.56 ± 2.69[c]	12.22 ± 4.90[c]	24.32 ± 4.29[c]

[a]Exposure occurred throughout gestation.
[b]Substantial maternal toxicity was noted at this dose.
[c]Significantly different from controls by Student's t-test, $p < 0.05$.

trioxide/m^3 a NOAEL in this study (Grin et al. 1987). However, the study is of limited use for quantitative toxicity assessment purposes because of the lack of information on the purity and particle size of the antimony trioxide used and the fact that maternal toxicity was seen. Therefore, the subcommittee decided not to use the study by Grin et al. (1987) for the determination of a critical level.

As summarized in Reprotox (1999), studies with antimony compounds other than the trioxide have shown that, although antimony can enter the fetus (Gerber et al. 1982), antimony compounds are not teratogenic in chicks (Ridgway and Karnofsky 1952), rats (Rossi et al. 1987), or sheep (James et al. 1966). However, antimony trichloride (0.1 and 1 mg/dL in drinking water for 38 d) did decrease pup body weight and had some effects on cardiovascular responses to noradrenaline, isoprenaline, and acetylcholine (Rossi et al. 1987).

Cancer

Three epidemiological studies have evaluated the potential carcinogenicity of antimony following occupational exposure (Jones 1994; Potkonjak and Pavlovich 1983; Schnorr et al. 1995). Jones (1994) studied a cohort of 2,508 smelter workers and reported that antimony exposure was associated with an increased risk of lung cancer, with a standardized mortality ratio (SMR) of 2.18 ($p < 0.001$) in workers employed prior to 1961, but not in those employed after 1960. No data on cigarette smoking were provided, and many possible confounding exposures existed in the workplace, including exposure to arsenic, sulfur dioxide, and polycyclic aromatic hydrocarbons. Schnorr et al. (1995) conducted a retrospective cohort study of 1,014 smelter workers and reported a lung cancer SMR of 1.39; the 90% confidence interval was 1.01-1.88, indicating that even at the 90% confidence level, this SMR was only marginally statistically significant. No data on cigarette smoking were reported and workplace exposures levels were not measured. Potkonjak and Pavlovich (1983) evaluated 51 workers exposed to 5.5-64 mg antimony trioxide/m^3 (particle size < 5 μm) for an average of 18 yr. No malignancies were observed in that study.

Although the study by Jones (1994) suggests a correlation between antimony exposure and lung cancer risk, the use of this study is limited by the lack of an appropriate control population, and failure to control for bias and confounding factors.

Results from animal studies are also conflicting. Two animal studies reported that antimony trioxide induced lung cancers in two strains of rats (Groth et al. 1986; Watt 1983). However, additional studies in rats (Newton et al. 1994) and a study in pigs (Watt 1983) did not confirm this effect.

In a study by Groth et al. (1986), Wistar rats (90/sex/group) were exposed to antimony trioxide[8] for 1 yr (target concentration = 50 mg antimony trioxide/m^3, 7 hr/d, 5 d/wk, killed 20 wk after end of exposure). The incidence of lung tumors was increased in female rats only, with tumors occurring in 19 of the 70 exposed females compared to 0 of the 70 control females. Of the lung tumors, nine were squamous-cell carcinomas, five were scirrhous carcinomas,[9] and 11 were bronchioloalveolar adenomas and carcinomas. Some rats had more than one type of lung tumor. Rats were 8 mo of age at the beginning of exposure, and the first tumor was seen in a rat killed after wk 53.

In an unpublished study, conducted by Watt (1983), groups of 48-50 female Charles River CDF rats were exposed to antimony trioxide[10] (0, 1.9, or 5.0 mg antimony trioxide/m^3, 6 hr/d, 5 d/wk) for 1 yr. Surviving rats were kept for up to 17 mo postexposure. Only 13-18 rats/group survived until 29 mo. Non-neoplastic lung lesions included focal fibrosis, pneumonocyte hyperplasia, cholesterol clefts,[11] and multinucleated giant cells. Adenomatous hyperplasia of the lung was evident at the high concentration. The most common lung tumor was scirrhous carcinoma (incidences of 0/41, 0/44, and 15/45 in the control, low-, and high-concentration groups, respectively). Bronchioloalveolar adenomas were also increased at the high-exposure concentration group (incidences of 1/41, 1/44, and 4/45 in the control, low-, and high-concentration groups, respectively). Squamous cell carcinomas (2/45 at the high concentration) were observed in the high-exposure concentration group, but not in the low-exposure concentration or control groups. The study authors noted that the neoplasms appeared to arise from the alveolar epithelial lining cells. The tumor incidence was not significantly increased in any other tissue. This study was limited in that only 13-18 rats remained in each dose group at the end of 29 mo. In addi-

---

[8]The authors noted difficulty in consistently generating 50 mg antimony trioxide/m^3, the mean daily TWA = 45.5 mg/m^3, range = 0-191.1 mg antimony trioxide/m^3; particle size: median circular area equivalent diameter = 0.347 μm, MMD = 1.23 μm, MMAD = 2.80 μm, analyzed using a scanning electron microscope and image analyzer; purity: 80% antimony by proton-induced X-ray emission, 0.04 mg arsenic/g, and 2.3 mg lead/g.

[9]An adenocarcinoma with a small number of tumor cells, in relation to an abundant amount of dense collagenous stroma, isolated and dispersed throughout the fibrous components (Becker et al. 1986).

[10]Generated by a modified hammer mill, average particle sizes were 0.44 μm (geometric SD = 2.23) and 0.40 μm (geometric SD = 2.13) for low and high concentrations, respectively; purity = 99.4% antimony, 0.02% arsenic, 0.2% lead.

[11]Elongated defects that represent the site of a cholesterol crystal that has been dissolved during the preparative procedures (Becker et al. 1986).

tion, continuing the study until 29 mo after study initiation increased the potential for age-related tumors, thus decreasing the study sensitivity. However, in light of the low background and clear increases seen in scirrhous carcinoma, these limitations do not affect the study conclusions.

Watt (1983) also examined the carcinogenicity of antimony trioxide in female pigs and found no neoplasms at the end of the 1-yr exposure (1.9 or 5.0 mg antimony trioxide/m^3, 6 hr/d, 5 d/wk). The negative response could be due to a low sensitivity of this species, or the lack of an appropriate observation period following exposure.

No increase in cancers was observed in F-344 rats (65/sex/group) exposed to antimony trioxide[12] (0, 0.06, 0.51, or 4.5 mg antimony trioxide/m^3, 6 hr/d, 5 d/wk) for 1 yr, and observed up to 1 yr after exposure (Newton et al. 1994). Extensive gross necropsy and histopathology were conducted. The lung tissue examined included the major bronchi. Elevated total leukocyte counts and atypical lymphocytes indicated leukemia in all groups. However, the authors noted that leukemia is a common finding in aged F-344 rats. Two males (one from the control group and one from the 4.5-mg/m^3 concentration group) and one female (0.51-mg/m^3 concentration group) had pulmonary carcinomas; the carcinomas were not considered to be treatment related. According to the authors, other neoplastic findings occurred sporadically or with an incidence similar to that of the controls (Newton et al. 1994).

Based on the pathological examination of the lungs from all three rat studies discussed above (Groth et al. 1986; Watt 1983; Newton et al. 1994), Newton et al. (1994) suggested that the differences in carcinogenesis are due to a different deposition pattern of antimony trioxide in the lungs. Newton et al. (1994) noted, however, that particle size could not explain these differences (see Table 10-4). Although the rats were reportedly exposed to similar concentrations in the Watt (1983) and Newton et al. (1994) studies, the rats in the Watt (1983) study had more damage and considerably more test material in the lung. It was suggested that the rats in the Watt (1983) study actually had a higher exposure than measured based on the extent of particle deposition (Newton et al. 1994). Newton et al. (1994) also concluded that the foreign body reaction cannot completely account for the tumors observed in the Groth et al. (1986) study, since females, but not males, were affected in that study.

---

[12]MMAD = 3.7 μm, geometric SD = 1.7; purity = 99.68 + 0.10%, contaminants not reported.

TABLE 10-4 Antimony Trioxide Particle Size (Micrometers)

Particle Sizing Technique	Sample	Count Median Diameter	Geometric Standard Deviation	Mass Median Diameter	Mass Median Aerodynamic Diameter	*N*	Reference
SEM: Feret diameter	High chamber	0.40	2.13	2.22[a]	5.06[b]	NA	Watt 1983
	Low chamber	0.44	2.23	3.03	6.9	NA	
SEM: equivalent area diameter	Chamber sample	0.347		1.23[c]	2.8[b]	1,948	Groth 1986
TSI Particle Sizer	Groups II-IV		1.79		3.76 ± 0.84	20	Newton et al. 1994
Cascade Impactor	Group IV		1.80		4.55	2	

CMD, count median diameter; GSD, geometric standard deviation; MMAD, mass median aerodynamic diameter; MMD, mass median diameter; NA, not available; SEM, scanning electron microscopy; sqrt, square root; TEM, transmission electron microscopy; TSI, TSI Aerodynamic Particle Sizer AP340.

[a]Calculated using MMD = CMD exp ($3\ln^2$GSD).

[b]Calculated using MMAD = MMD × sqrt (density).

[c]Calculated using the equivalent area diameters and an assumed spherical particle with density of 5.2 g/cm^2.

Source: Modified from Newton et al. 1994.

The International Agency for Research on Cancer (IARC) classifies antimony trioxide as a possible carcinogen to humans, group 2B, based on sufficient evidence for the carcinogenicity in experimental animals (by inhalation), but inadequate evidence for the carcinogenicity in humans (IARC 1989). That assessment was completed before the publication of the negative study by Newton et al. (1994).

In summary, based on the weight of evidence, the subcommittee concluded that there is suggestive evidence that antimony trioxide is carcinogenic and a quantitative cancer risk assessment was performed based on the study by Watt (1983) (see Cancer section under Quantitative Toxicity Assessment).

Other Systemic Effects

No studies were identified that investigated the immunological or neurological effects of antimony trioxide following inhalation exposure.

Oral Exposure

Systemic Effects

There are no data on the health effects of antimony trioxide in humans following oral exposure.

Oral exposure studies conducted in animals are summarized in Table 10-5. An oral LD_{50} of >20 g/kg body weight has been reported in rats for antimony trioxide (Smyth and Carpenter 1948, as cited in ATSDR 1992; Ebbens 1972). Diarrhea has been reported in rats administered 16.7 g/kg body weight antimony trioxide in oil by gavage (Myers et al. 1978). The same dose in water given by gavage (Gross et al. 1955a), or provided in food (Smyth and Thompson 1945) did not produce any observable toxicity. Rats gavaged with 8.6-29 g/kg body weight antimony trioxide exhibited hypoactivity and ruffed fur within 1 hr after dosing, but returned to normal after 2 d (Ebbens 1972). No gross pathologic alterations were observed upon necropsy in that study.

No significant treatment-related effects were seen in rats following gavage with 134-501 mg/kg-bw/d antimony trioxide when administered in either 0.4% hydrochloric acid or 4% citric acid/0.4% hydrochloric acid for 20 d. Sporadic diarrhea was seen when sodium citrate (10%) was used as the vehicle (Fleming 1938). In a 21-d study, two dogs were gavaged daily with 1,000 mg antimony trioxide (79 mg/kg-bw/d) in water (Fleming 1938). The animals developed

TABLE 10-5 Toxic Effects of Antimony Trioxide Following Oral Exposure

Species, Strain, Sex, Number	Dose	Duration, Route	Effects	NOAEL/LOAEL (mg/kg-d)	Reference
Rat, Sherman, NS, 6/dose	NS	Single dose, oral	$LD_{50} > 20$ g/kg	ND	Smyth and Carpenter 1948, as cited in ATSDR 1992
Rat, Charles River, M/F, 4/dose	10.25, 15.38, 23.07, 34.6 g/kg in water	Single dose, gavage	$LD_{50} > 34.6$ g/kg; hypoactivity and ruffed fur at all doses	ND	Ebbens 1972
Rat	16.7 g/kg in oil	Single dose, gavage	Diarrhea	ND	Meyers et al. 1978
Rat	16.7 g/kg in water	Single dose, gavage	No effects observed	ND	Gross et al. 1955a
	1.3 g/kg-d	240 d, feed	Decreased weight gain	ND	
Rat, Albino, M, 10/dose	16.7 g/kg	Single dose, feed		ND	Smyth and Thompson 1945
	60, 270, 1,070 mg/kg-d	30 d, feed	Decreased food consumption, decreased weight gain, increased red blood cell count in high-dose group	ND	
Dog, NS, NS, 2	79 mg/kg-d	21 d, gavage	Diarrhea, reversible	ND	Fleming 1938
	7.7 mg/kg-d in 21% citric acid, second treatment in same dogs	11 d, gavage	Diarrhea, weight loss, gastrointestinal and liver lesions	ND	

Rat	670 mg/kg-d	12 wk, feed	Decreased weight gain, spleen weight, heart weight; increased lung weight	ND	Hiraoka 1986, as cited in ATSDR 1992
Rat, Wistar, M, 5/dose	500, 1,000 mg/kg-d	24 wk, feed	Decreased red blood cell count (500, 1,000), increased serum glutamic oxaloacetic transaminase; no macroscopic changes	LOAEL: 500	Sunagawa 1981
Rat, Wistar, M/F, 12/sex/dose	M: 84, 421, 1,686 mg/kg-d F: 97, 494, 1,879 mg/kg-d	90 d, feed	Increased red blood cell count (high-dose M/F), increased urine volume (high-dose M/F), decreased urine specific gravity (high-dose F), increased serum cholesterol (high-dose F), increased triglycerides (high-dose M), decreased alkaline phosphatase activity (high-dose M, mid- and high-dose F), increased aspartate and alanine aminotransferase (high-dose F)	Based on enzyme changes in F: NOAEL: 494 LOAEL: 1,879	Hext et al. 1999

F, female; LOAEL, lowest-observed-adverse-effect level; M, male; ND, not determined; NOAEL, no-observed-adverse-effect level; NS, not specified.

severe diarrhea, which lasted 6 or 7 d, but resolved prior to completion of the treatment, suggesting either that the diarrhea was not a severe response or that the animals adapted to the treatment. Following the 21-d treatment with antimony trioxide dissolved in water, the same dogs were administered 7.7 mg/kg-bw/d antimony trioxide dissolved in 5% citric acid for 11 d. Diarrhea, weight loss, and gastrointestinal and liver lesions were observed. Although the usefulness of this study is limited by the small number of animals used and the lack of control group, the results suggest that solubility plays a role in the toxicity of orally administered antimony trioxide (Fleming 1938).

In a short-term exposure toxicity study of antimony trioxide in which groups of 10 male albino rats received antimony trioxide in their diet (0%, 0.1%, 0.45%, 1.8%; corresponding to 0, 60, 270, 1,070 mg antimony trioxide/kg-d) for 30 d, rats in the high-dose group (1,070 mg/kg-d) had significantly decreased food consumption (41%) and decreased body weight gain (43%) compared with controls (Smyth and Thompson 1945). Hematological examination indicated that rats in the high-dose group had an increased red blood cell count but no change in hemoglobin concentration compared to controls. The high dose of 1,070 mg antimony trioxide/kg-d was considered a NOAEL for the derivation of the oral reference dose (RfD) because the subcommittee did not consider an increase in red blood cell count to be an adverse effect and because the other effects are probably related to decreased food consumption.

Rats fed with 670 mg antimony trioxide/kg bw-d in the diet for 12 wk had a decrease in overall weight gain, spleen and heart weight, and an increase in lung weight (Hiraoka 1986; as cited in ATSDR 1992). Reduced weight gain was also seen in rats given approximately 1.3 g antimony trioxide/kg bw-d in food for 240 d. No gross or microscopic changes were seen in those animals (Gross et al. 1955a).

Sunagawa (1981) fed male Wistar rats (five animals/group) 0%, 1.0%, or 2.0% antimony trioxide (calculated to be 0, 500, or 1,000 mg antimony trioxide/kg bw-d) in the diet for 24 wk. Exposure to antimony trioxide had no effect on gross appearance or behavior, and did not affect body weight, food and water intake, or organ weights in the rats. Red blood cell count was significantly decreased (not dose-dependent) in both treated groups compared with controls. No changes were observed in white blood cell count, hematocrit, or hemoglobin concentration. Serum glutamic oxalacetic transaminase (SGOT) was significantly increased ($p < 0.05$) in both dose groups. Histopathological evaluation of the liver indicated some (not statistically significant) disorders of hepatic laminae, cloudy swelling in hepatic cords, and vacuolar degeneration in hepatic cells. Based on the suggestion of liver toxicity and the decreased red blood cell count, a LOAEL of 500 mg antimony trioxide/kg bw-d was identified from this study. However, this study is of limited usefulness for the derivation of an RfD for antimony trioxide because of the small number of animals used and the fact that only the abstract and data tables were available in English.

Hext et al. (1999) fed male and female Wistar rats (12/sex/dose) diets containing 0, 1,000, 5,000, or 20,000 ppm antimony trioxide for 90 d (0, 84, 421, and 1,686 mg antimony trioxide/kg-d for males, and 0, 97, 494, and 1,879 mg antimony trioxide/kg-d for females; based on measured food consumption and body weights). Urine volume was significantly increased, and specific gravity was significantly decreased in high-dose females. Serum cholesterol and urine volume in high-dose females (dose-related trend), and triglycerides and red blood cell count in high-dose males were increased. Alkaline phosphatase (AlkP) activity was significantly decreased in high-dose males and mid- and high-dose females (dose-related trend). SGOT and serum glutamic aminotransferase (SGPT) were significantly increased in high-dose females. Absolute and relative liver weights were increased by approximately 10% in high-dose males and females. No other treatment-related effects were seen. The subcommittee concluded that the effects seen in this study are adverse when considered together with the data from Sunagawa (1981) and Smyth and Thompson (1945). Based on the increase in serum enzymes (statistically significant only in females), and the liver weight, 1,879 mg/kg-d is identified as a LOAEL for this study; the NOAEL is 494 mg/kg-d.

Other Systemic Effects

No studies were identified that investigated immunological, neurological, reproductive, developmental, or carcinogenic effects of antimony trioxide following oral exposure.

Genotoxicity

Although a single oral gavage of antimony trioxide (400, 666.67, and 1,000 mg/kg) did not cause chromosome aberrations in mouse bone marrow cells, aberrations were observed following repeated administration of those doses (Gurnani et al. 1992). Repeated oral doses of antimony trioxide, however, did not cause unscheduled DNA synthesis in the liver cells of rats, or an increase in the micronucleated polychromatic erythrocytes in the mouse bone marrow micronucleus assay (Elliott et al. 1998).

Antimony trioxide was not mutagenic in *Salmonella typhimurium* or *E. coli* strains (Kanematsu et al. 1980; Kuroda et al. 1991), but it did cause sister chromotid exchange (SCE) in V79 Chinese hamster cells (Kuroda et al. 1991). DNA damage occurred following antimony trioxide treatment in *Bacillus subtilis* in Rec assays (Kanematsu et al. 1980; Kuroda et al. 1991).

QUANTITATIVE TOXICITY ASSESSMENT

Noncancer

Dermal Assessment

There are inadequate dermal toxicity data on antimony trioxide to derive a reference dose for dermal exposure.

Inhalation RfC

In 1995, the EPA derived a reference concentration (RfC) value for antimony trioxide (EPA 1999) based on the study by Newton et al. (1994). The subcommittee agrees that the Newton et al. (1994) study is the critical study for the derivation of an inhalation RfC, and that the critical end points chosen by the EPA are appropriate. The subcommittee, therefore, used EPA's benchmark concentration (BMC) analysis to determine their recommended level for antimony trioxide. The BMC was calculated for chronic pulmonary inflammation,[13] granulomatous inflammation, and fibrosis in males, females, and both sexes combined. The lower 95% confidence level on the concentration corresponding to a 10% extra risk of pulmonary inflammation (i.e., a 10% increase in the incidence of pulmonary inflammation) (the $BMCL_{10}$) was determined. The most sensitive end point was chronic inflammation in female rats, for which the $BMCL_{10}$ was 0.87 mg antimony trioxide/m^3. Adjusted for intermittent exposure of 6 hr/d, 5 d/wk, the BMC_{10}(ADJ) was 0.16 mg antimony trioxide/m^3. The human equivalent concentration, BMC_{10} (ADJ, HEC), of that exposure was calculated to be 0.074 mg/m^3 (using a regional deposited dose ratio [RDDR] for the thoracic region of 0.46). That value is similar to the HEC of 0.042 mg/m^3 calculated from the NOAEL of 0.51 mg antimony trioxide/m^3. The derivation of the RfC is shown in Table 10-6. To derive the RfC from the BMC_{10} (ADJ, HEC) of 0.16 mg antimony trioxide/m^3, a composite uncertainty factor of 300 was used, which included a factor of 3 for interspecies extrapolation, a factor

[13]The lung tissue examined included the right lobes and the major bronchi.

TABLE 10-6 Inhalation Reference Concentration for Antimony Trioxide

Critical Effect	Species	Effect Level (mg/m^3)	Uncertainty Factors		RfC ($\mu g/m^3$)	Reference
Pulmonary toxicity	Rats	BMC_{10}: (ADJ, HEC) 0.074	UF_A:	3	0.2	Newton et al. 1994
			UF_H:	10		
			UF_S:	3		
			UF_D:	3		
			Total:	300		

BMC_{10}, concentration corresponding to an extra risk of 10% (adjusted for intermittent exposure and calculated as the human equivalent concentration); RfC, reference concentration; UF_A, extrapolation from animals to humans; UF_H, intraspecies variability; UF_S, extrapolation from a study of less-than-lifetime duration; UF_D, inadequate or deficient toxicity database.

of 10 for intraspecies variation, a factor of 3 for database inadequacies, and a factor of 3 for a less-than-lifetime exposure that was longer than the standard subchronic study. Division of BMC_{10} (ADJ, HEC) by the composite uncertainty factor resulted in an RfC of 0.2 µg antimony trioxide/m^3.

The key study used for the derivation of the inhalation RfC was assigned a medium confidence level. Although it was well conducted and well documented, it is not a lifetime exposure study. Confidence in the database is medium because of the absence of adequate developmental or reproductive toxicity studies. Therefore, confidence in the RfC is medium.

Oral RfD

The database for developing an oral reference dose (RfD) for antimony trioxide is limited to one high-quality subchronic feeding study in rats (Hext et al. 1999). That study is supported by data from a subchronic study in rats (Sunagawa 1981), and short-term studies in rats (Smyth and Thompson 1945) and dogs (Fleming 1938). Overall, those data indicate that the hematological system (increased serum enzymes), the liver (increased liver weights), and the gastrointestinal tract are the target organs for antimony trioxide. Based on the weight of evidence, the subcommittee considered the increases in serum enzymes in females and the increase in liver weight in males and females at 1,879 mg Sb_2O_3/kg bw-d to be adverse effects (Hext et al. 1999). Therefore, the LOAEL for that study is 1,879 mg antimony trioxide/kg-d and the NOAEL is 494 mg antimony trioxide/kg-d. A composite uncertainty factor of 3,000 is applied to that NOAEL to yield an RfD of approximately 0.2 mg antimony trioxide/kg-d. The composite uncertainty factor comprises a factor of 10 for interspecies extrapolation; a factor of 10 to for intraspecies variability; a factor of 10 for extrapolation from a subchronic to a chronic study; and a factor of 3 for data base deficiencies (i.e., lower than the default of 10 because there is some data that indicate there is no progression in severity of effects). A summary of the derivation of that oral RfD is provided in Table 10-7.

The key study used for the derivation of the RfD was conducted according to current testing guidelines and is well documented; therefore, confidence in the key study is high. However, confidence in the overall database is low, because there are no adequate data on developmental or reproductive effects. Several additional studies are needed to complete the database, including a bioassay in a second species, a multigeneration reproduction study, and developmental toxicity studies in two species. Longer-term assays would also be informative. As a result, the confidence for the derived RfD is low to medium.

Cancer

Dermal

The carcinogenicity of antimony trioxide by the dermal route of exposure cannot be determined because of lack of data.

Inhalation

Based on the weight of evidence (from animal studies), the subcommittee concludes that the data are suggestive of carcinogenicity following inhalation

TABLE 10-7 Oral Reference Dose for Antimony Trioxide

Critical Effect	Species	Effect Level (mg/kg-d)	Uncertainty Factors	RfD (mg/kg-d)	Reference
Increases in serum enzymes; increased liver weight	Female rats	NOAEL = 494	UF_A: 10 UF_H: 10 UF_S: 10 UF_D: 3 Total: 3,000	0.2	Hext et al. 1999

NOAEL, no-observed-adverse-effect level; RfD, reference dose; UF_A, extrapolation from animals to humans; UF_H, intraspecies variability; UF_S, extrapolation from a study of less-than-lifetime duration; UF_D, inadequate or deficient toxicity database.

exposure to antimony trioxide. The cancer risk from antimony trioxide following inhalation exposure was estimated based on the study by Watt (1983). It should be noted, however, that the study by Watt (1983) is not published in the peer review literature and the results are controversial. A linear extrapolation from the observable region to the low-dose region is appropriate because there are insufficient data to suggest a nonlinear mode of action.

The data by Watt (1983) was modeled using the linear multistage model. The three tumor end points from the Watt (1983) study that were modeled were bronchioalveolar adenomas, scirrhous carcinomas, and squamous-cell carcinomas.[14] Concentrations were adjusted for discontinuous exposure (multiplied by 6 hr/24 hr × 5 d/7 d), converted to a HEC using the regional deposited dose ratio (RDDR = 1.8342) of particles for the thoracic region (MMAD = 0.4 microns, sigma g = 2.2), and adjusted for the less-than-lifetime exposure. The modeling results are listed in Table 10-8.

Because all the tumors occurred in the bronchioalveolar region and appeared to be arising from the alveolar epithelial lining cells, the three tumor types were combined, and total bronchioalveolar tumors were also modeled for the LED_{10}. Using the combined bronchioalveolar tumor incidence yields the most conservative (health-protective) estimate of the risk, with an LED_{10} of 0.14 mg antimony trioxide/m^3. Based on a linear extrapolation, the unit risk (cancer potency factor) of lung cancer is 7.1×10^{-4}/µg antimony trioxide/m^3.

Oral

The carcinogenicity of antimony trioxide by the oral route of exposure cannot be determined because of lack of data.

EXPOSURE ASSESSMENT AND RISK CHARACTERIZATION

Noncancer

Dermal

The assessment of noncancer risk by the dermal route of exposure is based on the scenario described in Chapter 3. This exposure scenario assumes that an

---

[14]The incidence of tumors in all animals was used in the hazard identification. However, since the concentration-response modeling is based on tumors following a lifetime exposure, and there were several interim kills in this study, only tumors in the animals sacrificed at study termination were used in the modeling.

TABLE 10-8 Results of Modeling for the Watt (1983) Study on Antimony Trioxide

	Multistage model (mg/m^3)		Risk by EPA methods $(mg/m^3)^{-1}$	
End point	ED_{10}	LED_{10}	$0.1/ED_{10}$	$0.1/LED_{10}$
Adenomas	0.51	0.24	0.185	0.417
Scirrhous carcinomas	0.35	0.21	0.286	0.476
Squamous-cell carcinomas	0.83	0.43	0.120	0.233
Bronchioalveolar tumor (combined)	0.24	0.14	0.417	0.714

ED_{10}, effective dose corresponding to a 10% tumor response in test animals; LED_{10}, lower 95% bound on the effective dose corresponding to a 10% tumor response in test animals.

adult spends 1/4th of his or her time sitting on furniture upholstery treated with antimony trioxide, that 1/4th of the upper torso is in contact with the upholstery, and that clothing presents no barrier. Antimony trioxide is considered to be ionic, and is essentially not absorbed through the skin. However, to be conservative, the subcommittee assumed that ionized antimony trioxide permeates the skin at the same rate as water, with a permeability rate of 10^{-3} cm/hr (EPA 1992). Using that permeability rate, the highest expected application rate for antimony trioxide (2.5 mg/cm^2), and Equation 1 in Chapter 3, the subcommittee calculated a dermal exposure level of 2.0×10^{-2} mg/kg-d. The oral RfD for antimony trioxide (0.2 mg/kg-d; see Oral RfD in Quantitative Toxicity section) was used as the best estimate of the internal dose for dermal exposure. Dividing the exposure level by the oral RfD yields a hazard index of 0.1. Thus it was concluded that antimony trioxide used as a flame retardant in upholstery fabric is not likely to pose a noncancer risk by the dermal route.

Inhalation

Particles

The assessment of the noncancer risk by the inhalation route of exposure is based on the scenario described Chapter 3. This scenario corresponds to a person spending 1/4th of his or her life in a room with low air-change rate

(0.25/hr) and with a relatively large amount of fabric upholstery treated with antimony trioxide (30 m^2 in a 30-m^3 room), with this treatment gradually being worn away over 25% of its surface to 50% of its initial quantity over the 15-yr lifetime of the fabric. A small fraction, 1%, of the worn-off antimony trioxide is released into the indoor air as inhalable particles and is breathed by the occupant. Equations 4 through 6 in Chapter 3 were used to estimate the average concentration of antimony trioxide in the air. The highest expected application rate for antimony trioxide is 2.5 mg/cm^2. The estimated release rate is 2.3×10^{-7}/d. Using those values, the estimated time-averaged exposure concentration for antimony trioxide is 0.24 μg/m^3.

Division of that exposure concentration (0.24 μg/m^3) by the inhalation RfC (2×10^{-4} mg/m^3; see Quantitative Toxicity Assessment section) results in a hazard index of 1.2, indicating that under the worst-case exposure scenario, antimony trioxide might possibly pose a noncancer risk via inhalation of particles.

Vapors

In addition to the possibility of release of antimony trioxide in particles worn from upholstery fabric, the subcommittee considered the possibility of its release by evaporation. However, because of antimony trioxide's negligible vapor pressure at ambient temperatures, the subcommittee considered antimony trioxide not likely to pose a noncancer risk by exposure to vapors.

Oral Exposure

The assessment of the noncancer risk by the oral exposure route is based on the scenario described in Chapter 3. That exposure assumes that a child sucks on 50 cm^2 of fabric backcoated with antimony trioxide daily for two yr, one hr/d. The highest expected application rate (per unit time) for antimony trioxide is about 2.5 mg/cm^2. The fractional release rate of antimory trioxide is estimated as 0.001/d, based on the leaching of antimony from polyvinyl chloride cot mattresses (Jenkins et al. 1998). Using those assumptions and Equation 15 in Chapter 3, the average oral dose rate is estimated to be 0.00052 mg/kg-d. Division of that exposure estimate (0.00052 mg/kg-d) by the oral RfD (0.2 mg/kg-d; see Quantitative Toxicity Assessment section) results in a hazard index of 2.6×10^{-3}. Therefore, under the worst-case exposure assumptions, antimony trioxide, used as a flame retardant in upholstery fabric, is not likely to pose a noncancer risk by the oral route of exposure.

Cancer

There are inadequate data to assess the carcinogenicity of antimony trioxide from dermal or oral exposures.

Inhalation (Particles)

The average room-air concentration and average exposure concentration for antimony trioxide were obtained as described for the noncancer risk assessment of particles. The estimated time-averaged exposure concentration is 0.24 μg/m^3. Using the inhalation unit cancer risk (cancer potency factor) of 7.1×10^{-4}/μg antimony trioxide/m^3, the lifetime excess cancer risk estimate from exposure to antimony trioxide as particles is 1.7×10^{-4}. However, the inhalation unit risk (cancer potency factor) of antimony trioxide is itself suspect (see Hazard Identification Section). Furthermore, even if the reservations concerning the study by Watt (1983) are discounted and the inhalation unit risk is considered to be accurate, better exposure assessment is required before any definitive conclusions can be drawn about the carcinogenic risk from antimony trioxide via inhalation in the particulate phase.

Inhalation (Vapors)

Antimony trioxide has negligible vapor pressure at ambient temperatures, so antimony trioxide used as a flame retardant in upholstery fabric is not likely to pose a cancer risk for exposure to vapors.

RECOMMENDATIONS FROM OTHER ORGANIZATIONS

The American Conference of Governmental Industrial Hygienists (ACGIH) has established a Threshold Limit Value (TLV) for antimony trioxide of 0.5 mg antimony/m^3 (AGCIH 1999).

The Occupational Safety and Health Administration (OSHA) and the National Institute for Occupational Safety and Health (NIOSH) do not have standards for exposure to antimony trioxide.

EPA's inhalation RfC of 0.2 μg antimony trioxide/mg^3 is the same as that of the subcommittee.

DATA GAPS AND RESEARCH NEEDS

There are little data on the toxicity of antimony trioxide following dermal exposure. The hazard index of 0.1 indicates that antimony trioxide is not likely to pose a non-cancer risk from dermal exposure. Therefore, the subcommittee does not recommend further research on the effects of antimony trioxide from dermal exposure for the purposes of flame-retarding upholstery furniture.

The subcommittee's risk characterization indicates that antimony trioxide might possibly pose a risk for noncancer and cancer end points via inhalation in the particulate phase. Therefore, better exposure information is essential to accurately assess the risks of antimony trioxide use as a flame retardant in upholstery fabric. If that research shows that actual exposures are substantially lower than the subcommittee's estimated levels, there will be a reduced need to perform toxicity studies. One study indicated that there are reproductive effects following inhalation of antimony trioxide. However, the purity of the antimony trioxide in that study is not known, and studies of other antimony compounds show no reproductive effects (Reprotox 1999). The study on which the quantitative toxicity assessment for cancer is based is suspect, and further studies would clarify if antimony indeed poses a cancer risk following inhalation exposure.

There are no studies that evaluated the chronic toxicity of antimony trioxide from the oral route of exposure. There are no studies that have measured exposure from the oral route. The hazard index of 2.6×10^{-3} indicates that antimony trioxide is not likely to pose a noncancer risk from oral exposure. Therefore, the subcommittee does not recommend further studies of antimony trioxide following oral exposure for the purposes of its use as a flame retardant in upholstery furniture fabric.

With respect to cancer, the effects following inhalation exposure are portal-of-entry specific (i.e., only occur in the lung), and, therefore, the subcommittee does not recommend carcinogenic studies following other routes of exposure.

Based on an inhalation hazard index greater than one and a potential cancer risk following inhalation exposure, the subcommittee recommends that the potential for particle release from treated fabric be investigated.

REFERENCES

ACGIH (American Conference of Government Industrial Hygienists). 1999. Threshold Limit Values and Biological Exposure Indices. Cincinnati, OH: American Conference of Government Industrial Hygienists, Inc.

ATSDR (Agency for Toxic Substances and Disease Control). 1992. Toxicological Profile for Antimony. Agency for Toxic Substances and Disease Registry, U.S. Department of Health and Human Services, Public Health Service, Atlanta, GA.

Bailly, R., R. Lauwerys, J.P. Buchet, P. Mahieu, and J. Konings. 1991. Experimental and human studies on antimony metabolism: Their relevance for the biological monitoring of workers exposed to inorganic antimony. Br. J. Ind. Med. 48(2):93-97.

Becker, E.L., Landau, S.I., A. Manuila, W.J.H. Butterfeild, A.M.Harvey, R.H. Heptinstall, and L. Thomas, eds. 1986. International Dictionary of Medicine and Biology. New York: Wiley.

Belyaeva, A.P. 1967. The effect produced by antimony on the generative function. [Abstract]. Gig. Truda. Prof. Zabol. 11(1):32-37.

Bio/dynamics (Bio/dynamics, Inc.). 1990. A One-year Inhalation Toxicity Study of Antimony Trioxide in the Rat (With a One-year Recovery Period). Project No. 83-7647. Bui/dynamics, Inc., East Millstone, N.J.

Budavari, S., M.J. O'Neil, A. Smith, and P.E. Heckelman. 1989. The Merck Index, 11th Ed. S. Budavari, M.J. O'Neil, A. Smith, and P.E. Heckelman, eds. Rahway, N.J.: Merck & Co., Inc.

Cooper, D.A., E.P. Pendergrass, A.J. Vorwald, R.L. Mayock, and H. Brieger. 1968. Pneumoconiosis among workers in an antimony industry. Am. J. Roentgenol. Radium Ther. Nucl. Med. 103(3):495-508.

Dernehl, C.U., C.A. Nau, and H.H. Sweets. 1945. Animal studies on the toxicity of inhaled antimony trioxide. J. Ind. Hyg. Toxicol. 27(9):256-262.

Ebbens, K. 1972. Acute Toxicity Studies with Antimony Oxide 2996-30. Industrial Bio-Test Labs, Inc., Northbrook, IL. EPA/OTS Doc #88-920007957.

Elliott, B.M., J.M. Mackay, P. Clay, and J. Ashby. 1998. An assessment of the genetic toxicology of antimony trioxide. Mutat. Res. 415(1-2):109-117.

EPA (U.S. Environmental Protection Agency). 1992. Dermal Exposure Assessment: Principles and Applications. EPA/600/8-91-011B. Office of Health and Environmental Assessment, U.S. Environmental Protection Agency, Washington, D.C.

EPA (U.S. Environmental Protection Agency). 1999. Integrated Risk Information System (IRIS). [Online]. Available: http://www.epa.gov/iris/subst/0676.htm Office of Research and Development. National Center for Environmental Assessment, Cincinnati, OH. Downloaded June, 1999; Antimony file dated 1991, antimony trioxide file dated 1995.

Fleming, A.J. 1938. The Toxicity of Antimony Trioxide. E.I. du Pont de Nemours and Co, Wilmington, Del. EPA/OTS Document #87220297.

Gerber, G.B., J. Maes, and B. Eykens. 1982. Transfer of antimony and arsenic to the developing organism. Arch. Toxicol. 49(2):159-168.

Gerhardsson, L., D. Brune, G.F. Nordberg, and P.O. Wester. 1982. Antimony in lung, liver and kidney tissue from deceased smelter workers. Scand. J. Work Environ. Health 8(3):201-208.

Grin, N.V., N.N. Govorunova, A.N. Bessemrnyi, and L.V. Pavlovich. 1987. Embryotoxic action of antimony oxide in an experiment. [Article in Russian]. Gig. Sanit. 10:85-86.

Gross, P., J.H.U. Brown, M.L. Westrick, R.P. Srsic, N.L. Butler, and T.F. Hatch. 1955a. Toxicologic study of calcium halophosphate phosphors and antimony trioxide. I. Acute and chronic toxicity and some pharmacologic aspects. Arch. Ind. Health 11:473-478.

Gross, P., M.L. Westrick, J.H.U. Brown, R.P. Srsic, H.H. Schrenk, and T.F. Hatch. 1955b. Toxicologic study of calcium halophosphate phosphors and antimony trioxide. II. Pulmonary studies. Arch. Ind. Health 11:479-486.

Groth, D.H., L.E. Stettler, J.R. Burg, W.M. Busey, G.C. Grant, and L. Wong. 1986. Carcinogenic effects of antimony trioxide and antimony ore concentrate in rats. J. Toxicol. Environ. Health 18(4):607-26.

Gurnani, N., A. Sharma, and G. Talukder. 1992. Comparison of the clastogenic effects of antimony trioxide on mice in vivo following acute and chronic exposure. BioMetals 5(1):47-50.

Haskell Laboratory (Haskell Laboratory for Toxicology and Industrial Medicine). 1970a. Primary Skin Irritation and Sensitization Tests in Guinea Pigs. U.S. EPA OTS document ID# 878220294.

Haskell Laboratory (Haskell Laboratory for Toxicology and Industrial Medicine). 1970b. Primary Skin Irritation and Sensitization Tests. U.S. EPA OTS document ID# 878220307.

Hatlelid, K. 1999. Toxicity Review for Calcium Molybdate and Zinc Molybdate. Memorandum, dated March 2, 1999, from Kristina Hatlelid, Toxicologist, Division of Health Sciences, to Ronald Medford, Assistant Executive Director for Hazard Identification and Reduction, U.S. Consumer Product Safety Commission, Washington, DC.

Hext, P.M., P.J. Pinto, and B.A. Rimmel. 1999. Subchronic feeding study of antimony trioxide in rats. J. Appl. Toxicol. 19(3):205-209.

Hiraoka, N. 1986. The toxicity and organ distribution of antimony after chronic administration to rats. J. Kyoto Prefect Univ. Med. 95:997-1017.

IARC (International Agency for Research on Cancer). 1989. Antimony trioxide and antimony trisulfide. Pp. 291-305 in IARC Monographs on the Evaluation of Carcinogenic Risks to Humans, Vol. 47, Some Organic Solvents, Resin Monomers and Related Compounds, Pigments and Occupational Exposures in Paint Manufacture and Painting. Lyon, France: World Health Organization, International Agency for Research on Cancer.

ICRP (International Commission on Radiological Protection). 1981. Metabolic data for antimony. Pp. 46-49 in Radiation Protection, ICRP Publication 30, Part 3, Including Addendum to Parts 1 and 2, Limits for Intakes of Radionuclides by Workers. International Commission on Radiological Protection. Elmsford, N.Y.: Pergamon Press.

Industrial Bio-Test Laboratories, Inc. 1973. Human Repeated Insult Patch Test with TS-162. U.S. EPA OTS document I.D.# 878211206.

James, L.F., V.A. Lazar, and W. Binns. 1966. Effects of sublethal doses of certain minerals on pregnant ewes and fetal development. Am. J. Vet. Res. 27(116):132-135.

Jenkins, R.O., P.J. Craig, W. Goessler, and K.J. Irgolic. 1998. Antimony leaching from cot mattresses and sudden infant death syndrome (SIDS). Hum. Exp. Toxicol. 17(3):138-139.

Jones, R.D. 1994. Survey of antimony workers: Mortality 1961-1992. Occup. Environ. Med. 51(11):772-776.

Kanematsu, N., M. Hara, and T. Kada. 1980. Rec assay and mutagenicity studies on metal compounds. Mutat. Res. 77(2):109-116.

Kim, K.-W., B.-S. Choi, S.-K. Kang, H.-Y. Kim, S.S. Park, Y.-S. Cho, M.-G. Song, and Y.-H. Moon. 1997. Assessment of workers' exposure to antimony trioxide in Korea. J. Occup. Health 39(4):345-348.

Kuroda, K., G. Endo, A. Okamoto, Y.S. Yoo, and S. Horiguchi. 1991. Genotoxicity of beryllium, gallium, and antimony in short-term assays. Mutat. Res. 264(4):163-170.

Leffler, P, L. Gerhardsson, D. Brune, and G.F. Nordberg. 1984. Lung retention of antimony and arsenic in hamsters after the intratracheal instillation of industrial dust. Scand. J. Work Environ. Health 10(4):245-251.

Lide, D.R. 1991-1992. Handbook of Chemistry and Physics, 72nd Ed. Boca Raton, FL: CRC Press.

Lüdersdorf, R., A. Fuchs, P. Mayer, G. Skulsuksai, and G. Schäcke. 1987. Biological assessment of exposure to antimony and lead in the glass-producing industry. Int. Arch. Occup. Environ. Health 59(5):469-474.

McCallum, R.I. 1963. The work of an occupational hygiene service in environmental control. Ann. Occup. Hyg. 6(2):55-64.

McCallum, R.I. 1967. Detection of antimony in process workers' lungs by x-radiation. Trans. Soc. Occup. Med. 17(4):134-138.

Muhle H., B. Bellmann, O. Creutzenberg, U. Henrich, M. Ketkar, and R. Mermelstein. 1990. Dust overloading of lungs after exposure of rats to particles of low solubility: Comparative studies. J. Aerosol. Sci. 21(3):374-377.

Myers, R.C., E.R. Homan, C.S. Weil, and G.A. Webb. 1978. Antimony Trioxide Range-Finding Toxicity Studies. Carnegie-Mellon Institute of Research, Carnegie-Mellon University, Pittsburgh, PA. Sponsored by Union Carbide Corp., Danbury, CT. EPA/OTS Document #878210813.

Newton, P.E., H.F. Bolte, I.W. Daly, B.D. Pillsbury, J.B. Terrill, R.T. Drew, R. Ben-Dyke, A.W. Sheldon, and L.F. Rubin. 1994. Subchronic and chronic inhalation toxicity of antimony trioxide in the rat. Fundam. Appl. Toxicol. 22(4):561-576.

Potkonjak, V., and M. Pavlovich. 1983. Antimoniosis: A particular form of pneumoconiosis. I. Etiology, clinical and X-ray findings. Int. Arch. Occup. Environ. Health 51(3):199-207.

Renes, L.E. 1953. Antimony poisoning in industry. Arch. Ind. Hyg. Occup. Med. 7:99-108.

Reprotox. 1999. Subscription information service. [Online]. Available: http://reprotox.org/

Ridgway, L.P., and D.A. Karnofsky. 1952. The effects of metals on the chick embryo: Toxicity and production of abnormalities in development. Ann. N.Y. Acad. Sci. 55:203-215.

Rossi, F., R. Acampora, C. Vacca, S. Maione, M.G. Matera, R. Servodio, and E. Marmo. 1987. Prenatal and postnatal antimony exposure in rats: Effect on vasomotor reactivity development of pups. Teratog. Carcinog. Mutagen. 7(5):491-496.

Schnorr, T.M., K. Steenland, M.J. Thun, and R.A. Rinsky. 1995. Mortality in a cohort of antimony smelter workers. Am. J. Ind. Med. 27(5):759-770.

Smyth, H.F., and C.P. Carpenter. 1948. Further experience with range finding test in the industrial toxicology laboratory. J. Ind. Hyg. Toxicol. 30:63-68.

Smyth, H.F., Jr., and W.L. Thompson. 1945. The Single Dose and Subacute Toxicity of Antimony Trioxide (Sb_2O_3). Mellon Institute of Industrial Research, University of Pittsburgh, Pittsburgh, PA. Sponsored by Union Carbide Corp. EPA/OTS Doc #878210812.

Sunagawa, S. 1981. Experimental studies on antimony poisoning. [Article in Japanese]. Igaku Kenkyu 51(3):129-142.

Watt, W.D. 1983. Chronic Inhalation Toxicity of Antimony Trioxide: Validation of the Threshold Limit Value, Dissertation, Wayne State University, Detroit, MI.

White, G.P., Jr., C.G.T. Mathias, and J.S. Davin. 1993. Dermatitis in workers exposed to antimony in melting process. J. Occup. Med. 35(4):392-395.

Witschi, H.R., and J.A. Last. 1996. Toxic responses of the respiratory system. Pp. 443-462 in Casarett & Doull's Toxicology: The Basic Science of Poisons, 5th Ed. C.D. Klaassen, M.O. Amdur, and J. Doull, eds. New York: McGraw-Hill.

11

Antimony Pentoxide and Sodium Antimonate

THIS chapter reviews the physical and chemical properties, toxicokinetics, toxicological, epidemiological, and exposure data on pentavalent antimony (antimony pentoxide and sodium antimonate). The subcommittee reviewed the available data on these compounds and determined that toxicological information necessary for risk assessment was mostly available for antimony pentoxide. The subcommittee used that information to characterize the health risk from exposure to pentavalent antimony. The subcommittee also identified data gaps and recommended research relevant for determining the health risk from exposure to pentavalent antimony.

PHYSICAL AND CHEMICAL PROPERTIES

The physical and chemical properties of antimony pentoxide and sodium antimonate are summarized in Table 11-1.

OCCURRENCE AND USE

Antimony pentoxide (Sb_2O_5) and sodium antimonate ($NaSbO_3$) are the pentavalent forms of antimony most widely used as flame retardants. Pentavalent

TABLE 11-1 Physical and Chemical Properties of Antimony Pentoxide and Sodium Antimonate

Property	Value	Reference
Antimony pentoxide		
Chemical formula	Sb_2O_5, Sb(V)	Budavari et al. 1989
CAS Registry #	1314-60-9	Budavari et al. 1989
Synonyms	Antimonic oxide, stibic anhydride, antimonic acid, antimonic (V) acid	Budavari et al. 1989
Molecular weight	323.52	Budavari et al. 1989
Physical state	Yellow powder	Budavari et al. 1989
Solubility	Slightly soluble in water; practically insoluble in HNO_3; slowly dissolves in warm HCl or in warm KOH	Budavari et al. 1989
Melting point	380°C	Budavari et al. 1989
Density	3.78 g/cm^3	Budavari et al. 1989
Sodium antimonate		
Chemical formula	$NaSbO_3$, Sb(V)	Lide 1991-1992
CAS Registry #	15432-85-6	Lide 1991-1992
Synonyms	Sodium (meta) antimonate	Lide 1991-1992
Molecular weight	192.74	Lide 1991-1992
Physical state	White, granular powder	Lide 1991-1992
Solubility	Slightly soluble in water; practically insoluble in HNO_3; soluble in tartaric acid	Lide 1991-1992

antimonates function primarily as a stable colloid or synergist with halogenated flame retardants.

Antimony pentoxide retards flammability by forming halogenated antimony compound which excludes oxygen from the front of the flame (Gerhartz et al. 1985; ATSDR 1992). Sodium antimonate ($NaSbO_3$) is used in industrial applications where special colors are required or when antimony trioxide may produce unwanted chemical reactions (IPCS 1997).

TOXICOKINETICS

Absorption

Dermal

No studies were found that examined the absorption of pentavalent antimonates following dermal exposure.

Inhalation

Elevated blood and urine antimony levels were reported in workers occupationally exposed to antimony compounds (pentoxide and trioxide) suggesting that antimony is absorbed by the inhalation route (Cooper et al. 1968).

Inhalation studies in Syrian hamsters indicate that antimony is absorbed and distributed to various tissues following exposure to pentavalent aerosols derived from an antimony-tartrate complex (Felicetti et al. 1974).

Oral

The International Commission on Radiological Protection (ICRP 1981) has recommended that a 1% absorption rate of antimony compounds (including antimony pentoxide and sodium antimonate) be assumed when estimating exposure from the gastrointestinal tract. This recommendation is based on studies of various organic and inorganic antimony compounds, some of which have a pentavalent valence state.

Oral exposure studies in Syrian hamsters with an antimony-tartrate complex found only minute amounts of antimony were absorbed through the gastrointestinal tract (Felicetti et al. 1974).

Distribution

The distribution of antimony in the body is affected by the valence states of the particular antimony species. Pentavalent antimony has less affinity for liver tissue and accumulates to a greater extent in the spleen of hamsters than do the trivalent forms (Gellhorn et al. 1946, as cited in ATSDR 1992). Human (Otto et al. 1947) and hamster (Felicetti et al. 1974) erythrocytes concentrate the trivalent form but not the pentavalent form of antimony. Additionally, skeletal uptake in hamsters is greater for pentavalent antimony than the trivalent form.

In a retrospective study on deceased smelter workers exposed to a number of metals (including antimony), elevated levels of antimony were found in their lung tissues as compared with those in non-occupationally exposed individuals (Gerhardsson et al. 1982). Neither the form nor the valence state of the antimony exposure was reported.

High levels of antimony were found in the liver, skeleton, and pelt (in nonlung tissue) of hamsters following inhalation exposure to pentavalent antimony as a tartrate (Felicetti et al. 1974).

Metabolism and Excretion

Few data are available on the metabolism of pentavalent antimony. It is known that inorganic antimony is not methylated in vivo but can covalently bind to sulfhydryl groups and phosphate (Bailly et al. 1991). Pentavalent antimony can form conjugates with glutathione and be excreted in the bile (Bailly et al. 1991).

Excretion of absorbed antimony in smelter workers exposed to pentavalent antimony occurs primarily in the urine. Rees et al. (1980) reported that more than 80% of pentavalent antimony (administered as sodium stibogluconate) is excreted in 6 hr in humans following intravenous or intramuscular injection.

Following a single intramuscular injection, pentavalent antimony had a half-life of about 1 hr in hamsters (Berman et al. 1988). Experimental animals excrete trivalent antimony in the feces and to a lesser extent in the urine (Gross et al. 1955).

HAZARD IDENTIFICATION[1]

There are inadequate toxicity data on antimony pentoxide and sodium antimonate from any route of exposure. Chapter 10 provides a detailed discussion of toxicity of antimony trioxide and other antimony compounds. This chapter contains discussion only on antimony pentoxide and sodium antimonate.

Dermal Exposure

Male smelter workers (n = 51) exposed to airborne dusts containing antimony trioxide and pentoxide were diagnosed with skin changes. Conjunctivitis was identified in 27% of the workers while dermatosis was identified in 63% of the 51 workers (Potkonjak and Pavlovich 1983). No actual exposure levels for these workers were given. The authors note that the affected persons worked in the factory from 9 to 31yr (mean = 17.91), almost exclusively as smelters.

No other dermal toxicity data were identified for any pentavalent antimonates.

---

[1]In this section, the subcommittee reviewed toxicity data on antimony pentoxide and sodium antimonate, including the toxicity assessment prepared by the U.S. Consumer Product Safety Commission (Hatlelid 1999).

Inhalation Exposure

Systemic Effects

Male smelter workers (n = 51) exposed to airborne dust containing antimony, arsenic, silica, and ferric oxides were evaluated for symptoms of antimoniosis (pneumoconiosis caused by antimony exposure) over a 2-yr reporting period (Potkonjak and Pavlovich 1983). All subjects had worked exclusively as smelters for a period of 9 to 31 yr (mean = 17.91) and had shown positive X-ray findings of antimoniosis over a range of severity. Exposure was primarily to antimony trioxide (5.5-64 mg/m^3) while exposure to the pentoxide form ranged from 0.27 to 5.0 mg/m^3. Other pulmonary symptoms included chronic coughing (60.8%), upper airway inflammation (35.3%), and conjunctivitis (27.5%). Pulmonary function tests did not find an increased incidence of any obstructive changes. Fibrosis was not detected in any of the subjects, and no other symptoms of systemic antimony poisoning were reported. However, the results of this study cannot be used to attribute the observed effects to antimony exposure because of the confounding exposure to other contaminants.

In an acute inhalation exposure study, the LC_{50} for male and female Sprague-Dawley rats (5/sex) exposed to colloidal antimony pentoxide for 4 hr at concentrations of 2.64, 5.01, or 8.62 mg/L was found to be 6.14 mg/L for males and 8.62 mg/L for females. The generated particles had a mass median aerodynamic diameter (MMAD) of 2.95-3.54 microns with a geometric standard deviation of 2.15 to 2.39. Survivors did not have any findings at necropsy that were not incidental or spontaneous postmortem changes (Hazelton Laboratories 1989).

No F-344 rats died or developed abnormal gross pathology that were intratracheally instilled with a suspension of antimony pentoxide (3.5 µg /kg body weight) for 6 mo (American Biogenics Corp. 1987). No significant differences in body weight were seen when compared to control animals. It is not known whether histopathological analysis was conducted on these animals.

Macrophage cytotoxicity was also determined for cultured F-344 rat alveolar macrophages (American Biogenics Corp. 1986). Antimony trioxide, antimony pentoxide-A, and antimony pentoxide-ZTA concentrations tested were 0.02, 0.2, 2, and 20 mM for 20 hr. The viability indices, expressed as an EC_{50}, were 0.55, 1.14, and 25.44, respectively, for the trioxide, pentoxide-A, and pentoxide-ZTA.

Other Systemic Effects

There are no data on the immunological, neurological, reproductive, or developmental effects of pentavalent antimonates following inhalation exposure.

Cancer

Potkonjak and Pavlovich (1983) did not find any malignancies among 51 smelter workers exposed to antimony (as antimony trioxide and pentoxide). Workers had occupational exposure in the smelter industry from 9 to 31yr (mean = 17.91).

There are no data on the carcinogenicity of pentavalent antimonates in animals following inhalation exposure.

Oral Exposure

There are no carcinogenicity data on antimony pentoxide or sodium antimonate in humans or animals following oral exposure.

Genotoxicity

Antimony pentoxide did not induce chromosome damage or damage to chromosomal spindle apparatus in male or female CD-1 mice (five/sex/group) treated by gavage with antimony pentoxide at doses of 2,500, 5,000, or 10,000 mg/kg (Nissan Chemical Industries, Ltd. 1985).

Antimony pentoxide was negative in the Bacillus subtilis assay (60 μg/disk) and did not induce sister chromatid exchanges (40 μg/mL) in Chinese hamster ovary cells (Kuroda et al. 1991). Antimony pentoxide was also negative for mutagenicity in various *Salmonella* strains at concentrations up to 200 μg/plate. Antimony trioxide was positive in the rec assay and induced SCE at doses 100 times lower than the pentoxide (Kuroda et al. 1991).

QUANTITATIVE TOXICITY ASSESSMENT

Noncancer

Dermal, inhalation, and oral toxicity data on pentavalent antimonates are insufficient to derive RfDs or an RfC. Acute oral toxicity data for the pentavalent forms of antimony suggest that it is less toxic than the trivalent forms of antimony.

Cancer

There are insufficient cancer data on pentavalent antimonates from any route of exposure to calculate cancer potency factors.

EXPOSURE ASSESSMENT AND RISK CHARACTERIZATION

Noncancer

Dermal exposure to pentavalent antimony was estimated using the dermal exposure scenario described in Chapter 3. This exposure scenario assumes that an adult spends 1/4th of his or her time sitting on furniture upholstery treated with antimony pentoxide or sodium antimonate and also assumes 1/4th of the upper torso is in contact with the upholstery and clothing presents no barrier.

The subcommittee believes that pentavalent antimony is an ionic substance and, therefore, is essentially not absorbed through the skin. However, to be conservative, the subcommittee assumed that ionized antimony pentoxide and sodium antimonate permeate the skin at the same rate as water, with a permeability rate of 10^{-3} cm/hr (EPA 1992). Using that permeability rate, the highest expected application rate for zinc borate of 2.5 mg/cm^2 , and Equation 1 in Chapter 3, the subcommittee calculated a worst-case dermal exposure level of 2.0×10^{-2} mg/kg-d. The dermal or oral RfDs for antimony pentoxide or sodium antimonate were not derived because of a lack of adequate toxicity data. As a result, the noncancer risk associated with dermal exposure to antimony pentoxide or sodium antimonate, used as a flame retardant, cannot be characterized at this time.

Inhalation Exposure

Particles

The assessment of the noncancer risk from inhalation of upholstery particles containing pentavalent antimony is based on the inhalation exposure scenario described in Chapter 3. In this scenario, a person is exposed to upholstery particles containing pentavalent antimony. It is assumed that particles are formed from the wear of the upholstery and 50% of the pentavalent antimony present in 25% of the treated surface is released as particles over the 15-yr lifetime of the fabric. It is also assumed that only 1% of the worn-off pentavalent antimony is released into the indoor air as particles that may be inhaled and that a person spends 1/4th of his or her lifetime in a 30-m^3 room that contains 30 m^2 of treated upholstery with an air-change rate of 0.25/hr.

Particle exposure was estimated using Equations 4 and 5 in Chapter 3. The subcommittee estimated an upholstery application rate (S_a) for pentavalent antimony of 2.5 mg/cm^2. The release rate (μ_r) for pentavalent antimony from upholstery fabric was estimated to be 2.3×10^{-7}/d yielding a room airborne particle concentration (C_p) of 0.95 μg/m³ and a short time-averaged exposure concentration of 0.24 μg/m³. The time-averaged exposure concentration for particles was calculated using Equation 6 in Chapter 3.

An inhalation RfC is currently not available for pentavalent antimony. As a result, the noncancer risk associated with the inhalation of particles containing antimony pentoxide cannot be characterized at this time. However, a structurally similar compound, antimony trioxide, was found to be a possible concern for noncancer effects. Therefore, the subcommittee recommends that exposure levels for these compounds be measured.

Vapors

Antimony pentoxide has negligible vapor pressure at ambient temperatures. Therefore, inhalation of antimony pentoxide vapor is not anticipated to pose a noncancer toxic risk when incorporated into furniture upholstery.

Oral Exposure

The assessment of noncancer risk from oral exposure to antimony pentoxide or sodium antimonate is based on the oral exposure scenario described in Chapter 3. This scenario assumes a child is exposed to pentavalent antimony by sucking on 50 cm² of fabric treated with pentavalent antimony, 1 hr/d for 2 yr. The subcommittee estimated an upholstery application rate (S_a) for pentavalent antimony of 2.5 mg/cm² and a fractional rate of pentavalent antimony extraction (μ_a) by saliva of 0.001/d based on levels reported by Jenkins et al. (1998). Oral exposure was calculated using Equation 15 in Chapter 3.

Using the above equation, the worst-case average oral daily dose for antimony pentoxide was estimated as 0.00052 mg/kg-d. An oral RfD is not currently available for antimony pentoxide; therefore the noncancer risk associated with the estimated worst-case daily dose cannot be characterized at this time.

Cancer

There are inadequate data to characterize carcinogenic risk from the use of antimony pentoxide as a flame retardant from the dermal, inhalation, or oral routes of exposure.

RECOMMENDATIONS FROM OTHER ORGANIZATIONS

The Threshold Limit Value recommended by the American Conference of Governmental Industrial Hygienists for antimony and its compounds is 0.5 mg/m^3 (ACGIH 1999). The subcommittee is not aware of any other regulatory exposure limits for antimony pentoxide or sodium antimonate.

DATA GAPS AND RESEARCH NEEDS

There are no subchronic or chronic toxicity studies in the literature on pentavalent forms of antimony from any route of exposure. Additionally, there are no studies on the toxic effects of pentavalent antimony on reproduction or development. There are no exposure measurements from any route of exposure.

Based on the lack of toxicological data and possible concern from exposure to a structurally similar compound, antimony trioxide, the subcommittee recommends that the release rates into saline solution and air from fabrics treated with antimony pentoxide or sodium antimonate be investigated.

REFERENCES

ACGIH (American Conference of Government Industrial Hygienists). 1999. Threshold Limit Values and Biological Exposure Indices. Cincinnati, OH: American Conference of Government Industrial Hygienists, Inc.

American Biogenics Corp. (American Biogenics Corporation). 1986. Evaluation of the Cytotoxicity of Antimony Trioxide and Two Antimony Pentoxides Using Cultured Alveolar Macrophages. Study No. 420-2630. Decatur, IL.

American Biogenics Corp. (American Biogenics Corporation). 1987. A Six-month Intratracheal Toxicity Study of Antimony Oxides in Albino Rats. Study no. 420-2211 (Draft). Decatur, IL.

ATSDR (Agency for Toxic Substances and Disease Control). 1992. Toxicological Profile for Antimony. Agency for Toxic Substances and Disease Registry, U.S. Department of Health and Human Services, Public Health Service, Atlanta, GA.

Bailly, R., R. Lauwerys, J.P. Buchet, P. Mahieu, and J. Konings. 1991. Experimental and human studies on antimony metabolism: Their relevance for the biological monitoring of workers exposed to inorganic antimony. Br. J. Ind. Med. 48(2):93-97.

Berman, J.D., J.F. Gallalee, and J.V. Gallalee. 1988. Pharmacokinetics of pentavalent antimony (Pentostam) in hamsters. Am. J. Trop. Med. Hyg. 39(1):41-45.

Budavari, S., M.J. O'Neil, A. Smith, and P.E. Heckelman. 1989. The Merck Index, Eleventh Ed. S. Budavari, M.J. O'Neil, A. Smith, and P.E. Heckelman, eds. Rahway, NJ.: Merck & Co., Inc.

Cooper, D.A., E.P. Pendergrass, A.J. Vorwald, R.L. Mayock, and H. Brieger. 1968.

Pneumoconiosis among workers in an antimony industry. Am. J. Roentgenol. Radium Ther. Nucl. Med. 103(3):495-508.

EPA (U.S. Environmental Protection Agency). 1992. Dermal Exposure Assessment: Principles and Applications. EPA/600/8-91-011B. Office of Health and Environmental Assessment, U.S. Environmental Protection Agency, Washington, DC.

Felicetti, S.A., R.G. Thomas, and R.O. McClellan. 1974. Metabolism of two valence states of inhaled antimony in hamsters. Am. Ind. Hyg. Assoc. J. 35(5):292-300.

Gellhorn, A., N.A. Tupikova, and H.B. van Dyke. 1946. The tissue distribution and excretion of four organic antimonials after single or repeated administration to normal hamsters. J. Pharmacol. Exp. Ther. 87:169-180.

Gerhardsson, L., D. Brune, G.F. Nordberg, and P.O. Wester. 1982. Antimony in lung, liver and kidney tissue from deceased smelter workers. Scand. J. Work Environ. Health 8(3):201-208.

Gerhartz, W., Y.S. Yamamoto, F.T. Campbell, R. Pfefferkorn, and J.F. Rounsaville, eds. 1985. Pp. 124-125 in Ullman's Encyclopedia of Industrial Chemistry, Vol. A1. Weinheim, Germany: VCH.

Gross, P., J.H.U. Brown, M.L. Westrick, R.P. Srsic, N.L. Butler, and T.F. Hatch. 1955. Toxicologic study of calcium halophosphate phosphors and antimony trioxide. I. Acute and chronic toxicity and some pharmacologic aspects. Arch. Ind. Health 11:473-478.

Hatlelid, K. 1999. Toxicity Review for Calcium Molybdate and Zinc Molybdate. Memorandum, dated March 2, 1999, from Kristina Hatlelid, Toxicologist, Division of Health Sciences, to Ronald Medford, Assistant Executive Director for Hazard Identification and Reduction, U.S. Consumer Product Safety Commission, Washington, DC.

Hazleton Laboratories (Hazelton Laboratories America, Inc.). 1989. Acute Inhalation Toxicity Study with Colloidal Antimony Pentoxide in the Rat. HLA Study No. 2096-169. Hazelton Laboratories America, Inc., Rockville, MD.

ICRP (International Commission on Radiological Protection). 1981. Metabolic data for antimony. Pp. 46-49 in Radiation Protection, ICRP Publication 30, Part 3, Including Addendum to Parts 1 and 2, Limits for Intakes of Radionuclides by Workers. International Commission on Radiological Protection. Elmsford, NY: Pergamon Press.

IPCS (International Programme on Chemical Safety). 1997. Environmental Health Criteria 192 Flame Retardants: A General Introduction. Geneva: World Health Organization.

Jenkins, R.O., P.J. Craig, W. Goessler, and K.J. Irgolic. 1998. Antimony leaching from cot mattresses and sudden infant death syndrome (SIDS). Hum. Exp. Toxicol. 17(3):138-139.

Kuroda, K., G. Endo, A. Okamoto, Y.S. Yoo, and S. Horiguchi. 1991. Genotoxicity of beryllium, gallium, and antimony in short-term assays. Mutat. Res. 264(4):163-170.

Lide, D.R. 1991-1992. Handbook of Chemistry and Physics, 72nd Ed. Boca Raton, FL.: CRC Press.

Nissan Chemical Industries, Ltd. 1985. Micronucleus Test with Sb_2O_5 in Mice. Report No. 85-32-02803. Saitama-ken, Japan.

Otto, G.F., T.H. Maren, and H.W. Brown. 1947. Blood levels and excretion rates of antimony in persons receiving trivalentand pentavalent antimonials. Am. J. Hyg. 46:193-211.

Potkonjak, V. and M. Pavlovich. 1983. Antimoniosis: A particular form of pneumoconiosis. I. Etiology, clinical and X-ray findings. Int. Arch. Occup. Environ. Health 51(3):199-207.

Rees, P.H., M.I. Keating, P.A. Kager, and W.T. Hockmeyer. 1980. Renal clearance of pentavalent antimony (sodium stibogluconate). Lancet 2(8188):226-229.

12

Ammonium Polyphosphates

THIS chapter reviews the physical and chemical properties, toxicokinetics, toxicological, epidemiological, and exposure data on ammonium polyphosphates (APPs). The subcommittee used that information to characterize the health risk from exposure to APPs. The subcommittee also identified data gaps and recommended research relevant for determining the health risk from exposure to APPs.

PHYSICAL AND CHEMICAL PROPERTIES

Ammonium polyphosphates (APPs) are a class of ammonium salts of linearly condensed polyphosphoric acids of the general chemical formula $(NH_4)_kH_{(n+2-k)}P_nO_{(3n+1)}$. The physical properties of this class of chemicals vary as a function of the degree of phosphate condensation. Short-chain APPs are water soluble, while longer chain APPs have lower water solubility. Toxicity data were available for three commercial APPs: ANTIBLAZE® LR2 (LR2), ANTIBLAZE® LR4 (LR4), and ANTIBLAZE® MC(M). The physical and chemical properties of these commercial APPs are summarized in Table 12-1.

Based on information provided by the manufacturer (Stewart Miller, Albright and Wilson, pers. commun., Nov. 1, 1999), a typical species distribution of polyphosphates in LR2 is 20% orthophosphate, 40% pyrophosphate,

TABLE 12-1 Physical and Chemical Properties of Ammonium Polyphosphates

Property	Value	Reference
ANTIBLAZE® LR2		
Chemical formula[a]	$(NH_4)_k H_{(n+2-k)}P_nO_{(3n+1)}$	Albright and Wilson 1998a
Components	APP (CAS # 68333-79-9), urea, (57-13-6), water	Albright and Wilson 1998a
Synonyms	AMGARD® LR2	Albright and Wilson 1998a
Physical State	Clear, colorless, odorless liquid	Albright and Wilson 1998a
Vapor Pressure	Not volatile at ambient temperature	Albright and Wilson 1998a
Solubility	Completely miscible in water	Albright and Wilson 1998a
pH	~7.3	Albright and Wilson 1998a
Boiling Point	105° C	Albright and Wilson 1998a
Flash Point	Non-flammable	Albright and Wilson 1998a
Thermal Decomposition	Begins at 80°C	Albright and Wilson 1998a
Density	1.26 g/mL at 20°C	Albright and Wilson 1998a
ANTIBLAZE® LR4		
Chemical formula[a]	$(NH_4)_k H_{(n+2-k)}P_nO_{(3n+1)}$	Albright and Wilson 1998a
Chain length	200	Albright and Wilson 1998a
Components	APP (CAS # 68333-79-9)	Albright and Wilson 1998b
Synonyms	AMGARD® LR4	Albright and Wilson 1998b
Physical State	White, odorless powder	Albright and Wilson 1998b
Solubility	4.0 g/100 g water Maximum solubility 10%[a]	Albright and Wilson 1998b
pH	5.0-7.0	S. Miller, Albright and Wilson, pers. commun., Nov. 1, 1999
Flash Point	Non-flammable	Albright and Wilson 1998b
Thermal Decomposition	Begins at 300°C	Albright and Wilson 1998b
Density	1.8 g/cm^3 at 20°C	Albright and Wilson 1998b
ANTIBLAZE® MC(M)		
Chemical formula[a]	$(NH_4)_k H_{(n+2-k)}P_nO_{(3n+1)}$	Albright and Wilson 1998a
Chain length	300	S. Miller, Albright and Wilson, pers. commun., Nov. 1, 1999
Components	APP (CAS # 68333-79-9)	S. Miller, Albright and Wilson, pers. commun., Nov. 1, 1999
Molecular weight	30,000	
Physical state	Solid	S. Miller, Albright and Wilson, pers. commun., Nov. 1, 1999

[a]In this formula, "k" represents the degree of replacement of hydrogen ions with ammonium ions. The value of "k" was not available; however, based on the reported molecular weight of ANTIBLAZE® MC(M), it appears that the value of "k" is approximately equal to n, the degree of polymerization.

40% tripolyphosphate, and 20% tetrapolyphosphate plus higher-level polyphosphates (Stewart Miller, Albright and Wilson, pers. commun., Nov. 1, 1999).

OCCURRENCE AND USE

APP-based flame retardants have been sold in the U.S., Europe, and Asia for several years. In the U.S., they are used in the treatment of commercial furniture upholstery, automotive interior fabrics, draperies, and in other applications (Albright and Wilson 1998c). Outside the U.S., APPs are also used as flame retardants in commercial furniture upholstery. Water-soluble forms of APPs are approved for use in food as a sequestrant and emulsifier (JECFA 1982).

Both LR2 and LR4 are used for semi-durable, flame-retardant (FR) applications (Albright and Wilson 1999). Water-soluble LR2 is applied to cellulose-rich upholstery fabrics. Less-soluble LR4 is applied to fabrics as a latex back-coating. Information on the uses of ANTIBLAZE® MC(M) was not available.

Phosphorus is essential in human physiology. Phosphate is a structural component of bones and teeth and is essential in many enzymatic processes.

TOXICOKINETICS

Absorption

JECFA (1974) reported that gastrointestinal absorption of higher polyphosphates is probably low. Polyphosphates are most likely hydrolyzed by stomach acids to phosphate and ammonium ions, which could then be absorbed. Ebel (1958, as cited by JECFA 1974) reported that the rates of hydrolysis and absorption of polyphosphates decrease with increasing size.

There are some limited data on the uptake of phosphate from highly polymeric polyphosphates. Approximately 10-30% of an orally administered dose of sodium hexametaphosphate (a cyclic polyphosphate) or Kurrol's salt[1] was absorbed as monophosphate (Lang et al. 1955, as reported by JECFA 1974; Lang 1958, as reported by JECFA 1974). Overall, these data suggest that about

---

[1]Kurrol's salt is a high-molecular-weight polymer with the chemical formula $(KPO_3)_n \cdot H_2O$, with n = 400-5000. Kurrol's salt is nearly insoluble in water but, as noted below, the mixture of Kurrol's salt and diphosphate could be dissolved, and a 1% solution had a pH of 7.6. It is unclear how these authors administered Kurrol's salt without the addition of diphosphate.

one-half of the ingested dose of LR2 or LR4 would be absorbed as monophosphate.

No data were located on the absorption of APPs following dermal or inhalation exposure.

Metabolism

No data were located on the metabolism of APPs following exposure by the dermal, inhalation, or oral routes of exposure.

It is known that APPs are hydrolyzed by stomach acids and dissociate to the ammonium and phosphate ions following ingestion. Bacteria located in the intestinal tract may also contribute to the hydrolysis of polyphosphates following ingestion (Schreier and Noller 1955, as reported by JECFA 1974).

Distribution

Normal adult serum phosphorus levels are about 0.87-1.41 mmol/L (2.5-97.5 percentiles) (IOM 1997). Phosphorous intake estimated to result in the upper boundary of normal adult serum phosphorus is 3.5 g/d. Serum phosphorus levels during infancy range from 1.88 to 2.2 mmol/L. The adult intake of phosphorus that would result in the serum phosphorus levels observed in infants would be 10.2 g/d.

No data were located on the distribution of APPs following dermal or inhalation exposure.

Excretion

Grosselin et al. (1952) found that various polyphosphates were excreted in trace amounts in urine of rodents following oral exposure to hexametaphosphate. Particularly, orthophosphate was found in trace amounts representing a small fraction of the dose. About 40% of an oral dose of sodium trimetaphosphate was excreted in urine (38% as orthophosphate) within 24 hr. About 22% of a dose of sodium tetrametaphosphate was excreted in urine (18% as orthophosphate). These data suggest that large polymers of polyphosphate are excreted less efficiently than smaller polymers.

About one-half of the radiolabeled Kurrol's salt (primarily as polyphosphate) was found in the feces of rats following gavage, while a small percentage of monophosphate was excreted in the urine as monophosphate.

No data were located on the excretion of APPs following dermal or inhalation exposure.

HAZARD IDENTIFICATION[2]

Dermal Exposure

Irritation

LR2 (5% or 10% dissolved in distilled water, (Inveresk 1996) was found to be non-irritating in 36 human volunteers (26 female, 10 male) in a dermal patch test. Volunteers (age: 18-65 yr) were treated with approximately 20 μL of 0%, 5%, and 10% w/w LR2 in petrolatum, which was placed in Finn chambers and applied to their backs[3]. Forty-eight hr after application, the patches were removed and the application sites were rinsed. Application sites were evaluated 2 hr, 1 d, 2 d, and 5 d after patch removal. Two subjects showed what were reported as doubtful skin reactions to 10% LR2 at the 2-hr evaluation time point. No skin reactions to 5% or 10% LR2 were observed in the volunteers at later time points.

LR2 and LR4 were also non-irritating when applied to rabbits (Inveresk 1989, Safepharm 1993a). A single application of 0.5 mL (650 g) of LR2 in neat form (Inveresk 1989) or of 0.5 g of LR4 in 0.5 mL distilled water (Safepharm 1993a) was applied to the dorsum of three male New Zealand White rabbits and covered for 4 hr. The patch was removed and the skin was evaluated for reactions after 1, 24, 48, and 72 hr. No erythema or edema was observed from LR2 exposure. Very slight erythema was observed from LR4 exposure in 2 of the 3 animals 1 hr after patch removal. There were no skin reactions to treatment with LR4 when the animals were evaluated after 24 and 72 hr.

No irritation was observed in one male and two female New Zealand White rabbits exposed topically to fabric treated with LR2 (Inveresk 1990a). The rabbits were exposed to fabric (approximately 2.5 cm × 2.5 cm) "impregnated" with LR2 five times for 23 hr each. The amount of LR2 applied to or present

---

[2]In this section, the subcommittee reviewed the toxicity data on ammonium polyphosphates, including the toxicity assessment prepared by the U.S. Consumer Product Safety Commission (Ferrante 1999).

[3]Each individual was treated with seven patches: two concentrations of LR2, two concentrations of each of two other flame retardants, and the control using a randomized sequence.

in the fabric was not reported. The application sites were observed for skin reactions 1 hr after each application. No erythema or edema was observed following any of the exposures to the fabric.

Sensitization

LR2 was found to be a poor skin-sensitizing agent in the Magnusson and Kligman maximization test (Safepharm 1993b). Initially, 20 female Dunkin-Hartley guinea pigs were treated with three 0.1-mL injections containing (1) Freund's complete adjuvant plus water in a 1:1 ratio; (2) a 1% w/v dilution of test material in distilled water; and (3) a 1% w/v dilution of test material in Freund's complete adjuvant plus distilled water, respectively. Topical induction was attempted on d 7 with undiluted LR2 (0.2-0.3 mL) applied to filter paper placed over the injection sites and covered for 48 hr. Scattered mild redness was observed at the injection sites in 20/20 animals treated with LR2 1 hr after removal of the filter paper and in 2/20 animals after 24 hr of removal of the filter paper. The animals were then challenged on d 21 for 24 hr with filter paper patches containing 0.1-0.2 mL of 50% or 75% (v/v) solutions of LR2. None of the 20 animals developed a tissue reaction to 50% or 75% LR2 24 or 48 hr after challenge.

LR4 was also found to be a poor skin-sensitizing agent in the Magnusson and Kligman maximization test (Safepharm 1993c). Twenty female guinea pigs were initially injected intradermally with a 25% (w/v) solution of LR4. Topical induction was then attempted on d 7 with filter paper patches containing 75% (w/w) LR4 in distilled water. Only 1 of 20 animals had skin changes (scattered mild redness) at the application site 1 hr after removal of the patches. No animals had any visible skin reactions 24 hr after patch removal. None of the animals showed any tissue reaction either 24 or 48 hr after topical challenge with filter paper patches containing 50% or 75% solutions of LR4.

Systemic Effects

Acute dermal toxicity studies in Sprague-Dawley rats estimated the LD_{50}s for LR2 and LR4 to be >2,000 mg/kg (Safepharm 1994, 1993d). Animals were treated with 2,000 mg/kg of neat LR2 or LR4 (1.59 mL/kg) to a skin area of approximately 10% of the rat total body surface area and covered for 24 hr. The bandage was then removed and the test sites were observed for skin reactions for 14 d. There were no deaths, signs of systemic toxicity, or skin irritation, and weight gain was not affected in animals from either test group. No abnormalities were observed at necropsy.

Other Systemic Effects

No studies were identified that examined the effect of dermal exposure to APPs, ammonium ions, or polyphosphates on immunological, neurological, reproductive, or developmental parameters in humans or experimental animals. Additionally, no studies were identified that investigated the carcinogenicity of these compounds in humans or animals following dermal exposure.

Inhalation Exposure

Systemic Effects

Only one inhalation toxicity study was located for APPs. Five male and five female Sprague-Dawley rats were exposed to Amgard MC(M) at 5,090 mg/m^3 (nose-only exposure) for 4 hr and monitored for 14 d (Safepharm 1993e). The mass median aerodynamic diameter (MMAD) of the APP particles was 5.8 μm, and their geometric standard deviation was 0.35 μm. No deaths occurred among the exposed animals and weight gain was normal over the 14-d observation period. Exposed animals developed a hunched posture, decreased respiratory rate, lethargy, and tiptoe gait 4 hr after exposure, but these symptoms were not evident after 3 d of exposure. The 4-hr-inhalation LC_{50} for Amgard MC(M) was estimated to be >5,090 mg/m^3 (>5.09 mg/L) for male and female rats.

Material Safety Data Sheets (MSDSs) report that exposure to LR2 or LR4 may cause nose and upper respiratory tract inflammation (Albright and Wilson 1998a, 1998b, 1998d, 1998e). The MSDSs for both LR2 and LR4 (Albright and Wilson 1998d, 1998e) report a 1-hr-inhalation LC_{50} >20 mg/L (species and sex not reported). It is not known whether this is an experimentally determined or an estimated value.

No data on the inhalation toxicity of ammonium ions or of other polyphosphates were identified.

Other Systemic Effects

No studies were identified that examined the toxic effects of inhalation exposure to APPs, ammonium, or polyphosphates on immunological, neurological, reproductive, or developmental parameters in humans or laboratory animals. Additionally, no studies were identified that investigated the carcinogenic effects of these compounds following inhalation exposure.

Oral Exposure

Systemic Effects

No oral toxicity data for APPs were located for humans. Typical human dietary phosphorous levels are not harmful, especially in the presence of adequate calcium and vitamin D intake. The mean daily phosphorus dietary intake for adult males and females is estimated to be 1,500 mg/d and 1,000 mg/d, respectively. However, if the intake of phosphorus from processed foods was included in these values, the estimated dietary intake of phosphorous would be up to 20% higher.

It is known that high doses of ammonium ions can cause metabolic acidosis; persons with compromised liver function are at highest risk. Exposure to oral doses of 3.2 or 4.8 g ammonium/d as ammonium chloride for 5 d caused metabolic acidosis in two humans with compromised liver function (Sartorius et al. 1949). Effects secondary to acidosis include renal enlargement and demineralization of bone (ATSDR 1990).

The oral LD_{50} for LR2 in rats was estimated to be >5,000 mg/kg (Inveresk 1990b). No deaths or clinical signs of toxicity were observed among five male and five female Sprague-Dawley rats given a single gavage dose of 5,000 mg/kg LR2 in distilled water and observed for 14 d. Weight-gain was normal in females during the second wk of the 2-wk observation period, but weight-gain in exposed males was reduced during the second wk. Gross post-mortem analysis of all dosed animals revealed no abnormalities.

No deaths or toxicity symptoms were observed among five male and five female Sprague-Dawley rats treated with a single gavage dose of 2,000 mg/kg LR4 in distilled water (Safepharm 1993f). Body weight gain was normal for all animals during both wk of the observation period and no gross abnormalities were detected at necropsy. Therefore, it was concluded that the LD_{50} for LR4 in rats is >2,000 mg/kg.

JECFA (1974) summarized a number of toxicity studies on phosphates and polyphosphates (see Table 12-2). The primary effect identified in these studies is kidney calcification (nephrocalcinosis), resulting from the precipitation of calcium phosphate due to an upset in phosphate homeostasis. JECFA (1974) noted that it is difficult to identify an effect level for nephrocalcinosis in toxic-

TABLE 12-2 Summary of Oral Toxicity Studies on Polyphosphates

Species, Strain, Sex, Number	Dietary Concentrations Tested (%)	Duration	Test Material	Effect and Lowest Level in Diet[a]	Reference
Rat, NS, M, 5/dose	0, 0.2, 2, 10	1 mo	Sodium hexametaphosphate	Kidney inflammation at 2%; no kidney effects at 0.2%	Hodge 1964c[b]
Rat, NS, M, 5/dose	0, 0.2, 2, 10	1 mo	Sodium tripolyphosphate	Kidney inflammation at 2%; no kidney effects at 0.2%	Hodge 1964c[b]
Rat, NS, M/F, 36/sex/dose	0, 3, 5	24 wk	Sodium tripolyphosphate	Nephrocalcinosis at 3%	Hahn and Seifen 1959[c], as cited in JECFA 1974; Hahn et al. 1958[c], as cited in JECFA 1974
Rat, NS, M/F, 36/sex/dose	0, 1.8, 3, 5	24 wk	Graham's salt	Nephrocalcinosis at 3%; slight or absent nephrocalcinosis at 1.8%	Hahn and Seifen 1959[c], as cited in JECFA 1974; Hahn et al. 1958[c], as cited in JECFA 1974
Rat, NS, M/F, 10/sex/dose	0, 0.5, 1, 2.5, 5	12 wk	1/3 Kurrol's salt and 2/3 disodium/ tetrasodium diphosphate mixture	Nephrocalcinosis at 2.5%; data on kidney at 1% not available	van Esch et al. 1957[c], as cited in JEFCA 1974; van Genderen 1958[c], as cited in JECFA 1974
Rat, NS, M, 12/dose	0, 0.9, 3.5 (Mineral levels equalized)	150 d	Sodium hexametaphosphate	No kidney histopathology up to 3.5%	Dymsza et al. 1959[c]
Rat, NS, 30M/dose, 10F/dose	0, 0.5, 1, 2.5, 5 (Control diet contained 0.47% phosphorus)	Life span, two generations of offspring	1/3 Kurrol's salt and 2/3 disodium/ tetrasodium diphosphate mixture	Renal calcification at 1%	van Esch et al. 1957[c], as cited in JECFA 1974

TABLE 12-2 *(Continued)*

Species, Strain, Sex, Number	Dietary Concentrations Tested (%)	Duration	Test Material	Effect and Lowest Level in Diet[a]	Reference
Rat, NS, M/F, 50/sex/dose	0, 0.05, 0.5, 5	2 yr	Sodium tripolyphosphate	Increased kidney weight, intertubular calcification and fibrosis at 5%; no apparent effects at 0.5%	Hodge 1964a[b]
Rat, NS, M/F, 50/sex/dose	0, 0.05, 0.5, 5	2 yr	Sodium hexametaphosphate	Increased kidney weight and calcification at 5%; no apparent effects at 0.5%	Hodge 1964b[b]

F, female; M, male; NS, not specified.

[a]NOAELs not identified because of the difficulty in determining (based on the available data) whether the next lower dose is really a NOAEL.

[b]Unpublished study.

[c]Published study.

ity studies because renal calcification occurs naturally to some extent in control rats which is determined by dietary intake of calcium and vitamin D.

Studies by van Esch et al. (1957, as cited in JECFA 1974) and Hodge (1964a, 1964b) suggest that chronic exposure to 0.5% polyphosphates in the diet may cause increased kidney weight but no kidney histopathology, while higher concentrations may cause kidney calcification when mineral levels are not equalized.

Reproductive and Developmental Effects

No information was found regarding the reproductive or developmental effects of APPs following oral exposure. Information on reproductive and developmental effects of ammonium ions was also not located.

A study conducted by van Esch et al. (1957, as cited in JECFA 1974) found decreased fertility in rats that were treated with a mixture of one-third Kurrol's salt and two-thirds diphosphate at a dietary concentration of 5%. No reproductive effects were reported at lower concentrations.

Lang (1959, as cited in JECFA 1974), found no effects on reproduction in three generations of rats, each fed diets containing 0.4% or 0.75% phosphoric acid for 90 wk.

No effects on fertility, litter size, neonate growth, or neonate survival was observed in a three-generation reproduction study in groups of rats administered 0.5% sodium tripolyphosphate or 0.5% sodium hexametaphosphate (Hodge 1964a, 1964b, and BIBRA 1964). There was also no apparent effect on the histopathology of major organs of the third generation.

Other Systemic Effects

No studies were identified that examined the immunological, neurological, or carcinogenic effects of oral exposure to APPs, ammonium ions, or polyphosphates. Ammonium ions and polyphosphates are unlikely to be carcinogenic, in light of the physiological roles of ammonium ions and phosphate in the body.

Genotoxicity

LR2 was not found to be mutagenic in 5 strains of *Salmonella typhimurium* (TA1535, TA1537, TA1538, TA98, TA100) exposed at concentrations of 25,

75, 250, 750, 2,500, or 5,000 μg LR2/plate in the presence or absence of exogenous metabolic activation (rat liver S9) (Safepharm 1995). No other genotoxicity data were located for APPs or for other polyphosphates.

QUANTITATIVE TOXICITY ASSESSMENT

Noncancer

Dermal Assessment

There are no appropriate APP toxicity data available for deriving a dermal RfD.

Inhalation RfC

Only acute LC_{50} data are available for APPs; these data are insufficient for the derivation of an RfC.

Oral RfD

There are no subchronic or chronic toxicity data for LR2 and LR4, therefore it is not possible to derive an oral RfD for these compounds.

IOM (1997) has developed a recommended dietary allowance (RDA) for phosphorous of 460 mg for children 1-3 yr old, 500 mg for children 4-8 yr old, 1,250 mg for ages 9-18 yr old, and 700 mg for ages 18 and older. The importance of the calcium:phosphate ratio was also noted; a ratio of less than 1:2 can cause resorption and loss of bone in animals.

The Joint Expert Committee on Food Additives (JECFA) of the Food and Agriculture Organization of the World Health Organization (FAO/WHO) has established a maximum tolerable daily intake for phosphates, including APP, of 70 mg phosphorus/kg (JECFA 1974). This value includes all phosphates including those in food and applies to diets that are sufficient in calcium.

The JECFA value of 70 mg/kg was derived from the lowest dietary concentration that produced nephrocalcinosis in the rat (1% phosphorous) as described by van Esch et al. (1957, as cited in JECFA 1974). Extrapolation to the lowest phosphorous level that might conceivably cause nephrocalcinosis in humans was based on a daily food intake of 2,800 calories and a dietary phosphate level of 6,600 mg/d (JECFA 1974). Further details on the calculation were not provided.

The subcommittee chose to use the JECFA (1974) acceptable daily intake for phosphorous of 70 mg phosphorus/kg-d for deriving an oral RfD for APPs. The oral RfD was calculated by multiplying the approximate amount of phosphorous present, by weight, in LR2 and LR4. The solids portion of LR2 contains approximately 26% phosphorus, therefore, the oral RfD for LR2 is estimated to be 270 mg/kg-d (≈300 mg/kg-d). LR4 contains a minimum of 27% phosphorus by weight (Stewart Miller, Albright and Wilson, pers. commun., Nov. 1, 1999), corresponding to a minimal oral RfD of 260 mg/kg-d (≈300 mg/kg-d).

The U.S. EPA has developed an RfD for ammonium sulfamate (EPA 1999), but the subcommittee did not use this RfD for APPs because the primary toxic moiety of ammonium sulfamate may be the sulfamate ion, rather than the ammonium ion.

Cancer

No data are available on the carcinogenic effects of APPs by any route of exposure. Because of the absence of carcinogenicity data, the subcommittee concluded that the carcinogenic potential of APPs cannot be determined. Carcinogenicity data on ammonium ion or polyphosphates were not located. However, based on the physiological roles of these compounds they would not be expected to be carcinogenic.

EXPOSURE ASSESSMENT AND RISK CHARACTERIZATION

Noncancer

Dermal Exposure

The assessment of noncancer risk by the dermal route of exposure is based on the scenario described in Chapter 3. This exposure scenario assumes that an adult spends 1/4th of his or her time sitting on furniture upholstered with APPs, that 1/4th of the upper torso is in contact with the upholstery, and that clothing presents no barrier. APPs are considered to be ionic and are essentially not absorbed through the skin. However, to be conservative, the subcommittee assumed that ionized APPs permeate the skin at the same rate as water, with a permeability rate of 10^{-3} cm/hr (EPA 1992). Using that permeability rate, the highest expected application rate for APPs of 7.5 mg/cm^2, and Equation 1 in Chapter 3, the subcommittee calculated a dermal exposure level of 2.2 mg/kg-d. The oral RfD for APPs (300 mg/kg-d; see Oral RfD in Quantitative Toxicity section) was used as the best estimate of the internal dose from dermal exposure. Dividing the exposure level by the oral RfD yields a hazard index of 7.3 $\times$ 10^{-3}. Thus it was concluded that APPs used as FRs in upholstery fabric are not likely to pose a noncancer risk from dermal exposure.

Inhalation Exposure

Particles

Inhalation exposure estimates for APPs were calculated using the exposure scenario described in Chapter 3. This scenario assumes that a person spends 1/4th of his or her life in a 30-m^3 room containing 30 m^2 of APP-treated fabric and the room is assumed to have an air-change rate of 0.25/hr. It is also assumed that 50% of the APP present in 25% of the surface area of the treated fabric is released over 15 yr and that 1% of released particles are small enough to be inhaled.

Particle exposure was estimated using Equations 4 and 5 in Chapter 3. The subcommittee estimated an upholstery application rate (S_a) for APPs of 7.5 mg/cm^2. The release rate (μ_r) for APPs from upholstery fabric was estimated to be 2.3 $\times$ 10^{-7}/d (by using Equation 5 in Chapter 3) yielding a room airborne particle concentration (C_p) of 2.9 $\mu g/m^3$ and a short time-averaged exposure concentration of 0.71 $\mu g/m^3$. The time-averaged exposure concentration for particles was calculated using Equation 6 in Chapter 3.

In the absence of relevant inhalation exposure data, the subcommittee chose to estimate inhalation RfCs from oral RfDs. The subcommittee, however, recognizes that it is not an ideal approach and also recognizes that the estimated RfC levels might be considerably different than actual levels (if inhalation data were available). Extrapolating from one route of exposure (oral) to another (inhalation) requires specific knowledge about the uptake kinetics into the body by each exposure route, including potential binding to cellular sites. The subcommittee believes that its extrapolation of oral RfDs to inhalation RfCs is highly conservative; it assumes that all of the inhaled compound is deposited in the respiratory tract and completely absorbed into the blood. The NRC committee on Toxicology (NRC 1985) has used this approach when inhalation exposure data were insufficient to derive inhalation exposure levels. The subcommittee believes that such an approach is justified for conservatively estimating the toxicological risk from exposure to APPs. This RfC should be used as an interim or provisional level until relevant data become available for the derivation of an inhalation RfC for the calculation of a hazard index.

In order to calculate a hazard index for the inhalation route, a provisional inhalation RfC of 1,050 mg/m^3 was derived using the oral RfD for ANTIBLAZE® LR2 and ANTIBLAZE® LR4 of 300 mg/kg-d, respectively and Equation 7 in Chapter 3.

Division of the time-average exposure concentration of 0.71 μg/m^3 by the provisional RfC for APPs of 1,050 mg/m^3 yields a hazard index of 6.8×10^{-7}. These findings suggest that under this worst-case exposure scenario, inhalation of APP particles from furniture upholstery is not likely to pose a noncancer risk to humans.

Vapors

APPs are ionic with negligible vapor pressure at ambient temperatures. Therefore, inhalation of APP vapor is not likely to pose a noncancer risk when incorporated into furniture upholstery.

Oral Exposure

The assessment of noncancer toxicological risk for oral exposure to APPs is based on the oral exposure scenario described in Chapter 3. This scenario assumes a child is exposed to APPs by sucking on 50 cm^2 of fabric backcoated with APPs, 1 hr/d for 2 yr. The subcommittee estimated an upholstery application rate (S_a) for APPs of 7.5 mg/cm^2. Oral exposure was calculated using Equation 15 in Chapter 3. The extraction rate (μ_w) for APPs was estimated to be 0.038 based on extraction data for organic phosphates in polyester fiber (McIntyre et al. 1995). The release rate from the fiber for estimating extraction was 0.06/d at 28°C calculated using the equation $2d/2\pi R$ (d = film thickness, R = fiber radius) with a correction from fiber to film of a factor of 0.63.

The worst-case average oral daily dose for APPs was estimated to be 0.059 mg/kg-d. Division of the oral dose estimate by the oral RfD for APPs of 300 mg/kg-d yields a hazard index of 2.0×10^{-4}. These results indicate that under the given worst-case exposure scenario, oral exposure to APPs is not likely to pose a health risk to humans.

Cancer

There are no adequate data available to assess the carcinogenicity of APPs by the dermal, inhalation, or oral routes.

RECOMMENDATIONS FROM OTHER ORGANIZATIONS

The Joint FAO/WHO Expert Committee on Food Additives (JECFA) established a maximum tolerable daily intake for phosphates, including APPs, of 70 mg phosphorus/kg-d (JECFA 1974).

The Food and Nutrition Board of the Institute of Medicine (IOM 1997) derived a tolerable upper intake level (UL) for phosphorous of 4 g/d for adults 19-70 yr of age. The UL for persons over 70 yr is 3 g/d and accounts for the higher prevalence of impaired renal function among older persons.

DATA GAPS AND RESEARCH NEEDS

Key information, such as data on the leaching of APPs from furniture upholstery and data on dermal penetration of APPs, is not available. There are no data on subchronic or chronic toxicity, developmental or reproductive effects, and little information on the genotoxicity of APPs. However, acute studies indicate that these compounds are probably not very potent toxicants and current use of these compounds as food additives further supports this opinion.

Hazard indices calculated using the given exposure scenarios were less than one for the dermal, inhalation, and oral routes of exposure, indicating that these chemicals are not likely to pose health risks. Therefore, the subcommittee concludes that no further research is needed for assessing health risks from exposure to ammonium polyphosphates.

REFERENCES

Albright and Wilson (Albright & Wilson UK Limited). 1998a. Safety Data Sheet (To requirements of Directive 91/155/EEC) ANTIBLAZE® LR2. Ref. FR/LR2/010, Revision 1. West Midlands, UK.

Albright and Wilson (Albright & Wilson UK Limited). 1998b. Safety Data Sheet (To requirements of Directive 91/155/EEC) ANTIBLAZE® LR4. Ref. FR/LR4/012, Revision 1. West Midlands, UK.

Albright and Wilson. 1998c. Product Summary for ANTIBLAZE® LR2, LR3, and LR4, Ammonium Polyphosphates. Glen Allen, VA.

Albright and Wilson (Albright & Wilson Americas, Inc.). 1998d. Material Safety Data Sheet. ANTIBLAZE® LR2. Ref. PSMSD-30B, Revision 4. Glen Allen, VA.

Albright and Wilson (Albright & Wilson Americas, Inc.). 1998e. Material Safety Data Sheet. ANTIBLAZE® LR4. Ref. PSMSD-30C, Revision 3. Glen Allen, VA.

Albright and Wilson. 1999. Product Summary. [Online]. Available: http://www.albright-wilson.com/feb99/products/pda/flame/flame.htm

ATSDR (Agency for Toxic Substances and Disease Registry). 1990. Toxicological

Profile for Ammonia. TP-90-03. Agency for Toxic Substances and Disease Registry, U.S. Public Health Service.

BIBRA (British Industrial Biological Research Association). 1964. Summaries of toxicological data: Toxicity studies on phosphates. Food Cosmet. Toxicol. 2:147-154.

Dymsza, H.A, G. Reussner, Jr., and R. Thiessen, Jr. 1959. Effect of normal and high intakes of orthophosphate and metaphosphate in rats. J. Nutr. 69:419-428.

Ebel, J.P. 1958. Generalités sur la structure des phosphates condensés. Annales de la Nutrition et de L Alimentaion 12:57-97.

EPA (U.S. Environmental Protection Agency). 1992. Dermal Exposure Assessment: Principles and Applications. EPA/600/8-91-011B. Office of Health and Environmental Assessment, U.S. Environmental Protection Agency, Washington, DC.

EPA (U.S. Environmental Protection Agency). 1999. Integrated Risk Information System (IRIS). National Center for Environmental Assessment. [Online]. Available: http://www.epa.gov/iris U.S. Environmental Protection Agency, Washington, DC. Downloaded August, 1999.

Ferrante, J. 1999. Toxicity review for ammonium polyphosphates. Memorandum, from Jacqueline Ferrante, Pharmacologist, Division of Health Sciences, to Ronald Medford, Assistant Executive Director for Hazard Identification and Reduction. U.S. Consumer Product Safety Commission. Washington, DC.

Gosselin, R.E., A. Rothstein, G.J. Miller, and H.L. Berke. 1952. The hydrolysis and excretion of polymeric phosphate. J. Pharmacol. Exp. Ther. 106:180-191.

Hahn, F., and E. Seifen. 1959. Further studies on the effect of chronic feeding of phosphates. Arzneimittel-Forschung 9:501-503.

Hahn, F., H. Jacobi, and E. Seifen. 1958. Do ortho- and polyphosphates show variable compatibilities on chronic feeding? Arzneimittel-Forschung 8:286-289.

Hodge, H.C. 1964a. Chronic oral toxicity studies of sodium tripolyphosphate in rats. [Summary]. Food Cosmet. Toxicol. 2:147-154.

Hodge, H.C. 1964b. Chronic oral toxicity studies of sodium hexametaphosphate in rats. [Summary]. Food Cosmet. Toxicol. 2:147-154.

Hodge, H.C. 1964c. Short-term oral toxicity tests of condensed phosphates in rats and dogs. [Summary]. Food Cosmet. Toxicol. 2:147-154.

Inveresk (Inveresk Research International). 1989. AMGARD LR2: Acute Dermal Irritation Test in Rabbits. Project 243137. Report No. 5915. Musselburgh, Scotland.

Inveresk (Inveresk Research International). 1990a. Fabric Treated with AMGARD LR1 and LR2: Dermal Irritancy Study in Rabbits - 5 Day Repeat Application. Project 243137. Report No. 5920. Musselburgh, Scotland.

Inveresk (Inveresk Research International). 1990b. AMGARD LR2: Acute Oral Toxicity (LD50) Test in Rats. Project 243137. Report 5910. Musselburgh, Scotland.

Inveresk (Inveresk Research International). 1996. Three Flame-Retardant Formulations a Human Skin Patch Test (46 h Exposure). Project 591521. Report No. 14158. Tranent, Scotland.

IOM (Institute of Medicine). 1997. Dietary Reference Intakes for Calcium, Phosphorus, Magnesium, Vitamin D, and Fluoride. Food and Nutrition Board, Institute of Medicine. Washington, DC.: National Academy Press.

JECFA (Joint FAO/WHO Expert Committee on Food Additives). 1974. Toxicological Evaluation of Some Food Additives Including Anticaking Agents, Antimicrobials, Antioxidants, Emulsifiers and Thickening Agents. WHO Food Additives Series No. 5. Seventeenth report of the Joint FAO/WHO Expert Committee on Food Additives, Technical Report Series 539; FAO Nutrition Meetings Report Series, No. 53.

JECFA (Joint FAO/WHO Expert Committee on Food Additives). 1982. Evaluation of Certain Food Additives and Contaminants. Twenty-sixth report of the Joint FAO/WHO Expert Committee on Food Additives. WHO Technical Report Series, No. 683.

Lang, K. 1958. Kondensierte Phosphate in Lebensmitteln, Berlin: Springer.

Lang, K. 1959. Z. Lebensmitt-Untersuch. 110:450.

Lang, K. et al. 1955. Der stoffwechsel von polyphosphaten. Biochemische Zeitschrift. 327(8):118-125.

McIntyre, J.E., I. Holme, and O.K. Sunmonu. 1995. The desorption of model compounds from poly(ethylene terephthalate) fibre. Colourage 41(13)77-81.

NRC (National Research Council). 1985. Emergency and Continuous Exposure Limits for Selected Airborne Contaminants, Volume 5. Committee on Toxicology. Board on Toxicology and Environmental Health Hazards, National Research Council. Washington, DC.: National Academy Press.

Safepharm (Safepharm Laboratories, Limited). 1993a. AMGARD LR4: Acute Dermal Irritation Test in the Rabbit. Project 71/163. Derby, UK.

Safepharm (Safepharm Laboratories, Limited). 1993b. AMGARD LR2: Magnusson & Kligman Maximisation Study in the Guinea Pig. Project 71/154. Derby, UK.

Safepharm (Safepharm Laboratories, Limited). 1993c. AMGARD LR4: Magnusson & Kligman Maximisation Study in the Guinea Pig. Project 71/162. Derby, UK.

Safepharm (Safepharm Laboratories, Limited). 1993d. AMGARD LR4: Acute Dermal Toxicity (Limit Test) in the Rat. Project 71/161. Derby, UK.

Safepharm (Safepharm Laboratories, Limited). 1993e. AMGARD MC(M): Acute inhalation toxicity study. Four-hour exposure (nose-only) in the rat. Project 71/167.

Safepharm (Safepharm Laboratories, Limited). 1993f. AMGARD LR4: Acute Oral Toxicity (Limit Test) in the Rat. Project 071/160. Derby, UK.

Safepharm (Safepharm Laboratories, Limited). 1994. AMGARD LR2: Acute Dermal Toxicity (Limit Test) in the Rat. Project 71/320. Derby, UK.

Safepharm (Safepharm Laboratories, Limited). 1995. AMGARD LR2: Reverse Mutation Assay "Ames Test" Using Salmonella Typhimurium. Project 071/385. Derby, UK.

Sartorius, O.W., J.C. Roemmelt, and R.F. Pitts. 1949. The renal regulation of acid-base balance in man. IV. The nature of the renal compensation in ammonium chloride acidosis. J. Clin. Invest. 28:423-439.

Schreier, K., and H.G. Nöller. 1955. Naunyn-Schmiedeberg's Arch. Exp. Pathol. Pharmakol. 227:199.

van Esch, G.J., H.H. Vink, S.J. Vit, and H. van Genderen. 1957. The physiologic action of polyphosphates. Arzneimittel Forsch. 7:172-175.

van Genderen, H. 1958. Kondensierte Phosphate in Lebensmitteln. Berlin: Springer.

13

Phosphonic Acid, (3-{[Hydroxymethyl]amino}-3-Oxopropyl)-Dimethyl Ester

THIS chapter reviews the physical and chemical properties, toxicokinetics, toxicological, epidemiological, and exposure data on phosphonic acid, (3-{[hydroxymethyl]amino}-3-oxopropyl)-dimethyl ester, or PA. The subcommittee used that information to characterize the health risk from exposure to PA. The subcommittee also identified data gaps and recommended research relevant for determining the health risk from exposure to PA.

PHYSICAL AND CHEMICAL PROPERTIES

The physical and chemical properties of PA (CAS No. 20120-33-6) are presented in Table 13-1.

OCCURRENCE AND USE

PA flame retardants have been used in various applications in more than 30 countries for almost 40 yr (Ciba 1998a). Ciba-Geigy (formally American

TABLE 13-1 Physical and Chemical Properties of Phosphonic Acid, (3-{[Hydroxymethyl]amino}-3-oxopropyl)-Dimethyl Ester

Property	Value	Reference
Chemical formula	$C_6H_{14}NO_5P$	ChemID 1999
Structure	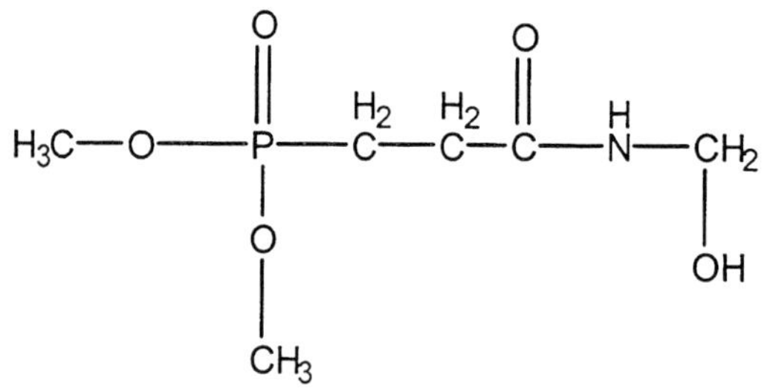	
CAS Registry #	20120-33-6	ChemID 1999
Synonyms	Dimethylphosphono-N-hydroxymethyl-3-propionamide; dimethyl (3-{[hydroxymethyl)]-amino}-3-oxopropyl)phosphonate; phosphonic acid, (2-{[hydroxymethyl)carbamamoyl}ethyl)dimethyl ester; N-methylol dimethylphosphonopropionamide	ChemID 1999
Trade names	Pyrovatex 3805®; Pyrovatex CP Neu/Special®; Pyrovatex 7620®; Amgard TFR1®; Spolapret OS®	ChemID 1999
Molecular weight	211.15	ChemID 1999
Physical state	White powder	Ciba 1998a
Solubility	Miscible	Ciba 1998a
Partition coefficient	- 1.68 (pH 6.9; 25°C)	Ciba 1998a
Density	1.27-1.29 g/cm³	Ciba 1998a

Cyanamide) is the leading manufacturer of PA-based flame retardants in the United States. PA flame retardants are produced by Ciba-Geigy include Pyrovatex CP Neu (or New)/Special® and Pyrovatex® 7620. Thor Chemical (South Africa) and Albright & Wilson (United States) have produced PA-based flame retardants similar to Pyrovatex CP®. Pax Enterprise Pvt. Ltd., of India, advertises a flame retardant for cellulose-rich fabrics that has chemical characteristics similar to those of the Pyrovatex CP® (see Appendix B, Table 2 for additional flame-retardant products containing PA).

PA-based flame retardants have been used in residential furniture in the United Kingdom to meet flammability standards imposed in 1988. In both the United Kingdom and the United States, PA flame retardants have been used to treat commercial furniture, draperies, and work clothing (FRCA 1998). In addition, Pyrovatex CP® was used as a flame retardant in children's sleepware until 1998, but was voluntarily withdrawn by Ciba from this market in 1999 because of questions about the durability of its flame-retardent properties after repeated washing.

The volume of PA flame retardants used in the United States is not known. Estimated use of PA-based flame retardants in the United Kingdom is about 600 tons/yr, and about 80 tons are used for treating textiles used in the manufacture of residential furniture (A.R. Horrocks, Bolton Institute, pers. commun., August 24, 1999).

PA and its derivatives are affixed directly to cellulose fibers in fabrics by the application of high heat in the presence of cross-linking agents such as melamine and glyoxal. Cross-linking imparts resistance to laundering through several washings (Ciba 1998b). Applied PA can undergo slow ambient hydrolysis and release from the fabric over a period of years (see Appendix B, Case 3). The interaction of PA, cellulose, and resin compounds prevents the formation of flammable gases during pyrolysis and severely retards combustion. If flame or high heat comes into contact with the PA-resin finish, the cellulose dehydrates, forms a carbon scaffold with PA and the cross-linking resin, and counteracts the penetration of heat and the spread of fire (Ciba 1998b). Production and marketing guidelines state that the weight of Pyrovatex CP New/Special®/ unit fabric area should not be less than 150 g/m^2 (Ciba 1998b).

The process of affixing PA-based flame retardants to textiles takes place at the end of the textile-milling process. Textiles are treated with PA flame retardants by dipping the fabric in vats that contain Pyrovatex CP Neu®[1] or other PA flame retardant, which typically contains a cross-linking resin, an acid catalyst for the cross-linking reaction (such as phosphoric acid), fabric softeners, and fabric-wetting agents. The fabric is then heated to 160 °C for 90 sec to cure the cross-linking resin. The treated fabric is then neutralized in an alkaline solution and excess flame retardant is removed. It has been acknowledged by one manufacturer of PA-based flame retardants that this process increases the handling of fabrics and the time required to bring a milled textile to market. Therefore, treatment of upholstery with PA is less economically appealing as

[1]Pyrovatex CP Neu® contains (by weight) 40% PA, 8% glyoxal resin, 2% melamine resin, 4% fabric softener (silicone-polyethylene blend), and 2.5% phosphonic acid. Competing PA-containing flame retardants are expected to contain similar components.

compared with use of latex back-coating. In the United Kingdom, back-coated fabrics are generally cheaper than PA-treated fabrics and, as a consequence, dominate the upholstery market.

TOXICOKINETICS

Absorption

Dermal

The dermal permeability of PA was evaluated in an in vitro pig-ear skin assay (CCR 1993a), which has been shown to be an adequate model of human dermal absorption. PA was extracted from a 100% cotton textile treated with Pyrovatex CP Neu® using a synthetic perspiration extract (the concentration of PA was not reported). Diffusion chambers were constructed with freshly shaved pig skin taken from the outer ear (2-3 mm thick). The surface area of skin separating the donor and receptor chambers was about 1.13 cm^2. No PA was detected in the receptor chamber after 0, 0.5, 1, 2, 4, 6, 8, and 24 hr. However, it is not known whether PA was extracted from the treated textile in sufficient quantities to create a concentration gradient large enough to drive diffusion of PA across the dermal barrier.

The subcommittee found no studies that measured the absorption of PA through the lung epithelium or gut wall following inhalation or oral exposure.

Distribution, Metabolism, and Excretion

The subcommittee found no studies that addressed the metabolism, distribution, or excretion of PA in animals or humans following dermal, oral, or inhalation exposure.

HAZARD IDENTIFICATION[2]

All PA-containing commercial FR preparations contain PA as the main ingredient and also contain melamine or a similar polymeric resin or resins, phosphoric acid, detergents, fabric-wetting agents, and possibly other com-

---

[2]In this section, the subcommittee reviewed data on toxicity of phosphonic acid, (3-{[hydroxymethyl]amino}-3-oxopropyl)-dimethyl ester, including the toxicity assessment prepared by the U.S. Consumer Product Safety Commission (Bittner 1999).

pounds including known impurities (Kapura 1994). All studies discussed below are on PA, unless otherwise indicated.

Dermal Exposure

There are no peer-reviewed, published studies on the dermal effects of PA flame retardants in humans or animals. Several unpublished dermal-toxicity studies of PA were conducted for Ciba-Geigy by the contract laboratory Research & Consulting Company AG. These are described below.

Irritation and Sensitization

PA was found to be nonirritating to the skin of male and female rabbits (RCC 1992a). PA (0.5 mL/animal) was applied to the shaved skin of New Zealand rabbits (one male and two females) and covered with a dressing. The site was uncovered after 4 hr and evaluated for dermal toxicity after 1, 24, 72 hr, and 7 d. Very slight (nearly imperceptible) erythema was observed in 2 of the 3 rabbits after 1 hr and in 1 of the 3 rabbits after 24 hr. Erythema was not observed in any of the animals 48, 72 hr, or 7 d after application. Local edema was not observed in any of the animals.

Fifty volunteers were exposed to PA or a commercial preparation (Muster 2, containing PA at 380 g/L) in combination with other fabric-finishing chemicals (Tronnier 1989a,b). PA or Muster 2 was spotted onto a bandage and applied to the arms of the volunteers, and the test site was assessed for irritation and sensitization. Some volunteers in each study had irritation 48 hr after application, but these reactions subsided within 72 hr. No sensitization reactions were observed in an 11-person subset known to have allergies. No other details are provided.

PA produced moderate skin sensitization when tested in guinea pigs using the maximization test (RCC 1992b). Initially, three groups of 20 female guinea pigs were given three, 0.1 mL intradermal injections of (a) a 50:50 mixture of Freund's complete adjuvant and physiological saline, (b) PA diluted to 5% with physiological saline, and (c) PA diluted to 5% in a 50:50 emulsion of Freund's complete adjuvant and physiological saline. Control animals were given similar injections but without PA. The skin was treated with 10% sodium lauryl sulfate in petrolatum oil on d 7 to enhance skin sensitization. On d 8, PA was applied to the test sites and covered. Controls were similarly treated with physiological saline. The application sites were then uncovered and assessed for erythema and edema 1 and 2 d after the removal of the dressing. At 2 wk, test and control animals were challenged with undiluted PA on the left and right flanks and covered. The sites were then uncovered 1 d after challenge and assessed for erythema and edema reactions. Ten of 19 animals had skin reactions the next day after treatment. Six of 19 animals had skin reactions 2 d after challenge. PA was of moderate allergenic potency in female albino guinea pigs according to the criteria developed by Magnusson and Kligman (1969). PA did not induce sensitization in the albino guinea pig maximization test when administered as a 25% solution dissolved in distilled water using identical protocols (RCC 1993).

Systemic Effects

No systemic effects were observed in Wistar rats treated topically with a single application of PA at a concentration of 2,000 mg/kg (RCC 1992c). Rats (5 male, 5 female) were topically treated with 2 mL of PA and the sites were covered for 24 hr. The dressing was removed and the sites were assessed for skin reactions four times during d1 and once on d 2-15 using methods developed by Noakes and Sanderson (1969). Animals were also evaluated for clinical signs of toxicity. All animals were killed and necropsied on d 15. Over the 15-d observation period, no deaths occurred, and no signs of systemic toxicity were observed. Skin changes were observed at the application site (scaling, yellowing) that disappeared 11-14 d post exposure.

Neurological Effects

No changes in general behavior or motor susceptibility (such as spasms, clonic muscle spasms, tremor, and muscle twitching) were observed among male and female Wistar rats 15 d after a single, 2 mL dermal application of PA at 2,000 mg/kg (RCC 1992c).

Other Systemic Effects

No studies were found on the immunological, reproductive, developmental, or carcinogenic effects of PA flame retardants following dermal exposure of humans and animals.

Inhalation Exposure

Systemic Effects

There are no published studies on the toxic effects of PA in humans or animals following inhalation exposure. There is one unpublished acute inhalation study of PA (RCC 1992d). No deaths occurred and there were no clinical signs of toxicity among Wistar rats (5 male and 5 female) exposed (nose-only) to PA at a target concentration of 4.82 mg/L for 4 hr. Animals were monitored for 15 d after exposure. Inhalation exposure was conducted using the methods of Cannon et al. (1983). The nominal concentration of PA used was 4.83 mg/L, and the gravimetric concentration was 2.71 ± 1.44 mg/L (mean ± standard deviation). Ninety-four percent of the PA particles were smaller than 3 μm. No significant changes in body weight occurred in treated animals. Isolated dark red foci were observed in the caudal lung lobes of 2 males and 2 females exposed to PA. However, the authors did not comment on the significance of this finding.

Other Systemic Effects

No studies were found on the immunological, neurological, reproductive, developmental, or the carcinogenic effects of PA following inhalation of PA by humans or animals.

Oral Exposure

The subcommittee found no studies on toxic effects of PA following oral exposure in humans.

Suzuki et al. (1983) reported that the oral LD_{50} of PA administered to rats as a 50% solution in water was 13,000 mg/kg (95% confidence interface [C.I.], 11,400-14,800 mg/kg) in males and 13,200 mg/kg (95% C.I., 12,300-14,200 mg/kg) in females. Five rats of each sex were given a single dose of PA by oral gavage of 9,100, 10,400, 12,000, 13,600, 15,900, or 18,300 mg/kg and observed for 7 d. Death occurred within 1 hr to 3 d. Toxic signs included depression of spontaneous movement, piloreaction, diarrhea, hypothermia, and bloody lacrimation. Eldefrawi et al. (1977) reported that the LD_{50} of PA in rats was greater than 10,000 mg/kg, but provided no reference for this statement finding. Ishizu (1975) reported that no deaths were observed in mice 72 hr after they received a dose of 10,000 ppm PA.

The subcommittee found a number of unpublished oral-toxicity studies of PA[3] conducted for Ciba-Giegy (CCR 1992a; RCC 1992e, f). No deaths, clinical signs of toxicity, or structural organ changes were observed among male and

[3]PA is referred to as FAT 80'001/I in these studies.

female Wistar rats over a 15-d period after treatment with a single dose of PA[4] of 2,000 mg/kg by oral gavage (RCC 1992e). In a 5-d range-finding study, no deaths occurred among male or female adult Wistar rats given PA by gavage doses of 0, 50, 200, or 1,000 mg/kg (RCC 1992f).

The acute toxicity of PA was assessed in NMRI mice in a range-finding study (CCR 1992a). Mice (2/sex/dose) given PA by gavage doses of 2,000, 3,000, 4,000, or 5,000 mg/kg exhibited apathy, a reduction in spontaneous activity, and eyelid closure 1 to 72 hr after dosing.

PA was found to be a low potency toxicant when administered to Wistar rats over 28 d (RCC 1992g). Male and female Wistar rats (five/sex-dose group) were given PA by gavage once a d for 28 d at 0, 50, 200, or 1,000 mg/kg body weight. A second group of rats (five/sex) was given PA by gavage once a d for 28 d at 1,000 mg/kg and was sacrificed 15 d after dosing was ended. All animals were necropsied and assessed for pathological changes. There were no treatment-related deaths or effects on clinical signs, blood or clinical biochemistry, urinary excretion, eye chemistry or pathology, gross pathology, or histopathology (adrenals, heart, kidneys, liver, spleen, and stomach) in any of the treatment groups. However, heart weights and heart-body weight ratios were significantly higher among female rats in the 50-mg/kg dose group than in female controls. However, there was no dose-response associated with these increases, therefore the subcommittee concluded that these findings were not treatment-related. Females in the 200-mg/kg dose group had a slightly but statistically significantly higher pituitary:body weight ratio than female controls. High-dose (1,000 mg/kg) males had significantly higher body weight and food consumption over d 22-28 than male controls, but these changes were considered to be coincidental. High-dose males from the recovery group had slightly but significantly increased ($p \leq 0.05$) spleen:brain and decreased thyroid:body weight ratios ($p \leq 0.001$). These changes were minor, and no pathological findings accompanied these weight changes. Pathological examination of other organs and tissues did not reveal the number or types of lesions encountered to be in excess of those expected in animals of this strain and age.

Immunological Effects

Saliva leachates of PA-treated textiles were tested in the female mouse lymphocyte-blastogenesis assay and were not effective in inducing immunotoxicity (CCR 1994a). The amount of PA in the extracts was not reported. Splenocytes were exposed to the extracts for 1 hr at subcytotoxic doses and

---

[4]Dissolved in deionized water unless noted otherwise.

then they were exposed to mitogens Con A and LPS for 72 hr. Exposure to saliva extracts at concentrations of 80-84% reduced splenocyte viability. Lower concentrations did not induce a mitogenic response (proliferation). Concommitant assays with positive control substances produced the expected dose-dependent suppression of T-lymphocyte blastogenesis.

Neurological Effects

No changes in general behavior or motor function (spasms, clonic muscle spasms, tremor, muscle twitching) were observed in male and female Wistar rats 15 d after a single dose of PA by oral gavage of 2,000 mg/kg (RCC 1992e).

Apathy, a reduction in spontaneous activity, and increased incidence of eyelid closure was observed in mice 72 hr after being given a single dose of PA by oral gavage of 2,000, 3,000, 4,000, or 5,000 mg/kg (CCR 1992a). These findings could be attributed to either a direct or indirect high-dose effect on the nervous system.

Other Systemic Effects

No studies were found on the reproductive, developmental, or carcinogenic effects of PA following oral exposure of humans and experimental animals.

Genotoxicity

PA was mutagenic in the Ames test (*Salmonella typhimurium* TA98, TA100, and TA1537); however, the authors did not report the concentration of PA tested or whether the testing was conducted in the presence of exogenous metabolic activation (Ishidate and Yoshikawa 1980). PA (at a concentration of 1 mg/L) induced chromosomal aberrations in Chinese hamster lung fibroblast cells (CHL) in the presence of exogenous metabolic activation (Ishidate and Yoshikawa 1980). PA (450 mg/L) caused a statistically significant increase in chromosomal breaks (excluding gaps) in Chinese hamster cells (Don-6) and human fibroblast cells (HE 2144) (Sasaki et al. 1980).

An unpublished study found PA to be nonmutagenic when tested at concentrations of 10, 33, 100, 333, 1,000, or 5,000 μg/plate in the Ames test (*S. typhimurium* TA98, TA100, TA1535, TA1537, and TA1538) with or without metabolic activation (from Wistar rat S9 microsomal fraction) (CCR 1992b). PA was toxic to *S. typhimurium* at 5,000 μg/plate without metabolic activation. Repeat assays confirmed these findings. The subcommittee concluded that this study adequately characterized the lack of genotoxicity of PA in the Ames assay.

Saliva leachates of textile samples treated with PA were judged to be nonmutagenic in *S. typhimurium* strains TA98 and TA100 with or without the incorporation of rat S9 microsomal fraction (CCR 1994b). The amount of PA in these extracts was not reported.

Two unnpublished in vivo studies did not find evidence of the genotoxicity of PA (CCR 1992a, 1993b). PA did not induce DNA damage in hepatocytes of male Wistar rats treated orally with PA either at doses of 200 or 2,000 mg/kg for 16 hr or 2,000 mg/kg for 2 hr (CCR 1993b). The subcommittee notes that there was no evidence that the maximum tolerated dose of PA was used in this experiment. Treatment with positive control compounds produced genotoxicity. The number of polychromatic erythrocyte micronuclei were not increased at doses of 24, 48, or 72 hr in male or female NMRI mice treated with a single dose of PA (at 5,000 mg/kg) compared with controls (CCR 1992a). Clinical signs of toxicity were observed in all treated animals and treatment of the mice with positive-control chemicals produced genotoxicity.

QUANTITATIVE TOXICITY ASSESSMENT

The current toxicity database for PA and the subcommittee's corresponding risk assessments do not take into consideration the presence of polymer resins, PA-polymer resin conglomerates, formaldehyde, or other unidentified compounds that might be present in upholstery treated with PA.

Noncancer

Dermal Assessment

The available data on PA are inadequate to derive a dermal reference dose (RfD). Only one unpublished study for PA is available and involves only acute exposure (see RCC 1992c). Human data suggest that PA is not a potent skin irritant.

Inhalation RfC

The toxicity data on PA are inadequate for deriving an inhalation reference concentration (RfC). Only one inhalation toxicity study is available for PA, but involves only acute exposure in rats. No human toxicity data are available for PA.

Oral RfD

There are no adequate toxicity data for deriving an oral RfD for PA. No human toxicity data are available on PA, and no subchronic or chronic toxicity studies have been done in laboratory animals.

Cancer

There are no studies available on the carcinogenicity of PA. The available data are not sufficient for making conclusive judgments on the genotoxic potential of PA. The subcommittee found one in vitro mutagenicity study of PA in the peer-reviewed literature, and its results were inconclusive. A well-conducted but unpublished Ames assay found PA to be nonmutagenic when tested in the presence or absence of exogenous metabolic activation (CCR 1992b). The in vivo data for the clastogenicity of PA are equivocal (Ishidate and Yoshikawa 1980; Sasaki et al. 1980).

EXPOSURE ASSESSMENT AND RISK CHARACTERIZATION

The following exposure and risk assessment assumes that PA is not chemically altered during the application and curing processes and that leachates from treated upholstery are unreacted PA. The subcommittee notes that this assumption may be inaccurate, but currently there are no data on the chemical forms and amounts of cured PA that is present in PA-treated upholstery following the curing process. Some unreacted PA is left on the upholstery surface after curing, but in reality much of the free PA present on treated upholstery evaporates from the fabric or is washed away in the alkaline wash after curing. Therefore, there is very little free PA present on newly treated upholstery that is available for uptake by end-users. It is known that some PA can be released from the treated fabric by ambient hydrolysis over a period of years (see Appendix B, Case 3).

Noncancer

Dermal Exposure

Dermal exposure to PA was estimated using the dermal exposure scenario described in Chapter 3. This exposure scenario assumes that an adult spends 1/4th of his or her time sitting on furniture upholstery backcoated with PA and also assumes 1/4th of the upper torso is in contact with the upholstery and clothing presents no barrier. Exposure to other chemicals present in the backcoating was not included in this assessment.

As a first estimate of exposure, it was assumed that skin, clothing, and the upholstery did not impede dermal exposure to PA. It was also assumed that there would be sufficient water present from sweat to facilitate dissolution of PA from the material and absorption through the skin. In this scenario, only the dissolution rate of PA from the material is assumed to be the limiting factor in absorption by the body. It is assumed that all of the PA that dissolves is immediately absorbed into the body by the sitting person.

Dermal exposure was estimated using Equation 1 in Chapter 3. For this calculation, the subcommittee estimated an upholstery application rate (S_a) for PA of 3.6 mg/cm^2. The extraction rate (μ_w) for PA was estimated to be 0.001 based on laundering data (Horrocks et al. 1992).

Using these assumptions, an estimated absorbed daily dose of 0.028 mg/kg was calculated for PA. No adequate data are available to calculate dermal RfD for PA. In addition, PA is likely to polymerize after application to the upholstery fabric; therefore, exposure to PA is not likely to occur and it should not pose a toxic risk by the dermal route of exposure when used as a FR in furniture upholstery.

Inhalation Exposure

Particles

Inhalation exposure estimates for PA were calculated using the exposure scenario described in Chapter 3. This scenario assumes that a person spends 1/4th of his or her life in a 30-m^3 room containing 30 m^2 of PA-treated fabric and the room is assumed to have an air-change rate of 0.25/hr. It is also assumed that 50% of the PA present in 25% of the surface area of the treated fabric is released over 15 yr and 1% of released particles are of the size that can be inhaled.

Particle exposure was estimated using Equations 4 and 5 in Chapter 3. The subcommittee estimated an upholstery application rate (S_a) for PA of 3.6 mg/cm^2. The release rate (μ_r) for PA from upholstery fabric was estimated to be 2.3 × 10^{-7}/d (using Equation 5 in Chapter 3) yielding a room airborne particle concentration (C_p) of 1.4 µg/m^3 and a short time-averaged exposure concentration of 0.35 µg/m^3. The time-averaged exposure concentration for particles was calculated using Equation 6 in Chapter 3. No inhalation RfC is available for calculating a hazard index. Therefore, no conclusions can be drawn about the noncancer toxicological risks posed by the inhalation of PA-containing particles.

Vapors

In addition to the possibility of release of PA in particles from worn upholstery fabric, the subcommittee considered the possibility of the release of PA by evaporation. This approach is described in Chapter 3, and uses an exposure scenario similar to that described above for exposure to PA particles.

The rate of flow of PA vapor from the room is calculated using Equations 8-11 in Chapter 3. A saturated vapor concentration (C_v) of 200,000 mg/m^3 was estimated for PA. The application density (S_a) for PA in the treated upholstery was estimated as 3.6 mg/cm^2.

Using the parameters described, the equilibrium room-air concentration of PA was estimated to be 170,000 mg/m^3. The short-term time-average exposure concentration for PA was estimated as 42,500 mg/m^3 using Equation 12 in Chapter 3 and the equilibrium room-air concentration for PA. It was estimated that this concentration could be maintained for approximately 39 minutes. These results suggest that the vapor inhalation scenario is unrealistic for PA-treated furniture in a residential setting. This conclusion refers to the worst case scenario of a fully hydrolyzed treatment which is a worst-case assumption. In practice, and with an acceptable fabric cleaning regime (see Appendix B, Case 3), the flame retardant will most likely be in a fully polymerized form and so have zero vapor pressure.

No inhalation RfC is available for calculating a hazard index, therefore no conclusions can be made about the noncancer toxicological risks posed by the inhalation of PA vapors.

Oral Exposure

The assessment of noncancer toxicological risk for oral exposure to PA is based on the oral exposure scenario described in Chapter 3. This scenario assumes a child is exposed to PA by sucking on 50 cm^2 of fabric treated with PA, 1 hr/d for two yr. The subcommittee estimated an upholstery application rate (S_a) for PA of 3.6 mg/cm^2. Oral exposure was calculated using Equation 15 in Chapter 3. The extraction rate (μ_w) for PA was estimated to be 0.001 based on laundering data (Horrocks et al. 1992-see Appendix B, Case 2).

The worst-case average oral daily dose for PA was estimated as 0.00075 mg/kg-d. Therefore, the level of exposure to PA by the oral route is anticipated to be small, given the worst-case parameters and conditions used in the exposure calculation. No oral RfD is available for calculating a hazard index; therefore, no conclusions can be drawn concerning noncancer toxicological risks associated with oral exposure to PA in upholstered fabric.

Cancer

No data are available on the carcinogenicity of PA by the dermal, oral, or inhalation routes of exposure.

RECOMMENDATIONS FROM OTHER ORGANIZATIONS

The subcommittee is not aware of any exposure limits that have been recommended by any regulatory agency or other organizations.

DATA GAPS AND RESEARCH NEEDS

Key information on the types and amounts (and ratios) of PA derivatives present in treated upholstery are not available. Information on exposure levels to PA from the dermal, inhalation, and oral routes are also not available. Data on the dermal penetration of PA, as well as the amounts of PA leached from treated fabric by simulated body fluids, are also not available. Subchronic or chronic toxicity data are not available for the dermal, inhalation, or oral routes of exposure. There are no data on the effects of PA on reproduction or development.

It is important to note that PA polymerizes within the fiber and fabric structure and may also react with other FR formulation components present; therefore, it will likely undergo other chemical changes that could alter its chemical properties and toxicity. It is also likely that oxidized forms of this FR will be present in or on the aged FR-treated fabric. The subcommittee recommends that research be conducted to determine whether new chemical species are formed and, if so, to identify them.

REFERENCES

Bittner, P. 1999. Toxicity review for phosphonic acid, (3-{[hydroxymethyl]amino}-3-oxopropyl)-dimethyl ester. Memorandum, dated March 3, 1999, from Patricia Bittner, Toxicologist, Division of Health Sciences, to Ronald

Medford, Assistant Executive Director for Hazard Identification and Reduction, U.S. Consumer Product Safety Commission, Washington, DC.

Cannon, W.C., E.F. Blanton, and K.E. McDonald. 1983. The flow-past chamber: An improved nose-only exposure system for rodents. Am. Ind. Hyg. Assoc. J. 44(12):923-928.

CCR (Cytotest Cell Research GmbH & Co. KG). 1992a. Micronucleus Assay in Bone Marrow Cells of the Mouse with FAT 80'001/I. CCR Project 269537. Study conducted by Cytotest Cell Research GmbH & Co. KG for Ciba-Geigy AG.

CCR (Cytotest Cell Research GmbH & Co. KG). 1992b. Salmonella Typhimurium Reverse Mutation Assay with FAT 80'001/I. CCR Project 269515. Study conducted by Cytotest Cell Research GmbH & Co. KG for Ciba-Geigy AG.

CCR (Cytotest Cell Research GmbH & Co. KG). 1993a. Skin permeability *in vitro* absorption through pig ear skin with Pyrovatex® CP neu. CCR Project 310228. Study conducted by Cytotest Cell Research GmbH & Co. KG for Ciba Speciality Chemicals Corp.

CCR (Cytotest Cell Research GmbH & Co. KG). 1993b. In vivo/in vitro Unscheduled DNA Synthesis in Rat Hepatocytes with FAT 80'001/I. CCR Project 404818. Study conducted by Cytotest Cell Research GmbH & Co. KG for Ciba-Geigy AG.

CCR (Cytotest Cell Research GmbH & Co. KG). 1994a. Lymphocyte Blastogenesis Assay on Spleenocytes of Mice In Vitro with Pyrovatex CP New (Saliva Extract of the Textile Sample). CCR. Projects 477104, 477204, 477304. Study conducted by Cytotest Cell Research GmbH & Co. KG for Pfersee Chemie GmbH.

CCR (Cytotest Cell Research GmbH & Co. KG). 1994b. Salmonella Typhimurium Reverse Mutation Assay with Pyrovatex CP New (Saliva Extract of the Textile Sample). CCR. Projects 477101, 477201, 477301. Study conducted by Cytotest Cell Research GmbH & Co. KG for Pfersee Chemie GmbH.

ChemID. 1999. Chemical ID Data Base. National Library of Medicine, National Toxicology Information Program, Bethesda, MD.

Ciba (Ciba Specialty Chemicals Corporation). 1998a. Pyrovatex® Flame Retardants for Use in Upholstered Furniture. Presentation to the United States Consumer Product Safety Commission, May 5-6, 1998 by Carl D'Ruiz and D. Parkes. Ciba Specialty Chemicals, Consumer Care Division. High Point, NC.

Ciba (Ciba-Geigy Corporation). 1998b. Fabric Processing and Finishing: Cotton-Pyrovatex®. Production and Marketing Guidelines. Ciba-Geigy Consumer Care.

Eldefrawi, A.T., N.A. Mansour, L.B. Brattsten, V.D. Ahrens, and D.J. Lisk. 1977. Further toxicologic studies with commercial and candidate flame-retardant chemicals. Part II. Bull. Environ. Contam. Toxicol. 17(6):720-726.

FRCA (Fire Retardants Chemicals Association). 1998. Textile Flame-Retardant Applications by Product Classes for 1997 Within and Outside of the United States. Submitted to the U.S. Consumer Product Safety Commission, prepared by Fire Retardants Chemicals Association. Lancaster, PA.

Horrocks, A.R., J. Allen, S. Ojinnaka, and D. Price. 1992. Influence of laundering on durable flame-retarded cotton fabrics-Part 1. Effect of oxidant concentration and detergent type. J. Fire Sci. 10:335-351.

Ishizu, S. 1975. Toxicity of organophosphorus retardants. Kobunshi 24:788-92.

Ishidate, M., Jr., and K. Yoshikawa. 1980. Chromosome aberration tests with Chinese hamster cells in vitro with and without metabolic activation—A comparative study on mutagens and carcinogens. Arch. Toxicol. Suppl. 4:41-44.

Kapura, A.A. 1994. Chemistry of flame retardants. II. NMR and HPLC analysis of Pyrovatex CP. J. Fire Sci. 12:3-12.

Magnusson, B., and A.M. Kligman. 1969. The identification of contact allergens by animal assay. The guinea pig maximization test. J. Invest. Dermatol. 52(3):268-76.

Noakes, D.N., and D.M. Sanderson. 1969. A method for determining the dermal toxicity of pesticides. Br. J. Ind. Med. 26(1):59-64.

RCC (Research & Consulting Company AG). 1992a. Primary Skin Irritation Study with Fat 80'001/i in Rabbits (4-hour Semi Occlusive Application). RCC Project 316697. Study conducted by RCC Research & Consulting Company AG for Ciba Speciality Chemicals Corp.

RCC (Research & Consulting Company AG). 1992b. Contact Hypersensitivity to FAT 80'001/I in Albino Guinea Pigs Maximization-Test. RCC Project 316710. Study conducted by RCC Research & Consulting Company AG for Ciba-Geigy AG.

RCC (Research & Consulting Company AG). 1992c. Acute Dermal Toxicity Study With FAT 80'001/I in Rats. RCC Project 316686. Study conducted by RCC, Research & Consulting Company AG for Ciba-Geigy AG.

RCC (Research & Consulting Company AG). 1992d. 4-Hour acute inhalation toxicity study (limit test) with FAT 80'001/I in rats. RCC Project 317002. Study conducted by RCC Research & Consulting Company AG for Ciba Speciality Chemicals Corp.

RCC (Research & Consulting Company AG). 1992e. Acute Oral Toxicity Study with FAT 80'001/I in Rats. RCC Project 316675. Study conducted by RCC Research & Consulting Company AG for Ciba-Geigy AG.

RCC (Research & Consulting Company AG). 1992f. 5-Day Oral Toxicity (Range-Finding) Study With FAT 80'001/I in Rats. RCC Project 316721. Study conducted by RCC Research & Consulting Company, Ltd. for Ciba-Geigy AG.

RCC (Research & Consulting Company AG). 1992g. Subacute 28-Day Oral Toxicity (Gavage) Study With FAT 80'001/I in the Rat. RCC Project 316732. Study conducted by RCC Research & Consulting Company, Ltd. for Ciba-Geigy AG.

RCC (Research & Consulting Company AG). 1993. Contact Hypersensitivity to Pyrovatex CP Neu in Albino Guinea Pigs Maximization-Test. RCC Project 347692. Study conducted by RCC Research & Consulting Company AG for Pfersee Chemie GmbH.

Sasaki, M., K. Sugimura, M.A. Yoshida, and S. Abe. 1980. Cytogenetic effects of 60 chemicals on cultured human and Chinese hamster cells. Kromosomo (Tokyo) 2:574-584.

Suzuki, Y., K. Naito, and M. Tobe. 1983. Acute toxicity of chemicals used in the household [English abstract]. Eisei Shikensho Hokoku 101:152-6.

Tronnier, H. 1989a. Dermal irritation/sensitization study of parent compound in human volunteers. [Report in German]. Study conducted by Prof. Dr. H. Tronnier, Dortmund, Germany, for Ciba Speciality Chemicals Corp.

Tronnier, H. 1989b. Dermal irritation/sensitization study of Muster 2 in human volunteers. [Report in German]. Study conducted by Prof. Dr. H. Tronnier, Dortmund, Germany, for Ciba Speciality Chemicals Corp.

14

Organic Phosphonates

THIS chapter reviews the physical and chemical properties, toxicokinetics, toxicological, epidemiological, and exposure data on dimethyl hydrogen phosphite (DMHP),[1] an organic phosphonate. The subcommittee used that information to characterize the health risk from exposure to DMHP. The subcommittee also identified data gaps and recommended research relevant for determining the health risk from exposure to DMHP.

PHYSICAL AND CHEMICAL PROPERTIES

As a pure substance, DMHP is a mobile, colorless liquid with a mild odor (Hawley 1981, as cited in IARC 1990; HSDB 1999). The physical and chemical properties of DMHP are summarized in Table 14-1.

OCCURRENCE AND USE

According to IPCS (1997), both halogenated and non-halogenated phosphonate esters are the predominant phosphorous-based flame retardants in use.

[1]In this chapter, the toxicity data were discussed for DMHP because most of the relevant data for organic phosphates were available for this compound.

TABLE 14-1 Physical and Chemical Properties of Dimethyl Hydrogen Phosphite

Property	Value	Reference
Chemical formula	$C_2H_7O_3P$	Howard and Meylan 1997
Structure	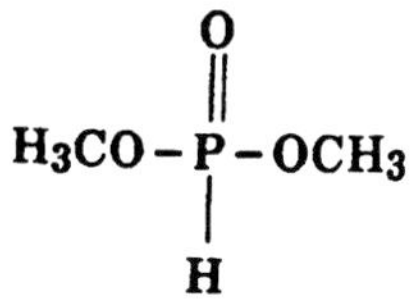	NTP 1985
CAS Registry #	868-85-9	Howard and Meylan 1997
Synonyms	DMHP, dimethyl phosphonate (Chem. Abstr. Name; IUPAC Systematic Name), Bis(hydroxy-methyl) phosphine oxide; dimethoxyphosphine oxide; dimethylhydrogenphosphite; dimethyl phosphite; dimethyl acid phosphite; *O,O'*-dimethyl phosphonate; dimethyl phosphorous acid; DMHP; hydrogen dimethyl phosphite; methyl phosphonate; NCI-C54773; phosphonic acid dimethyl ester; phosphorous acid dimethyl ester; TL 585	IARC 1990; NTP 1985; HSDB 1999
Molecular weight	110.05	Howard and Meylan 1997
Physical state	Mobile liquid	Hawley 1981; as cited in IARC 1990
Solubility	Solubility in water = 1×10^6 mg/L; miscible in most organic solvents; soluble in pyrimidine	Howard and Meylan 1997; CRC Press 1992
Color	Colorless	Hawley 1981; as cited in IARC 1990
Vapor pressure	4.52 mm Hg at 25°C	Howard and Meylan 1997
log K_{OW}	-1.13	Howard and Meylan 1997
pH	Not available	—
Melting point	Not available	—
Boiling point	170.5 °C	Howard and Meylan 1997
Flash point	96 °C	Hawley 1981, as cited in IARC 1990

(Continued)

Decomposition	When heated to decomposition, can emit highly toxic fumes of phosphorous oxides	Sax 1984
Reactivity	Hydrolyzes in water with half-life of approximately 10 d at 25°C and 19 d at 20°C; basic conditions accelerate hydrolysis	Bel'skii et al. 1969; Vilceanu and Schulz 1972; all as cited in HSDB 1999
Density (water = 1)	1.195	NTP 1985

DMHP is commercially produced by the reaction of phosphorous trichloride and methanol or sodium methoxide (HSDB 1999; EPA 1985 as cited in IARC 1990). In 1982, U.S. production was estimated to be about 3 million pounds/yr (W. Smithey, Jr., pers. commun. to J. Dunnick 1982, as cited in NTP 1985).

DMHP is used in the manufacture of adhesives, pesticides, and is used to impart flame resistance to textiles (Hatlelid 1999; IARC 1990; HSDB 1999; Siemer 1980; Lewis 1975 as cited in NTP 1985; and Lewis 1993).

TOXICOKINETICS

Dermal Exposure

There are no studies that have investigated the absorption, distribution, metabolism, or excretion of DMHP in humans or animals following dermal exposure. Dermal LD_{50} studies suggest that DMHP is systemically available following dermal application.

Inhalation Exposure

No data describing toxicokinetics of DMHP from the inhalation route were identified during the course of this review.

Oral Exposure

Nomeir and Matthews (1997) examined the metabolism and disposition of ^{14}C-labeled DMHP in F-344/N rats and $B6C3F_1$ mice. After gavage administration (10-200 mg/kg), the radio-labeled compound was almost completely absorbed from the gastrointestinal tract in both rats and mice and was primarily eliminated as expired CO_2 (44-57%) within 24 hr. Radioactivity was primarily distributed to the liver, kidneys, spleen, lungs, and forestomach. The authors concluded that absorption, metabolism, and disposition of DMHP were linear in both species within the dose range that was examined. The rate of clearance was twice as fast in the mouse than in the rat. The metabolite monomethyl hydrogen phosphite (MMHP) was excreted in the urine in both species and indicates that DMHP is demethylated in vivo. Within the first 24 hr of exposure, elimination via urine (28-49%) greatly exceeded elimination by the fecal route (1-2%) or as volatile organic compounds (2-3%). Repeat administration of labeled compound over a period of 5 consecutive days (once/d) had little effect on metabolism to CO_2 or elimination in urine.

In vitro tests indicate that DMHP is metabolized to formaldehyde (CH_2O) by microsomes prepared from the liver, lungs, kidneys, forestomach, and glandular stomach of rats (Nomeir and Mathews 1997).

HAZARD IDENTIFICATION[2]

Dermal Exposure

Irritation

No signs of dermal irritation were observed in rabbits receiving dermal applications of DMHP at concentrations of up to 3,160 mg/kg (details discussed below under Systemic Effects) (Keller 1961).

Systemic Effects

DMHP (undiluted, purity not described) was applied to the occluded skin of albino rabbits at doses of 100, 316, 1,000, and 3,160 mg/kg for 24 hr (Keller 1961). After the occlusion period, the treated site was rinsed and the animals were examined for toxic effects and mortality periodically during the first 24 hr post-exposure and every day thereafter for a total of 7 d. During the first 24 hr, all animals had normal appearance and behavior. Between 24 and 48 hr post-exposure, mortality occurred at doses of 1,000 and 3,160 mg/kg (three of four rabbits at each dose died). By 72 hr post-exposure, the fourth animal receiving 1,000 mg/kg died. It exhibited systemic effects of depression, ptosis,

[2]In this section, the subcommittee reviewed toxicity data on organic phosphonates, including the toxicity assessment prepared by the U.S. Consumer Product Safety Commission (Hatlelid 1999).

labored respiration, ataxia, and placidity before dying. Necropsy results included hemorrhagic lungs, red-tinged fluid in the pleural cavity, thymus and kidney congestion, and stomach mucosal edema. The remaining rabbit in the 3,160 mg/kg group exhibited slight depression and labored respiration after 48 hr of exposure, but recovered by d 3. Thereafter, it appeared normal and gained weight for the duration of the study. Autopsy results for it were also normal. No mortality occurred at the two lowest doses. The calculated LD_{50} for this study was 681 mg/kg.

Neurological Effects

Keller (1961) reported depression, ptosis, labored respiration, ataxia, and placidity in a rabbit that received a dermal dose of DMHP of 1,000 mg/kg. These findings could indicate neurotoxicity, but are not adequate to conclude that DMHP causes neurotoxicity following dermal exposure.

Other Systemic Effects

No data were found regarding the immunological, reproductive, developmental, or carcinogenic effects of DMHP following dermal exposure.

Inhalation Exposure

A summary of the inhalation toxicity data of DMHP is presented in Table 14-2.

Systemic Effects

Rusch (1980) exposed Sprague-Dawley rats (5 males and 5 females/dose) to DMHP at concentrations of 0, 431, 843, and 934 ppm (0, 1940, 3794 and 4203 mg/m^3), 6 hr/d for 5 consecutive days. Degradation of the test article to phosphoric acid occurred with deposition on the rats as well as exposure chamber surfaces. Skin, eye, and mucous membrane irritation was observed at all exposure levels, but was more severe at the higher concentrations. Attempts to minimize "wall losses" and hydrolytic degradation were made by dehumidifying chamber air and increasing chamber temperature to 79 °F. Nevertheless, the author considered that "generation of [test article] levels . . . in excess of 500 ppm [2250 mg/m^3] would be difficult" (Rusch 1980). After 4 d of exposure, all

TABLE 14-2 Inhalation Exposure Studies on Dimethyl Hydrogen Phosphite

Species, Strain Sex, Number	Purity	Concentration	Duration	Effects	Comments	Reference
Rats, Sprague-Dawley, M/F, 5/sex/dose	As received from Mobil Chemical Co., Edison, NJ	0, 431, 843, 934 ppm (0, 1940, 3794, 4203 mg/m^3)	6 hr/d, 5 consecutive days	All HC animals died or killed in extremis after d 3 with red stained CSF under brain meninges, gastrointestinal tract vascular lesions, discoloration of trachea and fluid retention in small intestine; labored breathing and neuromuscular impairment at HC; skin, eye, and mucous membrane irritation at all exposures; significant body weight decrease at HC; significant increase in M lung weight at MC and HC; concentration-related lung discoloration; corneal changes	Range-finding study for Bio/dynamics 1980; test material degraded to phosphoric acid; degradation product considered responsible for irritant responses	Rusch 1980[a]
Rat, Sprague-Dawley, M/F, 20/sex/dose	As received from Mobil Chemical Co., Edison, NJ	0, 12, 35, 119, 198 ppm (0, 54, 158, 536, 891 mg/m^3)	6 hr/d, 5 d/wk for up to 4 wk; left untreated for next 4 wk	Excessive mortality at HC; significant body weight decrease at ≥536 mg/m^3 (M/F); significant total WBC count at ≥536 mg/m^3 (M) and at 891 mg/m^3 (F); significant decrease in HCT at ≥536 mg/m^3 (M); decreasing trend in HGB ≥536 mg/m^3 (M) and 891 mg/m^3 (F); significant decrease in blood GLU ≥536 mg/m^3 (M/F); increasing trend in blood SGPT, AP, and BUN at 891 mg/m^3 (M/F); lenticular opacities at ≥158 mg/m^3 (4 wk; M/F); cataracts at ≥536 mg/m^3 (8 wks; M/F); concentration-related increase in absolute and relative kidney weight at all levels (M/F); at necropsy, red discoloration lung/turbinates in HC (irritation observed in vivo)	LOAEL: 12 ppm (54 mg/m^3) based on absolute and relative kidney weight changes in M	Bio/dynamics 1980[a,b]

AP, alkaline phosphatase; BUN, blood urea nitrogen; CSF, cerebrospinal fluid; F, female; GLU, glucose; HC, high concentration; HCT, hematocrit; HGB, hemoglobin; LOAEL, lowest-observed-adverse-effect level; M, male; MC, mid-concentration; SGPT, serum glutamic pyruvic transaminase; WBC, white blood cell.

[a] Lab report from Bio/Dynamics, Inc., to Mobil Oil Corp.

[b] Submittal from Mobil Oil Corp. to EPA (1981).

animals in the high-dose group were dead or were killed. These animals exhibited neuromuscular impairment (slowed righting reflex, loss of toe-pinch reflex, prostration, lack of response to sound stimuli, muscular contractions and splayed stance) and significant body weight depression. Significant body weight depression (males and females) was also noted at concentrations of 3,794 mg/m^3. At the 1,940- and 3,794-mg/m^3 concentration levels, statistically significant increases in lung weight were noted in males. No necropsy results were reported for animals dead or dying at the highest concentration tested.

Based on the results of Rusch (1980), Bio/dynamics (1980) exposed groups of male and female Sprague-Dawley rats (20 males and 20 females/dose) to DMHP at concentrations of 0, 12, 35, 119, and 198 ppm (0, 54, 158, 536, and 891 mg/m^3) 6 hr/d, 5 d/wk for 4 wk. The duration-adjusted concentrations were 0, 10, 28, 96, and 159 mg/m^3. After the 4-wk period of exposure was completed, Bio/dynamics (1980) maintained the treated and control populations for an additional 4 wk with no further exposure.

The most significant effects observed where dose-related increases in absolute and relative kidney weight (that persisted during the 4-wk recovery period), and lenticular opacities that progressed to cataracts during the recovery period (see Table 14-2). According to Bio/dynamics (1980), reduced body weights at the higher exposure levels somewhat obscured the increase in absolute and relative kidney weights. The differences were significant at all exposure concentrations greater than 158 mg/m^3 for males and greater than 891 mg/m^3 for females. The findings of cataracts in male rats receiving DMHP by gavage supports this finding as a treatment-related effect (NTP 1985).

Neurological Effects

Neuromuscular impairment (in the form of slowed righting reflex, loss of toe-pinch reflex, prostration, etc.) was observed in rats at the highest concentration level tested by Rusch (1980). Bio/dynamics (1980) observed neurological impairment in all rats of the highest exposure group and in some rats exposed to 536 mg/m^3. These effects were usually reversed after cessation of exposure. No studies have been specifically carried out to determine the the neurotoxicity of DMHP following inhalation exposure.

Other Systemic Effects

No data were found on the immunological, reproductive, developmental, or carcinogenic effects of inhaled DMHP.

Oral Exposure

A summary of the noncancer effects from oral exposure to DMHP is presented in Table 14-3.

Systemic Effects

Groups of two albino rats were given single oral doses of DMHP (in a 0.5% aqueous methyl cellulose solution) ranging from 10 to 3,160 mg/kg. The animals were observed for 48 hr. No explanation regarding the basis for selecting methyl cellulose as a vehicle was provided. A single death occurred at the highest dose tested of 3,160 mg/kg; this animal also exhibited signs of toxicity (prostration, labored breathing, tremors) 4 hr after exposure and shortly before death. Pathological results on this rat revealed lung hemorrhage, congested kidneys, and gastrointestinal tract inflammation. Analyses of these data resulted in an estimated LD_{50} of 3,160 mg/kg.

The NTP (1985) administered DMHP by gavage to F-344/N rats and $B6C3F_1$ mice for 1 d, 15 consecutive days, 13 wk (5 d/wk), and 103 wk (5 d/wk). DMHP was administered in corn oil for all doses except the 15-d 3,000 mg/kg dose, which was administered as undiluted DMHP. Doses, number of rats, dose volume, and toxic response are summarized in Table 14-3. The estimated LD_{50}s were 3,283 mg/kg for males and 3,040 mg/kg for females. In general, females were less sensitive than males, and mice were less sensitive than rats.

Dose-related testicular atrophy was observed in $B6C3F_1$ mice given doses of 375 mg/kg-d or greater for 13 wk (NTP 1985). All male mice given doses of 750 mg/kg-d or greater died by wk 4. The NOAEL for this study was 190 mg/kg-d based on testicular atrophy. In the 13-wk study in rats, the NOAEL for DMHP was determined to be 100 mg/kg-d based on body weight depression in female rats.

In the 103-wk study, dose-related histopathological changes were observed in tissues of the lung, forestomach, eye, cerebellum, and hematopoietic system. Malignancies of the lung and forestomach occurred in high-dose males (200 mg/kg-d), who also exhibited increased incidences of mononuclear cell leukemia and cataracts. Body weight depression was noted in the high-dose groups, with the greatest difference observed in the high-dose male rats. Survival of the high-dose males was also significantly shorter when compared to the vehicle controls ($p = 0.008$ by life table pairwise comparison). Survival rates for males were 78% (vehicle control), 58% (100 mg/kg), and 46% (200 mg/kg). For females, survival was 80% (vehicle control), 66% (50 mg/kg), and 64% (100 mg/kg). An increased incidence of chronic interstitial pneumonia occurred

TABLE 14-3 Noncancer Effects from Oral Exposures to Dimethyl Hydrogen Phosphite

Species, Strain, Sex, Number	Purity	Dose	Duration, Route	Effects	Comments	Reference
Rat, albino, NS, 2/dose	As received from Hooker Chem. Co. ("considered to be free of impurities")	10.0, 31.6, 100, 316, 3,160 mg/kg in 0.5% (v/v) aqueous methyl cellulose solution; untreated control not reported	Single dose, intubation	No deaths at 10.0-100 mg/kg; single death at 3,160 mg/kg at 4 hr, exhibited prostration, labored respiration, and tremors. Lung hemorrhage, kidney congestion, gastrointestinal inflammation	LD_{50}: ~3,160 mg/kg	Keller 1961[a]
Rat, F-344/N, M/F, 5/sex/dose	~96-98%	0, 1,470, 2,150, 3,160, 4,640, 6,810 mg/kg in corn oil (dose volume: 5.675 mL/kg)	Single dose, gavage	Mortality; inactivity, weakness, shallow breathing on d 1 at ≥3,160 mg/kg; gas in stomach and intestines at necropsy	LD_{50}: 3,283 mg/kg (M); 3,040 mg/kg (F)	NTP 1985[b]
Rat, F-344/N, M/F, 5/sex/dose	~96-98%	0, 250, 500, 1,000, 2,000 mg/kg-d in corn oil; 3,000 mg/kg undiluted (dose volume: 2.5 mL/kg)	15 d (consecutive), gavage	Deaths at ≥500 mg/kg; inactivity after dosing	No other clinical signs or necropsy findings reported	NTP 1985[b]
Rat, F-344/N, M/F, 10/sex/dose	~96-98%	0, 25, 50, 100, 200, 400 mg/kg-d in corn oil (dose volume: 3.33 mL/kg)	13 wk (5 d/wk), gavage	Deaths at ≥200 mg/kg in M and at ≥100 mg/kg in F; body weight depression at ≥400 mg\kg in M and at ≥200 mg/kg in F; lens degeneration at 400 mg/kg in M and F; corneal inflammation at 400 mg/kg in F; urinary bladder calculi at 400 mg/kg in M	3/5 deaths at 100 and 200 mg/kg perhaps due to accidental lung gavage. NOAEL: 100 mg/kg (based on body-weight depression in F)	NTP 1985[b]

TABLE 14-3 *(Continued)*

Species, Strain, Sex, Number	Purity	Dose	Duration, Route	Effects	Comments	Reference
Rat, F-344/N, M/F, 50/sex/dose	~96-98%	0, 100, 200 mg/kg-d in corn oil for M; 0, 50, 100 mg/kg in corn oil for F (dose volume: 4.0 mL/kg)	103 wk (5 d/wk), gavage	Mortality at 200 mg/kg in M; dose-related decreased weight gain at all doses for M and high dose F; dose-related alveolar and adenomatous hyperplasia, forestomach hyperplasia and hyperkeratosis, cerebellum mineralization in high-dose M; chemical pneumonia	NOAEL: 50 mg/kg-d (based on hyperplasia of lung tissue and forestomach in F) Evidence of dose-related cancers, especially M rats	NTP 1985[b]; Dunnick et al. 1986[c]
Mouse, B6C3F, M/F, 5/sex/dose	~96-98%	0, 1,470, 2,150, 3,160, 4,640, 6,810 mg/kg in corn oil (dose volume: 5.675 mL/kg)	Single dose, gavage	Deaths by d 2 at ≥3,160 mg/kg; inactivity, prostration; shallow breathing on d 2 at >2,150 mg/kg; white opaque eyes in M at necropsy	LD_{50}: 2,815 mg/kg (M) Survival curve too steep for estimating LD_{50} for F	NTP 1985[b]
Mouse, $B6C3F_1$, M/F, 5/sex/dose	~96-98%	0, 250, 500, 1,000, 2,000, 3,000 mg/kg-d in corn oil (dose volume: 10 mL/kg)	15 d (consecutive), gavage	Deaths by d 9 at ≥2,000 mg/kg; inactivity at ≥1,000 mg/kg, thickening and nodules in squamous stomach region at ≥250 mg/kg at necropsy		NTP 1985[b]
Mouse, $B6C3F_1$, M/F, 10/sex/dose	~96-98%	0, 95, 190, 375, 750, 1500 mg/kg-d in corn oil (dose volume: 3.33 mL/kg)	13 wk (5 d/wk), gavage	Deaths, tremors, and decreased activity in first 4 wk at ≥375 mg/kg; testicular atrophy in M at ≥375 mg/kg; some lung congestion, cardiac mineralization in M; hepatocellular vacuolization in F	NOAEL: 190 mg/kg-d (M, testicular atrophy)	NTP 1985[b]

Mouse, B6C3F$_1$, M/F, 50/sex/dose	~96-98%	0, 100, 200 mg/kg-d in corn oil (dose volume: 4.0 mL/kg)	103 wk (5 d/wk), gavage	Decreased body weight gain and survival at 200 mg/kg in M; focal calcification in testes of M at both doses	LOAEL: 100 mg/kg-d (M, testicular calcification)	NTP 1985[b]; Dunnick et al. 1986[c]
Rat, F-344/N, M, 18/dose	As received from Aldrich Chem. Co., Milwaukee, WI	0, 200 mg/kg-d in corn oil, (dose volume: 4 mL)	4, 5, or 6 wk (5 d/wk), gavage	Increase in forestomach weight for all doses; forestomach tissue exhibited hyperplasia, hyperkeratosis, subepithelial inflammation, and edema at 6 wk; significantly increased angiotensin converting enzyme in serum at 4-6 wk but returned to near control levels after 1 wk recovery; Increased level nonprotein soluble sulfhydryls in forestomach at 6 wk; decreased active soluble carboxylesterase in lungs and forestomach at 6 wk		Nomeir and Uraih 1988[c]
Rat, F-344/N, M, NS/dose Mouse, B6C3F$_1$, M, NS/dose ("at least three animals . . . for each time point")	Radio-chemical purity of 97% for labeled compounds	10, 100, 200 mg/kg [^{14}C]DMHP in corn oil (dose volume: 4 mL/kg); untreated control not reported	Rats: single dose, gavage at 10, 100,200 mg/kg; 5 d (consecutive), gavage at 200 mg/kg-d Mice: single dose, gavage at 200 mg/kg	Highest activity in liver, kidneys, spleen, lungs, forestomach; lowest activity in brain, skeletal muscle, adipose tissue. Elimination as CO_2 in expired air (44-57%) and urine (28-49%); little activity in feces (1-2%), volume organic compounds (2-3%)	Rate of labeled DMHP clearance in mice approximately 2 times greater than in rats	Nomeir and Matthews 1997[c]

F, female; LD_{50}, lethal dose to 50% of test animals; LOAEL, lowest-observed-adverse-effect level; M, male; NOAEL, no-observed-adverse-effect level; NS, not specified.

[a]Lab report from Hazleton Labs to Hooker Chemical Corporation (1961); TSCA Submittal from Occidental Chemical to EPA (1992).

[B]Report of the National Toxicology Program.

[C]Published study.

among males and appeared to be dose-related. This was considered a chemical pneumonia, evidenced by the fact that all assays were negative for infection (NTP 1985, Appendix L, p. 170). All 24 male rats at the high dose also had lung neoplasms, and pneumonia was widespread in this group (43/50). The authors did not observe an association between pneumonia and these lesions (NTP 1985). The NOAEL for this study was considered to be 50 mg/kg-d based on hyperplasia of lung and forestomach tissue in females. The LOAEL was determined to be 100 mg/kg-d based on testicular calcification in males.

Immunological Effects

No data were found regarding immunological effects after oral exposure to DMHP.

Neurological Effects

Oral administration of DMHP at doses of 200 mg/kg to male F-344/N rats for 103 wk resulted in an increase in the incidence of focal mineralization in the granular layer of the cerebellum (observed in 12 of 49 rats) (NTP 1985). Multiple basophilic concretions up to 1-mm diameter were observed in clusters but were not associated with cell damage or the presence of blood vessels. This effect was not observed in any other treatment group of male or female rats and was not noted in the $B6C3F_1$ mice.

Reproductive and Developmental Effects

No data on reproductive and developmental toxicity of DMHP were located. However, there are some studies on a closely related chemical—dimethyl methyl phosphonate (DMMP), which is discussed here. DMMP has been shown to be a reproductive toxicant in 13-wk gavage studies of male F-344 rats and male $B6C3F_1$ mice (Dunnick et al. 1984a, 1984b) and in 12-wk gavage studies with male F-344 rats (Chapin et al. 1984). Dunnick et al. (1984a, 1984b) reported dose-related decreases in rat sperm count, motility, and male fertility at all dose levels tested (250, 500, 1,000, and 2,000 mg/kg). Although reproductive function was altered, histological changes were noted only in tissues from rats in the 2,000 mg/kg dose group, characterized by a lack of spermatogenesis and necrosis of cells in the spermatogenic tubules (Dunnick et al. 1984a). Gavage treatment with 1,750 mg/kg DMMP in tap water for periods of 5-12 wk

(5 d/wk) produced morphological alterations in Sertoli cells and elongated spermatids, as well as functional defects in spermatozoa (Chapin et al. 1984). Eighty percent of the rats had normal seminiferous tubules at the end of a 14 wk recovery period. However, all recovered tubules displayed a loss of normal epithelial organization (Chapin et al. 1984).

Administration of DMMP to gestating Tif/RAI rats and CD-1 mice in their drinking water or by oral gavage (Hardin et al. 1987; Fritz 1978) did not result in reproductive or developmental toxicity (doses for rats were 2 g/kg-d on gestation d 6-15; doses for mice were 4.2 g/kg-d on gestation d 6-13). However, the high dose rats in the Fritz (1978) study exhibited maternal toxicity.

Cancer

DMHP administered by gavage to male F-344/N rats at a dose of 200 mg/kg, 5 d/wk for 103 wk induced alveolar/bronchiolar adenomas or carcinomas in 48% (24 of 50) of the animals. At a dose of 100 mg/kg, alveolar/bronchiolar carcinomas occurred in one of 50 animals while none occurred among 50 vehicle controls. There was also a dose-related increase in the incidence of squamous cell carcinomas in the lungs of the male rats (0 of 50 in vehicle control, 0 of 50 at 100 mg/kg, 5 of 50 at 200 mg/kg; $p = 0.020$, life table test) (NTP 1985). Further, the combined incidences of squamous cell papillomas and carcinomas of the forestomach were significantly increased in male rats when compared to the vehicle controls (0 of 50 in control, 1 of 50 at 100 mg/kg, 6 of 50 at 200 mg/kg; $p = 0.006$ life table test). The low-dose (100 mg/kg) male rats exhibited a significantly increased incidence of mononuclear cell leukemia when compared to the vehicle control (NTP 1985). Time-to-first tumor for males occurred at wk 92 (NTP 1985, Appendix A, p. 68). The 15% decrease in average body weight of the high-dose males indicates that the maximum tolerated dose for DMHP was reached in this study.

The authors note that DMHP caused the highest incidence of lung tumors in the male rat of all chemicals studied by the National Toxicology Program (Dunnick et al. 1986). Historical incidence data for lung squamous cell carcinomas in male rats is 0 of 50 for tests with diallyphthalate, tris (2-ethyl hexyl) phosphate, and toluene diisocyanate; while the historical incidence of alveolar/bronchiolar carcinomas was 1 of 50 for each of the above 3 compounds (NTP 1985, Appendix F, p. 128). For alveolar/bronchiolar adenomas, the historical incidence was 2 of 50, 1 of 50, and 2 of 50 for the above 3 compounds, respectively (NTP 1985, Appendix F, p. 128).

Female F-344/N rats were tested under the same protocol but using a different dosing regimen (50 and 100 mg/kg-d DMHP) (NTP 1985; Dunnick et al.

1986). Alveolar/bronchiolar carcinomas were observed in 6% of the animals in the 100 mg/kg dose group (3 of 50) while none occurred in the vehicle controls (0 of 50) and one occurred in the 50 mg/kg dose group (1 of 49). Female rats displayed a significant ($p < 0.05$) positive trend for alveolar-bronchiolar carcinoma, but the high-dose effect was not found to be statistically significant when compared to controls. There was no evidence of DMHP carcinogenicity in male or female B6C3F_1 mice administered doses of 100 or 200 mg/kg, 5 d/wk for 2 yr (Dunnick et al. 1986). NTP concluded that there was "clear evidence of carcinogenicity" in male F-344/N rats, "equivocal evidence of carcinogenicity" in female F-344/N rats, and "no evidence of carcinogenicity" in male or female B6C3F_1 mice (NTP 1985, Dunnick, et al. 1986).

IARC (1990) concluded that there was limited evidence for the carcinogenicity of DMHP in experimental animals and that DMHP is not classifiable as to its carcinogenicity to humans (i.e., it is a Group 3 carcinogen).

Genotoxicity

In vivo genotoxicity studies of DMHP include tests of sex-linked lethal mutagenicity (Woodruff et al. 1985; NTP 1985), the bithorax test of Lewis, Y-chromosome loss test, dominant lethality test, and somatic reversion test of white-ivory in male Canton-S *Drosophila melanogaster* (Bowman 1980). Exposure routes included injection, ingestion, and inhalation (see Table 14-4 for summaries).

Male Canton-S *Drosophila melanogaster* fruit flies were tested for the presence of sex-linked recessive mutations after feeding (600 or 650 ppm) or injection (1,500 ppm) of DMHP (Woodruff 1985; NTP 1985). The results were negative for this effect. In the Woodruff (1985) study, 30% mortality occurred within 72 hr after feeding or 24 hr after injection.

Exposure of Canton-S males to DMHP aerosol (0.07 mL/25 mL air for 5 min) gave positive results for the sex-linked lethal and dominant lethal tests but negative results for the bithorax, Y-chromosome loss, and somatic reversion tests (Bowman 1980).

In 13-wk oral (gavage) studies of dimethyl methylphosphonate, a compound structurally similar to DMHP, there was evidence of dominant lethal mutagenicity (Dunnick et al. 1984a, 1984b) . There were increased resorptions in dams mated to male F-344 rats administered DMHP at doses of 250, 500, 1,000, and 2,000 mg/kg (Dunnick et al. 1984a). In B6C3F_1 mice, Dunnick et al. (1984b) reported dominant lethal effects at doses of 1,000 mg/kg and greater (dose regimen mirrored that of the Dunnick et al. 1984a rat study); after a 15-wk recovery period, resorptions associated with those doses declined to the control group rate (Dunnick et al. 1984b).

TABLE 14-4 Genotoxicity Studies of Dimethyl Hydrogen Phosphite

Species, Sex	Purity	Dose	Duration	Effects	Comments	Reference
Canton-S *Drosophila melanogaster*, M; Basc virgin, F	NA	150 ppm injection; 600 ppm feeding	Single injection; feeding-NA	Sex-linked lethality:negative. 30% mortality at 24 hr post-injection or 72 hr post-feeding	M mated individually to 3 harems of Basc F	Woodruff et al. 1985[a]
Canton-S *D. melanogaster*, M; Basc virgin, F	~96-98%	0, 1500 ppm injection (dose vol. = 0.03 µL); 0, 650 ppm in feed	Single injection with 24 hr recovery before mating; 3 d in feed	Sex-linked recessive lethal mutations: negative.	M mated with 3 harems of Basc F	NTP 1985[b]
Canton-S *D. melanogaster*, M	As received from sponsor	0, 0.07 mL undiluted aerosol in 25 mL air	5 min	Sex-linked lethal-positive; bithorax-negative; Y chromosome loss-negative; dominant lethal-positive; somatic reversion of white ivory:negative		Bowman 1980[c]
Salmonella typhimurium strains TA1535, 1537, 1538, 98, 100; *Saccharomyces cerevisiae* strain D4	As received from sponsor	0.001, 0.01, 0.10, 1.0, 5.0 µL/plate, with and without S9 activation	48 hr incubation	Reversion mutation w/o activation-negative; reversion mutation w/activation: negative	Positive controls	Brusick 1977[d]
S. typhimurium strains TA1535, 1537, 98, 100	~96-98%	0, 100, 333, 1000, 3333, 10,000 µg/plate; with and without S9 activation	20 min	Mutagenicity with and without activation: negative		NTP 1985[b]

TABLE 14-4 *(Continued)*

Species, Sex	Purity	Dose	Duration	Effects	Comments	Reference
S. typhimurium strains TA1535, 1537, 1538, 98, 100	As received from sponsor	0.1, 5.0, 10.0, 15.0 µL/plate; with and without S9 activation	20 min	*Salmonella*/mammalian microsome pre-incubation mutagenicity assay: negative for all strains, with and without S9 activation	Positive controls	Haworth 1979b[c]
Mouse lymphoma L5178Y TK +/- cells	As received from sponsor	0.13, 0.18, 0.24, 0.32, 0.42, 0.56, 0.75, 1.0, 1.3, 1.8 µL/plate; with and without S9 activation	4 hr	Mouse lymphoma mutagenesis assay without activation: negative; with activation: positive	Significant dose-dependent effects with S9 activation	Kirby 1979[c]
Mouse lymphoma L5178Y TK +/- cells	From NTP repository; same purity as used in NTP 1985 study	Without activation: 125, 250, 500, 1000, 2000 µg/plate; with activation: 1700, 1900, 2100, 2300, 2500 µg/plate	4 hr	Mouse lymphoma mutagenesis assay without activation: negative; with activation: positive at ≥2100 µg	LOED: 2100 µg/mL with S9 activation	McGregor et al. 1988[a]
Chinese Hamster Ovaries	Where possible, same chemical lot as NTP study	NA; with and without activation	NA	Chromosomal aberrations at 1,600 µg/mL and sister chromatid exchange at 250 µg/mL both with and without activation		Tennant et al. 1987[a]
$B6C3F_1$ mouse, M	From NTP repository; same purity as used in NTP 1985 study	Intraperitoneal injection of 0, 250, 500 mg/kg in corn oil (dose vol. = 0.4 mL); 1 injection/d	3 d	Mouse bone marrow micronucleus assay-increasing trend; no significant increase in micronucleated percentage of polychromated erythrocytes	Data judged by authors to be adequate evidence of effect; additional tests needed	Shelby et al. 1993[a]

S. typhimurium strains TA1978 and 1538; *E. coli* strains WP2 and 100	As received from sponsor	0.3, 3.0, 30.0, 50 µL/plate; with and without S9 activation	90 min	DNA damage/repair suspension test for *E. coli* strain WP100: positive at 30 µg with activation and positive at 50 µgL/plate without activation; same test for *Salmonella* strain TA1538: negative with activation and positive at 30 µgL/plate without activation	No clear dose response. Strains with positive reaction are repair-deficient	Haworth 1979c[c]

F, female; LOED, Lowest Effective Dose; M, male; NA, not applicable.
[a]Published study.
[b]Study of the National Toxicology Program.
[c]Lab report from EG&G Mason Research Institute, Rockville, MD, to sponsor, Mobil Chemical Company of Princeton, NJ.
[d]Lab report from Litton Bionetics, Inc., Kensington, MD, to Mobil Chemical Company (1977); TSCA submittal by Mobil Research and Development Corporation of NY, to EPA Office of Toxic Substances (1981).

DMHP was studied in several in vitro genotoxicity assays, which include Ames tests, with and without S9 activation, in various tester strains of *Salmonella typhimurium* and *Saccharomyces cerevisiae* (Brusick 1977; Haworth 1979b; NTP 1985); mouse lymphoma assays (Kirby 1979; McGregor et al. 1988); a mouse bone marrow micronucleus assay (Shelby et al. 1993); Chinese hamster ovary (CHO) assay for chromosomal aberrations and sister chromatid exchange (Tennant et al. 1987); and DNA damage/repair tests of various *Escherichia coli* and *S. typhimurium* tester strains (Haworth 1979c) (See Table 14-4).

DMHP was tested in Ames *Salmonella* tester strains TA98, TA100, TA1535, TA1537, and TA1538 at concentrations ranging from 0.001-15.0 μL/plate (Brusick 1977; Haworth 1979b) or 100-10,000 μg/plate (NTP 1985) and the results were negative, with and without S9 activation. Mouse lymphoma assay results were positive at high concentrations (lowest observed effective dose [LOED] = 2,100 μg/plate) with S9 activation (McGregor et al. 1988) but negative without activation (Kirby 1979). The Kirby (1979) test results indicate dose dependency.

In the CHO cell line, DMHP at 1,600 μg/mL caused chromosomal aberrations and, at 250 μg/mL, caused sister chromatid exchanges (both with and without S9 activation) (Tennant et al. 1987). Mouse bone marrow micronucleus tests revealed a positive, but not significant, increasing trend in percentage of micronucleated, polychromated erythrocytes (Shelby et al. 1993). The authors concluded that there was adequate evidence of clastogenic effect, but that additional tests are needed. DNA damage and repair have been tested in *E. coli* strains WP2 and WP100 with and without activation (Haworth et al. 1979b).

QUANTITATIVE TOXICITY ASSESSMENT

Noncancer

Dermal Assessment

There are insufficient dermal toxicity data from which to develop an estimate of a dermal reference dose (RfD) for DMHP.

Inhalation RfC

There are inadequate toxicity data for deriving an RfC for DMHP. No chronic inhalation toxicity studies area available for DMHP. There is one subchronic inhalation study for DMHP in rats by Bio/dynamics (1980), however, the subcommittee concluded that this study was not adequate for use in deriving an RfC for DMHP.

Oral RfD

The 2-yr chronic gavage exposure study performed by NTP (1985) in F-344/N female rats was selected by the subcommittee as the critical study for the development of a chronic oral RfD for DMHP based on treatment-related hyperplasia of the lung and forestomach. Similar results were observed in mice, but these lesions occurred at greater incidence in rats. EPA does not usually establish RfDs on the basis of hyperplasia where cancer is also seen. However, it is not clear that all forms of observed hyperplastic response is associated only with carcinogenesis. Adenomatous hyperplasia in the female rats is not known to be clearly linked to malignant tumor formation.

The NOAEL for hyperplasia of the lung (alveolar epithelium hyperplasia and adenomatous hyperplasia) and forestomach in female F-344/N rats was identified as 50 mg/kg-d. The NOAEL was adjusted for discontinuous exposure by multiplying by the ratio of (5/7) to accommodate the 5-d dosing regimen employed (NTP 1985; Dunnick et al. 1986) yielding an adjusted NOAEL of 35.7 mg/kg-d. A composite uncertainty factor of 300 was applied to the NOAEL giving an oral RfD of 0.12 mg/kg-d (see Table 14-5). A factor of 10 was applied for intraspecies variation, a factor of 10 for extrapolating from animals to humans and a factor of 3 for the adequacy of the toxicity database of DMHP (availability of chronic toxicity results in 2 species). No mammalian multigenerational reproductive toxicity studies and no mammalian developmental toxicology studies on DMHP were found and their absence leaves a critical gap in the toxicological characterization of this compound (Cicmanec et al. 1996).

Cancer

Dermal

No data were found regarding the carcinogenicity of the dermal application of DMHP.

TABLE 14-5 Oral Reference Dose for Dimethyl Hydrogen Phosphite

RfD (mg/kg-d)	Critical effect	Species	Effect level (mg/kg-d)	Uncertainty factors	Reference
0.12	Increases in hyperplasia in the lung (alveolar epithelium hyperplasia, adenomatous hyperplasia)	Female rats	NOAEL: 35.7	$UF_A = 10$ $UF_H = 10$ $UF_D = 3$ Total = 300	NTP1985

NOAEL, no-observed-adverse-effect level; RfD, reference dose; UF_A, extrapolation from animals to humans; UF_H, intraspecies variability; UF_D, inadequate or deficient toxicity database.

Inhalation

No data were found regarding the carcinogenicity of DMHP following inhalation exposure. For the calculating a hazard index for this route, an inhalation unit risk of $1.54 \times 10^{-6}/\mu g/m^3$ was estimated using Equation 16 in Chapter 3 and the oral cancer potency factor for DMHP (see proceeding Oral section).

Oral

The subcommittee believes that there are adequate oral carcinogenicity data for deriving a cancer potency estimate for DMHP. As previously discussed, DMHP induced alveolar/bronchiolar adenomas or carcinomas when administered by gavage to male F-344/N rats at doses of 200 mg/kg, 5 d/wk for 103 wk. There was also a dose-related and increased incidence of squamous cell carcinomas of the lung and squamous cell papillomas and carcinomas of the forestomach combined in male rats exposed to DMHP, 5 d/wk for 103 wk (NTP 1985). There was also a significantly increased incidence of mononuclear cell leukemia in male rats exposed to DMHP at a dose of 100 mg/kg-d, 5 d/wk, for 103 wk as compared to the vehicle controls (NTP 1985). An increased incidence of alveolar/bronchiolar carcinomas occurred in female F-344/N rats exposed to DMHP at dose levels of 50 or 100 mg/kg-d for 5 d/wk, for 103 wk, as compared with vehicle controls (NTP 1985; Dunnick et al. 1986). The female rats displayed a significant ($p < 0.05$) positive trend for alveolar-bronchiolar carcinomas, but the high-dose effect was not found to be statistically significant when compared to controls (Dunnick et al. 1986). No evidence was observed for the carcinogenicity of DMHP in $B6C3F_1$ mice (NTP 1985, Dunnick, et al. 1986). NTP concluded that these studies provide clear evidence of carcinogenicity for DMHP in male F-344/N rats, equivocal evidence of carcinogenicity in female F-344/N rats, and no evidence of carcinogenicity in male or female $B6C3F_1$ mice (NTP 1985, Dunnick, et al. 1986).

In its evaluation of the carcinogenicity of DMHP, IARC (1990) concluded that there is limited evidence for the carcinogenicity of DMHP in experimental animals and DMHP is not classifiable as to its carcinogenicity to humans and assigned DMHP a Group 3 rating. The subcommittee concluded that available data suggests that DMHP might be carcinogenic.

Cancer slope factors (SF) were derived for DMHP using lung and forestomach tumor data for male and female F-344/N rats. SF calculations were performed using a computerized program of the Global 86 linearized multistage model and several other benchmark dose models. These programs provided LED_{10} values for derivation of cancer SFs. Dose levels used in the calculations were adjusted for discontinuous exposure and scaled to $bw^{3/4}$ to estimate human equivalent doses. The estimated oral SFs for lung and forestomach tumors is summarized in Table 14-6. The subcommittee believes that the use of cancer SF for male lung tumors is appropriate because of the high incidence of these tumors and because the NTP classification for DMHP (clear evidence of carcinogenicity in male rats) was based on these data. Therefore, the subcommittee used the oral cancer SF of 5.4×10^{-3}/mg/kg-d for calculating cancer risk estimates for DMHP

EXPOSURE ASSESSMENT AND RISK CHARACTERIZATION

Noncancer

Dermal Exposure

Dermal exposure to DMHP was estimated using the dermal exposure scenario described in Chapter 3. This exposure scenario assumes that an adult spends 1/4th of his or her time sitting on furniture upholstery backcoated with DMHP and also assumes 1/4th of the upper torso are is in contact with the upholstery and clothing presents no barrier. Exposure to other chemicals present in the backcoating was not included in this assessment.

First Iteration

As a first estimate of exposure, it was assumed that skin, clothing, and the upholstery did not impede dermal exposure to DMHP present in the backcoating. It was also assumed that there would be sufficient water present from sweat to facilitate dissolution of DMHP from the backcoating and absorption through the skin. In this scenario, only the dissolution rate of DMHP from the backcoating is assumed to be the limiting factor in absorption by the body. It is assumed that all of the DMHP that dissolves is immediately absorbed into the body by the sitting person.

Dermal exposure was estimated using Equation 1 in Chapter 3. For this calculation, the subcommittee estimated an upholstery application rate (S_a) for DMHP of 7.5 mg/cm^2. The extraction rate (μ_w) for DMHP was estimated to be 0.038 based on extraction data for organic phosphates in polyester fiber (McIntyre et al. 1995). The release rate from the fiber for estimating extraction was 0.06/d at 28°C calculated using the equation $2d/2\pi R$ (d = film thickness, R = fiber radius) with a correction from fiber to film of a factor of 0.63.

TABLE 14-6 Cancer Risk Values for Dimethyl Hydrogen Phosphite Determined from Different Curve-fitting Models (Data from NTP 1985)

Model[a]	MLE or BMD (mg/kg-d)	LED_{10}[b] (mg/kg-d)	Slope Factors ($0.1/LED_{10}$) (per mg/kg-d)	Comment
DATA SET: F-344/N male rats, alveolar/bronchiolar adenoma/carcinoma; body weight$^{3/4}$ utilized for calculation of the human equivalent dose				
Global 86 (linearized multistage)	21.51	18.51	5.4×10^{-3} (2.5×10^{-3})[c]	The p-values for the Monte Carlo test were ≥0.05
EPA's BMD: Gamma	25.27	22.68	4.4×10^{-3}	p-value for χ^2 test was 1.00; graphical representation showed a good fit for the data
EPA's BMD: Logistic	27.58	22.73	4.4×10^{-3}	p-value for χ^2 test was 0.88; graphical representation showed a good fit for the data
EPA's BMD: Probit	26.33	21.75	4.6×10^{-3}	p-value for χ^2 test was 0.97; graphical representation showed a good fit for the data
EPA's BMD: Weibull	26.41	21.27	4.7×10^{-3}	p-value for χ^2 test was 0.98; graphical representation showed a good fit for the data
DATA SET: F-344/N female rats, alveolar/bronchiolar adenoma/carcinoma; body weight$^{3/4}$ utilized for calculation of the human equivalent dose; tumor incidence showed a statistically significant dose-related trend, but not a statistically significant pair-wise comparison of incidence at the high-dose with that of control.				
GLOBAL 86 (linearized multistage)	25.54	16.92	5.9×10^{-3} (5.7×10^{-3})[c]	p-value for the Monte Carlo test was 1.00
EPA's BMD: Gamma	26.18	16.87	5.9×10^{-3}	p-value for χ^2 test was 1.00; graphical representation showed a good fit for the data
EPA's BMD: Log-logistic	26.11	16.97	5.9×10^{-3}	p-value for χ^2 test was 1.00; graphical representation showed a good fit for the data; logistic model run as log-logistic with slope parameter restricted to ≥1

EPA's BMD: Log-probit	23.86	16.63	6.0×10^{-3}	p-value for χ^2 test was 0.95; graphical representation showed a good fit for the data; probit model run as log-probit with slope parameter restricted to ≥ 1
EPA's BMD: Quantal quadratic	24.26	16.90	5.9×10^{-3}	p-value for χ^2 test was 0.99; graphical representation showed a good fit for the data
DATA SET: F-344/N male rats, forestomach squamous cell papilloma/carcinoma; body weight$^{3/4}$ utilized for calculation of the human equivalent dose				
GLOBAL 86 (linearized multistage)	35.46	27.02	3.7×10^{-3} (3.8×10^{-3})[c]	p-value for the Monte Carlo test was 1.00
EPA's BMD: Gamma	35.17	27.97	3.6×10^{-3}	p-value for χ^2 test was 0.00; graphical representation showed a good fit for the data
EPA's BMD: Log-logistic	35.26	17.93	3.6×10^{-3}	p-value for χ^2 test was 1.00; graphical representation showed a good fit for the data; the logistic model run was as log-logistic with slope parameter restricted to ≥ 1
EPA's BMD: Log-probit	34.94	26.51	3.8×10^{-3}	p-value for χ^2 test was 1.00; graphical representation showed a good fit for the data; probit model run as log-probit with slope parameter restricted to ≥ 1
EPA's BMD: Quantal quadratic	35.91	27.09	3.7×10^{-3}	p-value for χ^2 test was 0.95; graphical representation showed a good fit for the data
EPA's BMD: Weibull	35.34	26.74	3.7×10^{-3}	p-value for χ^2 test was 1.00 graphical representation showed a good fit for the data

BMD, benchmark dose; LED_{10}, lower 95% bound on the effective dose corresponding to a 10% tumor response in test animals; MLE, maximum likelihood estimate

[a]All other models were rejected because p-values for χ^2 tests were <0.05, graphical representation did not show a good fit for the data, and/or the benchmark dose computation failed.

[b]LED_{10} is the same as the BMDL (lower confidence limit on the BMD) in EPA's BMD software program.

[c]The number in parenthesis is the q_1* derived from GLOBAL 86.

Using these assumptions, an estimated absorbed daily dose of 2.2 mg/kg was calculated for DMHP. A hazard index of 18.3 was calculated for this first iteration by dividing the estimated daily dermal dose of 2.2 mg/kg-d by the oral RfD for DMHP of 0.12 mg/kg-d. At this time, the oral RfD is used as the best estimate of the internal dose associated with dermal exposure to DMHP. These results suggest that DMHP could pose a toxic risk from dermal exposure.

Alternative Iteration

The estimated dermal daily dose for DMHP is also calculated using an estimate of the dermal penetration rate for DMHP (Chapter 3: Equations 2 and 3). Instead of assuming that all dissolved DMHP immediately penetrates the skin and enters systemic circulation, it is assumed that the skin slows the absorption of DMHP to a specific amount of chemical absorbed/unit of time. This estimate can be measured experimentally and is referred to as the skin permeability coefficient K_p. However, the dermal penetration constant for DMHP has not been measured experimentally. However, K_p can be estimated from a correlation between the octanol-water partition coefficient (K_{ow}) and molecular weight (mass/unit amount of substance) using Equation 2 in Chapter 3 yielding an alternate K_p of 1.46×10^{-3} cm/d.

In the absence of a dermal RfD, the subcommittee believes it is appropriate to use the oral RfD for DMHP of 0.12 mg/kg-d as the best estimate of the internal dose from dermal exposure

Using Equation 3 in Chapter 3 and the alternate K_p, the dermal daily dose rate for DMHP was estimated to be 11.4 mg/kg-d. A hazard index of 95 was calculated by dividing the estimated daily dermal dose of 11.4 mg/kg-d by the oral RfD for DMHP of 0.12 mg/kg-d. These results suggest that under the given exposure conditions, dermal exposure to DMHP could pose a noncancer toxic risk to humans and should be investigated further.

Inhalation Exposure

Particles

Inhalation exposure estimates for DMHP were calculated using the exposure scenario described in Chapter 3. This scenario assumes that a person spends 1/4th of his or her life in a 30 m^3 room containing 30 m^2 of DMHP-treated fabric and the room is assumed to have a air-change rate of 0.25/hr. It is also assumed that 50% of the DMHP present in 25% of the surface area of the treated fabric is released over 15 yr and 1%, of released particles are of size that can be inhaled.

Particle exposure was estimated using Equations 4 and 5 in Chapter 3. The subcommittee estimated an upholstery application rate (S_a) for DMHP of 7.5 mg/cm^2. The release rate (μ_r) for DMHP from upholstery fabric was estimated to be 2.3×10^{-7}/d (see Chapter 3, Equation 5) yielding a room airborne particle concentration (C_p) of 2.9 µg/m^3 and a short time-averaged exposure concentration of 0.725 µg/m^3. The time-averaged exposure concentration for particles was calculated using Equation 6 in Chapter 3.

In the absence of relevant inhalation exposure data, the subcommittee chose to estimate inhalation RfCs from oral RfDs. The subcommittee, however, recognizes that it is not an ideal approach and also recognizes that the estimated RfC levels might be considerably different than actual levels (if inhalation data were available). Extrapolating from one route of exposure (oral) to another (inhalation) requires specific knowledge about the uptake kinetics into the body by each exposure route, including potential binding to cellular sites. The subcommittee believes that its extrapolation of oral RfDs to inhalation RfCs is highly conservative; it assumes that all of the inhaled compound is deposited in the respiratory tract and completely absorbed into the blood. The NRC committee on Toxicology (NRC, 1985) has used this approach when inhalation exposure data were insufficient to derive inhalation exposure levels. The subcommittee believes that such an approach is justified for conservatively estimating the toxicological risk from exposure to DMHP. That RfC should be used as interim or provisional level until relevant data becomes available for the derivation of inhalation RfC.

In order to calculate a hazard index for the inhalation route, a provisional inhalation RfC of 0.42 mg/m^3 was derived using the oral RfD for DMHP and Equation 7 in Chapter 3.

Division of the time-average exposure concentration of 0.725 µg/m^3 by the provisional RfC for DMHP of 0.42 mg/m^3 yields a hazard index of 1.73×10^{-3} indicating that inhalation of DMHP-containing particulate from treated upholstery is not likely to pose a noncancer toxic risk to humans based on worst-case estimates in the given exposure scenario.

Vapors

In addition to the possibility of release of DMHP in particles from worn upholstery fabric, the subcommittee considered the possibility of the release of DMHP by evaporation. This approach is described in Chapter 3, and uses an exposure scenario similar to that just described for exposure to DMHP particles.

The rate of flow of DMHP vapor from the room is calculated using Equations 8-11 in Chapter 3. A saturated vapor concentration (C_v) of 26,800 mg/m^3

was estimated for DMHP. The application density (S_a) for DMHP in the treated upholstery was estimated as 7.5 mg/cm^2.

Using the parameters described, the equilibrium room-air concentration of DMHP was estimated to be 22,600 mg/m^3. The short-term time-average exposure concentration for DMHP was estimated as 5,650 mg/m^3 using Equation 12 in Chapter 3 and the equilibrium room-air concentration for DMHP. It was estimated that concentration could be maintained for approximately 10 hr. These results clearly show that the model for this scenario is substantially incorrect for DMHP if it is a useful FR, since any such material would have to be sufficiently well bound to the fabric to stay in place for years. However, the subcommittee has no further information on plausible rates of evaporation of DMHP from treated fabrics, and these calculations suggest that further information is required.

Oral Exposure

The assessment of noncancer toxicological risk for oral exposure to DMHP is based on the oral exposure scenario described in Chapter 3. This scenario assumes a child is exposed to DMHP by sucking on 50 cm^2 of fabric back-coated with DMHP, 1 hr/d for two yr. The subcommittee estimated an upholstery application rate (S_a) for DMHP of 7.5 mg/cm^2. Oral exposure was calculated using Equation 15 in Chapter 3. The extraction rate (μ_w) for DMHP was estimated to be 0.038 based on extraction data for organic phosphates in polyester fiber (McIntyre et al. 1995). The release rate from the fiber for estimating extraction was 0.06/d at 28°C calculated using the equation $2d/2\pi R$ (d = film thickness, R = fiber radius) with a correction from fiber to film of a factor of 0.63.

The worst case average oral daily dose for DMHP was estimated as 0.059 mg/kg-d. Division of the dose estimate by the oral RfD for DMHP of 0.12 mg/kg-d gives a hazard index of 0.49. This suggests that under the subcommittee's worst-case exposure assumptions, DMHP is not anticipated to be a noncancer toxic risk to children when incorporated into furniture upholstery at the given concentration level.

Cancer

Dermal

Human cancer risk from dermal exposure to DMHP was calculated by multiplying the oral cancer potency factor for DMHP by the most conservative lifetime average dermal dose rate of 11.4 mg/kg-d. The subcommittee believes that the use of the oral cancer potency factor for DMHP was acceptable for the calculation of cancer risk for dermal exposure since the oral cancer potency factor is based on carcinogenic effects following near-complete systemic absorption and the appearance of tumors not at the site of DMHP application.

Using the dose rate obtained in the alternate iteration, a lifetime average daily dose of 11.4 mg/kg-d was estimated for DMHP. Multiplication of 11.4 mg/kg-d times the cancer potency estimate of 5.4×10^{-3}/mg/kg-d, the lifetime risk estimate is 6.1×10^{-2}. This estimate suggests that the dermal route of exposure may pose a carcinogenic hazard for persons exposed to DMHP incorporated into residential furniture upholstery at the indicated concentration levels and under the given worst-case exposure scenario. Further evaluation of the cancer risk associated with dermal exposure to DMHP should be conducted.

Inhalation

Particles

The average room-air concentration and average exposure concentration to DMHP particles estimated in the previous sections were used for the cancer assessment. An inhalation cancer potency value was not available for DMHP, therefore a provisional inhalation cancer potency value of $1.54 \times 10^{-6}/\mu g/m^3$ was derived from oral cancer potency data for DMHP. Multiplication of the exposure estimates of 0.725 $\mu g/m^3$ for particulate times the provisional cancer potency value yields an estimated lifetime cancer risk of 1.1×10^{-6} and suggests that the cancer risk associated with the inhalation of DMHP particulates is negligible at the given upholstery concentrations and the exposure parameters in the worst-case exposure scenario. However, the subcommittee concluded that exposure to DMHP by this route needs further evaluation.

Vapors

For DMHP vapors, the equilibrium concentration of vapor-phase DMHP in room air was estimated as described in the Noncancer Inhalation Exposure section. The long-term time-average vapor exposure concentration for DMHP was estimated using Equation 14 in Chapter 3.

Using the estimated inhalation unit risk for DMHP of $1.54 \times 10^{-6}/\mu g/m^3$, the upper bound on lifetime cancer risk for inhalation exposure to DMHP in the vapor phase is 6.6×10^{-4}. This risk estimate indicates that further investigation of cancer risks associated with DMHP vapors should be considered.

Oral

As discussed previously, DMHP is judged by the subcommittee to be a rodent carcinogen. Therefore, the conservative approach for risk assessment purposes is to assume that DMHP represents a carcinogenic risk to humans.

Using Equation 16 in Chapter 3, the lifetime average dose rate for DMHP by the oral exposure route was calculated to be 1.7×10^{-3} mg/kg-d. Lifetime cancer risk for this exposure scenario was then estimated by multiplying the oral lifetime daily dose rate times the oral cancer potency factor for DMHP of 5.4×10^{-3}/mg/kg-d yielding a cancer risk estimate of 9.1×10^{-6}. This suggests that under the subcommittee's worst-case exposure assumptions, DMHP could be a carcinogenic hazard by the oral route of exposure.

RECOMMENDATIONS FROM OTHER ORGANIZATIONS

There is a single report documenting a workplace maximum air concentration for DMHP of 0.5 mg/m^3, established by the Ukrainian Ministry of Health (Kuz'minov et al. 1992).

In 1990, IARC reported that no regulatory standards or guidelines had been established for this compound. Further, DMHP is not listed on IRIS or HEAST, in the online version of the NIOSH Pocket Guide to Chemical Hazards, or in the online version of listings available from the American Conference of Governmental Industrial Hygienists. It has not been addressed in publications of the CRAVE Work Group or the EPA Office of Pesticide Programs.

DATA GAPS AND RESEARCH NEEDS

There are no data on the subchronic or chronic toxicity of DMHP by the dermal or inhalation routes of exposure. No information is available on human exposure to DMHP from treated furniture upholstery. No studies have been conducted on the leaching of DMHP from treated materials.

The hazard indices of greater than one were calculated for DMHP for the dermal route of exposure. Cancer risk estimates were greater than 1×10^{-6} for the dermal, inhalation, and oral routes of exposure. Therefore, the subcommittee concluded that future research for TDCPP should focus on determining the actual amounts of DMHP leached from treated furniture and the dermal penetration of these compounds through human skin.

An oral RfD for DMHP (the representative compound of the organic phosphonate and cyclic phosphonate ester flame retardants) is available based on a

2-yr chronic gavage study. An inhalation RfC was calculated by the subcommittee based on a 4-wk study. Cancer potency slope factors were available for oral and inhalation. Because DMHP is soluble in water, there is concern about noncancer effects after dermal absorption and concern about cancer risk by all three routes of exposure. The subcommittee recommends that the potential for release of vapor and particles into air and DMHP release into saline from treated fabric be investigated. Because of a dermal hazard index of greater than 1, the subcommittee also recommends that the dermal absorption of DMHP from treated fabric be investigated.

REFERENCES

Bel'skii, V.E., et al. 1969. Izv. Akad. Nauk. SSSR Ser. Khim. 6:1297-1300.

Bio/dynamics (Bio/dynamics, Inc.). 1980. A Four Week Inhalation Toxicity Study of MCTR-242-79 in the Rat. Bio/Dynamics Inc., East Millstone, NJ. EPA OTS Doc. #8EHQ-0182-0366.

Bowman, J.T. 1980. *Drosophila* Mutagenicity Assays of Mobil Chemical Company Compound MCTR-111-79. MRI Report No. 344. EG&G Mason Research Institute, Rockville, MD. EPA OTS Doc #88-8100187.

Brusick, D.J. 1977. Mutagenicity Evaluation of MCTR-214-77, Final Report. LBI Project No. 2683. Litton Bionetics, Kensington, MD. EPA/OTS Doc. #88-8100187.

Chapin, R.E., S.L. Dutton, M.D. Ross, B.M. Sumrell, and J.C. Lamb, IV. 1984. Development of reproductive tract lesions in male F344 rats after treatment with dimethyl methylphosphonate. Exper. Molec. Pathol. 41:126-140.

Cicmanec, J.L., M.L. Dourson, and R.C. Hertzberg. 1995. Noncancer risk assessment: Present and emerging issues. Pp. 293-310 in Toxicology and Risk Assessment: Principles, Methods and Applications. A.M. Fan, and L.E. Chang, eds. New York: Marcel Dekker.

Dunnick, J.K., B.N. Gupta, M.W. Harris, and J.C. Lamb, IV. 1984a. Reproductive toxicity of dimethyl methyl phosphonate (DMMP) in the male Fischer 344 rat. Toxicol. Appl. Pharmacol. 72(3):379-387.

Dunnick, J.K., H.A. Solleveld, M.W. Harris, R. Chapin, and J.C. Lamb, IV. 1984b. Dimethyl methyl phosphonate induction of dominant lethal mutations in male mice. Mutat. Res. 138(2-3):213-218.

Dunnick, J.K., G.A. Boorman, J.K. Haseman, J. Langloss, R.H. Cardy, and A.G. Manus. 1986. Lung neoplasms in rodents after chronic administration of dimethyl hydrogen phosphite. Cancer Res. 46(1):264-270.

EPA (U.S. Environmental Protection Agency). 1985. Chemical Hazard Information Profile: Dimethyl Hydrogen Phosphite. TSCA Interagency Testing Committee, Washington, DC.

EPA (U.S. Environmental Protection Agency). 1986. Guidelines for Carcinogen Risk Assessment. Fed. Regist. 51(185):33992-34003.

EPA (U.S. Environmental Protection Agency). 1999. Guidelines for Carcinogen Risk Assessment. Review Draft. NCEA-F-0644, Risk Assessment Forum, EPA, Washington, DC (July1999). [Online] Available: http://www.epa.gov/nceawww1/raf/crasab.htm

Fritz, H. 1978. Reproduction Study-FAT 80'021/B, Rat, Seg. II (Test for Teratogenic or Embryotoxic Effects). Ciba-Geigy Ltd., Basle, Switzerland. EPA/OTS Doc. #FYI-OTS-0784-0242-SU.

Hardin, B.D., R.L. Schuler, J.R. Burg, G.M. Booth, K.P. Hazelden, K.M. McKenzie, V.J. Piccirillo, and K.N. Smith. 1987. Evaluation of 60 chemicals in a preliminary developmental toxicity test. Teratog. Carcinog. Mutagen. 7(1):29-48.

Hatlelid, K. 1999. Memorandum from Kristina Hatlelid, Toxicologist, Division of Health Sciences, to Ronald L. Medford, Assistant Executive Director for Hazard Identification and Reduction, Subject: Toxicity Review for Organic Phosphonates and Cyclic Phosphonate Esters, March 25, 1999. U.S. Consumer Product Safety Commission, Washington, DC.

Hawley, G.G. 1981. Hawley's Condensed Chemical Dictionary, 10th ed. New York: Van Nostrand Reinhold.

Haworth, S.R. 1979a. *Salmonella*/mammalian-microsome Pre-incubation Mutagenicity Assay. Study No. 009-639-341-2. EG&G Mason Research Institute, Rockville, MD. EPA/OTS Doc #88-8100187.

Haworth, S.R. 1979b. Bacteria DNA Damage/Repair Suspension Assay. Study No. 009-639-342-6. EG&G Mason Research Institute, Rockville, MD. EPA/OTS Doc #88-8100187.

Howard, P.H., and W.M. Meylan, eds. 1997. Handbook of Physical Properties of Organic Chemicals. Bocat Raton, FL: Lewis Publishers.

HSDB (Hazardous Substance Data Bank). 1999. Hazardous Substance Data Bank. [Online]. Avaliable: http://sis.nlm.nih.gov/cgi-bin/sis/htmlgen?HSDB National Library of Medicine (18 Aug 1999).

IARC (International Agency for Research on Cancer). 1990. IARC Monographs on the Evaluation of Carcinogenic Risks to Humans, Some Flame Retardants and Textile Chemicals, and Exposures in the Textile Manufacturing Industry, Vol. 48. Lyon, France: World Health Organization.

IPCS (International Programme on Chemical Safety). 1997. Pp. 13; 85 in Environmental Health Criteria 102: Flame Retardants: A General Introduction. Geneva: World Health Organization.

Keller, J.G. 1961. 3 Studies on Phosphoric Acid-dimethyl Ester. Hazleton Labs, Falls Church, VA. EPA/OTS/ Doc #88-920001146.

Kirby, P.E. 1979. Evaluation of Test Article MCTR-110-79 (MRI #343) for Mutagenic Potential Employing the L5178y TK +/- Mutagenesis Assay. EG&G Mason Research Institute, Rockville, MD. EPA/OTS Doc #88-8100187.

Kuz'minov, B.P., V.R. Kokot, T.I. Sharova, S.S.H. Zhuk, and M.M. Vus. 1992. Experimental data on the hygienic regulation of dimethy phosphite in the air of a working area. [Abstract]. Gig. Tr. Prof. Zabol. 4:22-23.

Lewis, M. 1975. In Situ Soil Fumigation for Combating Nematodes Using Aminomethylphosphonates. U.S. Patent No. 3,865,936, Feb. 11.

Lewis, R.J., Sr. 1993. Pp. 419 in Hawley's Condensed Chemical Dictionary, 12th Ed. New York: Van Nostrand Reinhold.

Lide, D.R. 1991-1992. CRC Handbook of Chemistry and Physics, 72nd Ed. Boca Raton, FL: CRC Press.

McGregor, D.B., A. Brown, P. Cattanach, I. Edwards, D. McBride, and W.J. Caspary. 1988. Responses of the L5178Y tk^+/tk^- mouse lymphoma cell forward mutation assay. II: 18 coded chemicals. Environ. Mol. Mutagen. 11(1):91-118.

McIntyre, J.E., I. Holme, and O.K. Sunmonu. 1995. The desorption of model compounds from poly(ethylene terephthalate) fibre. Colourage 41(13)77-81.

Nomeir, A.A., and L.C. Uraih. 1988. Pathological and biochemical effects of dimethyl hydrogen phosphite in Fischer 344 rats. Fundam. Appl. Toxicol. 10(1):114-124.

Nomeir, A.A., and H.B. Matthews. 1997. Metabolism and disposition of dimethyl hydrogen phosphite in rats and mice. J. Toxicol. Environ. Health 51(5):489-501.

NRC (National Research Council). 1985. Emergency and Continuous Exposure Limits for Selected Airborne Contaminants, Volume 5. Committee on Toxicology. Board on Toxicology and Environmental Health Hazards, National Research Council. Washington, DC: National Academy Press.

NTP (National Toxicology Program). 1985. Toxicology and Carcinogenesis Studies of dimethyl Hydrogen Phosphite (CAS No. 868-85-9) in F344/N Rats and $B6C3F_1$ Mice (Gavage Studies). Technical Report Series No. 287. National Toxicology Program, U.S. Department of Health and Human Services, Public Health Service, National Institutes of Health, Research Triangle Park, NC. NIH Publication No. 86-2543.

Rusch, G.M. 1980. A Five-day Inhalation Toxicity Study of MCTR-174-79 in the Rat. Project No. 79-7344. Bio/Dynamics Inc., East Millstone, NJ. EPA/OTS/ Doc. #88-8100187.

Sax, N.I. 1984. Dangerous Properties of Industrial Materials, 6th ed. New York: Van Nostrand Reinhold.

Shelby, M.D., G.L. Erexson, G.J. Hook, and R.R. Tice. 1993. Evaluation of a three-exposure mouse bone marrow micronucleus protocol: Results with 49 chemicals. Environ. Mol. Mutagen. 21(2):160-179.

Siemer, S. 1980. Phosphorus Compounds as Sugarcane Ripeners. U.S. Patent No. 4,229,203. Oct. 21.

Tennant, R.W., B.H. Margolin, M.D. Shelby, E. Zeiger, J.K. Haseman, J. Spalding, W. Caspary, M. Resnick, S. Stasiewicz, B. Anderson, and R. Minor. 1987. Prediction of chemical carcinogenicity in rodents from in vitro genetic toxicity assays. Science 236(4804):933-941.

Vilceanu, R., and P. Schulz. 1972. Rev. Roum. Chim. 17:361-366.

Woodruff, R.C., J.M. Mason, R. Valencia, and S. Zimmering. 1985. Chemical mutagenesis testing in *Drosophila*. V. Results of 53 coded compounds tested for the National Toxicology Program. Environ. Mutagen. 7(5):677-702.

15

Tris Monochloropropyl Phosphates

THIS chapter reviews the physical and chemical properties, toxicokinetics, toxicological, epidemiological, and exposure data on tris monochloropropyl phosphates (TMCPPs). The subcommittee used that information to characterize the health risk from exposure to TMCPPs. The subcommittee also identified data gaps and recommended research relevant for determining the health risk from exposure to TMCPPs.

PHYSICAL AND CHEMICAL PROPERTIES

Tris monochloropropyl phosphates (TMCPPs) are four isomers having the molecular formula $C_9H_{18}Cl_3O_4P$ (MW 327.55) and belonging to the class of chlorinated alkyl phosphate esters. The most abundant isomer in commercial products is the completely branched isomer, tris (1-chloro-2-isopropyl) phosphate and the least abundant form is the completely linear isomer, tris (2-chloropropyl) phosphate. The chemical and physical properties of TMCPPs are summarized in Table 15-1. Variations in manufacturing methods result in commercial formulations that contain different proportions of the four TMCPPs listed in Table 15-1. Mixtures in which the linear forms are above trace levels tend to be pale yellow; other mixtures are colorless (Albright and Wilson 1980a; Courtaulds Chemicals 1988).

Although tris(1-chloro-2-isopropyl) phosphate (TCIP, CAS RN 13674-84-5) is the most abundant isomer, companies have tended to refer to their product

TABLE 15-1 Physical and Chemical Properties of Tris Monochloropropyl Phosphates

Property	Value	Reference
Chemical formula	$C_9H_{18}Cl_3O_4P$	RTECS 1999
Structures	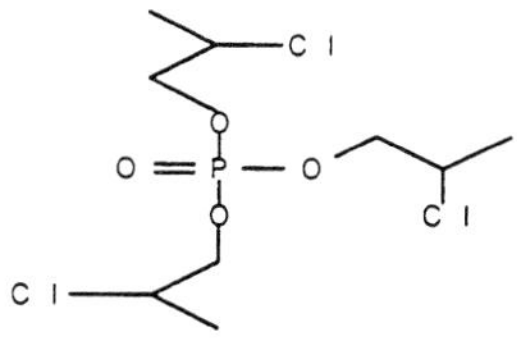 13674-84-5 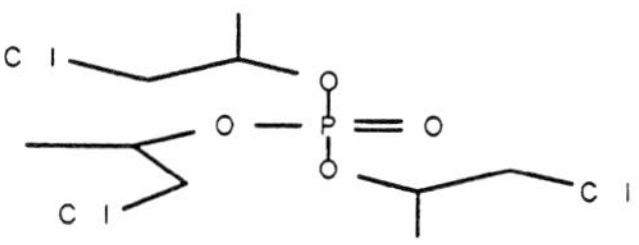 6145-73-9	
CAS Registry #'s	13674-84-5 [tris(1-chloro-2-isopropyl phosphate] 76025-080-6 [bis(1-chloro-2-isopropyl) (2-chloropropyl) phosphate] 76649-15-5 [bis(2-chloropropyl) (1-chloro-2-isopropyl) phosphate] 6145-73-9 [tris(2-chloropropyl) phosphate]	RTECS 1999
Synonyms	13674-84-5: tris(2-chloroisopropyl) phosphate; 2-propanol, 1-chlorophosphate (3:1); 1-chlor-2-propyl phosphate (3:1); tris(beta-chloropropyl) phosphate; phosphoric acid, tris(2-chloro-1-methylethyl) ester; Fyrol PCF; Amgard TMCP; Antiblaze 80; TCPP; TMCP; TCIP 76025-08-6: bis(1-chloro-2-isopropyl) (2-chloropropyl) phosphate; bis(2-chloro-1-methylethyl) 2-chloropropyl phosphate 76649-15-5: bis (2-chloropropyl) (1-chloro-2-isopropyl) phosphate; bis (2-chloropropyl) (1-chloro-2-propyl) phosphate; 2-chloro-1-methylethyl; bis (2-chloropropyl) phosphate	TSCA Interagency 1988; IPCS 1998; RTECS 1999

(Continued)

	6145-73-9: tris (2-chloropropyl) phosphate; 1-propanol, 2-chloro-phosphate (3:1); tris (beta-chloropropyl) phosphate, TCPP	
Molecular weight	327.59	RTECS 1999
Physical State	Liquid; sold commercially as mixture of isomers	Albright and Wilson 1980a; Bayer 1993, 1996
Solubility	In water, 1.1-1.2 g/L at 25°C In DMSO, 95% ethanol, acetone, ≥100 g/L	Albright and Wilson 1980a; Courtaulds Chemicals 1988; NTP 1999
Melting point	380°C	
Vapor Pressure	<2 mm Hg at 25°C	Courtaulds chemicals 1988
Specific Gravity	1.29 at 25°C	Akzo Nobel 1999
Boiling Point:	235-248°C	Courtaulds chemicals 1988; IPCS 1998
Flash Point:	218-220°C	Courtaulds chemicals 1988; IPCS 1998
Conversion Factor:	mg/m^3 = 13.39 ppm	IPCS 1998
Thermal Decomposition	Begins at ~70°C for the yellow product; at ~190°C for the colorless product	Albright and Wilson 1980a; Courtaulds chemicals 1988
Reactivity	Hydrolyzes slowly under alkaline or acidic conditions	Courtaulds chemicals 1988

by the name tris(2-chloropropyl) phosphate (TCPP), even though that name refers to CAS RN 6415-73-9 (e.g., Albright and Wilson 1999). This has led to a considerable degree of uncertainty in the literature and toxicity databases as to the identity of the substance that has undergone toxicity testing (see Ferrante 1999; Saltzman and Babich 1999). All TMCPP toxicity testing has been carried out on commercial mixtures containing variable amounts of TMCPP isomers, which in addition to those previously mentioned, include bis(1-chloro-2-isopropyl) (2-chloropropyl) phosphate (CAS RN 76025-08-6) and bis(2-chloropropyl) (1-chloro-2-isopropyl) phosphate (CAS RN 76649-15-5). Therefore, the designation TMCPPs in this document refers to the commercial mixtures containing variable amounts of TMCPP isomers.

OCCURRENCE AND USE

TCMPPs are not known to occur naturally but are manufactured from propylene oxide and phosphorus oxychloride (IPCS 1998). Although TMCPPs may be used for textile (non-apparel) finishes, they are also used as flame retardants in rigid and flexible polyurethane foams (Akzo Nobel 1999; Albright and Wilson 1999). According to the IPCS (1998), the annual worldwide demand for TMCPPs exceeded 40,000 metric tons in 1997. TMCPP mixtures are sold under various trade names, including Antiblaze® 80, Amgard® TMCP, Fyrol® PCF, and Hostaflam® OP 820.

TOXICOKINETICS

Absorption

Minegishi et al. (1988) administered radiolabeled ^{14}C-tris (1-chloro-2-propyl) phosphate (^{14}C-TMCPP, presumably CASRN 13674-84-5, 99% pure by gas chromatography), in a single oral dose of 16.4 mg/kg (0.05 mmol/kg; 0.213 mCi/mmol) in olive oil, to groups of three to five male Wistar rats. Absorption was calculated from the radioisotopic measurements of cumulative urinary excretion and cumulative exhalation (performed for 3 or 4 d post-administration), and the amount in blood and tissues of rats sacrificed at 3, 6, 12, 24, 72, and 168 hr following administration.

Absorption of radiolabeled TMCPP was rapid; radiolabel was detected throughout the body as early as 3 hr post-administration. At 168 hr, 75.6% of the administered oral dose had been excreted in the urine and expired air, or remained in the carcass. Another 22% was excreted in the feces; however, TMCPP undergoes enterohepatic circulation (see Excretion section), and it is not clear if this proportion includes unabsorbed TMCPP, TMCPP excreted in the bile, or both. Therefore, at least 75.6% is absorbed following oral administration to rats.

Distribution

Three hr after oral administration of a 16.4 mg/kg dose of ^{14}C-TMCPP in rats, the concentration of radiolabel per gram of tissue was highest in the liver (28.6 nmol/g) and kidney (27.3 nmol/g), and to a lesser extent, in the lung (9.4 nmol/g), but did not accumulate at high levels in other tissues (Minegishi et al. 1988). The concentrations of the radiolabel in blood, heart, spleen, brain, testis, adipose tissue, and muscle were all <4 nmol/g tissue. The radiolabel concentration remained highest in the liver during subsequent timepoints up to d 7. Radiolabel reached its maximum concentration in various tissues 3-6 hr after administration.

Metabolism

The metabolic transformation of TMCPPs has not been investigated.

Excretion

Approximately 97% of radiolabeled TMCPP (67% in urine, 22% in feces, and 7.7% in expired air) was excreted within 7 d of administration in rats, as reported by Minegeshi et al. (1988). Only 0.7% of the administered dose was recovered in the carcass after 168 hr, and approximately 2.5% was not recovered. Removal of TMCPP from the tissues followed a biphasic pattern. Excretion was fairly rapid from all compartments ($t_{1/2}$ = 5.2-13.5) for the first 24 hr. During the subsequent 6 d, tissue half-lives ranged from 45 hr in the liver to 103.4 hr in adipose tissue.

Minegishi et al. (1988) also measured biliary excretion of ^{14}C-TMCPP in cannulated rats for 48 hr following a single oral dose (as above). Biliary radioactivity peaked within 2 hr of administration, and after 48 hr, biliary excretion represented approximately 45% of the administered dose. The ratio of biliary excretion to fecal excretion was calculated to be 2.23, suggesting to the authors that TMCPP is subject to enterohepatic circulation.

HAZARD IDENTIFICATION[1]

No studies were found that investigated the toxicity of TMCPPs in humans. Animal studies were either acute or subacute in duration. Studies were not done according to GLP guidelines unless noted in the text.

Dermal Exposure

Irritation

A product safety data sheet published by Couraulds specifically warns that skin irritation may result from prolonged contact with liquid TMCPP (Courtaulds Chemicals 1988). A data sheet for Fyrol PCF recommends that it should not be used for fabrics intended for apparel uses (Akzo Nobel 1999).

---

[1]In this section, the subcommittee reviewed toxicity data on tris monochloropropyl phosphates, including the toxicity assessments prepared by the U.S. Consumer Product Safety Commission (Bittner 1999; Ferrante 1999; Soltzman and Babich 1999).

Albright and Wilson (1980a) carried out dermal irritation studies on two different production batches of their TMCPP flame retardant Antiblaze® 80. In both tests, 0.5 mL of TMCPP liquid was applied neat to two test sites (abraded and intact) on the back skin of six New Zealand white rabbits (three male and three female). For product A, the exposure sites were wiped after 24 hr and were evaluated after 24 and 72 hr (Albright and Wilson 1980a). Erythema was observed at low levels after 24 hr, was not increased by abrasion of the epidermis, and was reduced after 72 hr. For product B, the test sites were wiped after 24 hr, and scored after 0.5 and 72 hr (Albright and Wilson 1980a). The results of this study were available only in a preliminary report. The test material was assigned in a low irritancy score for intact skin (0.8/8.0) and it was only slightly higher for abraded skin (1.1/8.0); signs of irritation were reduced after 72 hr. The reports concluded that neither compound was a primary skin irritant.

Eye irritation tests were conducted on two production batches of the TMCPP flame retardant Antiblaze® 80 (Albright and Wilson 1980a). In both tests, six New Zealand white rabbits (three male and three female) received 0.1 mL of Antiblaze® 80, neat, in one eye and were evaluated for irritation at 1, 24, 48, and 72 h. Treatment with product A (Albright and Wilson 1980a) resulted in minor, transient irritation and the Draize test score (3/110) was considered to be negative. Similarly, treatment with product B resulted in low average irritation scores (Albright and Wilson 1980a, available only as a summary). The report concluded that product B was not a primary eye irritant.

Systemic Effects

Acute dermal toxicity studies were conducted on two different production batches of the TMCPP flame retardant Antiblaze® 80 (Albright and Wilson 1980a). In both studies, 0.5 mL of neat commercial TMCPP was administered dermally to six New Zealand white rabbits (240 cm^2 average exposure area). Two males and one female received TMCPP on abraded skin, and two females and one male received TMCPP on intact skin. The test areas were wiped after 24 h, and the animals were observed for 14 d. After treatment with product A (Albright and Wilson 1980a, Study 462-80), no mortality or other consistent clinical signs resulting from treatment were observed, other than anorexia in five of six animals on d 1. The exposure site in all animals exhibited some erythema and edema formation at 24 hr, but not in the following 2 wk, after which animals were sacrificed. Scaling of the test site was observed at necropsy in two animals. In all three female rabbits (two intact and one abraded) and one male rabbit (abraded), the lungs exhibited diffuse redness and/or white edges. In the females, the liver appeared to be pale, but no other gross pathological lesions were observed. Treatment with product B yielded similar results

(Albright and Wilson 1980a, Study 2426-80). The dermal median lethal doses for both product batches of Antiblaze® 80 were above 2 g/kg.

Other Systemic Effects

No studies were located that investigated the sensitization, immunological, neurological, reproductive, developmental, or carcinogenic effects of TMCPPs following dermal exposure.

Inhalation Exposure

Systemic Effects

In an acute inhalation study, five male and five female Sprague-Dawley rats were exposed (whole-body) to an aerosol of TMCPPs (Antiblaze® 80, product A) at a nominal concentration of 17,800 mg/m^3 (Albright and Wilson 1980a, Study 465-80) for 1 hr, and then observed for 14 d. The average integrated aerosol concentration was not stated. No animals died during the study period. Treatment effects included: decreased activity, partially closed eyes, swollen eyelids, lacrimation, soft stool, excessive salivation, yellow or orange discoloration of the fur in the ano-genital area, and oily and/or matted fur, a condition that persisted in all rats for 7 d and in 9/10 rats for 10 d. Most rats exhibited dry rales, wet or dry material around the facial area and excessive salivation. In a few animals, dry rales persisted throughout the observation period. Body weight decreased in four out of five rats of both sexes on the first day of treatment; the condition was transient in males, but three out of five females had lost weight by d 14, compared to their initial body weight.

In an acute inhalation study, five male and five female Sprague Dawley rats were exposed to an aerosol of TMCPPs (Antiblaze® 80, product B) at an average integrated aerosol concentration of 5,000 mg/m^3 for 4 hr and observed for 14 d (Albright and Wilson 1980a, Study 2425-80). The nominal concentration of 26,300 mg/m^3, comparable to that used in the product A study, is a less accurate measure of exposure. The mass median diameter of the particles was 4.1 µm, and >87% of the particles had a cumulative mass aerodynamically equivalent diameter of ≤ 10 µm. The animals were observed prior to exposure, at 15-min intervals during exposure and during the subsequent 0.5-hr recovery period, upon removal from the chamber, and hourly for 4 hr. Body weights were recorded on d 0, 1, 2, 4, 7, and 14, prior to sacrifice. Because of the design of the test chamber, no more than five to seven rats could be observed simultaneously during exposure. Treatment effects are listed in Table 15-2. In the subsequent 14-d period, no males, but two of four females died. Necropsy

TABLE 15-2 Inhalation Toxicity Studies on Tris Monochloropropyl Phosphates

Chemical, Purity	Species, Strain, Sex, Number	Concentration (mg/m^3)	Duration	Effects	Comments	References
Nominally IV-tris (2-chloropropyl) phosphate, 100% Probable mixture: I > II > III > IV; likely that I is lower than 75%	Rat, Sprague-Dawley, M/F, 5/sex/dose	17,800, nominal	1 hr (14-d observation)	No deaths; inactivity; swollen eyelids; increased lacrimation, salivation, dry rales; decreased body weight		Albright and Wilson 1980a[a]
Nominally IV- tris (2-chloropropyl) phosphate, lot 060815000; Mixture of I- 75%, II- 16%, III-1%, IV-<0.1%	Rat, Sprague-Dawley, M/F, 5/sex/dose	5,000 average, integrated	4 hr (14-d observation)	Mortality; inactivity, weakness, salivation; one F exhibited prostration, convulsions, dyspnea before death 0.5 hr after end of exposure; lacrimation, lethargy, alopecia, decreased body weight	LC_{50}(F): ~5,000 mg/m^3 LC_{50}(M): >5,000 mg/m^3	Albright and Wilson 1980a[b]

I: tris (1-chloro-2-isopropyl phosphate), CAS Registry # 13674-84-5. II: bis (1-chloro-2-isopropyl) (2-chloropropyl) phosphate, CAS Registry # 76025-08-6. III: bis (2-chloropropyl) (1-chloro-2-isopropyl) phosphate, CAS Registry # 76649-15-5. IV: tris (2-chloropropyl) phosphate, CAS Registry # 6145-73-9.

[a]Laboratory report from Bio/dynamics to Mobil Oil Corporation (1980); TSCA Submittal from Albright & Wilson to EPA (1989).

[b]Laboratory report from Gulf South Research Institute to Mobil Environmental Health Sciences Laboratory (1981); TSCA Submittal from Albright and Wilson to EPA (1989).

revealed no gross abnormalities, except for opacity of both corneas in one female that died during testing. The 4-hr LC_{50} was >5,000 mg/m^3 for males and approximately 5,000 mg/m^3 for females.

Other Systemic Effects

No studies were located that investigated the immunological, neurological, reproductive, developmental, or carcinogenic effects of TMCPPs following inhalation exposure.

Oral Exposure

Systemic Effects

Bayer (1996) administered a single dose of TMCPPs in corn oil by gavage to male and female Wistar rats (200, or 500 mg/kg for males; only 2,000 mg/kg in females) and found that at or below 500 mg/kg, animals exhibited no clinical signs of toxicity over the subsequent 2-wk observation period. No gross pathological signs of toxicity were observed in treated animals during necropsy. In females given 2,000 mg/kg, all developed clinical signs of toxicity with increasing intensity over time and died within 3-6 hr (see Table 15-3). Necropsy revealed red, mottled lungs. The "approximate median lethal dose" for females was calculated by the authors to be 632 mg/kg (the geometric mean of the 0% and 100% response concentrations); the LD_{50} for males was above 500 mg/kg.

In a range-finding study, groups of five male Wistar rats received TMCPPs at doses of 0, 1, 10, 100, or 1,000 mg/kg-d by gavage in peanut oil for 7 d (Bayer 1993). Treatment had no effect on mortality, clinical signs, body weight, or food intake, but water intake was significantly higher (>30%) in the 1,000 mg/kg-d group than in controls. Absolute testicular weight was significantly reduced in one rat in the 100 mg/kg-d group.

In a preliminary trial, groups of male and female CD rats (two/sex) were administered 2,000 or 4,000 mg/kg of neat TMCPPs (neat) by gavage (Gardner 1987) and observed for 5 d. All animals died within 24 hr at the high dose, and both males died at the lower dose. One of two females died at the lower dose. In a second trial, groups of male and female rats (two/sex) were gavaged with 400, 800, 2,000, or 4,000 mg/kg of TMCPP dissolved in corn oil. In the 5-d observation period, there was no mortality at or below the 2,000-mg/kg dose, but 2/2 males and 1/2 females died at the 4,000-mg/kg dose within the first d. In another study (Gardner 1987), groups of five male and five female CD rats received a single dose of TMCPPs (2,500, 3,200, 4,000, or 5,000 mg/kg) by gavage, diluted with corn oil. Clinical signs are described in Table 15-3. Mor-

TABLE 15-3 Oral Toxicity Studies on Tris Monochloropropyl Phosphates

Chemical, Purity	Species, Strain, Sex, Number	Dose (mg/kg or mg/kg-d)	Duration, Route	Effects	Comments	Reference
Tris (chloropropyl) phosphate, TCPP: I; 2 other components, possibly II and II, but different names stated in text; (IV at trace levels)	Rat, Wistar, F, 5/dose	945, 1,137, 1,349, 1,633, 2,000, 2,400 (dose vol. = 10 mL/kg)	Single dose (96-hr observation), gavage	Mortality: no deaths at 945 mg/kg; 100% deaths at 2,400 mg/kg within 2-5 hr); clinical signs: tremors, wheezing, oronasal bleeding, clonic convulsions	LD_{50}: 1,017 mg/kg	Kawasaki et al. 1982[a]
Trichloropropyl phosphate (later document identifies as TCIP, mixture 90-95% isopropyl, 5-10% n-propyl phosphates)	Rat, CD, M/F, 2/sex/dose	2,000, 4,000	Single dose (5-d observation), gavage	Mortality: 4 of 4 died at 4,000 mg/kg; 2 of 2 M and 1 of 2 F died at 2,000 mg/kg		Courtaulds Chemicals 1988[b]
Trichloropropyl phosphate (later document identifies as TCIP, mixture 90-95% isopropyl, 5-10% n-propyl phosphates)	Rat, CD, M/F, 4/sex/dose	400, 800, 2,000, 4,000 (dose vol. = 10 mL/kg)	Single dose (5-d observation), gavage	Mortality: no deaths at 400-2,000 mg/kg; 2 of 2 M and 1 of 2 F died at 4,000 mg/kg		
Trichloropropyl phosphate (later document identifies as TCIP, mixture 90-95% isopropyl, 5-10% n-propyl phosphates)	Rat, CD, M/F, 5/sex/dose	2,500, 3,200, 4,000, 5,000 (dose vol. = 10 mL/kg)	Single dose (14-d observation), gavage	Mortality: deaths in M at ≥3,200 mg/kg, in F at ≥4,000 mg/kg; clinical signs: piloerection, hunched posture, abnormal gait, increased salivation; followed by lethargy, pallor, ptosis, decreased respiratory rate; clonic convulsions at ≥4,000 mg/kg; diarrhea at ≤3,200 mg/kg; decreased body weight gain in wk 1	LD_{50}: 3,600 mg/kg (3,800 mg/kg in M; 3,400 mg/kg in F)	

TABLE 15-3 *(Continued)*

Chemical, Purity	Species, Strain, Sex, Number	Dose (mg/kg or mg/kg-d)	Duration, Route	Effects	Comments	Reference
Antiblaze ®80: product A; lot PP-2B	Rat, Sprague-Dawley, M/F, 1-10/dose	M: 320, 630, 1,000, 1,250, 1,300, 1,700, 2,300, 2,500, 3,000, 5,000 F: 320, 500, 630, 700, 1,000, 1,400, 2,000, 2,500, 5,000	Single dose (14-d observation), gavage	Mortality: no deaths in F at ≤700 mg/kg or in M at ≤1,250 mg/kg; clinical signs: altered activity, discharge from orifices, hunching, rough coat, anorexia, diarrhea, dehydration, decreased body temperature, alopecia, aggression, teeth chattering, emaciation, clonic convulsions, ataxia, sensitivity to touch	LD_{50} (M): 1,546 mg/kg (95%CI: 1,066-2,241 mg/kg) LD_{50} (F): 017 mg/kg (95% CI: 727-1,423 mg/kg)	Albright and Wilson 1980b[c]
Antiblaze ®80: product B; sample 11108005; tris (2-chloro-propyl) phosphate	Rat, NS, M/F, 4-5 M/dose, 10-12 F/dose	M: 1,000, 1,208, 1,450, 2,085, 3,000, 5,000 F: 700, 840, 1,000, 1,450, 2,085, 5,000	Single dose (14-d observation), NS	Mortality: 20% for lowest dose in M and F; clinical signs: hunched posture, inactivity, oral discharge; convulsions at 5,000; body weight loss in one F at 5,000.	LD_{50} (M): 2,085-3,000 mg/kg LD_{50} (F): 840-1,000 mg/kg	Albright and Wilson 1980a[c]
Fyrol ®PCF: "Tri (2-chloropropyl phosphate)": probable mixture	Hen, white leghorn, 17 (10 control)	13,200 mg/kg (dose vol = 10 mL/kg)	2 doses 3 wk apart (3-wk observation after last dose), gavage	No deaths; no egg production, decreased food intake and body weight; loss of feathers; no effect on walking or central nervous histology		Sprague et al. 1981[a]
Tris-chloroisopropyl phosphate: I- 63.2%; II- 27.2%; III- 4.4%; IV- 0.5%; other ≤.8%	Rat, Wistar, M/F, 5/sex/dose	M: 200, 500 F: 200, 500, 2,000 (dose vol. = 10 mL/kg)	Single dose (14-d observation), gavage	Mortality: no deaths in dose range 200-500 mg/kg; 100% mortality at 2,000 mg/kg; clinical signs: apathy, palmospasms, crusted blood on snout; reddened and mottled lungs	LD_{50} (M): >500 mg/kg LD_{50} (F): 632 mg/kg	Bayer 1996[d]

Tris-chloroisopropyl phosphate: 97.85% = (I plus unspecified isomers)	Rat, Wistar, M/F, 5/sex/dose	0, 1, 10, 100, 1,000 (dose vol. = 2 mL/kg)	7 consecutive days, gavage	No effect on mortality, clinical signs, body weight, or food intake; water intake increased at 1,000 mg/kg-d; no effect on gross pathology, weight of testes (but decreased size and weight of testes in one animal at 100 mg/kg-d)		Bayer 1993[e]
Tris (chloropropyl) phosphate, TCPP: I ; 2 other components, possibly II and III, but different names stated in text; (IV at trace level)	Rat, Wistar, F, 5/dose	0, 8, 40, 200, 1,000	7 consecutive days (14-d observation)	Mortality: one rat died at 1,000 mg/kg-d; no clinical signs; no effect on body weight; relative liver weight increased at 1,000 mg/kg-d; relative kidney weight increased at ≥40 mg/kg-d		Kawasaki et al. 1982[a]
Tris (chloropropyl) phosphate, TCPP: I ; 2 other components, possibly II and III, but different names stated in text; (IV at trace level)	Rat, Wistar, F (pregnant), 11-14/dose	0, 6, 70, 625 (0, 0.01, 0.1, 1% in feed)	Gestation d 0-20 (30-d postnatal observation)	In dams: no effect on body weight; no clinical signs; no effect on implantation or resorption. In fetus: no effect on mortality, body weight, or sex ratio; no significant increase in external, visceral, or skeletal malformations; at weaning, no effect on survival, body weight	Dose-related increase in cervical ribs and missing 13th rib (not statistically significant); no necropsy reported for dams	Kawasaki et al. 1982[a]

F, females; M, males. I: tris (1-chloro-2-isopropyl phosphate), CAS Registry # 13674-84-5. II: bis (1-chloro-2-isopropyl) (2-chloropropyl) phosphate, CAS Registry # 76025-08-6. III: bis (2-chloropropyl) (1-chloro-2-isopropyl) phosphate, CAS Registry # 76649-15-5. IV: tris (2-chloropropyl) phosphate, CAS Registry # 6145-73-9.

[a]Published study.

[b]Laboratory report from Huntingdon Research Centre to Courtaulds Chemical Co. (1987); TSCA submittal from Aceto Chem. Co. to EPA.

[c]Incomplete laboratory report by Mobil Environmental Health Sciences Laboratory (1980); TSCA submittal from Albright and Wilson to EPA.

[d]Good Laboratory Practice laboratory report of the Institute of Industrial Toxicology Bayer, A.G. (1996); TSCA submittal from Miles, Inc. to EPA.

[e]Good Laboratory Practice laboratory report of the Institute of Industrial Toxicology Bayer, A.G. (1993); TSCA submittal from Bayer, A.G. to EPA.

tality occurred at and above 3,200 mg/kg, during the first 2 d. The authors estimated the acute LD_{50} of TMCPPs to be 3,600 mg/kg for rats (3,400 mg/kg for females and 3,800 mg/kg for males).

According to a preliminary report (Albright and Wilson 1980b), groups of Sprague Dawley rats (from one to five of each sex/dose) were given TMCPPs (Antiblaze® 80, product A) by mouth and observed for up to 14 d. Females received 320, 500, 630, 700, 1,000, 1,250, 1,400, 2,000, 2,500, or 5,000 mg/kg and males received 320, 630, 1,000, 1,250, 1,300, 1,700, 2,300, 2,500, 3,000, or 5,000 mg/kg. The authors did not indicate whether the material was delivered neat or in a vehicle. Clinical signs observed are noted in Table 15-3. Most deaths occurred on d 1; the LD_{50} for females was calculated to be 1,017 mg/kg (727-1,423 mg/kg), and for males, it was 1,546 mg/kg (1,066-2,241 mg/kg).

In another preliminary report (Albright and Wilson 1980a), groups of rats (strain unspecified) were given a single dose of TMCPPs (Antiblaze® 80, product B) by mouth, and observed for up to 14 d. Groups of five to seven female rats received 700, 840, 1,000, 1,450, 2,085, or 5,000 mg/kg, and groups of four to five male rats received 1,000, 1,208, 1,450, 2,085, 3,000, or 5,000 mg/kg. Treatment related effects are listed in Table 15-3. Mortality was 20% at the lowest dose and 100% at the highest dose for both males and females. The median lethal dose was between 2,085 and 3,000 mg/kg for males, and between 840 and 1,000 mg/kg for females.

Kawasaki et al. (1982) gave groups of five female Wistar rats a single dose of 945, 1,137, 1,349, 1,633, 2,000, or 2,400 mg/kg of TMCPPs by gavage in olive oil. Treatment-related effects are listed in Table 15-3. No mortality occurred at the lowest dose, but was 100% at the highest dose and occurred within 2-5 hr after administration. The estimated 96-hr LD_{50} for this study is 1,017 mg/kg.

Kawasaki et al. (1982) gave groups of five female Wistar rats 0, 8, 40, 200, or 1,000-mg/kg of TMCPPs by gavage in olive oil for 7 d. No abnormalities or mortality was observed except for one rat that died in the 1000 mg/kg group. The only treatment-related effects were significant increases in the relative weights of the liver (at 1,000 mg/kg) and of the kidneys ($\geq$40 mg/kg).

Neurological Effects

Sprague et al. (1981) administered a single neat dose of 13.2 g/kg of TMCPPs (Fyrol® PCF) by gavage to 18 adult White Leghorn hens; 10 hens in the control group received a 10-mL/kg dose of corn oil. No significant effects on the activities of brain neurotoxic esterase or plasma cholinesterase were observed. In a delayed neurotoxicity study, a group of 17 hens received a neat dose of 13.2 g/kg of TMCPPs and 10 hens received 10 mL corn oil; hens received a second dose 3 wk later, and were observed for another 3 wk. Body weight and food consumption were measured every 3-4 d, and walking behavior was observed weekly. Surviving hens were sacrificed, and the brains and spinal cord were examined histologically. Effects of TMCPP treatment included cessation of egg production, feather loss, and a significant reduction in food consumption and body weight. However, TMCPPs had no significant effect on walking behavior or histology of the central nervous system.

Reproductive and Developmental Effects

Kawasaki et al. (1982) fed groups of 11-14 pregnant female Wistar rats a diet containing 0%, 0.01%, 0.1%, or 1% TMCPPs during gestation d 0-20. The daily intake of TMCPPs was estimated at 0, 6, 70, or 625 mg/kg-d[2]. Some dams were sacrificed on gestation d 20, and the fetuses were examined for skeletal or visceral abnormalities. Litters born naturally from groups of five to seven dams were culled to eight neonates each and monitored for 4 wk. TMCPPs had no effect on food consumption or body weight gain in dams. TMCPPs did not affect the number of implants, the number of resorptions, or the number or weight of live fetuses. No fetuses died or had obvious external malformations. The incidence of skeletal or visceral malformations in fetuses from treated litters was not statistically different from controls. However, dose-related increases in the incidence of missing 13th ribs and cervical ribs were observed in treated fetuses; delayed ossification of sternebra was observed in all groups. Dilatation of the renal pelvis was observed in one fetus in the 68 mg/kg-d group. Treatment in utero had no effect on growth or survival of weanlings up to postnatal d 28.

Other Systemic Effects

No studies were located that investigated immunological effects to TMCPPs following oral exposure.

---

[2]Doses calculated from the following values reported by the authors: total 21-d intake, total body weight gain, and the initial average body weight of the females used for toxicity tests, described in Section 4.1.1; (21-d mg "TMCPPs")/21 d = mg "TMCPPs"/d divided by the time-weighted average body weight in kg). For example, the 21-d intake of TMCPPs for the 0.1% dose group was 380 mg; it was divided by 21 to yield 18 mg TMCPP/d. The average body weight gain for that group was 0.117 kg; using 0.20 kg as an initial body weight (from the acute study in the same paper), the average body weight was (initial + final)/2 = (0.20 kg + 0.317 kg)/2 = 0.259 kg. Therefore, rats in the 0.1% dose group received 18 mg TMCPP/ 0.259 kg = 70 mg/kg-d.

Genotoxicity

No studies were located regarding the in vivo genotoxicity of TMCPPs. All of the in vitro tests of genotoxicity yielded negative results.

TMCPPs (TCIP, CASRN 13674-84-5), at concentrations ranging from 0 to 1,000 µg/plate, yielded negative results in mutagenicity tests using *Salmonella typhimurium* strains TA97, TA98, TA100, TA1535, and TA1537, with or without metabolic activation (Zeiger et al. 1992). TMCPPs, at concentrations of 0, 0.1, 1, 10, 100, 500, or 2,000 µg/plate were not mutagenic to *S. typhimurium* strains TA98, TA100, TA1535, TA1537, or TA1538, with or without metabolic activation (BIBRA 1977). TMCPPs, at concentrations of 0, 98, 328, 980, or 3,300 µg/plate were not mutagenic in tests using *S. typhimurium* strains TA100 or TA1535, with or without metabolic activation (Nakamura et al. 1979). TMCPPs (Antiblaze®80, product A), in concentrations ranging from 38 to 425 µg/plate yielded negative results in mutagenicity tests using *S. typhimurium* strains TA98, TA100, TA1535, TA1537, and TA1538, with or without metabolic activation (Albright and Wilson 1980a).

QUANTITATIVE TOXICITY ASSESSMENT

Noncancer

Dermal RfD

The database on dermal toxicity of TMCPPs is inadequate for developing a dermal RfD. The duration of all available dermal studies was only 24 hr.

Inhalation RfC and Oral RfD

No chronic or subchronic studies in humans or animals were located that could provide the basis for developing an inhalation RfC or oral RfD for TMCPPs.

Cancer

No long-term studies in humans or animals were located that could be used to assess the carcinogenic potency of TMCPPs by any route of exposure.

EXPOSURE ASSESSMENT AND RISK CHARACTERIZATION

Noncancer

Dermal Exposure

Dermal exposure to TMCPP was estimated using the dermal exposure scenario described in Chapter 3. This exposure scenario assumes that an adult spends 1/4th of his or her time sitting on furniture upholstery treated with TMCPP and also assumes 1/4th of the upper torso is in contact with the upholstery and clothing presents no barrier.

As a first estimate of exposure, it was assumed that skin, clothing, and the upholstery did not impede dermal exposure to TMCPP. It was also assumed that there would be sufficient water present from sweat to facilitate dissolution of TMCPP from the upholstery and absorption through the skin. In this scenario, only the dissolution rate of TMCPP from upholstery is assumed to be the limiting factor in absorption by the body. It is assumed that all of the TMCPP that dissolves is immediately absorbed into the body by the sitting person.

Dermal exposure was estimated using Equation 1 in Chapter 3. For this calculation, the subcommittee estimated an upholstery application rate (S_a) for TMCPP of 5 mg/cm^2. The extraction rate (μ_w) for TMCPP was estimated to be 0.038 based on extraction data for organic phosphates in polyester fiber (McIntyre et al. 1995). The release rate from the fiber for estimating extraction was 0.06/d at 28°C calculated using the equation $2d/2\pi R$ (d = film thickness, R = fiber radius) with a correction from fiber to film of a factor of 0.63.

Using these assumptions, an estimated absorbed daily dose of 1.5 mg/kg was calculated for TMCPP. Insufficient data are available for deriving a dermal or oral RfD for TMCPP. Therefore, no conclusions can be drawn concerning the noncancer health risks associated with dermal exposure to TMCPP in furniture upholstery. Since there was no appropriate data for estimating TMCPP dermal penetration, a second dermal exposure iteration was not performed for this compound.

Inhalation Exposure

Particles

Inhalation exposure estimates for TMCPP were calculated using the exposure scenario described in Chapter 3. This scenario assumes that a person spends 1/4th of his or her lifetime in a 30 m^3 room containing 30 m^2 of

TMCPP-treated fabric and the room is assumed to have a air-change rate of 0.25/hr. It is also assumed that 50% of the TMCPP present in 25% of the surface area of the treated fabric is released over 15 yr and 1% of released particles are a size that can be inhaled.

Particle exposure was estimated using Equations 4 and 5 in Chapter 3. The subcommittee estimated an upholstery application rate (S_a) for TMCPP of 5 mg/cm^2. The release rate (μ_r) for TMCPP from upholstery fabric was estimated to be 2.3×10^{-7}/d (see Chapter 3, Equation 5) yielding a room airborne particle concentration (C_p) of 1.9 µg/m^3 and a short time-averaged exposure concentration of 0.48 µg/m^3. The time-averaged exposure concentration for particles was calculated using Equation 6 in Chapter 3.

No inhalation RfC has been derived for TMCPP. Therefore, no conclusions can be drawn concerning the noncancer health risks associated with inhalation of particles from furniture upholstery containing TMCPP.

Vapors

In addition to the possibility of release of TMCPP in particles from worn upholstery fabric, the subcommittee considered the possibility of the release of TMCPP by evaporation. This approach is described in Chapter 3, and uses an exposure scenario similar to that described above for exposure to TMCPP particles.

The rate of flow of TMCPP vapor from the room is calculated using Equations 8-11 in Chapter 3. A saturated vapor concentration (C_v) of 35,300 mg/m^3 was estimated for TMCPP. The application density (S_a) for TMCPP in the treated upholstery was estimated as 5 mg/cm^2.

Using the parameters described, the equilibrium room-air concentration of TMCPP was estimated to be 30,000 mg/m^3. The short-term time-average exposure concentration for TMCPP was estimated as 7,500 mg/m^3 using Equation 12 in Chapter 3 and the equilibrium room-air concentration for TMCPP. It was estimated that concentration could be maintained for approximately 5 hr. These results suggest that the vapor inhalation scenario is unrealistic for TMCPP-treated furniture in a residential setting. Nevertheless, no inhalation RfC is available for calculating a margin of exposure for TMCPP. Therefore, no conclusions can be drawn about the noncancer health risks posed by the inhalation of TMCPP vapors.

Oral Exposure

The assessment of noncancer toxicological risk for oral exposure to TMCPP is based on the oral exposure scenario described in Chapter 3. This scenario assumes a child is exposed to TMCPP by sucking on 50 cm^2 of fabric treated with TMCPP, 1 hr/d for 2 yr. The subcommittee estimated an upholstery application rate (S_a) for TMCPP of 5 mg/cm^2. Oral exposure was calculated using Equation 15 in Chapter 3. The extraction rate (μ_w) for TMCPP was estimated to be 0.038 based on extraction data for organic phosphates in polyester fiber (McIntyre et al. 1995). The release rate from the fiber for estimating extraction was 0.06/d at 28°C calculated using the equation $2d/2\pi R$ (d = film thickness, R = fiber radius) with a correction from fiber to film of a factor of 0.63.

The worst case average oral daily dose for TMCPP was estimated to be 0.04 mg/kg-d. However, a margin of exposure could not be calculated for the oral route because there were insufficient data to derive an oral RfD for this compound.

Cancer

There are insufficient data for assessing the cancer risk from exposure to TMCPP by any route of exposure.

RECOMMENDATIONS FROM OTHER ORGANIZATIONS

The subcommittee is not aware of acceptable exposure levels recommended by any regulatory agency or other organizations.

DATA GAPS AND RESEARCH NEEDS

There are no data on the subchronic or chronic toxicity of TMCPPs for the dermal, inhalation, or oral routes of exposure. No studies have been conducted on the effects of TMCPP exposure on reproduction. Data on the effects of TMCPP exposure on reproduction are also not available. Data on the dermal absorption of TMCPPs as well as information on human exposure to TMCPPs from treated upholstery are also not available.

The subcommittee recommends that the potential for release of TMCPP vapor into air and TMCPP released into saline from treated fabric be investigated.

REFERENCES

Akzo Nobel. 1999. Fyrol® PCF: product information sheet. [Online]. Available: http://www.akzo-nobel.com/phosphorus/TDS/fyrol/fyrolpcf.PDF (Retrieved September 8, 1999).

Albright and Wilson (Albright & Wilson Americas, Inc.). 1980a. Eighteen health and safety studies on Antiblaze® 80. TSCATS 311056. EPA/OTS Doc. # 86-890000114. OTS # 0517715.

Albright and Wilson (Albright & Wilson Americas, Inc.). 1980b. Oral LD_{50} of Tris (2-Chloropropyl) Phosphate, Lot PP-28, in Sprague-Dawley Rats After a Single Administration. TSCATS 311058. EPA/OTS Doc. # 86-0000115. OTS 0517716.

Albright and Wilson (Albright & Wilson Americas, Inc.). 1988. Memorandum from B.E. Johnson to H.H. King, Re: Composition of Antiblaze 80 and Equivalents, dated December 21, 1988.

Albright and Wilson. 1999. Flame Retardants. [Online]. Available: http://www.albright-wilson.com/feb99/products/pda/flame/flame.htm (Retrieved September 8, 1999).

Bayer (Bayer AG). 1993. Tris-chlorisopropylphosphat, Vorversuch zur Dosisfindung fur eine subakute toxikologische Studie an mannlichen Wistar-Ratten (Verabreichung mit der Magensonde uber 7 Tage). [Article in German]. TSCATS 442098. EPA/OTS Doc. # 86-94-0000032. OTS # 0556628.

Bayer (Bayer AG). 1996. Tris(2-chlorisopropyl)phosphat (13674-84-5): Acute Oral Toxicity Study in Male and Female Wistar Rats. Study No. T0060914. Bayer AG, Institute of Toxicology, Wuppertal, Germany. TSCATS 444651. EPA/OTS Doc. #86-960000566. OTS# 0558767.

BIBRA (British Industrial biological Research Association). 1977. Microbial Mutagenicity Test With Trichloroethyl Phosphate and Trichloropropyl Phosphate. Report No. 214/1/77. Surrey, UK.

Bittner, P. 1999. Toxicity Review for Tris(2-chloropropyl) Phosphate. Memorandum, dated March 3, 1999, from Patricia Bittner, Toxicologist, Division of Health Sciences, to Ronald Medford, Assistant Executive Director for Hazard Identification and Reduction, U.S. Consumer Product Safety Commission, Washington, DC.

Courtaulds Chemicals. 1988. Hazard Data Sheet, Tris-(2-Chloroisopropyl) Phosphate (TCPP). Derby, UK.

Ferrante, J. 1999. Memorandum from Jacqueline Ferrante, Pharmacologist, Division of Health Sciences to Ronald L. Medford, Assistant Executive Director for Hazard Identification and Reduction, Subject: Toxicity Review for Tris (1-chloro-2-propyl) phosphate and bis (2-chloropropyl) 1-(chloro-2-propyl) phosphate. U.S. Consumer Product Safety Commission, Washington, DC.

Gardner, J.R. 1987. Acute Oral Toxicity to Rats of Trichloropropyl Phosphate. HRC Report No. 871285D/CLD 23/AC. Huntingdon Research Centre, Ltd., Cambridgeshire, UK.

IPCS (International Programme on Chemical Safety). 1998. Environmental Heath Criteria 209: Flame Retardants: Tris(chloropropyl) Phosphate and Tris(2-chloroethyl) Phosphate. Geneva: World Health Organization.

Kawasaki, H. et al. 1982. Studies on the Toxicity of Insecticides and Food Additives in Pregnant Rats- (5) Foetal Toxicity of Tris-(Chloropropyl) Phosphate. Oyo Yakuri 24(5):697-702. TSCATS 442991. EPA/OTS Doc. # 86-950000008. OTS# 0557521.

McIntyre, J.E., I. Holme, and O.K. Sunmonu. 1995. The desorption of model compounds from poly(ethylene terephthalate) fibre. Colourage 41(13)77-81.

Minegishi, K.-I., H. Kurebayashi, S. Nambaru, K. Morimoto, T. Takashi, and T.

Yamaha. 1988. Comparative studies on absorption, distribution, and excretion of flame retardants halogenated alkyl phosphate in rats. Eisei Kagaku 34(2):102-114.

Nakamura, A., N. Tateno, S. Kojima, M.A. Kaniwa, and T. Kawamura. 1979. The mutagenicity of halogenated alkanols and their phosphoric acid esters for Salmonella typhimurium. Mutat. Res. 66(4):373-380.

NTP (National Toxicology Program). 1999. [Online]. Available: http://ntp-server.niehs.nih.gov/. Last updated 08/13/99

RTECS (Registry of Toxic Effects of Chemical Substances). 1999. Micromedex, Inc.

Sprague, G.L., L.L. Sandvik, M.J. Brookins-Hendricks, and A.A. Bickford. 1981. Neurotoxicity of two organophosphorus ester flame retardants in hens. J. Toxicol. Environ. Health 8(3):507-518.

Saltzman, L., and M. Babich. 1999. Tris(chloropropyl) phosphate. Memorandum, dated August 4, 1999, from Lori Saltzman, Director, Division of Health Sciences, and Michael Babich, Chemist, Divison of Health Sciences, to Ronald Medford, Assistant Executive Director for Hazard Identification and Reduction, U.S. Consumer Product Safety Commission, Washington, DC.

TSCA Interagency (Toxic Substances Control Act Interagency Testing Committee). 1988. Information Review: Tris (Chloropropyl) Phosphates. IR-509. Toxic Substances Control Act Interagency Testing Committee. Prepared by Dynamac Corporation. December 9, 1988.

Williams, G.M., H. Mori, and C.A. McQueen. 1989. Structure-activity relationships in the rat hepatocyte DNA-repair test for 300 chemicals. Mutat. Res. 221(3):263-286.

Zeiger, E., B. Anderson, S. Haworth, T. Lawlor, and K. Mortelmans. 1992. Salmonella mutagenicity tests: V. Results from the testing of 311 chemicals. Environ. Mol. Mutagen. 19(Suppl. 21):2-141.

16

Tris(1,3-dichloropropyl-2) Phosphate

THIS chapter reviews the physical and chemical properties, toxicokinetics, toxicological, epidemiological, and exposure data on tris(1,3-dichloropropyl-2) phosphate, or TDCPP. The subcommittee used that information to characterize the health risk from exposure to TDCPP. The subcommittee also identified data gaps and recommended research relevant for determining the health risk from exposure to TDCPP.

PHYSICAL AND CHEMICAL PROPERTIES

TDCPP is a chlorophosphonate that is used as a fire retardant and plasticizer in various plastic foams, resins, and latexes in the U.S. and Europe (IPCS 1998). Chemical and physical properties of TDCPP are reported in Table 16-1. TDCPP is a viscous, colorless liquid; is not volatile; and is soluble in water and most organic solvents. It is manufactured from epichlorohydrin and phosphorus oxychloride and contains 49% chlorine and 7.2% phosphorus by weight (HSDB 1989). The commercial product consists mainly of 1,3-dichloro-2-propyl groups but can contain trace amounts of tris(2,3-dichloropropyl) phosphate (CAS 78-43-3). TDCPP has been mistakenly referred to as TCPP, which is tris(1-chloro-2-propyl) phosphate (CAS 13674-84-5).

TABLE 16-1 Chemical and Physical Properties of Tris(1,3-dichloropropyl-2) Phosphate

Properties	Value	Reference
Chemical formula	$C_9H_{15}Cl_6O_4P$	HSDB 1989; IPCS 1998
Structure	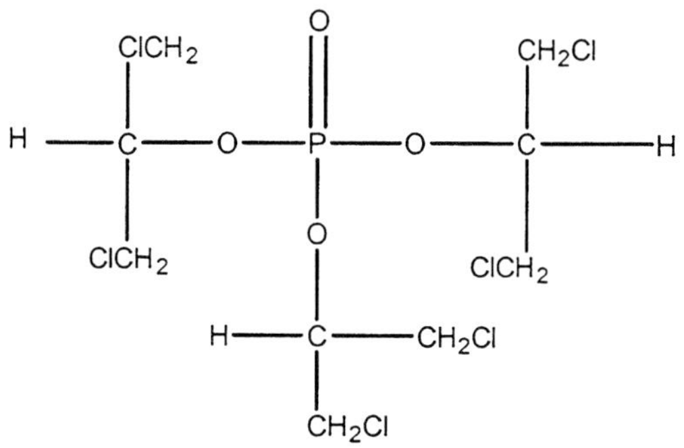	
CAS Registry #	13674-87-8	HSDB 1989; IPCS 1998
Synonyms	1,3-Dichloro-2-propanol phosphonate (3:1); Phosphoric acid tris (1,3-dichloro-2-propyl)ester; 2-Propanol, 1,3-dichloro-, phosphate (3:1); Tris(2-chloro-1-(chloromethyl)ethyl)phosphate; Tris(1-chloromethyl-2-chloroethyl)phosphate; Tris(1,3-dichloroisopropyl)phosphate; Tris(1,3-dichloro-2-propyl)-phosphate; chloroalkyl phosphate; TDCPP; TDCP; TCPP	IPCS 1998
Trade names	Fyrol FR2; PF 38; PF 38/3; Firemaster T33P; CRP; Apex Flame Proof Emulsion 197 or 212; Antiblaze 195; Amgard; Foforan Troj-(1,3-dwuchloroizopropylowy (Polish)	HSDB 1989; IPCS 1998
Molecular weight	430.88	HSDB 1989
Physical state	viscous liquid	IPCS 1998
Solubility	100 ppm or 7 mg/l in H_2O (temp. not given) 100 mg/l in H_2O (30°C)	HSDB 1989; IPCS 1998
Vapor pressure	0.01 mm Hg at 30°C	IPCS 1998
Partition coefficient	log K_{ow} = 3.76	HSDB 1989; IPCS 1998
Specific gravity	1.52 at 25°C	IPCS 1998
Boiling point	236-237°C at 5 mm Hg	HSDB 1989, IPCS 1998
Flash point	252°C (Cleveland open cup method)	IPCS 1998
Viscosity	1,800 cP (25°C)	IPCS 1998
Conversion factors	1 ppm= 0.0567 mg/m^3 1 mg/m^3 = 17.62 ppm	IPCS 1998

OCCURRENCE AND USE

TDCPP was first synthesized by chemists of the Stauffer Chemical Company in the 1950s. It was introduced as a flame retardant commercially in 1962 and was later given the commercial trade name Fyrol® FR2 (Sanders 1978). It has been produced by other chemical companies under the trade names Emulsion 212, PF38, and PF38/3.

Annual worldwide demand for TDCPP in 1997 was estimated at 8,000 tons and was growing (IPCS 1998). Current volume of use in the U.S. was not located, but TDCPP appears in the US Environmental Protection Agency's (EPA) 1990 list of high production volume chemicals (chemicals that are manufactured or imported into the U.S. at greater than 1 million lb/yr).

TDCPP is added as a flame retardant and plasticizer to polyurethane foam (both rigid and flexible), other plastics and resins, and latexes for textile back-coating and binding of nonwoven fabrics (IPCS 1998). It is also applied as a flame retardant in polyisocyanurate foams, automotive seating, and styrene-butadiene rubber (HSDB 1989). A 1997 survey of producers of flame retardants marketed to the textile industry found that TDCPP was used as a flame retardant in automobile and truck upholstery, draperies, and wall coverings in the U.S. and in commercial and residential furniture and "other transportation" upholstery outside the U.S. (Fire Retardant Chemicals Association 1998).

TDCPP was used as a flame retardant in childrens, and infants sleepware until May 1977, when it was withdrawn from sales to the apparel market after published reports that it was mutagenic in bacteria (Sanders 1978).

TOXICOKINETICS

Absorption

Dermal

Experimental animal data suggest that TDCPP is absorbed extensively from the skin, but the rate or extent of absorption has not been measured. Nomeir et al. (1981) reported that TDCPP was readily absorbed through the skin of rats treated topically with 60 µL of a methanol-TDCPP solution on a 4- cm^2 area at a concentration of 2.0 µmol/kg. Each animal was treated with 60 µL/kg solution. The 60 µL was applied to a 4-cm^2 area of skin. Ulsamer et al. (1980) reported that the dermal absorption of TDCPP in rats and rabbits was about twice that of tris(2,3-dibromopropyl) phosphate (Tris). The dermal penetration of Tris was reported to be about 3.0-15.0% in rabbits (Ulsamer et al. 1980).

Inhalation

No studies were identified that investigated the absorption of TDCPP by the dermal, inhalation, or oral routes in humans or laboratory animals.

Oral

TDCPP is readily absorbed through the GI tract in rodents. Matthews and Anderson (1979) reported that about 90% of an unstated oral dose of TDCPP was absorbed through the gut wall in rats. Nomeir et al. (1981) also found that about 90% of TDCPP given orally at 0.2, 2.0, and 20.0 µmol/kg was absorbed from the rat gut within 24 hr. Similar results were reported by Minegishi et al. (1988) for rats given TDCPP at 50 µmol/kg by gavage.

Distribution

Dermal

Distribution of TDCPP in rats after topical application of TDCPP at 2.0 µmol/kg was studied by Nomeir et al. (1981). Most of the dermal dose was found in the liver, lungs, skin, and kidneys at 4 hr after application. Smaller amounts were found in adipose tissue, muscle, and blood.

Inhalation

No studies were identified that investigated the distribution of TDCPP following inhalation exposure in humans or laboratory animals.

Oral

TDCPP is distributed primarily to the lungs, liver, and kidneys within 24 hr of oral administration in rats (Minegishi et al. 1988, Nomeir et al. 1981). TDCPP is also distributed to a lesser extent to the heart, adipose tissue, skin, brain, spleen, and gonads after oral administration in rats.

Other Routes of Exposure

Morales and Mathews (1980) reported that [^{14}C]-TDCPP readily bound with the DNA present in the liver of male CD-1 mice. Mice were given a single intravenous dose of TDCPP at 94.4 µmol/kg and sacrificed 6 hr later. TDCPP in liver DNA was estimated at 8.3 ± 2.3 pmol/mg. Traces of radioactivity were detected in kidney DNA and none in muscle; much larger amounts were detected in protein and low-molecular-weight RNA isolated from liver and kidney.

Metabolism

No studies were identified that have investigated the metabolism of TDCPP in humans or laboratory animals following dermal, inhalation, or oral exposure.

The major urinary metabolite of TDCPP formed in rats is bis(1,3-dichloro-2-propyl) phosphate 24 hr after following intraperitoneal injection (Lynn et al. 1981). Of radiolabeled TDCPP, 54% was excreted in the urine within 5 d and 62% of the urinary label was bis(1,3-dichloro-2-propyl) phosphate. Two other metabolites isolated were 1,3-dichloro-2-propyl phosphate and 1,3-dichloro-2-propanol.

TDCPP was rapidly metabolized in vitro by an NADPH-dependent microsomal mixed-function oxidase system and glutathione S-transferases from rat liver to 1,3-dichloro-2-propanol, 3-chloro-1,2-propanediol, bis (1,3-dichloro-2-propyl) phosphate, and one unidentified metabolite that was thought to be a glutathione conjugate (Nomeir et al. 1981).

Excretion

No studies were identified that have investigated the excretion of TDCPP in humans or laboratory animals following dermal or inhalation exposure.

Minegishi et al. (1988) reported that 43% of a single oral dose of TDCPP at 50 µmol/kg in rats was excreted in the urine within a week of exposure; 39% and 16% were excreted in the feces and air, respectively. About 2.5% of the administered dose remained in the carcass. Those studies also demonstrated that TDCPP is extensively excreted (40%) in the bile; this increases the retention time of TDCPP because of enterohepatic recirculation.

Nomeir et al. (1981) found that intravenous TDCPP at 2.0 µmol/kg in rats was eliminated primarily in the bile, feces, and urine; some small amounts were eliminated as CO_2. About 80% of the intravenous dose was eliminated within 24 hr, but traces of radioactivity were found in most tissues 10 d after exposure.

HAZARD IDENTIFICATION[1]

Dermal Exposure

Irritation

Citing industrial reports and an unpublished chronic dermal-toxicity study of TDCPP in rabbits, Ulsamer et al. (1980) stated that TDCPP had been shown not to be a primary irritant. Piotrowski et al. (1976), cited in the entry for TDCPP in the Hazardous Substances Data Bank (HSDB) reported that TDCPP did not produce irritation in rabbit skin or sensitization reactions in guinea pig skin; however, no details on the amounts of TDCPP applied were available to the subcommittee.

The International Programme on Chemical Safety (IPCS)(1998) reports that TDCPP irritated rabbit skin in a 1989 industry-sponsored study. The irritating properties of TDCPP were evaluated in three New Zealand white rabbits via a patch test at an unspecified dose. Well-defined erythema was recorded in two animals an hour after patch removal; slight erythema was observed in the third animal at the contact site. All exposed sites had returned to normal 48 hr after exposure.

Sensitization

No studies were identified that investigated the sensitizing potential of TDCPP following repeated dermal application in humans or laboratory animals.

Systemic Effects

Dermal LD_{50} s of TDCPP in rats and rabbits are consistently above 2,000 mg/kg (IPCS 1998). Ulsamer et al. (1980) reported that one unpublished study had found that the LD_{50} in male albino rabbits was greater than 23.9 g/kg when TDCPP was applied with an occlusive bandage. At that dose, rabbits were irritable and had diarrhea, miosis, and increased muscle tonus as evidenced by trembling; these effects persisted during most of the unstated observation period. Ulsamer et al. also reported that no deaths or toxic signs were observed 14 d after exposure in albino rabbits treated topically with TDCPP of 4.64 g/kg for 24 hr.

---

[1]In this section, the subcommittee reviewed toxicity data on tris(1,3-dichloropropyl-2) phosphate, including the toxicity assessment prepared by the U.S. Consumer Product Safety Commission (Ferrante 1999).

Ulsamer et al. (1980) summarized the results of an unpublished chronic dermal-toxicity study of TDCPP in rabbits. Rabbits were exposed to TDCPP or TRIS at 1.45 g/kg for 90 d (frequency of dermal application not stated). Kidney weights were significantly increased in animals exposed to either compound. Histopathological analysis showed that TDCPP did not produce any effects in any tissues (examined tissue not stated).

Immunological Effects

The effects of subcutaneous injection of TDCPP on immune functions and host susceptibility to infectious agents were investigated in groups of seven to 10 adult B6C3F_1 mice given TDCPP intravenously at 0, 0.25, 2.5, or 25 kg/d for 4 d (Luster et al. 1981). There were no clinical signs of toxicity in the treated mice, and no significant effects on body weight, organ weight, or histopathological status were detected. Immune measures evaluated in this study were bone marrow cellularity and colony formation, lymphoproliferative responses to mitogens, delayed hypersensitivity, and serum IgG, IgM, and IgA concentrations. TDCPP treatment induced minimal changes in immune functions and host susceptibility. The 25 mg/kg-d mice had lower proliferative responses to mitogens and higher tumor rates than nonexposed mice after tumor-cell implantation. No changes latent period for tumorigenicity were detected in the high-dose animals compared with controls.

Other Effects

No toxicity studies were found that investigated the toxic effects of topically-applied TDCPP on neurological, reproductive or developmental parameters in humans or laboratory animals. No studies were identified that investigated the carcinogenicity of TDCPP in humans or laboratory animals following dermal exposure.

Inhalation Exposure

Systemic Effects

The subcommittee located only one study that investigated the inhalation toxicity of TDCPP. IPCS (1998) cites a single industry-sponsored study that found that the LC_{50} of TDCPP was greater than 5,220 mg/m^3 in male and female Sprague-Dawley rats. There were no subchronic or chronic studies.

IPCS (1998) summarizes the results of a 1981 occupational-health survey by

Stauffer Chemical Company of 124 workers employed at a TDCPP-manufacturing plant in the U.S. The clinical and general health of 93 workers classified as having potential exposure to TDCPP was compared with that of 31 workers exposed to TDCPP and matched for age, alcohol consumption, and smoking. The investigation found no significant increase of morbidity in exposed workers compared with controls. Chest x-rays of the two groups were comparable. Abnormal electrocardiograms were twice as common among the exposed workers. No significance differences were seen in any clinical-chemistry measures investigated (measures not stated by IPCS). The prevalence of "minor respiratory disease" was slightly, but apparently not significantly, increased in exposed workers. Air samples taken in the plant within various area or job classifications all contained TDCPP at less than 0.4 $\mu g/m^3$.

Cancer

No studies were found that investigated the carcinogenicity of TDCPP by the inhalation route in humans or laboratory animals.

A 1981 mortality study at a Stauffer Chemical Company plant that manufactured TDCPP found an increased incidence of lung-cancer deaths in a cohort of 289 workers employed at the plant for at least 3 mo from 1956 to 1977 as compared to the incidence of lung cancer among the U.S. general population (comparison years not indicated). Three lung-cancer deaths occurred over this period for a standardized mortality ratio of 399. However, one of the three workers was identified as nonexposed, and another was employed for only 2 yr. All three were reported to be heavy smokers. No TDCPP was detected in air samples from various job or exposure areas of the plant (maximal detection limit, 0.13 mg/m^3).

Other Effects

No toxicity studies were identified that investigated the effects of TDCPP on the immune system, nervous system, the reproductive system, development, or behavior in humans or laboratory animals following inhalation exposure.

Oral Exposure

Systemic Effects

TDCPP does not appear to be a potent acute toxicant on the basis of reported oral LD_{50}s in rodents (see Table 16-2). Reported oral LD_{50}s for TDCPP are

TABLE 16-2 LD_{50} and LC_{50} Data for Tris(1,3-dichloropropyl-2) Phosphate

Route	Species	LD_{50}/LC_{50}	Study
Oral	Rat	3.16 g/kg	Stauffer Chemical Co. 1981
		2.83 g/kg	Celanese Corp. 1960, as cited in Ulsamer et al. 1980
		2.36 g/kg	Osterberg and Bierbower 1978, as cited in Ulsamer et al. 1980
		1.85 g/kg (frationated: small doses over 120 hr)	Piotrowski et al. 1976, as cited in HSDB 1989
		4.5 g/kg (frationated: large doses over 48 hr)	
	Mouse	4.99 mL/kg	Stauffer Chemical Company 1977-78, as cited in Ulsamer et al. 1980
		2.67 g/kg (males)	Kamata et al. 1989
		2.25 g/kg (females)	
	Rabbit	6.8 g/kg	Azko Nobel 1998, as cited in CPSC 1999
Dermal	Rabbit	>23.9 g/kg	Celanese Corp. 1960, as cited in Ulsamer et al. 1980
		>4.64 g/kg	Stauffer Chemical Company 1977-78, as cited in Ulsamer et al. 1980
		>15 mL/kg	Azko Nobel 1998, as cited in CPSC 1999
Inhalation	Rat	>9.8 mg/L	Azko Nobel 1998, as cited in CPSC 1999

1.85- 4.5 g/kg in rats, 2.25- 2.67 g/kg in mice, and 6.8 g/kg in rabbits. Typical clinical signs of toxicity in rats given high doses of TDCPP are hypokinesia, piloerection, soiled coats, ataxia, irritability, hyperactivity, convulsions, tetanus, rhinorrhea, salivation, and congestion of heart, lungs and liver (Celanese Corp. 1960, as cited in Ulsamer et al. 1980; Osterberg and Bierbower 1978, as cited in Ulsamer et al. 1980; Kamata et al. 1989). One study reported fatty degeneration and renal necrosis (Piotrowski et al. 1976, as cited in HSDB 1989); the remaining studies indicated no remarkable histopathological effects

(Celanese Corp. 1960, as cited in Ulsamer et al. 1980; Osterberg and Bierbower 1978, as cited in Ulsamer et al. 1980).

Increases in relative liver and kidney weights and in mortality, were observed in mice fed TDCPP for 3 mo in their diet (Kamata et al. 1989). Male mice were given TDCPP at 0, 13, 47, 171, 576, or 1,792 mg/kg-d in their diet, and females were given TDCPP at 0, 15, 62, 214, 598, or 1,973 mg/kg-d. Males and females in the high-dose groups (1,792 and 1,973 mg/kg-d) all died within a month; they were emaciated and had rough hair coats and tremor. Liver weights in males after 3 mo at 171 and 576 mg/kg-d were increases by 32% and 51%, respectively. Liver weight in females after 3 mo at 62, 214, and 598 mg/kg-d were statistically significantly increased by 16%, 29%, and 51%, respectively. Significant increases in kidney weights were observed—by an average of 39%—in male mice fed 576 mg/kg-d and significant increases in relative kidney weight of 34 and 40% were observed in female mice fed 214 and 598 mg/kg-d, respectively. Histopathological examination showed slight necrosis of liver tissue in two female mice fed 598 mg/kg-d for 3 mo. No other statistically significant findings were reported. Decreased hemoglobin concentration was observed in males given 576 mg/kg-d and females given 598 mg/kg-d. Increases in serum alkaline phosphatase and serum alanine aminotransferase were observed with increasing exposure but were not statistically significantly elevated after 3 mo. The no-observed-adverse-effect level (NOAEL) for the increase in liver weight was 47 mg/kg-d in males and 15 mg/kg-d for females. The lowest observed-adverse-effect level (LOAEL) for significantly increased liver weight was 171 mg/kg-d in males and 62 mg/kg-d for females (IPCS 1998).

Statistically significant increases in mortality; in liver, kidney, and thyroid weights; and in abnormal histopathological findings were observed in Sprague-Dawley rats fed TDCPP in their diet for 12 and 24 mo in an unpublished study conducted by Bio/dynamics for Stauffer Chemical Company (Bio/dynamics 1981). Male and female Sprague-Dawley rats (60/sex/group) were given TDCPP in the diet at 0, 5, 20, and 80 mg/kg-d. Ten animals/sex/dose group were sacrificed after 12 mo, and the remaining animals were sacrificed at 24 mo. Mortality among the high-dose males was significantly increased as compared with controls at 17 mo. Mortality was not significantly increased among males or females at 12 mo, and mortality was not significantly higher among females in the 80 mg/kg-d dose group or in males or females of the 5 or 20 mg/kg-d dose groups at 24 mo. Chronic exposure to TDCPP did not result in significant clinical signs representative of toxicity in any of the dose groups at any time during the study. Body weights were significantly decreased in high-dose male and females beginning at 7 wk. Mean organ weights and organ: body weight ratios were increased for liver, kidneys and thyroid at 12 and 24 mo in males and females given 80 mg/kg-d (see Table 16-3). At 24 mo, the incidence

TABLE 16-3 Changes in Organ Weights and Organ/Body Weight Ratios in Sprague-Dawley Rats Fed Tris(1,3-dichloropropyl-2) Phosphate (mg-kg-d)

	Males				Females			
Tissue	0	5	20	80	0	5	20	80
				12 Month Mean Values				
LIVER								
Organ weight (mg)	13.86	14.98	15.50	17.52[a]	8.17	8.73	8.70	10.10[a]
Organ/body weight ratio	2.39	2.49	2.89	3.56[a]	2.60	2.56	2.72	3.25[a]
KIDNEY								
Organ weight (mg)	3.19	3.57	3.74	4.70	2.03	2.18	2.27	2.84[a]
Organ/body weight ratio	0.56	0.60	0.70[a]	0.95[a]	0.65	0.64	0.73	0.92[a]
THYROID								
Organ weight (mg)	0.028	0.030	0.032	0.035	0.020	0.024	0.025	0.024
Organ/body weight ratio	0.0049	0.0050	0.0060	0.0072[a]	0.0062	0.0072	0.0078[a]	0.0077[a]
				24-Month Mean Values				
LIVER								
Organ weight (mg)	14.43	14.74	16.33[a]	16.77[a]	10.55	11.27	11.39	12.20
Organ/body weight ratio	2.40	2.58	3.27[a]	4.00[a]	2.75	2.80	3.29[a]	4.62[a]
KIDNEY								
Organ weight (mg)	3.74	3.94	5.49[a]	5.75[a]	2.64	2.96	3.42[a]	4.34[a]
Organ/body weight ratio	0.63	0.67	1.14[a]	1.34[a]	0.71	0.74	0.98[a]	1.40[a]
THYROID								
Organ weight (mg)	0.040	0.038	0.039	0.043	0.029	0.030	0.030	0.034[a]
Organ/body weight ratio	0.0066	0.0068	0.0079[a]	0.0102[a]	0.0077	0.0076	0.0086	0.0110[a]

[a]$p < 0.05$; chi square analysis.
Source: Adapted from Bio/dynamics 1981.

of altered liver foci was significantly increased ($p < 0.05$) in females given 80 mg/kg-d, and the incidence of convoluted tubule hyperplasia was significantly increased in males given 20 and 80 mg/kg-d and females given 80 mg/kg-d (Table 16-4). An increased incidence of bone marrow erythroid-myeloid hyperplasia was observed in males and females given 80 mg/kg-d, but no bone marrow samples were taken from males or females in the lower-dose groups.

Ulsamer et al. (1980) reported increased mortality in rats given TDCPP by gavage at 25.0 or 250.0 mg/kg-d for 90 d. Numbers of deaths were not reported. The authors noted that absolute and relative liver and kidney weights were significantly increased in both dose groups. Histopathological evaluation showed "no remarkable differences in any tissues."

Immunological Effects

No oral-toxicity studies of TDCPP were located that investigated effects on the immune system.

Neurological Effects

Exposure to TDCPP at high concentrations produces various clinical symptoms, such as convulsions and hyperactivity, that could be the result of direct interaction of TDCPP with nervous tissue (see Systemic Effects of Oral Exposure section under Hazard Identification).

Ulsamer et al. (1980) fed chickens TDCPP at 0.6, 1.2, 2.4, or 4.8 g/kg-d for 5 d. Chickens fed 1.2 mg/kg-d or more exhibited leg and wing weakness (flaccid paralysis); all the chickens given 4.8 mg/kg-d died. It was determined that TDCPP had about 5% of the paralyzed activity of tri-o-cresyl phosphate (TOCP), a known neurotoxicant in chickens.

No neurotoxicity was observed in 12-mo old white leghorn hens 21 d after being fed TDCPP at 420 mg kg-d for 5 d (Bullock and Kamienski 1972, as cited in IPCS 1998). TOCP fed to control hens induced inability to walk, hypertension, ataxia, and prostration.

Reproductive and Developmental Effects

Wilczynski et al. (1983) investigated the effects of TDCPP following oral exposure on reproductive measures in male rabbits. No adverse effects on various reproductive measures were observed in male rabbits treated with

TABLE 16-4 Histopathological Observations in Sprague-Dawley Rats Fed Tris(1,3-dichloropropyl-2) Phosphate (mg/kg-d)

	Males				Females			
Tissue	0	5	20	80	0	5	20	80
LIVER								
Altered foci	22/60	20/60	16/60	31/60	16/60	23/60	19/55	36/60[a]
KIDNEY								
Convoluted tubule hyperplasia	2/60	10/60	29/60[a]	24/59[a]	0/60	1/60	3/57	22/60[a]
Nephropathy	39/60	26/60	36/60	39/59	12/60	13/60	11/57	25/60
SPLEEN								
Erythroid/myeloid metaplasia	13/60	3/6	4/6	17/58	13/60	5/6	3/4	33/60[a]
PARATHYROID								
Hyperplasia	1/29	1/1	0/2	12/38[a]	6/29	—	—	9/28
TESTES								
Oligospermia	35/57	31/60	45/60	51/56	NA	NA	NA	NA
Eosinophilc material/lumen	2/57	4/60	12/60[a]	11/56[a]	NA	NA	NA	NA
Sperm stasis	5/57	5/60	11/60	14/56	NA	NA	NA	NA
Periarteritis nodosa	5/57	10/60	19/60[a]	16/56[a]	NA	NA	NA	NA
EPIDIDYMES								
Oligospermia	11/55	9/33	7/14	36/55[a]	NA	NA	NA	NA
Degenerated seminal product	8/55	7/33	3/14	22/55[a]	NA	NA	NA	NA
SEMINAL VESICLE								
Decreased secretory product	1/56	11/13[a]	17/20[a]	23/52[a]	NA	NA	NA	NA
Atrophy	0/56	4/13[a]	6/20[a]	10/52[a]	NA	NA	NA	NA

NA, not applicable.
[a]$p < 0.05$; chi square analysis.
Source: Adapted from Bio/dynamics 1981.

TDCPP by gavage at 0, 2, 20 or 200 mg/kg-d for 12 wk. Reproductive measures evaluated were mating behavior, fertility, and sperm quantity and quality.

Three studies were located that investigated the developmental and maternal toxicity of oral TDCPP in rats: Stauffer Chemical Company 1977-78; Tanaka et al. 1981; and Kawashima et al. 1983. Doses given to pregnant rats were as follows: Stauffer Chemical Company (1977-78) 0, 25, 100, or 400 mg/kg-d on d 6-15 of gestation; Tanaka et al. (1981) 0, 25, 50, 100, 200, or 400 mg/kg-d on d 7-15 of gestation; and Kawashima et al. (1983) 0, 25, 50, 200, or 400 mg/kg-d on d 7-15 of gestation. None of the studies found that TDCPP caused developmental abnormalities in surviving fetuses despite an increased incidence of fetal and maternal toxicity in the high-dose groups. Fetal toxicity was characterized by increased incidence of fetal death, decreased fetal weight, decreased fetal length, or increased incidence of resorption. Maternal toxicity observed at 100 mg/kg-d or greater was characterized by reduced body weight and reduced food consumption compared with negative controls (Stauffer Chemical Company 1977-78; Kawashima et al. 1983). Tanaka et al. (1981) observed significant increases in kidney weights in pregnant rats given 200 or 400 mg/kg-d; rats given 400 mg/kg-d also had a higher incidence of reduced body-weight gain and higher incidences of piloerection, salivation, and hematuria. Tanaka et al. (1981) concluded that the no-observed-adverse-effects level (NOAEL) for developmental effects in their study was 400 mg/kg-d although it was not determined whether the observed fetal effects were related to maternal toxicity. It was concluded that the NOAEL and lowest observed-adverse effects level (LOAEL) for maternal toxicity in this study were 100 and 200 mg/kg-d, respectively.

Cancer

An increase in the incidence of various tumors was observed in male and female Sprague-Dawley rats chronically exposed to TDCPP for 104 wk (Bio/dynamics 1981). The tumor incidence for this study at 104 wk is summarized in Table 16-5. Statistically significant increases in the incidence of liver adenomas and of liver adenomas and carcinomas combined were observed in males and females given 80 mg/kg-d. Although the increase in the incidence of liver carcinomas was not statistically significant at 104 wk, it is assumed that some adenomas would progress to carcinomas if the in-life phase of the study was extended. A statistically significant increase in the incidence of renal cortical tumors was also observed in males and females fed 20 and 80 mg/kg-d. The incidence of testicular interstitial-cell tumors was significantly increased in males fed 20 and 80 mg/kg-d. The incidence of adrenal cortical adenomas was increased significantly in female mice fed 80 mg/kg-d.

Genotoxicity

A number of genotoxicity studies on TDCPP have been completed. They are summarized in Table 16-6. TDCPP was uniformly positive for mutagenicity in

Table 16-5 Tumor Incidence Among Sprague-Dawley Rats Fed for Tris(1,3-dichloropropyl-2) Phosphate 104 wk (mg/kg-d)

	Males				Females			
Tissue	0	5	20	80	0	5	20	80
LIVER								
Adenoma	2/60	7/60	1/60	16/60[a]	1/60	1/60	4/55	9/60[a]
Carcinoma	1/60	2/60	3/60	7/60	0/60	2/60	2/55	4/60
Adenoma and carcinoma combined	3/60	9/60	4/60	23/60[a]	1/60	3/60	6/55	13/60[a]
KIDNEY								
Cortical tumor	1/60	3/60	9/60[a]	32/59[a]	0/60	1/60	8/57[a]	29/60[a]
TESTES								
Interstitial cell tumor	7/57	8/60	26/60[a]	39/56[a]	NA	NA	NA	NA
ADRENAL								
Cortical adenoma	5/59	3/14	5/16	5/57	13/59	5/27	2/33	20/59[a]

NA, not applicable.
[a]$p < 0.05$ as compared with controls.
Source: Adapted from Bio/dynamics 1981.

the Ames assay in the presence of rat S9 liver fraction. It was weakly clastogenic in mouse lymphoma cells in the presence or absence of mouse S9 liver fraction (Brusick et al. 1980). It was also weakly positive in the UDS assay with rat hepatocytes (Soderlund et al. 1985). In vivo studies have failed to demonstrate consistently that TDCPP is a genotoxicant in mammalian test systems. The TDCPP metabolite 1,3-dichloro-2-propanone has been shown to be a direct-acting mutagen in the Ames assay (Gold et al. 1978); The metabolite 1,3-dichloro-2-propanol was found to be weakly mutagenic in the same study. Similar results have been reported by Lynn et al. (1981). Overall, the data indicate that TDCPP might be mutagenic after metabolic activation.

QUANTITATIVE TOXICITY ASSESSMENT

Noncancer

Dermal Assessment

The dermal-toxicity data available on TDCPP are not adequate for developing a dermal RfD. One unpublished subchronic dermal-toxicity study in rabbits

TABLE 16-6 Summary of Genotoxicity Studies of Tris(1,3-dichloropropyl-2) Phosphate

Assay	Results (−S9)	Results (+S9)	Comments	Study
			IN VITRO STUDIES	
Gene Mutation: *S. typhimurium*				
Tris(1,3-dichloropropyl-2) Phosphate				
TA 100	NT	9	Mouse S9 - phenobarbital induced	Gold et al. 1978
TA 100	NT	-	Mouse and Rat S9 - PCB induced	Brusick et al. 1980
TA 100	NT	-	Rat S9 - phenobarbital induced	Brusick et al. 1980
TA 100	NT	+	Mouse S9 - phenobarbital induced	Brusick et al. 1980
TA 100	NT	-	Human S9	Brusick et al. 1980
TA 100	-	+	Rat S9 - PCB induced	Nakamura et al. 1979
TA 1535	±	+	Rat S9 - PCB induced	Nakamura et al. 1979
TA 100	NT	+	Rat S9 - phenobarbital induced	Soderlund et al. 1985
TA 100	NT	-	Metabolic activation by co-culture with hepatocyte monolayer from phenobarbital-treated rats	Soderlund et al. 1985
TA 100	NT	±	Rat S9 - PCB induced or mouse S9 - phenobarbital induced. Positive response only at high cytotoxicity (<3% survival). Unclear for which S9 system data were reported.	Majeska and Matheson 1983
TA 100	-	±	Mouse S9 - phenobarbital induced	Lynn et al. 1981
TA 1538	-	-	Rat S9 - PCB induced	Prival et al. 1977
Tris(1,3-dichloropropyl-2) Phosphate Metabolites				
TA 100	NT	0	Mouse S9 - phenobarbital induced	Gold et al. 1978
	+ +			Lynn et al. 1981
TA 100	+ +	NT		Gold et al. 1978

(Continued)

TABLE 16-6 *(Continued)*

Assay	Results (−S9)	Results (+S9)	Comments	Study
TA100, TA1535, TA98, TA1537	++		Urine from mice dosed by gavage for 4 d with tris(1,3-dichloroisopropyl) phosphate up to 0.5 mL/kg	Brusick et al. 1980
Gene Mutation: *E. coli*				
	−	−	Activation system not described	Ulsamer et al. 1980
Mammalian Systems				
L5178Y mouse lymphoma	−	−	Mouse S9 - phenobarbital induced	Brusick et al. 1980
V79 Chinese Hamster Lung	NT	−	Rat S9 - phenobarbital induced; tested to only 19% cytotoxicity, 0.02 mM high dose	Soderlund et al. 1985
DNA Damage Assays				
Unscheduled DNA synthesis in rat hepatocytes	±	−	Hepatocytes derived from untreated or phenobarbital-treated rats	Soderlund et al. 1985
Sister chromatid exchange in	±	±	Mouse S9 - PCB induced	Brusick et al. 1980
L5178Y mouse lymphoma cells	±	±	Mouse S9 - phenobarbital induced	Brusick et al. 1980
Chromosome Aberration				
Mouse lymphoma cells	±	+	Mouse S9 - PCB induced	Brusick et al. 1980
	±	+	Mouse S9 - phenobarbital induced	Brusick et al. 1980
Morphological Transformation				
Balb 3T3 cells	−	NT		Brusick et al. 1980
Syrian hamster embryo cells	+	NT		Soderlund et al. 1985
			IN VIVO STUDIES	
Sex-linked recessive lethal	−		In *Drosophila*	Brusick et al. 1980.
Chromosome Aberration				
Mouse bone marrow	−		Mice exposed by gavage to 0.05, 0.17 or 0.5 mL/kg for 1 or 5 d	Brusick et al. 1980.
Chick embryo	−		No information provided	Bloom 1982 (cited by CPSC 1999)

−, negative; ±, weakly positive; +, positive; ++, strongly positive; NT, not tested (or results not reported in the CPSC (1999) summary.

is summarized in the review article by Ulsamer et al. (1980), but data necessary for deriving a dermal RfD based on this study are not reported. The study found an increase in kidney weights in treated rabbits—a response also reported in rodents orally exposed to TDCPP. Short-term dermal assays indicate that TDCPP causes minimal or no irritation to the skin and is not an effective sensitizer.

In the absence of a dermal RfD, the subcommittee believes it is appropriate to use the oral RfD for TDCPP of 0.005 mg/kg-d as the best estimate of the internal dose from dermal exposure.

Inhalation RfC

EPA has not developed an inhalation reference concentration (RfC) for TDCPP. The subcommittee found no adequate human or animal studies that could be used for deriving an inhalation RfC for TDCPP. A series of occupational-health investigations are available for a cohort of workers employed in a plant that manufactured TDCPP (Stauffer Chemical Company 1983a, 1983b, both as cited in IPCS 1998). However, air sampling failed to detect measurable amounts of TDCPP in various job or exposure areas frequented by the cohort in these investigations. Therefore, there are insufficient inhalation data to derive an inhalation RfC for TDCPP.

In the absence of relevant inhalation exposure data, the subcommittee chose to estimate inhalation RfCs from oral RfDs The subcommittee, however, recognizes that it is not an ideal approach and also recognizes that the estimated RfC levels might be considerably different than actual levels (if inhalation data were available). Extrapolating from one route of exposure (oral) to another (inhalation) requires specific knowledge about the uptake kinetics into the body by each exposure route, including potential binding to cellular sites. The subcommittee believes that its extrapolation of oral RfDs to inhalation RfCs is highly conservative; it assumes that all of the inhaled compound is deposited in the respiratory tract and completely absorbed into the blood. The NRC committee on Toxicology (NRC 1985) has used this approach when inhalation exposure data were insufficient to derive inhalation exposure levels. The subcommittee believes that such an approach is justified for conservatively estimating the toxicological risk from exposure to FRs. These RfCs should be used as interim or provisional levels until relevant data becomes available for the derivation of inhalation RfCs.

In order to calculate a hazard index for the inhalation route, a provisional inhalation RfC of 0.018 mg/m^3 was derived using the oral RfD for TDCPP and Equation 7 in Chapter 3.

Oral RfD

EPA has not developed an oral reference dose (RfD) for TDCPP. The oral-toxicity database relevant for developing an oral RfD for TDCPP consists of one subchronic study in mice (Kamata et al. 1989), one unpublished chronic study in rats (Bio/dynamics 1981), three developmental studies in rats (Stauffer Chemical Company 1977-78; Tanaka et al. 1981; Kewashima et al. 1983) and one male reproductive study in rabbits (Wilczynski et al. 1983). The subcommittee identified the chronic study by Bio/dynamics (1981) as the key study for deriving an oral RfD for TDCPP because the exposure period is of sufficient length (24 mo) and toxic effects occurred at dose levels in the rat that are lower than NOAELs reported for various effects in the TDCPP toxicity database.

There was statistically significant atrophy and decreased secretory product of the seminal vesicles in male rats fed diets containing 5 mg TDCPP/kg-d or greater. Animals were not tested at dose levels lower than 5 mg TDCPP/kg-d, therefore the subcommittee considered this dose level to be the LOAEL for testicular and seminal vesicle effects in this study.

Calculation of a benchmark dose for testicular effects was not possible, because of a lack of complete dose-response data on seminal vesical atrophy. This condition was investigated in all control and all high-dose males but not in males in the low- and medium-dose groups.

Using testicular and seminal vesicle effects as the critical effect and the LOAEL for these effects of 5 mg/kg-d, the oral RfD was then derived by applying a composite uncertainty factor (UF) of 1,000 yielding an oral RfD of 0.005 mg/kg-d (Table 16-7). An uncertainty factor (UF) for extrapolation to humans (UF_A) of 10 was applied to the NOAEL because there are no data for comparing the toxicokinetic and dynamic characteristics of TDCPP in rodents and humans. A UF of 3 was applied for intraspecies variability (UF_H) since developmental toxicity data does not indicate that immature animals are more sensitive to TDCPP than adults. A UF of 10 (UF_L) was applied because an LOAEL for organ toxicity was used to derive the oral RfD. An uncertainty factor of 3 (UF_D) was also applied because of the limited database for the toxicity of TDCPP.

The subcommittee has moderate confidence that the Bio/dynamics (1981) study identified the most critical toxic effects for TDCPP. Although the available data do not state that the study was conducted according to GLP procedures, Bio/dynamics (1981) appears to have been a well conducted study. However, this study is not peer-reviewed and does not establish a NOAEL for testicular effects related to TDCPP exposure.

The subcommittee has moderate confidence that the derived oral RfD will protect against noncancer toxic effects in most persons. That is based on its moderate confidence in the completeness of the toxicity database on TDCPP,

TABLE 16-7 Oral Reference Dose for Tris(1,3-dichloropropyl-2) Phosphate

RfD (mg/kg-d)	Critical effect	Species	Effect level (mg/kg-d)	Uncertainty factors	Reference
0.005	Testicular atrophy and decreased seminal vesicle secretory product	Male rats	LOAEL: 5.0	UF_A: 10 UF_H: 3 UF_L: 10 UF_D: 3 Total: 1,000	Bio/dynamics 1981

LOAEL, lowest-observed-adverse-effect level; NOAEL, no-observed-adverse-effect level; RfD, reference dose; UF_A, extrapolation from animals to humans; UF_H, intra-species variability; UF_L, NOAEL not determined for critical effect; UF_D, inadequate or deficient toxicity database

its moderate confidence in the sensitivity of the Bio/dynamics (1981) bioassay to detect critical toxic effects, and its inherently conservative approach to deriving the oral RfD. The subcommittee notes the presence of some uncertainty in the threshold dose associated with testicular and seminal vesicle effects after lifetime exposure in rodents.

Cancer

The subcommittee is not aware of any scientific organizations or authoritative bodies that have evaluated the weight of evidence of the carcinogenicity of TDCPP. EPA has not developed a cancer assessment for TDCPP, and it has not been evaluated by the International Agency for Research on Cancer (IARC) or the National Toxicity Program (NTP).

Dermal

No studies were identified that investigated the carcinogenicity of TDCPP in humans or laboratory animals following dermal exposure.

Inhalation

No inhalation carcinogenicity data are available for TDCPP and an inhalation unit risk has not been derived. One epidemiological assessment was found of the potential carcinogenicity of TDCPP in workers with possible inhalation exposure to this compound (Stauffer Chemical 1983a, 1983b, both as cited in

IPCS 1998). Air samples failed to detect TDCPP in any of the potential job or exposure areas in the manufacturing plant investigated. Therefore, the study is inadequate for determining the potential carcinogenicity of TDCPP.

For the purposes of chracterizing cancer risk, an inhalation unit risk of 1.71 $\times 10^{-5}/\mu g/m^3$ was estimated using Equation 16 in Chapter 3 and the oral cancer potency factor for TDCPP.

Oral

The available animal data on TDCPP provide sufficient evidence of carcinogenicity in rats following chronic oral exposure. Statistically significant and dose-dependent increases in the incidence of renal cortical tumors and liver adenomas were observed in male and female Sprague-Dawley rats chronically exposed to TDCPP in the diet for 2 yr (Bio/dynamics 1981). In addition, the incidence of adrenal cortical tumors in females and the incidence of interstitial-testicular cell tumors were significantly increased. Also in this study, there was a dose-related but not statistically significant increase in the incidence of liver carcinomas in animals exposed to TDCPP in the diet for 2 yr.

In vitro data suggest that TDCPP is a mutagen in the presence of liver S9 fraction, initiates DNA repair, and is clastogenic. These findings suggest that TDCPP (or one or more of its major metabolites), is DNA-reactive. Therefore, the subcommittee concluded that linear extrapolation is appropriate for estimating cancer risk in the low dose range.

Doses associated with a 10% increase in cancer risk were calculated for TDCPP using data for various tumor types with increased incidence reported in the Bio/dynamics (1981) study (Table 16-8). Cancer-potency estimates ($0.1/LED_{10}$) for TDCPP range from 0.01 to 0.06/mg/kg-d with the highest risk calculated for testicular tumors. It is debatable whether this tumor-type is caused by a mode of action relevant to humans (Akzo Nobel 1998, as cited in CPSC 1999). However, the subcommittee concluded that use of the cancer-potency estimate of 0.06/mg/kg-d for calculating TDCPP cancer risk is the most health-protective approach.

EXPOSURE ASSESSMENT AND RISK CHARACTERIZATION

Noncancer

Dermal

Dermal exposure to TDCPP was estimated using the dermal exposure scenario described in Chapter 3. This exposure scenario assumes that an adult

TABLE 16-8 Calculated ED_{10}, LED_{10}, $0.1/ED_{10}$, and $0.1/LED_{10}$'s for Tris(1,3-dichloropropyl-2) Phosphate[a]

Tumor type	ED_{10} (mg/kg-d)	LED_{10} (mg/kg-d)	$0.1/LED_{10}$	$0.1/ED_{10}$
FEMALES				
Adrenal cortical tumors	14.74	9.21	0.011	0.0068
Kidney cortical tumors	4.21	2.90	0.035	0.024
Liver adenoma	13.95	8.42	0.048	0.0072
Liver adenoma/carcinoma combined	9.21	6.05	0.017	0.011
MALES				
Testes interstitial cell tumor	1.84	1.58	0.063	0.054
Kidney cortical tumors	3.95	2.37	0.042	0.025
Liver adenoma	13.95	8.95	0.011	0.0072
Liver adenoma/carcinoma combined	11.05	5.26	0.019	0.0090

ED_{10}, effective dose corresponding to a 10% tumor response in test animals; LED_{10}, lower 95% bound on the effective dose corresponding to a 10% tumor response in test animals.

[a]Values calculated using tumor data from Bio/dynamics (1981).

spends 1/4th of his or her time sitting on furniture upholstery backcoated with TDCPP and also assumes 1/4th of the upper torso is in contact with the upholstery and clothing presents no barrier. Exposure to other chemicals present in the backcoating was not included in this assessment.

First Iteration

As a first estimate of exposure, it was assumed that skin, clothing, and the upholstery did not impede dermal exposure to TDCPP present in the backcoating. It was also assumed that there would be sufficient water present from sweat to facilitate dissolution of TDCPP from the backcoating and absorption through the skin. In this scenario, only the dissolution rate of TDCPP from the backcoating is assumed to be the limiting factor in absorption by the body. It is assumed that all of the TDCPP that dissolves is immediately absorbed into the body by the sitting person.

Dermal exposure was estimated using Equation 1 in Chapter 3. For this calculation, the subcommittee estimated an upholstery application rate (S_a) for TDCPP of 5 mg/cm^2. The extraction rate (μ_w) for TDCPP was estimated to be 0.038 based on extraction data for organic phosphates in polyester fiber

(McIntyre et al. 1995). The release rate from the fiber for estimating extraction was 0.06/d at 28°C calculated using the equation $2d/2\pi R$ (d = film thickness, R = fiber radius) with a correction from fiber to film of a factor of 0.63.

Using these assumptions, an estimated absorbed daily dose of 1.5 mg/kg was calculated for TDCPP. A hazard index of 300 was calculated for this first iteration by dividing the estimated daily dermal dose of 1.5 mg/kg-d by the oral RfD for TDCPP of 0.005 mg/kg-d. At this time, the oral RfD is the best estimate of the internal dose associated with dermal exposure to TDCPP. These results suggest that TDCPP could be a toxic hazard if all available TDCPP is absorbed simultaneously.

Alternative Iteration

The estimated dermal daily dose for TDCPP can be calculated using an estimate of the dermal penetration rate for TDCPP (Chapter 3: Equations 2 and 3). Instead of assuming that all dissolved TDCPP immediately penetrates the skin and enters systemic circulation, it is assumed that the skin slows the absorption of TDCPP to a specific amount of chemical absorbed per unit of time. This estimate can be measured experimentally and is referred to as the skin permeability coefficient K_p. However, the dermal penetration constant for TDCPP has not been measured experimentally. However, K_p can be estimated from a correlation between the octanol-water partition coefficient (K_{ow}) and molecular weight (mass/unit amount of substance) using Equation 2 in Chapter 3 yielding an alternate K_p of 4.76×10^{-2} cm/d.

Using Equation 3 in Chapter 3 and the alternate K_p, the dermal daily dose rate for TDCPP was estimated to be 2.6×10^{-3} mg/kg-d. A hazard index of 0.52 was calculated by dividing the estimated daily dermal dose of 2.6×10^{-3} mg/kg-d by the oral RfD for TDCPP of 0.005 mg/kg-d. At this time, the oral RfD is the best estimate of the internal dose associated with dermal exposure to TDCPP. These results suggest that TDCPP is not anticipated to be a toxic risk by the dermal route at the stated application concentrations and under the given worst-case exposure conditions.

Inhalation Exposure

Particles

Inhalation exposure estimates for TDCPP were calculated using the exposure scenario described in Chapter 3. This scenario assumes that a person spends 1/4th of their life in a 30 m^3 room containing 30 m^2 of TDCPP-treated fabric and the room is assumed to have an air-change rate of 0.25/hr. It is also assumed that 50% of the TDCPP present in 25% of the surface area of the treated fabric is released over 15 yr and 1% of released particles are a size that can be inhaled.

Particle exposure was estimated using Equations 4 and 5 in Chapter 3. The subcommittee estimated an upholstery application rate (S_a) for TDCPP of 5 mg/cm^2. The release rate (μ_r) for TDCPP from upholstery fabric was estimated to be 2.3×10^{-7}/d (see Chapter 3, Equation 5) yielding a room airborne particle concentration (C_p) of 1.9 μg/m^3 and a short time-averaged exposure concentration of 0.48 μg/m^3. The time-averaged exposure concentration for particles was calculated using Equation 6 in Chapter 3.

Division of the time-average exposure concentration of 0.48 μg/m^3 by the provisional RfC for TDCPP of 0.018 mg/m^3 gives a hazard index of 2.7×10^{-2}. This suggests that under the subcommittee's worst-case exposure assumptions, TDCPP would not be considered a toxic hazard by the inhalation route of exposure.

Vapors

In addition to the possibility of release of TDCPP in particles from worn upholstery fabric, the subcommittee considered the possibility of the release of TDCPP by evaporation. This approach is described in Chapter 3, and uses an exposure scenario similar to that just described for exposure to TDCPP particles.

The rate of flow of TDCPP vapor from the room is calculated using Equation 8-11 in Chapter 3. A saturated vapor concentration (C_v) of 230 mg/m^3 was estimated for TDCPP. The application density (S_a) for TDCPP in the treated upholstery was estimated as 5 mg/cm^2.

Using the parameters described, the equilibrium room-air concentration of TDCPP was estimated to be 200 mg/m^3. The short-term time-average exposure concentration for TDCPP was estimated as 50 mg/m^3 using Equation 12 in Chapter 3 and the equilibrium room-air concentration for TDCPP. It was estimated that concentration could be maintained for approximately 1 mo.

These results indicate that if all of the TDCPP is released from the fabric into the air, TDCPP could be a toxic risk to persons entering the room. In reality, any flame retardant that evaporated so rapidly would be useless in preventing upholstery flammability. Either TDCPP is much more strongly bound to the fabric than assumed in this scenario (so that the parameter γ in the analysis above is substantially less than unity), or the chemical is transformed during the application process. In either case, the emission rate would likely be controlled by some process other than diffusion through a boundary layer of air, as assumed here. It is in the opinion of the subcommittee that this exposure scenario provides no useful information about the potential toxicity of TDCPP vapors to humans associated with the emission of TDCPP vapors from treated furniture upholstery. Therefore, further investigation should be carried out to determine if exposure to TDCPP by this route poses a toxic risk to humans.

Oral Exposure

The assessment of noncancer toxicological risk for oral exposure to TDCPP is based on the oral exposure scenario described in Chapter 3. This scenario assumes a child is exposed to TDCPP by sucking on 50 cm^2 of fabric back-coated with TDCPP, 1 hr/d for two yr. The subcommittee estimated an upholstery application rate (S_a) for TDCPP of 5 mg/cm^2. Oral exposure was calculated using Equation 15 in Chapter 3. The extraction rate (μ_w) for TDCPP was estimated to be 0.038 based on extraction data for organic phosphates in polyester fiber (McIntyre et al. 1995). The release rate from the fiber for estimating extraction was 0.06/d at 28°C calculated using the equation $2d/2\pi R$ (d = film thickness, R = fiber radius) with a correction from fiber to film of a factor of 0.63.

The worst case average oral daily dose for TDCPP was estimated as 0.04 mg/kg-d. Division of the dose estimate by the oral RfD for TDCPP of 0.005 mg/kg-d gives a hazard index of 8.0. This suggests that under the subcommittee's worst-case exposure assumptions, TDCPP could be a toxic hazard by the oral route of exposure.

Cancer

Dermal

Human cancer risk for dermal exposure to TDCPP was calculated by multiplying the oral cancer potency factor for TDCPP by the lifetime average dermal dose rates of 1.5 mg/kg-d or 2.6×10^{-3} mg/kg-d (see Noncancer Dermal Exposure section). The subcommittee felt that the use of the oral cancer potency factor for TDCPP based on testicular tumors in rats was acceptable for the calculation of cancer risk for dermal exposure since the oral cancer potency factor is based on carcinogenic effects following near-complete systemic absorption and the appearance of tumors not at the site of TDCPP application.

A lifetime cancer risk of 9.0×10^{-7} was obtained by multiplying the first iteration exposure estimate of 1.5 mg/kg-d times the TDCPP oral cancer potency factor of 0.06 mg/kg-d. Multiplication of the oral cancer potency factor times the average dermal daily dose of 2.6×10^{-3} mg/kg-d developed in the alternative dermal exposure iteration gives an estimated lifetime cancer risk of 1.6×10^{-4}. These estimates suggest that the dermal route of exposure may pose a carcinogenic hazard for persons exposed to TDCPP incorporated into residential furniture upholstery at the indicated concentration levels and under the given worst-case exposure scenario.

Inhalation

Particles

The average room-air concentration and average exposure concentration to TDCPP particles estimated in the previous sections were used for the cancer assessment. An inhalation cancer potency value was not available for TDCPP, therefore a provisional inhalation cancer potency value was derived from oral cancer potency data for TDCPP. Multiplication of the exposure estimates of 0.48 $\mu g/m^3$ for particles times the provisional cancer potency value of $1.71 \times 10^{-5}/\mu g/m^3$ produces estimated lifetime cancer risks of 8.2×10^{-6} and suggests that the cancer risk associated with the inhalation of TDCPP particles is negligible at the given upholstery concentrations and the exposure parameters in the worst-case exposure scenario. However, the subcommittee noted that exposure to TDCPP by this route may need further evaluation.

Vapors

For TDCPP vapors, the equilibrium concentration of vapor-phase TDCPP in room air was estimated as described in the Noncancer Inhalation Exposure section. The long-term time-average vapor exposure concentration for TDCPP was estimated using Equation 14 in Chapter 3.

Oral

As discussed previously, TDCPP is judged to be a rodent carcinogen. Therefore, the conservative approach for risk assessment purposes is to assume that TDCPP represents a carcinogenic risk to humans.

Using Equation 16 in Chapter 3, the lifetime average dose rate for TDCPP by the oral exposure route was calculated to be 1.1×10^{-3} mg/kg-d. Lifetime cancer risk for this exposure scenario was then estimated by multiplying the oral lifetime daily dose rate times the most conservative oral cancer potency factor for TDCPP (0.06/mg/kg-d) yielding a cancer risk estimate of 6.6×10^{-5}. This suggests that under the subcommittee's worst-case exposure assumptions, TDCPP could be a carcinogenic hazard by the oral route of exposure.

RECOMMENDATIONS FROM OTHER ORGANIZATIONS

The subcommittee is not aware of exposure limits proposed by regulatory agencies or other organizations.

DATA GAPS AND RESEARCH NEEDS

There are no data on the chronic toxicity of TDCPP by the dermal or inhalation routes of exposure. Data on the rate of dermal absorption of TDCPP are needed and there is no information on the metabolism of TDCPP in animal or human systems. No information is available on human exposure to TDCPP from treated furniture upholstery. No studies have been conducted on the leaching of TDCPP from treated materials.

Based on an oral hazard index of greater than 1 and potential cancer risk from all three routes of exposure, the subcommittee recommends that the potential for particle and vapor release and TDCPP release into saline from treated fabric be investigated.

REFERENCES

Akzo Nobel. 1998. Public Submission to the US CPSC, June 1998.

Bio/dynamics (Bio/dynamics, Inc.). 1981. A Two-year Oral Toxicity/Carcinogenicity Study on Fyrol FR-2 in Rats (Final Report). Study conducted for Stauffer Chemical Company. U.S. EPA/OPTS Public files, Fiche # OTS0204911-1.

Bloom, S.E. 1982. Cytogenet. Assays Environ. Mutagens:137.

Brusick, D., D. Matheson, D.R. Jagannath, S. Goode, H. Lebowitz, M. Reed, G. Roy, and S. Benson. 1980. A comparison of the genotoxic properties of tris (1,3-dichloro-2-propyl)phosphate in a battery of short-term bioassays. J. Environ. Pathol. Toxicol. 3:207-226.

Bullock, C.H., and Kamienski, F.X. 1972. Report No. T-4055. Stauffer Chemical Company. Western Research Center, Richmond, CA. As cited in IPCS (1998).

Celanese Corp. (Celanese Corporation of America). 1960. Mellon Institute Report, Cellulube FR-2, February 1960.

EPA (U.S. Environmental Protection Agency). 1996. Proposed Guidelines for Carcinogen Risk Assessment. Fed. Regist. 61(Apr. 23):17960-18011.

Ferrante, J. 1999. Toxicity Review for Tris (1,3-dichloropropyl)-2 phosphate. Memorandum updated from Jacqueline Ferrante, Pharmacologist, Division of Health Sciences, to Ronald L. Medford, Assistant Executive Director for Hazard Identification and Reduction. U.S. Consumer Product Safety Commission, Washington, DC.

Fire Retardant Chemicals Association. 1998. Tri(1,3-dichloropropyl-2) phosphate. Pp. 7 in Textile Flame Retardant Applications by Product Classes for 1997 Within and Outside of the United States. Fire Retardant Chemicals Association, Lancaster, PA.

Gold, M.D., A. Blum, and B.N. Ames. 1978. Another flame retardant, tris(1,3-dichloro-2-propyl)-phosphate, and its expected metabolites are mutagens. Science 200:785-787.

HSDB (Hazardous Substances Data Bank). 1989. Tris(1,3-dichloro-2-propyl) phosphate; CAS# 13674-87-8. Entry last reviewed 03/17/89. Entry retrieved from MEDLINE on September 21, 1999.

IPCS (International Programme on Chemical Safety). 1998. Environmental Health Criteria 209. Flame Retardants: Tris(chloropropyl) phosphate and Tris (2-chloroethyl) phosphate. International Programme on Chemical Safety, World Health Organization, Geneva.

Kamata, E., K. Naito, Y. Nakaji, Y. Ogawa, S. Suzuki, T. Kaneko, K. Takada, Y. Kurokawa, and M. Tobe. 1989. Acute and subacute toxicity studies of tris (1,3-dichloro-2-propyl) phosphate on mice. Eisei Shikenjo Hokoku 107:36-43. [Article in Japanese]. [Abstract in English].

Kawashima, K., S. Tanaka, S. Nakaura, S. Nagao, T. Endo, K. Onoda, A. Takanaka, and Y. Omori. 1983. Effect of phosphoric acid tri-esters flame retardants on the prenatal and postnatal developments of rats. J. Toxicol. Sci. 8(1):339.

Luster, M.I., J.H. Dean, G.A. Boorman, D.L. Archer, L. Lauer, L.D. Lawson, J.A. Moore, and R.E. Wilson. 1981. The effects of orthophenylphenol, tris (2,3-dichloropropyl)phosphate, and cyclophosphamide on the immune system and host susceptibility of mice following subchronic exposure. Toxicol. Appl. Pharmacol. 58(2):252-261.

Lynn, R.K., K. Wong, C. Garvie-Gould, and J.M. Kennish. 1981. Disposition of the flame retardant, tris (1,3-dichloro-2-propyl) phosphate, in the rat. Drug Metab. Dispos. 9(5):434-41.

Majeska, J.B., and D.W. Matheson. 1983. Quantitative Estimate of Mutagenicity of Tris-(1,3-dichloro-2-propyl)-phosphate (TCPP) and Its Possible Metabolites in Salmonella. [Abstract]. Environ. Mutagen. 5:478-479.

Matthews, H.B., and M.W. Anderson. 1979. Disposition of tris-(1,3-dichloro-2-propyl)-phosphate in the rat. Toxicol. Appl. Pharmacol. 48:A184.

McIntyre, J.E., I. Holme, and O.K. Sunmonu. 1995. The desorption of model compounds from poly(ethylene terephthalate) fibre. Colourage 41(13)77-81.

Minegishi, K.-I., H. Kurebayashi, S. Nambaru, K. Morimoto, T. Takahashi, and T. Yamaha. 1988. Comparative studies on absorption, distribution, and excretion of flame retardants halogenated alkyl phosphate in rats. Eisei Kagaku 34(2):102-114.

Morales, N.M., and H.B. Matthews. 1980. In vivo binding of the flame retardants tris (2,3-dibromopropyl) phosphate and tris (1,3-dichloro-2-propyl) phosphate to macromolecules of mouse liver, kidney and muscle. Bull. Environ. Contam. Toxicol. 25(1):34-38.

Nakamura, A., N. Tateno, S. Kojima, M.A. Kaniwa, and T. Kawamura. 1979. The mutagenicity of halogenated alkanols and their phosphoric acid esters for Salmonella typhimurium. Mutat. Res. 66(4):373-380.

Nomeir, A.A., S. Kato, and H.B. Matthews. 1981. The metabolism and disposition of tris (1,3-dichloro-2-propyl) phosphate (Fyrol FR-2) in the rat. Toxicol. Appl. Pharmacol. 57(3):401-413.

NRC (National Research Council). 1985. Emergency and Continuous Exposure Limits for Selected Airborne Contaminants, Volume 5. Committee on Toxicology. Board on Toxicology and Environmental Health Hazards, National Research Council. Washington, D.C.: National Academy Press.

Osterberg, R.E., and G.W. Bierbower. 1978. Unpublished data on Fyrol toxicity.

Piotrowski, J.K., B. Baranowska-Dutkiewicz, A. Szadowska, and W. Szaniawski. 1976. Acute toxicity of technical tris(1,3-dichloroisopropyl)phosphate. Bromatol. Chem. Toksykol 9(2):141-147.

Prival, M.J., E.C. McCoy, B. Gutter, and H.S. Rosendranz. 1977. Tris(2,3-dibromo-propyl) phosphate: mutagenicity of a widely used flame retardant. Science 195(4273):76-78.

Sanders, H.J. 1978. Flame retardants: Government regulations and public emphasis on safety provide the impetus for an expanding industry. C&EN April 24, 1978. pp. 22-36.

Soderlund, E.J., E. Dybing, J.A. Holme, J.K. Hongslo, E. Rivedal, T. Sanner, and S.D. Nelson. 1985. Comparative genotoxicity and nephrotoxicity studies of the two halogenated flame retardants tris(1,3-dichloro-2-propyl)phosphate and tris(2,3-dibromopropyl)phosphate. Acta Pharmacol. Toxicol. 56(1):20-9.

Stauffer Chemical Company. 1977-1978. Report to CPSC, 1977-1978.

Stauffer Chemical Company. 1981. A Two-year Oral Toxicity/Carcinogenicity Study (Final Report) on Fyrol FR-2 in Rats (Sanitized). EPA/OPTS Public Files, Fiche # OTS0204911-1.

Stauffer Chemical Company. 1983a. A Morbidity Survey of Workers Employed at a Fyrol FR-2 Manufacturing Plant. Stauffer Chemical Company, Westport ,CT.

Stauffer Chemical Company. 1983b. A Mortality Study of Workers Employed at a Fyrol FR-2 Manufacturing Plant.

Tanaka, S., S. Nakaura, K. Kawashima, S. Nagao, T. Endo, K-I. Onoda, Y. Kasuya, and Y. Omori. 1981. Effect of oral administration of tris(1,3-dichloroisopropyl) phosphate to pregnant rats on prenatal and postnatal developments. Eisei Shikenjo Hokoku. 99:50-5.

Ulsamer, A.G., R.E. Osterberg, and J. McLaughlin, Jr. 1980. Flame-retardant chemicals in textiles. Clin. Toxicol. 17(1):101-131.

Wilczynski, S.L., J.M. Killinger, G.M. Zwicker, and R.I. Freudenthal. 1983. Fyrol® FR-2 fertility study in male rabbits. The Toxicologist 3(1):22.

17

Aromatic Phosphate Plasticizers

THIS chapter reviews the physical and chemical properties, toxicokinetics, toxicological, epidemiological, and exposure data on tricresyl phosphate (TCP), an aromatic phosphate ester. TCP is one of several aromatic phosphate esters used commercially as flame retardants and plasticizers. TCP was chosen as the representative aromatic phosphate ester flame retardant for risk assessment because it has the most complete toxicity database. The subcommittee used that information to characterize the health risk from exposure to TCP. The subcommittee also identified data gaps and recommended research relevant for determining the health risk from exposure to TCP.

PHYSICAL AND CHEMICAL PROPERTIES

Tricresyl phosphate (TCP) is one of several aromatic phosphate esters used commercially as a flame retardant. The physical and chemical properties of TCP are summarized in Table 17-1.

Commercial TCP is a complex mixture containing the meta TCP (TMCP) and para TCP (TPCP) isomers and mixed tricresyl and dicresyl phosphate esters. The ortho TCP isomer (TOCP) occurs only as a contaminant in commercial mixtures and usually at very low concentrations (<0.1%). This risk assessment focuses primarily on the toxicity of commercial TCP and the meta and para isomers of TCP. Data for TOCP are extensive and only those studies that are relevant for this risk assessment are included in this chapter.

TABLE 17-1 Physical and ChemicalProperties for Tricresyl Phosphate (TCP), Mixed Isomers

Properties	Value	Reference
Chemical Formula	$C_{21}H_{21}O_4P$	HSDB 1999
Structure	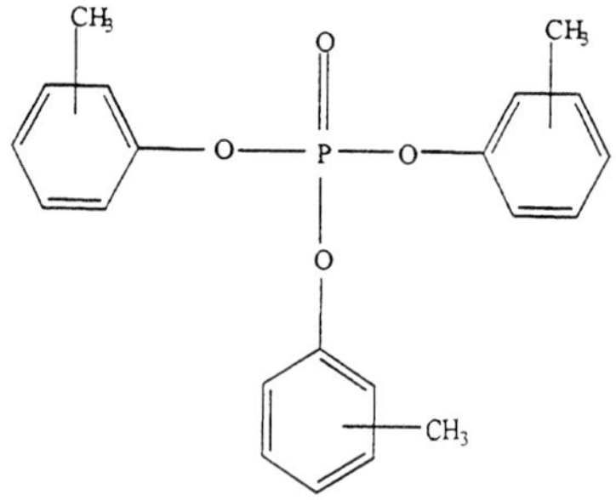	IPCS 1990
CAS Registry #	1330-78-5	ChemID 1999
Synonyms	Tritolyl phosphate; trimethylphenyl phosphate; phosphoric acid, tritolyl ester; phosphoric acid, tris(methylphenyl) ester; tris(tolyloxy)phosphine oxide	ChemID 1999; RTECS 1999
Trade Names	Celluflex 179C, Disflamoll TKP, Durad, Fyrquel 150, Flexol Plasticizer TCP, IMOL S 140, Lindol, Koflex 50[50], Kronitex-TCP, Phosflex 179, Pliabrac 521, PX.917, Santicizer 140	IPCS 1990; ChemID 1999; RTECS 1999
Molecular Weight	368.36	Budavari et al. 1989
Physical State	Colorless liquid	HSDB 1999
Solubility	0.36 mg/L in H_2O at 25 °C Miscible with all the common organic solvents and thinners, and also with vegetable oils	HSDB 1999 HSDB 1999
Vapor Pressure	1×10^{-4} mmHg at 20 °C	IPCS 1990
Partition Coefficients Log K_{ow}	5.11	HSDB 1999
Melting Point	-33 °C	IPCS 1990
Boiling Point	241-255 °C at 4 mm Hg	IPCS 1990
Flashpoint	257 °C	IPCS 1990
Henry's Law Constant	$1.1\text{-}2.8 \times 10^{-6}$ atm-m^3/mol	IPCS 1990
Refraction Index	1.556 at 25 °C	HSDB 1999
Viscosity	60 cSt at 25 °C, 4 cSt at 100 °C	IPCS 1990
Density	1.16 at 25 °C	HSDB 1999
Conversion Factor	1 ppm = 15.07 mg/m^3	IPCS 1990

OCCURRENCE AND USE

Currently, TCP is commercially manufactured by reacting phosphorous oxychloride with synthetically prepared cresols (IPCS 1990) to limit the formation of isomers and unwanted contaminants. Early manufacturing practices used petroleum- or coal tar-derived cresols.

Few data are available regarding the amount of TCP produced annually. Japanese production of TCP in 1984 was 33,000 tons, and that U.S. production in 1977 was 10,400 tons (IPCS 1990).

TCP is applied as a backcoating to upholstery fabrics when used as a flame retardant in upholstered furniture (Piccirillo 1999). It may be applied as backcoating on nylon, polyester, olefin, cotton, non-cotton cellulose, polyvinyl chloride (PVC) and blends of cotton/polyester, wool/nylon, wool/polyester, polyester/nylon, and nylon/olefin (R.Kidder, Fire Retardant Chemicals Association, pers. commun.,1998).

TOXICOKINETICS

Absorption

Dermal

No studies were identified that investigated the dermal absorption of TCP in humans.

It has been suggested that similarities with regard to structure and physical properties among the isomeric TCPs make it likely that the other isomeric TCPs would also be readily absorbed through the skin (NTP 1994). In the cat, 73% of the radioactivity from a 50-mg/kg dose of ^{14}C-TOCP was no longer present at the application site (intrascapular region)after 12 hr. Maximum concentrations of radioactivity were reached in the examined tissues within 24 hr. By d 10, at least 48% of the dose was absorbed as indicated by urinary and fecal excretion data (Nomeir and Abou-Donia 1986, as reviewed by NTP 1994).

Hodge and Sterner (1943), described by IPCS (1990), found that ^{32}P-TOCP (200 mg/kg) was poorly absorbed through dog abdominal skin. The absorption of 2 to 4 mg/kg TOCP by human palm skin was approximately 100 times faster than through the dog abdominal skin based on urinary excretion and surface-area data. Additional details were not provided.

Inhalation

No studies were identified that have investigated the absorption of TCP in humans or laboratory animals following inhalation exposure.

Oral

At least 41% of a single gavage dose of 7.8 mg/kg ^{14}C-labeled TPCP in rats was excreted in the urine over 7 d following administration (Kurebayashi et al. 1985). About 12% of a single gavage dose of 89.6 mg/kg in rats was excreted in the urine. Most of the urinary excretion occurred within the 24 hr after administration.

All three isomers of TCP (TMCP, TPCP, and TOCP) were administered by gavage to rats at doses of 2, 20, and 200 mg/kg in corn oil, were reported to be well absorbed by NTP (1994). The basis for this conclusion was not stated, but may have been based on comparisons of the area under the blood concentration versus time curves for intravenous (20 mg/kg) versus oral administration.

Distribution

Dermal

Distribution of radioactivity in the dog following a single 200-mg/kg application of ^{32}P-TOCP to the abdominal skin was highest in the liver followed by the blood, kidney, lung, muscle and spinal cord, brain and sciatic nerve at 24 hr post-exposure (Hodge and Sterner 1943, as reviewed by IPCS 1990). In cats, the highest levels of radioactivity occurred in the bile, gall bladder, urinary bladder, kidney, and liver at 1-10 d after application of 50 mg/kg of ^{14}C-TOCP (Nomeir and Abou-Donia 1986, as reviewed by IPCS 1990). In addition, low levels of radioactivity were found in the spinal cord and brain. Analysis showed that the parent compound was found primarily in the brain, spinal cord, and sciatic nerve, while metabolites were primarily found in the liver, kidney, and lung. It is not known if the patterns of distribution for TOCP and metabolites can be generalized to other TCP isomers.

Inhalation

No studies were identified that investigated the distribution of TCP in humans or laboratory animals following inhalation exposure.

Oral

Twenty-four hr after 89.6 mg/kg of ^{14}C-TPCP was administered by gavage to rats, the highest concentrations of radioactivity were found in the intestine (including contents), followed by the stomach, adipose tissue, liver, and kidneys (4-13-fold higher than blood concentrations). The lowest concentrations were found in heart, muscle, and brain (lower than blood concentrations) (Kurebayashi et al. 1985).

In rats, ^{14}C-TMCP, TPCP, and TOCP were rapidly distributed to muscle and liver following intravenous administration (NTP 1994). This was followed by a redistribution of radioactivity to adipose tissue and skin. The parent compounds were rapidly cleared rapidly from the tissues and did not bioaccumulate. Details of the study were not reported.

Metabolism

In rats, metabolism of TCP following oral gavage of 7.8 or 89.6 mg/kg was found to involve successive oxidations and hydrolysis resulting in the production of *p*-hydroxybenzoic acid (Kurebayashi et al. 1985). The major urinary metabolites identified were *p*-hydroxybenzoic acid, di-*p*-cresyl phosphate, and *p*-cresyl *p*-carboxyphenyl phosphate. The main biliary metabolites were di-*p*-cresyl phosphate, *p*-cresyl *p*-carboxyphenyl phosphate, and the oxidized triesters, di-*p*-cresyl *p*-carboxyphenyl phosphate, and *p*-cresyl *p*-carboxyphenyl phosphate. Fecal metabolites were similar to the biliary metabolites. $^{14}CO_2$ was found in expired air following administration and appeared to be formed probably through decarboxylation of *p*-hydroxybenzoic acid by intestinal microbes.

Many studies on the metabolism of TOCP are available. However, they might not be applicable to TMCP or TPCP. TOCP is metabolized to highly neurotoxic derivatives such as salingenin cyclic *o*-tolyl phosphate. However, there are no data to suggest that TMCP or TPCP is metabolized to neurotoxic metabolites. Exposure to TCP mixtures containing isomers with one or two ortho-cresol groups could result in the formation of neurotoxic metabolites (Johnson 1975; NTP 1994).

Excretion

Dermal

About 48% of a single dermal application of a 50 mg/kg dose was excreted by d 10 post-exposure with 28% of the dose excreted in the urine while 20% of the dose was excreted in the feces (Nomeir and Abou-Donia 1986, as reviewed by NTP 1994). Since the metabolism and excretion of orally administered TOCP appears to be different from TMCP and TPCP, the relevance of dermal excretion data for TOCP to other isomers is uncertain. No excretion data are available for other TCPs following dermal exposure.

Approximately 40-60% of an intravenous injection of 2 or 20 mg/kg of radiolabelled TMCP, TPCP, or TOCP underwent biliary excretion within 6 hr of administration (NTP 1994). For TPCP, biliary excretion increased with increasing dose from 2-20 mg/kg resulting in a doubling of biliary excretion. For all three TCPs, the percentage of administered radioactivity excreted in the feces was less than the percentage excreted in bile suggesting that the isomers underwent enterohepatic recirculation.

Inhalation

No studies were identified that investigated the excretion of TCP in humans or laboratory animals following inhalation exposure.

Oral

Excretion of radioactivity following oral administration of ^{14}C-TMCP, ^{14}C-TPCP, or ^{14}C-TOCP in rats at doses of 0.5 (^{14}C-TMCP and ^{14}C-TPCP only) 2, 20, and 200 mg/kg was investigated by NTP (1994). Radioactivity from TMCP was excreted primarily in the feces at all dose levels. Radioactivity from TPCP was excreted primarily in the urine at 0.5 and 2 mg/kg and primarily in the feces at 20 and 200 mg/kg. Radioactivity from TOCP was excreted primarily (70%) in the urine at all doses tested.

Rats that received ^{14}C-TPCP as a single gavage dose of 7.8 mg/kg excreted 41% of the dose of radioactivity in the urine, 44% in the feces, and 18% in the expired air within 7 d (Kurebayashi et al. 1985). A majority of the excretion occurred within 24 hr. Rats with cannulated bile ducts excreted 28% of the administered radioactivity in the bile during the first 24 hr. Rats treated in a similar manner with 89.6 mg/kg of ^{14}C-TPCP excreted 12% of the administered radioactivity in the urine, 77% in the feces, and 6% in the expired air. The radiolabeled material excreted in urine and bile was identified as metabolites of TPCP in high dose rats (see Metabolism section for details). Parent compound was the dominant isomer excreted in the feces with some lesser amounts of metabolites present.

HAZARD IDENTIFICATION[1]

Dermal Exposure

Irritation

Broadhurst et al. (1951) reported mild irritation following initial contact with TCP formulations in patch tests on volunteers. Rabbits studied by Broadhurst et al. (1951) did not develop any irritation to TCP.

The skin of rabbits exposed for 24 hr to commercial TCP mixtures at both lethal and sublethal doses showed mild inflammation, and in a few instances, focal acanthosis and slight hyperkeratosis (Treon et al. 1955). Repeated dermal exposure produced local skin inflammation. Inflammation was more severe in animals that died from treatment than among survivors (Treon et al. 1955).

Eastman Kodak Company (1978) reported moderate skin irritation for TOCP and TPCP, slight irritation for TMCP, and no irritation for a TCP mixture in guinea pigs. No study details were reported.

Sensitization

Sensitization by TCP has been reported, it is not common. Tarvainen (1995) found no positive responses among 839 patients at a dermatology clinic given patch tests with TCP. Broadhurst et al. (1951) also reported negative results for sensitization in patch testing of volunteers.

Pegum (1966) described the case of a housewife who became sensitized to PVC plasticized with TCP. Patch tests with TCP produced a similar sensitization reaction. However, there is evidence that triphenyl phosphate, which is present in TCP mixtures, is a sensitizing agent (Carlsen et al. 1986) and may account for some or all of the sensitizing potential of TCP mixtures.

Eastman Kodak Company (1978) reported moderate skin sensitizing activity for TMCP in guinea pigs. Study details were not available.

Systemic Effects

No studies were identified that investigated the systemic effects of TCP from dermal exposures in humans.

---

[1]In this section, the subcommittee reviewed toxicity data on aromatic phosphate plasticizers, including the toxicity assessment prepared by the U.S. Consumer Product Safety Commission (Ferrante 1999).

Treon et al. (1955) studied the effects in rabbits of 24-hr dermal exposure to seven commercial TCP mixtures. The minimum lethal dose varied between 0.4-0.6 mL/kg and 1.6-3.2 mL/kg for the different compounds. Lethal doses resulted in diffuse degenerative changes in the brain, liver, and kidney, and edema in the other viscera. Pathological changes were not observed in survivors. Treon et al. (1955) also conducted repeated dermal exposure studies in which groups of three to four female rabbits were exposed to 0.25, 0.5, 1, 2, or 5 mL of one of seven commercial TCP mixtures 2 hr/d, 5 d/wk, for several weeks. Death, which was produced by doses as low as 0.25 mL, was preceded by ataxia and tremors. Degenerative changes were seen in the brain, liver, and kidneys of rabbits that died. Mild changes were also seen in the liver and kidneys of survivors treated with one of the TCP formulations.

Broadhurst et al. (1951) reported death and clinical signs of delayed neuropathy (head drop, paralysis) in rabbits treated dermally with TCP. However, no changes in histopathology were reported.

Neurological Effects

Several cases have been reported of workers who developed polyneuropathy following occupational dermal exposure to TCP, and specifically TOCP (IPCS 1990). Percutaneous absorption was considered to be the most likely route of exposure in each of these cases, although some material may have been inhaled or ingested. The effects in these workers were similar to those observed in people who accidentally ingested TOCP.

No data were located regarding neurological effects in animals following dermal exposure to TCP.

Other Effects

No studies were identified that investigated the immunological, or reproductive, developmental, or carcinogenic effects of TCP following dermal exposure.

Inhalation Exposure

Systemic and Neurological Effects

Only one study was located that investigated the toxicity of TCP in humans following inhalation exposure. Bisesi (1994) reported that Tabershaw and

Kleinfeld (1957) found some inhibition of plasma cholinesterase, but no neuromuscular effects, in TOCP-manufacturing plant workers exposed to TOCP air concentrations ranging from 0.27 to 3.40 mg/m^3.

Animal inhalation toxicity data for TCP were also limited. Mortality was very high in rabbits exposed to TCP aerosols at concentrations of 5,900 mg/m^3 to 42,200 mg/m^3 for periods of 3 hr to 18 d (Broadhurst et al. 1951). Rabbits were observed to have considerably increased nasal and oral discharge during and immediately following exposure, and respiratory difficulties were noted. Diarrhea was also seen. Delayed neuropathy was also evident, progressing from hyperexciteability to tremors, gait impairment, and in several animals, paralysis of the hind legs. Serum cholinesterase was depressed. Histopathological evaluation of the respiratory tract revealed respiratory irritation, including bronchitis, inflammation of the larynx, and pulmonary edema.

Treon et al. (1955) reported only small, transitory changes in body weight, no deaths, no clinical signs of toxicity, and no treatment-related lesions (examined tissues not reported) in 3 rats, 5 mice, 2 rabbits, 2 guinea pigs, and 1 cat exposed 7 hr/d for 8 d to air containing 62 mg/m^3 of TCP vapor (unspecified mixture).

Other Effects

No studies were identified that investigated the immunological, reproductive, developmental, or carcinogenic effects of TCP in humans or animals following inhalation exposure.

Oral Exposure

Systemic Effects

NTP (1994) conducted 16-d gavage studies, 13-wk gavage and feed studies, and 2-yr feed studies of a commercial TCP mixture in rats and mice. A battery of chemical analyses revealed the test material to be a complex mixture containing 79% tricresyl phosphate esters and 18% dicresyl phosphate esters. The results of these and other studies are summarized in Table 17-2 and discussed below.

Acute Studies

The acute toxicity of TCP has been investigated in rodents by NTP (1994) and Chapin et al. (1988). NTP (1994) found that acute administration of TCP

TABLE 17-2 Summary of Oral Toxicity Dose-Response Data for Tricresyl Phosphate

Species, Strain, Sex, Number	Dose (mg/kg-d)	Duration, route	Effects	NOAEL/LOAEL (mg/kg-d)	Reference
Rat, F-344/N, M/F, 10/sex/dose	0, 360, 730, 1,450, 2,900, or 5,800	13-14 d, gavage	Death; diarrhea; decreased body weight; decreased neurobehavioral performance; increased liver weight; decreased thymus weight; testicular aspermatogenesis; necrosis in thymus, spleen, salivary gland and lymph node	LOAEL: 360 (increased liver weight)	NTP 1994
Mouse, $B6C3F_1$, M/F, 10/sex/dose	0, 360, 730, 1,450, 2,900, or 5,800	13-14 d, gavage	Death; decreased neurobehavioral performance; increased liver weight; decreased thymus weight; necrosis in thymus, spleen and lymph node	LOAEL: 360 (decreased neurobehavioral performance, increased liver weight)	NTP 1994
Rat, F-344/N, M/F, 10/sex/dose	0, 50, 100, 200, 400, or 800	13 wk, gavage	Decreased body weight; decreased serum cholinesterase; decreased neurobehavioral performance; increased liver weight; decreased thymus weight; atrophy of seminiferous tubules in testes; hypertrophy of ovarian interstitial cells; cytoplasmic vacuolization of the adrenal cortex	LOAEL: 50 (lesions in ovary and adrenals, decreased serum cholinesterase)	NTP 1994
Mouse, $B6C3F_1$, M/F, 10/sex/dose	0, 50, 100, 200, 400, or 800	13 wk, gavage	Decreased body weight; hind limb weakness and tremors; decreased neurobehavioral performance; decreased serum cholinesterase; increased liver weight; hypertrophy of ovarian interstitial cells; cytoplasmic vacuolization of the adrenal cortex; axonal degeneration in the spinal cord and sciatic nerve	LOAEL: 50 (lesions in ovary and adrenals, decreased serum cholinesterase)	NTP 1994
Rat, F-344/N, M/F, 10/sex/dose	M: 0, 55, 120, 220, 430, or 750 F: 0, 65, 120, 230, 430, or 770	13 wk, diet	Decreased food consumption; decreased body weight; emaciation; decreased neurobehavioral performance; decreased serum cholinesterase; increased liver weight; decreased testis weight; atrophy of seminiferous tubules in testes; hypertrophy of ovarian interstitial cells; cytoplasmic vacuolization of the adrenal cortex; edema and necrosis of the renal papilla; hypertrophy of pituitary basophils	LOAEL: 55/65 (lesions in ovary and adrenals, decreased serum cholinesterase)	NTP 1994

Mouse, B6C3F_1, M/F, 10/sex/dose	M: 0, 45, 110, 180, 380, or 900; F: 0, 65, 130, 230, 530, or 1,050	13 wk, diet	Decreased food consumption; decreased body weight; tremors; decreased neurobehavioral performance; decreased serum cholinesterase; hypertrophy of ovarian interstitial cells; cytoplasmic vacuolization of the adrenal cortex; axonal degeneration in the spinal cord and sciatic nerve; hyperplasia in gallbladder; renal tubule regeneration	LOAEL: 45/65 (adrenal lesions decreased serum cholinesterase)	NTP 1994
Rat, F-344/N, M/F, 95/sex/dose	M: 0, 3, 6, or 13; F: 0, 4, 7, or 15	104 wk, diet	Decreased neurobehavioral performance; decreased serum cholinesterase; hypertrophy of ovarian interstitial cells; cytoplasmic vacuolization of the adrenal cortex	Excluding decreased serum cholinesterase (see text): NOAEL: 7 LOAEL: 15 (lesions in ovary and adrenals)	NTP 1994
Mouse, B6C3F_1, M/F, 95/sex/dose	M: 0, 7, 13, or 27; F: 0, 8, 18, or 37	104 wk, diet	Decreased neurobehavioral performance; decreased serum cholinesterase; increased adrenal weight; lesions in the adrenal cortex and liver	Excluding decreased serum cholinesterase (see text): NOAEL: 7/8 LOAEL: 13/18 (lesions in liver and adrenals)	NTP 1994
Rat, Long-Evans, M/F, 12 M /dose, 24 F/dose	M: 0, 100, or 200; F: 0, 200, or 400	M: 56 d before breeding and 10 d during breeding, gavage F: 14 d before breeding through lactation, gavage	Decreased fertility; decreased litter size; decreased sperm concentration, motility and velocity; increased abnormal sperm; decreased epididymis weight; necrosis and degeneration of seminiferous tubules; hypospermia in epididymis; vacuolar cytoplasmic alteration of ovarian interstitial cells	LOAEL: 100/200 (reproductive effects)	Carlton et al. 1987

TABLE 17-2 *(Continued)*

Species, Strain, Sex, Number	Dose (mg/kg-d)	Duration, route	Effects	NOAEL/LOAEL (mg/kg-d)	Reference
Mouse, Swiss CD-1, M/F, 20/sex/dose	0, 62.5, 124 or 250	7 d before mating through 14 wk of breeding, diet	Decreased body weight; hind limb weakness; decreased litters per pair; decreased live pups per litter; decreased proportion of pups born alive; decreased pup body weight; decreased kidney and adrenal weight, testis weight and epididymis weight; decreased sperm concentration and motility; increased abnormal sperm; atrophy of seminiferous tubules; hypertrophy and degeneration of adrenals	LOAEL: 62.5 (reproductive effects, decreased body weight)	NTP 1984; Chapin et al. 1988
Rat, W star, M, 10/dose	0, 2.4, 6 or 12	6 wk, diet	Decreased humoral and cell-mediated immune response	NOAEL: 2.4 LOAEL: 6 (decreased immune response)	Banerjee et al. 1992

F, female; LOAEL, lowest-observed-adverse-effect level; M, male; NOAEL, no-observed-adverse-effect level.

by gavage to rats or mice caused increases in deaths, decreases in body weight, various neurological effects, and necrosis of various tissues. Diffuse aspermatogenesis was observed in high-dose male rats. Liver weights were significantly increased in mice of both sexes at dose levels ≥360 mg/kg-d.

Chapin et al. (1988) found that all mice treated with TCP at ≥2,280 mg/kg-d died before the end of the 14-d exposure period. Decreases in body weight were observed at doses of 570 and 1,140 mg/kg-d.

Subchronic Studies

The results of 90-d oral toxicity studies for TCP are summarized in Table 17-2. In a subchronic study conducted by NTP (1994), mortality was not increased in rats or mice treated by gavage with TCP. Terminal body weights were significantly reduced in male rats and mice at ≥200 mg/kg-d, and female mice at ≥400 mg/kg-d. Decreased hindlimb grip strength was noted in female rats at ≥400 mg/kg-d and various neurological parameters were considered abnormal at ≥200 mg/kg-d in both male and female mice. Liver and thymus weights were increased in male rats, mice and female rats at doses ≥400 mg/kg-d and ≥200 mg/kg-d in mice. Atrophy of the seminiferous tubules occurred ≥400 mg/kg-d in male rats. Hypertrophy of ovarian interstitial cells occurred in female mice and rats at ≥50 mg/kg-d. The adrenal cortex was also affected in both species (see Table 17-2).

NTP (1994) also conducted subchronic toxicity studies for TCP in which rats and mice (10 animals/sex/dose) were given TCP in their feed for 90 d (see Table 17-2). All animals survived to the end of the study. In rats, final body weight was significantly reduced in males in the 6,600 ppm group and in females in the 3,300 ppm and 6,600 ppm groups without large changes in food intake. In mice, final body weight was reduced in males at 4,200 ppm and in females at 2,100 ppm and 4,200 ppm. The reduction in body weight was accompanied by a reduction in feed consumption in the females. Hindlimb grip strength was reduced in male rats fed TCP at 13,000 ppm and in male and female mice at ≥2,100 ppm. Various pathological changes and increases in the weights of various organs were seen in both species for both sexes.

Chronic Studies

Groups of 95 male and female F-344/N rats were fed diets containing 0, 75, 150, or 300 ppm of TCP for 104 wk (estimated doses of 0, 3, 6, or 13 mg/kg-d in males and 0, 4, 7, or 15 mg/kg-d in females) (NTP 1994). A second group of rats was fed diets containing 600 ppm TCP for 22 wk and control feed thereafter. Fifteen rats of each sex from each dose group (including 600 ppm) were killed for interim evaluations after 3, 9, and 15 mo. Most of the 600 ppm group (those not included in the interim evaluation groups) were killed and discarded after the 3-mo interim evaluation. The results of this study are summarized in Table 17-2.

No treatment-related deaths occurred and no clinical signs of toxicity were reported. Food intakes and body weights of treated and controls were not significantly different. Hindlimb grip strength was significantly reduced in males at ≥300 ppm and females at 600 ppm in the 3-mo interim evaluation, but not in the subsequent 9- and 15-mo evaluations. A treatment-related reduction in serum cholinesterase was observed in both males and females at 3-, 9-, and 15-mo interim evaluations. Treatment-related adrenal cytoplasmic vacuolization occurred in males at 600 ppm at 3 mo, but not at later time periods when this group received control feed. In females, adrenal cytoplasmic vacuolization was significantly increased at 300 ppm at all time periods and at 600 ppm at the 3-mo evaluation. Absolute and relative adrenal weights were significantly increased at ≥300 ppm at the 3-mo interim evaluation. The incidence of minimal-to-mild interstitial cell hyperplasia of the ovary was significantly increased at ≥150 ppm in the 3-mo interim evaluation, but only at 300 ppm in subsequent evaluations.

In the mouse study, groups of 95 male and female B6C3F$_1$ mice were fed diets containing 0, 60, 125, or 250 ppm of TCP (estimated mg/kg-d dose for males: 7, 13, 27; for females: 8, 18, or 37) for 105 wk (NTP 1994). There were no effects on survival, feed consumption, body weight, or clinical findings. Hindlimb grip strength was significantly reduced in female mice in the 250 ppm group in the 3-mo evaluation, but not in the subsequent evaluations. Serum cholinesterase was significantly reduced in a dose-related manner in both males and females at ≥60 ppm at all time points. Non-neoplastic pathology findings were limited to the adrenal gland and liver. Absolute and relative weights of both the right and left adrenal glands were significantly increased in female mice at 250 ppm in the 15-mo interim evaluation. Microscopic examination revealed a dose-related increase in the severity of ceroid pigmentation, distension of epithelial cells, and macrophages of the adrenal cortex in female mice starting at the 9-mo interim evaluation and continuing through the end of the study. Male mice at the end of the study had significant increases in the incidences of liver lesions at ≥125 ppm and higher-dose groups, including clear cell foci (foci of enlarged cells with clear spaces in cytoplasm), fatty change (small vacuoles in hepatocytes throughout the liver), and ceroid pigmentation (cells, sometimes enlarged and clustered, with pigmented granules in cytoplasm).

Other Studies

Absolute and relative liver weights were significantly increased in male JCL-Wistar rats fed diets containing 0% (n = 18) or 0.5% (n = 8) TCP (estimated dose = 450 mg/kg-d) for 9 wk (Oishi et al. 1982). TCP was a mixture of unspecified composition. TCP had no effect on body and other organ weights as compared to controls. Serum cholesterol, bile acids, total protein, urea nitrogen, and alanine aminotransferase levels were significantly elevated. Mild histopathological changes were found in the liver characterized by cytoplasmic vacuolation, increased number of binucleated cells, and enlarged cell size.

Histopathological examination of unspecified tissues found no treatment-related effects in rats exposed for 3 mo to TCP (60-65% TMCP and 35-40% TPCP) suspended in water with 5% gum arabic at doses of 30, 100, 300, or 1,000 mg/kg-d (Saito et al. 1974 as reviewed by IPCS 1990). No further details of this study were reported.

Immunological Effects

No data were located regarding immunological effects of TCP in humans.

Banerjee et al. (1992) reported that dietary exposure to TCP caused a suppression of humoral and cell-mediated immune response in rats. Groups of 10 male Wistar rats were fed diets containing 0, 20, 50, or 100 ppm of TCP (90% ortho, meta, and para isomers) for 6 wk. Doses in the 0-, 20-, 50-, and 100-ppm groups were estimated to be 0, 2.4, 6, and 12 mg/kg-d of TCP, respectively. After 25 d of exposure, rats were immunized with tetanus toxoid. No clinical signs of toxicity were observed, and food and water intake, body weight and relative organ weights in treated rats were similar to controls. Serum antibody titres to tetanus toxoid were significantly reduced at 50 and 100 ppm TCP. Serum immunoglobulins (IgM and IgG) were significantly reduced at 100 ppm TCP while leucocyte and macrophage migration was inhibited at 50 and 100 ppm TCP. Although the results of this study suggest that the immune system may be a sensitive target for TCP, the study included only limited assessment of immune function and the test material was not well characterized.

Brinkerhoff et al. (1981) found little evidence of immunotoxicity in mice treated by oral gavage with TOCP at 0, 5, 50, or 500 mg/kg or 50 mg/kg of TMCP, once/wk, for 1-13 wk. Assays were conducted for splenic plaque formation, serum immunoglobulin levels (IgA, IgG, IgM), delayed hypersensitivity in response to sheep erythrocytes, and lymphocyte transformation in response to various mitogens in splenic cultures. No treatment-related changes in body or organ weights were detected in treated versus control animals. Lymphocyte proliferation to PWM (but not PHA or LPS) was reduced in all TOCP and TMCP dose groups at 13 wk. However, the decreases were small and there was no evidence of a dose-response relationship. No other evidence for immune effects of TOCP and TMCP were detected in other assays. The researchers concluded that this study found no significant alterations of immune function.

Neurological Effects

TCP, and especially TOCP, have been implicated in outbreaks of polyneuropathy that have affected tens of thousands of people around the world since the late 1890s (Morgan 1982; IPCS 1990). Most of these outbreaks were traced to contamination of food or cooking oil with lubricating oil, mineral oil, hydraulic fluid, or some similar material containing TCP. The first symptom of delayed neuropathy in affected persons, occurring 3-28 d after exposure, is sharp, cramp-like pain in the calves (IPCS 1990). This is followed within a few days by weakness of the leg muscles and unsteadiness. Symptoms progress over a period of days or weeks to partial or complete paralysis that may include the upper, as well as lower, extremities. These effects are associated with axonopathy of both motor and sensory distal axons characterized by transection of the axon and degeneration of the axon and myelin sheath distal to the transection (IPCS 1990; NTP 1994). The axonopathy is most prominent in long, large-diameter myelinated axons of peripheral nerves and long spinal tracts. Although the effects may regress over time in mild cases, many of the individuals affected in various outbreaks still showed severe effects many years after exposure.

Studies in laboratory animals have showed that TOCP produces delayed neuropathy to varying degrees in many species (IPCS 1990). Ability to inhibit brain neurotoxic esterase in animals is used as a marker of delayed neuropathy (referred to as organophosphorous-induced delayed neuropathy [OPIDN]). These studies have shown that TOCP is a much more potent inducer of OPIDN than TCP mixtures. Metabolism of the ortho-cresol residue of TOCP produces a cyclic phosphate that is thought to be the proximate toxicant for TOCP-induced OPIDN (NTP 1994). No evidence has been found that TMCP or TPCP can produce OPIDN (e.g., Aldridge and Barnes 1966; Johannsen et al. 1977; Sprague and Castles 1985). However, it is possible that cyclic phosphate could be formed from the metabolism of mixed, ortho-cresol-containing TCP isomers and produce neurotoxic effects if these isomers are present.

Reproductive and Developmental Effects

No studies were identified that evaluated the reproductive and developmental effects of TCP in humans.

There are a number studies that have investigated the effects of TCP on reproductive and developmental parameters; they are summarized in Table 17-2.

TCP, containing <9% TOCP, administered to breeding male and female rats was found to have an effect on various reproductive and developmental toxicity parameters (Carlton et al. 1987). Male rats (12/dose) were administered 0, 100, or 200 mg/kg-d of TCP in corn oil by daily gavage for 56 d prior to breeding and throughout a 10-d breeding period. Females rats (24/dose) were gavaged daily with 0, 200, or 400 mg/kg-d TCP for 14 d prior to breeding and throughout breeding, gestation, and d 21 of lactation. No clinical signs of toxicity or effects on body weight were observed in breeding males or females. Fertility and mean litter size were significantly reduced in a dose-dependent manner in breeding females. All high-dose pups died on lactation d 5. Pup weight, days of eye opening, and vaginal patency were not affected. In breeding adult males, a dose-related increase in the percentage of abnormal sperm was observed in males. Sperm parameters (concentration, motility, and velocity) and epididymis weight were markedly reduced or decreased in the high-dose group. High-dose males also had various reproductive tract lesions including necrosis and degeneration of seminiferous tubules (minimal to mild in severity), hypospermia in the epididymides, degenerated and immature spermatids in the seminiferous tubules and epididymides, and early sperm granulomas in the seminiferous tubules. A dose-related increase in the incidence of diffuse vacuolar cytoplasmic alteration of ovarian interstitial cells was observed in females.

Chapin et al. (1988) found that chronic administration of TCP in the diet resulted in decreased fertility among the F_0 generation of a continuous breeding study. Male and female mice were fed diets containing 0%, 0.05%, 0.1%, or 0.2% TCP (79% tricresyl phosphate esters, including 21% TMCP, 4% TPCP, and <0.1% TOCP, and 18% dicresyl phosphate esters) starting 7 d before mating and continuing for 14 wk. Doses were estimated by the researchers to be 0, 62.5, 124, and 250 mg/kg-d for the 0%, 0.05%, 0.1%, and 0.2% groups, respectively. After the 14-wk breeding period, the males and females were separated, but continued on treatment. Data were collected on all litters born before the end of 14 wk and then discarded. Litters delivered after the 14-wk period were weaned, treated until breeding age, and then mated to nonsiblings from the same treatment group to produce F_2 litters. F_2 mice from the high-dose group were then mated with controls in a cross-over study.

Hind limb weakness and decreased postpartum body weights were observed in F_0 females from the 0.2% dose group starting with delivery of a second litter. The number of pairs producing more than one litter was markedly reduced in the 0.2% group. A reduction in the number of live pups per litter and increase in number of dead pups per litter occurred in the 0.1% and 0.2% dose groups. The mean number of litters per pair, number of live pups per litter, proportion of pups born alive, and body weight of live pups were all significantly decreased in the 0.2% group. Pup body weight was also significantly reduced in the 0.1% group. At necropsy, body, and kidney and adrenal gland weights were decreased in F_0 females, while body, testis, and epididymis weights were decreased in males from the 0.2% dose groups. Sperm concentration and motility were significantly reduced, and the percentage of abnormal sperm was significantly increased in F_0 males from the 0.2% dose group. Examination of these animals showed atrophy of the seminiferous tubules in males and hypertrophy and degeneration in the adrenals in both males and females.

A dose-related decrease in fertility was observed in F_1 animals in the 0.05% and 0.1% dose groups. No F_1 matings were performed for the 0.2% group because of insufficient numbers of F_1 offspring produced in the 0.2% group. There was also a statistically significant, dose-related trend for decreased number of live pups per litter. Necropsy of F_1 mice showed an increased incidence and/or severity of adrenal lesions (hypertrophy, degeneration) in both sexes in both dose groups. Decreased body weights were observed in females from both dose groups and decreased testis and epididymis weight in males of the 0.1% group. Decreased sperm motility and increased abnormal sperm was observed in males in both dose groups. Bolon et al. (1997) examined the ovarian sections from both F_0 and F_1 females and found no effect on differential follicle count related to treatment with TCP.

The cross-over mating studies showed impaired reproductive performance whether treated males or females were mated with controls. However, the effect of TCP on reproductive performance was greater when treated females were used.

Morrissey et al. (1988) summarizes the effects of TCP on reproductive organ weight and sperm morphology from animals in the NTP (1994) study. Sperm motility and concentration were decreased, and the percentage of abnormal sperm was increased in male rats and mice treated with TCP for 90 d or 104 wk. These effects were generally accompanied by decreased absolute and/or relative weights of the cauda, epididymis, and/or testis. The dose levels at which these effects occurred within each study were not reported.

Latendresse et al. (1993, 1994a,b, 1995) used TCP as a positive control in a number of studies of reproductive toxicity in F-344 rats. In these studies, TCP was administered at a high-dose level (400 mg/kg-d) by daily gavage in sesame oil. The TCP used in these studies was analyzed by gas chromatography/mass spectrophotometry and was found to be a mixture of mostly the TPCP and TMCP isomers (62%) containing substantial amounts of cresyl-xylyl (18%) and cresyl-ethylphenyl (18%) phosphates and no detectable TOCP or other TCP species containing the ortho-isomer. It is possible that the ortho-cresol moiety was present in ethylphenyl and xylyl-substituted species. The initial study (Latendresse et al. 1994a) included light microscopic, morphometric, histochemical, and ultrastructural examination of reproductive tissues of male and female rats treated with TCP for 20, 40, or 60 d (3/sex/duration). Findings included cholesteryl lipidosis and hypertrophy of adrenocortical cells in rats of both sexes and ovarian interstitial cells in females, and degeneration of the seminiferous tubules in males. These lesions all occurred within 20 d of exposure and progressed as exposure continued. Testicular weight was significantly reduced in males after 60 d of exposure. Accumulation of cholesteryl ester in cytoplasmic lipid droplets in the adrenals and ovaries was associated with almost complete inhibition of neutral cholesteryl ester hydrolase, which converts cholesteryl ester to cholesterol, in both tissues (Latendresse et al. 1993). Cholesteryl lipidosis apparently did not result from inhibition of steroidogenesis, since serum concentrations of corticosterone and androstenedione were not decreased in treated rats, and serum concentrations of estradiol were significantly increased (Latendresse et al. 1995). TCP had no effect on the estrous cycle of treated rats (Latendresse et al. 1995).

In a continuous breeding study, TCP at 400 mg/kg-d slightly reduced body weight in F-344 rat dams, significantly reduced fertility (number producing at least one litter/number mated), eliminated the occurrence of second and third litters by breeding pairs, and significantly decreased the number of live pups per litter (Latendresse et al. 1994b). Cross-over mating trials showed that these reproductive findings were due to effects on the males and not the females.

The only teratogenicity study located for TCP and isomers was a study of TOCP in rats. Groups of 10-16 pregnant female Long-Evans Hooded rats were treated with 0, 87.5, 175, or 350 mg/kg-d of TOCP in corn oil on d 6-18 of gestation by gavage (Tocco et al. 1987). Deaths occurred among high-dose dams (28%). There were no treatment-related effects on preimplantation loss, resorptions, sex ratio, fetal body weight, or the incidence of external, visceral, or skeletal malformations or variations. The study did not include investigation of potential functional developmental deficits.

Cancer

No data were located regarding the carcinogenicity of TCP in humans. Extensive human experience with TCP, including poisonings of tens of thousands of people over the past 100 yr, has produced no evidence that oral exposure to TCP can cause cancer in humans.

NTP (1994) conducted a 2-yr cancer bioassay in rats and mice using a commercial TCP mixture. A battery of chemical analyses revealed the test material to be a complex mixture containing 79% tricresyl phosphate esters and 18% dicresyl phosphate esters. F-344/N rats (50/sex/dose) were fed diets containing 0, 75, 150, or 300 ppm of TCP for 104 wk (0, 3, 6, or 13 mg/kg-d in males; 0, 4, 7, or 15 mg/kg-d in females). $B6C3F_1$ mice (50/sex/dose) were fed diets containing 0, 60, 125, or 250 ppm of TCP for 105 wk (0, 7, 13, or 27 mg/kg-d in males and 0, 8, 18, or 37 mg/kg-d in females).

There were no effects on survival, feed consumption or body weight in either species. Treatment-related systemic effects were identified and the MTD (maximum tolerated dose) was achieved in both species (see Systemic Effects section). In female rats, there was an increased incidence of mononuclear cell leukemia. However, this effect was not considered to be treatment-related because of the unusually low tumor incidence in controls and low-dose groups, In mice, there was a nontreatment-related increase in the incidence of Harderian gland adenoma in males. NTP concluded that this study provided no evidence of carcinogenic activity for TCP in male or female rats or mice.

Genotoxicity

Few data were located regarding the genotoxicity of TCP. Results were negative for commercial TCP (<0.1% TOCP) and TMCP in the Ames assay (*Salmonella typhimurium* strains TA100, TA1535, TA1537, and TA98) with and without metabolic activation (Haworth et al. 1983; NTP 1994). The same TCP mixture was also negative in tests for sister chromatid exchange and chromosomal aberrations in Chinese hamster ovary cells with and without metabolic activation (NTP 1994). The TCP mixture used in these studies was the same as that used in the NTP toxicity and carcinogenicity studies. Mirsalis et al. (1983), in an abstract, reported negative results for TCP in an assay for unscheduled DNA synthesis in hepatocytes from Fischer-344 rats treated with TCP. Additional details regarding this study were unavailable.

QUANTITATIVE TOXICITY ASSESSMENT

Quantitative toxicity assessments of aromatic phosphate esters was estimated using toxicity data for TCP. Therefore, these assessments will be overly conservative for the toxicity of other aromatic phosphate esters.

Noncancer

Dermal Assessment

No studies were identified that could be used to derive a dermal RfD for TCP. In the absence of a dermal RfD, the subcommittee believes it is appropriate to use the oral RfD for TCP of 7×10^{-2} mg/kg-d as the best estimate of the internal dose from dermal exposure (derivation of the oral RfD for TCP is presented below).

Inhalation RfC

The available inhalation toxicity data are inadequate for the derivation of an RfC for TCP.

Oral RfD

The oral toxicity database for TCP contains several studies that are potentially useful for the derivation of an oral RfD, including chronic dietary studies in rats and mice (NTP 1994), subchronic feeding and gavage studies in rats and mice (NTP 1994), reproduction studies in rats and mice (Carlton et al. 1987; Chapin et al. 1988), and a study of immune function in rats (Banerjee et al. 1992). The subcommittee chose to use the results from the chronic feeding studies reported by NTP (1994) for deriving an oral RfD for TCP. These studies were chosen because they evaluated a broad range of toxicity end points in two species following lifetime exposure to an appropriate range of doses of TCP in the diet.

The most sensitive end point for toxicity in these studies was identified as changes in serum cholinesterase. However, the subcommittee concluded that this end point is not appropriate for setting an oral RfD for TCP because there is some question as to whether inhibition of serum cholinesterase by TCP in these studies constitutes an adverse effect. The subcommittee concluded that the neurobehavioral effects and neuropathology findings in the NTP (1994) studies are consistent with TCP-induced delayed neuropathy and not cholinesterase inhibition.

The subcommittee identified the adrenal gland and ovarian lesions in female rats and adrenal and liver lesions in female mice that occurred at 7 mg/kg-d to be the key critical effect for deriving an oral RfD for TCP. Application of a composite uncertainty factor (UF) of 100 (10 for interspecies variability and 10 for intraspecies variability) as summarized in Table 17-3, yields an oral RfD of 7×10^{-2} mg/kg-d.

Confidence in the oral toxicity database for TCP is medium. Supporting data were available from subchronic, reproductive, and developmental toxicity studies, but studies of reproductive function did not identify a NOAEL for TCP and only one developmental toxicity study was located. There was one report on immune-system effects associated with TCP exposure, but the available database was insufficient to evaluate this claim.

Confidence in the key studies is high. These studies included an appropriately identified range of doses, a large number of animals, lifetime exposure by a relevant route (diet), and evaluation of a broad array of systemic toxicity end points.

Cancer

Dermal

No studies were located that investigated the carcinogenicity of TCP in humans or animals following dermal exposure to TCP. The absence of route-specific data is a source of uncertainty with regard to a potential portal-of-entry effect, but there are no data to suggest that such an effect would be expected for TCP. Therefore, the weight-of-evidence classification from the oral data is expected to apply for dermal exposure as well.

Inhalation

No studies were located that investigated the carcinogenicity of TCP in humans or animals following inhalation exposure. The absence of route-specific data is a source of uncertainty with regard to a potential portal-of-entry effect, but there are no data to suggest that such an effect would be expected for TCP. Therefore, the weight-of-evidence classification from the oral data is expected to apply for inhalation exposure as well.

Oral

Extensive human experience with TCP, including poisonings of tens of thousands of people over the past 100 yr, has produced no evidence that oral exposure to TCP can cause cancer in humans. No evidence for the carcinogenicity of TCP was found in rats or mice chronically exposed to this compound for two yr in their diet (NTP 1994). Available genotoxicity studies, including assays for mutagenicity, cytogenetic effects, and DNA damage, found no evidence that TCP produces genotoxic effects. Therefore, TCP is considered not likely to be carcinogenic.

TABLE 17-3 Oral Reference Dose for Tricresyl Phosphate

Critical effect	Species	Effect level (mg/kg-d)	Uncertainty factors		RfD (mg/kg-d)
Adrenal and liver lesions	Female rats, male mice	NOAEL: 7.0	UF_A:	10	7×10^{-2}
			UF_H:	10	
			Total:	100	

NOAEL, no-observed-adverse-effect level; RfD, reference dose; UF_A, uncertainty factor for interspecies variability to humans; UF_H, uncertainty factor for intraspecies variability.

EXPOSURE ASSESSMENT AND RISK CHARACTERIZATION

Noncancer

Dermal

Dermal exposure to TCP was estimated using the dermal exposure scenario described in Chapter 3. This exposure scenario assumes that an adult spends 1/4th of his or her time sitting on furniture upholstery backcoated with TCP and also assumes 1/4th of the upper torso is in contact with the upholstery and clothing presents no barrier. Exposure to other chemicals present in the backcoating was not included in this assessment.

First Iteration

As a first estimate of exposure, it was assumed that skin, clothing, and the upholstery did not impede dermal exposure to TCP present in the backcoating. It was also assumed that there would be sufficient water present from sweat to facilitate dissolution of TCP from the backcoating and absorption through the skin. In this scenario, only the dissolution rate of TCP from the backcoating is assumed to be the limiting factor in absorption by the body. It is assumed that all of the TCP that dissolves is immediately absorbed into the body by the sitting person.

Dermal exposure was estimated using Equation 1 in Chapter 3. For this calculation, the subcommittee estimated an upholstery application rate (S_a) for TCP of 5 mg/cm^2. The extraction rate (μ_w) for TCP was estimated to be 0.038 based on extraction data for organic phosphates in polyester fiber (McIntyre et al. 1995). The release rate from the fiber for estimating extraction was 0.06/d at 28°C calculated using the equation $2d/2\pi R$ (d = film thickness, R = fiber radius) with a correction from fiber to film of a factor of 0.63.

Using these assumptions, an estimated absorbed daily dose of 1.5 mg/kg was calculated for TCP. In the absence of a dermal RfD, the subcommittee believes it is appropriate to use the oral RfD for TCP of 7×10^{-2} mg/kg-d as the best estimate of the internal dose from dermal exposure. A hazard index of 21.3 was calculated for this first iteration by dividing the estimated daily dermal dose of 1.5 mg/kg-d by the oral RfD for TCP of 0.07 mg/kg-d. These results suggest that TCP could be a toxic hazard if all applied TCP is absorbed into the body simultaneously. This is an impossible event.

Alternative Iteration

The estimated dermal daily dose for TCP can be calculated using an estimate of the dermal penetration rate for TCP (Chapter 3: Equations 2 and 3). Instead of assuming that all dissolved TCP immediately penetrates the skin and enters systemic circulation, it is assumed that the skin slows the absorption of TCP to a specific amount of chemical absorbed per unit of time. This estimate can be measured experimentally and is referred to as the skin permeability coefficient K_p. However, the dermal penetration constant for TCP has not been measured experimentally. However, K_p can be estimated from a correlation between the octanol-water partition coefficient (K_{ow}) and molecular weight (mass/unit amount of substance) using Equation 2 in Chapter 3 yielding an alternate K_p of 1.04 cm/d.

Using Equation 3 in Chapter 3 and the alternate K_p, the dermal daily dose rate for TCP was estimated to be 3.0×10^{-3} mg/kg-d. In the absence of a dermal RfD, the subcommittee believes it is appropriate to use the oral RfD for TCP of 7×10^{-2} mg/kg-d as the best estimate of the internal dose from dermal exposure. A hazard index of 4.3×10^{-2} was calculated by dividing the estimated daily dermal dose of 3.0×10^{-3} mg/kg-d by the oral RfD for TCP of 0.07 mg/kg-d. These results suggest that TCP is not anticipated to be a toxic risk by the dermal route at the stated application concentrations and under the worst-case exposure scenario.

Inhalation Exposure

Particles

Inhalation exposure estimates for TCP were calculated using the exposure scenario described in Chapter 3. This scenario assumes that a person spends 1/4th of his or her lifetime in a 30-m^3 room containing 30 m^2 of TCP-treated fabric and the room is assumed to have an air-change rate of 0.25/hr. It is also assumed that 50% of the TCP present in 25% of the surface area of the treated fabric is released over 15 yr and that 1% of released particles are small enough to be inhaled.

Particle exposure was estimated using Equations 4 and 5 in Chapter 3. The subcommittee estimated an upholstery application rate (S_a) for TCP of 5 mg/cm^2. The release rate (μ_r) for TCP from upholstery fabric was estimated to be 2.3×10^{-7}/d (see Chapter 3, Equation 5) yielding a room airborne particle concentration (C_p) of 1.9 μg/m^3 and a short time-average exposure concentration of 0.48 μg/m^3. The time-averaged exposure concentration for particles was calculated using Equation 6 in Chapter 3.

In the absence of relevant inhalation exposure data, the subcommittee chose to estimate inhalation RfCs from oral RfDs The subcommittee, however, recognizes that this is not an ideal approach and also recognizes that the estimated RfC levels might be considerably different than actual levels (if inhalation data were available). Extrapolating from one route of exposure (oral) to another (inhalation) requires specific knowledge about the uptake kinetics into the body by each exposure route, including potential binding to cellular sites. The subcommittee believes that its extrapolation of oral RfDs to inhalation RfCs is highly conservative; it assumes that all of the inhaled compound is deposited in the respiratory tract and completely absorbed into the blood. The NRC committee on Toxicology (NRC 1985) has used this approach when inhalation exposure data were insufficient to derive inhalation exposure levels. The subcommittee believes that such an approach is justified for conservatively estimating the toxicological risk from exposure to FRs, and the derived RfC value should be used as an interim or provisional level until relevant data become available for the derivation of an inhalation RfC.

In order to calculate a hazard index for the inhalation route, a provisional inhalation RfC of 0.245 mg/m^3 was derived using the oral RfD for TCP and Equation 7 in Chapter 3.

Division of the time-average exposure concentration of 0.48 μg/m^3 by the provisional RfC for TCP of 0.245 mg/m^3 gives a hazard index of 1.9×10^{-3}. This suggests that under the subcommittee's worst-case exposure assumptions, TCP would not be considered a toxic hazard by the inhalation route of exposure.

Vapors

In addition to the possibility of release of TCP in particles from worn upholstery fabric, the subcommittee considered the possibility of the release of TCP by evaporation. This approach is described in Chapter 3, and uses an exposure scenario similar to that described above for exposure to TCP particles. The rate of flow of TCP vapor from the room is calculated using Equations 8-11 in Chapter 3. A saturated vapor concentration (C_v) of 2.0 mg/m^3 was estimated for TCP. The application density (S_a) for TCP in the treated upholstery was estimated as 5 mg/cm^2.

Using the parameters described, the equilibrium room-air concentration of TCP was estimated to be 1.7 mg/m^3. The short-term time-average exposure concentration for TCP was estimated as 0.417 mg/m^3 using Equation 12 in Chapter 3 and the equilibrium room-air concentration for TCP. It was estimated that concentration could be maintained for approximately 10 yr. Division of the short-term inhalation vapor exposure concentration of 0.417 mg/m^3 by the provisional RfC of 0.245 mg/m^3 yields a hazard index of 1.7, which indicates that inhalation exposure at the worst-case levels might pose a noncancer risk.

Oral Exposure

The assessment of noncancer toxicological risk for oral exposure to TCP is based on the oral exposure scenario described in Chapter 3. This scenario assumes a child is exposed to TCP by sucking on 50 cm^2 of fabric backcoated with TCP, 1 hr/d for two yr. The subcommittee estimated an upholstery application rate (S_a) for TCP of 5 mg/cm^2. Oral exposure was calculated using Equation 15 in Chapter 3. The extraction rate (μ_w) for TCP was estimated to be 0.038 based on extraction data for organic phosphates in polyester fiber (McIntyre et al. 1995). The release rate from the fiber for estimating extraction was 0.06/d at 28°C calculated using the equation $2d/2\pi R$ (d = film thickness, R = fiber radius) with a correction from fiber to film of a factor of 0.63.

The worst-case average oral daily dose for TCP was estimated as 0.04 mg/kg-d. Division of the dose estimate by the oral RfD for TCP of 0.07 mg/kg-d gives a hazard index of 0.57. This suggests that under the subcommittee's worst-case exposure assumptions, TCP is not likely to pose a health risk by the oral route of exposure.

Cancer

Dermal

There are no studies available to evaluate the carcinogenicity of TCP in humans or laboratory animals following dermal exposure.

Inhalation

There are inadequate data to assess the carcinogenicity of TCP in humans or animals following inhalation exposure.

Oral

TCP is not likely to be a human carcinogen by the oral route of exposure. Therefore, TCP is not anticipated to cause cancer in humans from oral exposure to treated furniture upholstery.

RECOMMENDATIONS FROM OTHER ORGANIZATIONS

IPCS (1990) concluded that there is no safe level of ingestion for TCP, and that exposure through inhalation or dermal contact should be minimized.

There is an ACGIH Threshold Limit Value (TLV) for TOPCP (CAS RN 78-30-8) of 0.1 mg/m^3 for skin (ACGIH 1999). The NIOSH REL (skin) and OSHA PEL for TOCP is also 0.1 mg/m^3 (NIOSH 1996).

DATA GAPS AND RESEARCH NEEDS

There are no chronic toxicity data for TCP for the dermal and inhalation routes of exposure. There is no information on the types and amounts of TCP species that are present in upholstery backcoating. Data on the leaching of these species from upholstery backcoating are also not available. Information on the dermal penetration of TCP and its possible derivatives would be helpful.

Based on an inhalation hazard index greater than 1, the subcommittee recommends that the potential for vapor release from treated fabric should be investigated.

REFERENCES

ACGIH (American Conference of Government Industrial Hygienists). 1999. Threshold Limit Values and Biological Exposure Indices. Cincinnati, OH: American Conference of Government Industrial Hygienists.

Aldridge, W.N., and J.M Barnes. 1966. Esterases and neurotoxicity of some organophosphorus compounds. Biochem. Pharmacol. 15(5):549-554.

Banerjee, B.D., S. Saha, K.K. Ghosh, and P. Nandy. 1992. Effect of tricresyl phosphate on humoral and cell-mediated immune responses in albino rats. Bull. Environ. Contam. Toxicol. 49(2):312-317.

Bisesi, M.S. 1994. Esters. Pp. 2967-3118 in Patty's Industrial Hygiene and Toxicology, Fourth Ed., Vol. II, Part D, Toxicology. G.D. Clayton, and F.E. Clayton, eds. New York: John Wiley & Sons.

Bolon, B., T.J. Bucci, A.R. Warbritton, J.J. Chen, D.R. Mattison, and J.J. Heindel. 1997. Differential follicle counts as a screen for chemically induced ovarian toxicity in mice: results from continuous breeding bioassays. Fundam. Appl. Toxicol. 39(1):1-10.

Brinkerhoff, C.R., R.P. Sharma, and D.R. Bourcier. 1981. The effects of tri-*o*-tolyl phosphate (TOTP) on the immune system of mice. Ecotoxicol. Environ. Saf. 5(3):368-376.

Broadhurst, C.A., A.F. Grady, N. Jarvik, et al. 1951. A Toxicological Study of Tricresyl Phosphate. Institute of Industrial Medicine, New York University-Bellevue Medical Center.

Budavari, S., M.J. O'Neil, A. Smith, and P.E. Heckelman. 1989. The Merck Index, Eleventh Edition. S. Budavari, M.J. O'Neil, A. Smith, and P.E. Heckelman, eds. Rahway, NJ: Merck & Co., Inc.

Carlsen, L., K.E. Andersen, and H. Egsgaard. 1986. Triphenyl phosphate allergy from spectacle frames. Contact Dermatitis 15(5):274-277.

Carlton, B.D., A.H. Basaran, L.E. Mezza, and M.K. Smith. 1987. Examination of the reproductive effects of tricresyl phosphate administered to Long-Evans rats. Toxicology 46(3):321-328.

Chapin, R.E., J.D. George, and J.C. Lamb, 4th. 1988. Reproductive toxicity of tricresyl phosphate in a continuous breeding protocol in Swiss (CD-1) mice. Fundam. Appl. Toxicol 10(2):344-354.

ChemID. 1999. On-Line Database. [Online]. Available: http://igm.nlm.nih.gov/cgi-bin/doler?account=++&password=++&datafile=chemid National Library of Medicine, Bethesda, MD. Rev: May 10, 1999

Eastman Kodak Company. 1978. Toxicity and Health Hazard Summaries for Aryl Phosphates. OTS0206526.

Ferrante, J. 1999. Toxicity Review for Aromatic Phosphate Plasticizers. Memorandum updated from Jacqueline Ferrante, Pharmacologist, Division of Health Sciences, to Ronald L. Medford, Assistant Executive Director for Hazard Identification and Reduction. U.S. Consumer Product Safety Commission, Washington, DC.

Haworth, S., T. Lawlor, K. Mortelmans, W. Speck, and E. Zeiger. 1983. Salmonella mutagenicity test results for 250 chemicals. Environ. Mutagen. 5(Suppl. 1):3-142.

Hodge, H.C., and J.H. Sterner. 1943. The skin absorption of triorthocresyl phosphate as shown by radioactive phosphorus. J. Pharmacol. Exp. Ther. 79:225-234.

HSDB (Hazardous Substances Data Bank). 1999. [Online]. Available: http://sis.nlm.nih.gov/cgi-bin/sis/htmlgen?HSDB National Library of Medicine, Bethesda, MD.

IPCS (International Programme on Chemical Safety). 1990. Environmental Health Criteria 110: Tricresyl Phosphate. World Health Organization, Geneva.

Johannsen, F.R., P.L. Wright, D.E. Gordon, G.J. Levinskas, R.W. Radue, and P.R.

Graham. 1977. Evaluation of delayed neurotoxicity and dose-response relationships of phosphate esters in the adult hen. Toxicol. Appl. Pharmacol. 41(2):291-304.

Johnson, M.K. 1975. Organophosphorus esters causing delayed neurotoxic effects: Mechanism of action and structure activity studies. Arch. Toxicol. 34(4):259-288.

Kurebayashi, H., A. Tanaka, and T. Yamaha. 1985. Metabolism and disposition of the flame retardant plasticizer, tri-p-cresyl phosphate, in the rat. Toxicol. Appl. Pharmacol. 77(3):395-404.

Latendresse, J.R., S. Azhar, C.L.Brooks, and C.C. Capen. 1993. Pathogenesis of cholesteryl lipidosis of adrenocortical and ovarian interstitial cells in F344 rats caused by tricresyl phosphate and butylated triphenyl phosphate. Toxicol. Appl. Pharmacol. 122(2):281-289.

Latendresse, J.R., C.L.Brooks, and C.C. Capen. 1994a. Pathologic effects of butylated triphenyl phosphate-based hydraulic fluid and tricresyl phosphate on the adrenal gland, ovary, and testis in the Fischer-344 rat. Toxicol. Pathol. 22(4):341-352.

Latendresse, J.R., C.L. Brooks, C.D. Flemming, and C.C. Capen. 1994b. Reproductive toxicity of butylated triphenyl phosphate and tricresyl phosphate fluids in F344 rats. Fundam. Appl. Toxicol. 22(3):392-399.

Latendresse, J.R., C.L. Brooks, and C.C. Capen. 1995. Toxic effects of butylated triphenyl phosphate-based hydraulic fluid and tricresyl phosphate in female F344 rats. Vet. Pathol. 324:394-402.

McIntyre, J.E., I. Holme, and O.K. Sunmonu. 1995. The desorption of model compounds from poly(ethylene terephthalate) fibre. Colourage 41(13)77-81.

Mirsalis, J., K. Tyson, J. Beck, E. Loh, K. Steinmetz, C. Contreras, L. Austere, S. Martin, and J. Spalding. 1983. Induction of unscheduled DNA synthesis (UDS) in hepatocytes following *in vitro* and *in vivo* treatment. [Abstract]. Environ. Mutagen. 5:482.

Morgan, J.P. 1982. The Jamaica ginger paralysis. JAMA 248(15):1864-1867.

Morrissey, R.E., B.A. Schwetz, J.C. Lamb, 4th, M.D. Ross, J.L. Teague, and R.W. Morris. 1988. Evaluation of rodent sperm, vaginal cytology, and reproductive organ weight data from National Toxicology Program 13-week studies. Fundam. Appl. Toxicol. 11(2):343-358.

NIOSH (National Institute for Occupational Safety and Health). 1996. NIOSH Chemical Listing and Documentation of Revised IDLM Values. [Online]. Available: http://www.cdc.gov/niosh/idlh/intridl4.html (Retrieved March 8, 2000).

Nomeir, A.A., and M.B. Abou-Donia. 1986. Studies on the metabolism of the neurotoxic tri-*o*-cresyl phosphate. Distribution, excretion, and metabolism in male cats after a single, dermal application. Toxicology 38(1):15-33. (Cited in IPCS 1990 and NTP 1994)

NRC (National Research Council). 1985. Emergency and Continuous Exposure Limits for Selected Airborne Contaminants, Volume 5. Committee on Toxicology. Board on Toxicology and Environmental Health Hazards, National Research Council. Washington, DC: National Academy Press.

NTP (National Toxicology Program). 1994. Toxicology and Carcinogenesis Studies of Tricresyl Phosphate (CAS No. 1330-78-5) in F344/N Rats and B6C3F$_1$ Mice (Gavage and Feed Studies). NTP Technical Report 433. NIH Publication No. 94-3164.

Oishi, H., S. Oishi, and K. Hiraga. 1982. Toxicity of several phosphoric acid esters in rats. Toxicol. Lett. 13(1-2):29-34.

Pegum, J.S. 1966. Contact dermatitis from plastics containing tri-aryl phosphates. Br. J. Dermatol. 78(12):626-631.

Piccirillo, V.J. 1999. Flame-retardant product chemistries, toxicology and hazard evaluations. NPC Inc., Sterling, VA.

Saito, C., T. Kato, H. Taniguchi, et al. 1974. [Subacute toxicity of tricresylphosphate (TCP) in rats.] Pharmacometrics. 8: 107-118. [Article in Japanese].

Sprague, G.L., and T.R. Castles. 1985. Estimation of the delayed neurotoxic potential and potency for a series of triaryl phosphates using an *in vitro* test with metabolic activation. Neurotoxicology 6(1):79-86.

Tabershaw, I.R., and M. Kleinfeld. 1957. A.M.A. Arch. Ind. Health. 15: 541.

Tarvainen, K. 1995. Analysis of patients with allergic patch test reactions to a plastics and glue series. Contact Dermatitis 32(6):346-351.

Tocco, D.R., J.L. Randall, R.G. York, and M.K. Smith. 1987. Evaluation of the teratogenic effects of tri-*ortho*-cresyl phosphate in the Long-Evans hooded rat. Fundam. Appl. Toxicol. 8(3):291-297.

Treon, J.F., F.P. Cleveland, and J. Cappel. 1955. The toxicity of certain aromatic phosphate esters. University of Cincinnati, Wright Air Development Center. Technical Report 54-345.

18

Tetrakis(hydroxymethyl) Phosphonium Salts

THIS chapter reviews the physical and chemical properties, toxicokinetics, toxicological, epidemiological, and exposure data on tetrakis(hydroxymethyl) phosphonium chloride (THPC),[1] which is one of the tetrakis(hydroxymethyl) phosphonium salts. Although other tetrakis(hydroxymethyl) phosphonium salts have been used as flame retardants, the subcommittee chose to focus its assessment on THPC because it has a large toxicology database and is the most toxic of the phosphate salts. The subcommittee used the toxicity and exposure information on THPC to characterize the health risk from exposure to THPC. The subcommittee also identified data gaps and recommended research relevant for determining the health risk from exposure to THPC.

---

[1]It is important to note that THPC is polymerized onto fabric in combination with amine compounds and may undergo chemical changes that alter its chemical properties and toxicity. It is also reasonable to assume that oxidized forms of THPC may be present in or on aged THPC-treated fabric. However, the chemical species of THPC and the average ratio of these compounds present in treated fabric are not known at this time.

PHYSICAL AND CHEMICAL PROPERTIES

The physical and chemical properties of THPC are summarized in Table 18-1. THPC is a water-soluble tetrakis(hydroxymethyl) phosphonium salt (IARC 1990) that is a common ingredient in commercial flame-retardant (FR) formulations.

TABLE 18-1 Physical and Chemical Properties for Tetrakis(hydroxymethyl) Phosphonium Chloride

Properties	Value	Reference
Chemical formula	$C_4H_{12}O_4PCl$	HSDB 1999
Structure	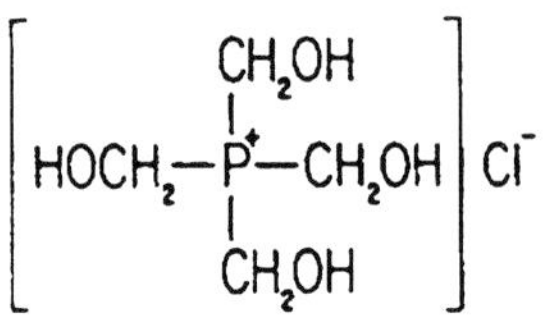	Loewengart and Van Duuren 1977
CAS Registry #	124-64-1	HSDB 1999
Synonyms	THPC	HSDB 1999
Molecular weight	190.58	HSDB 1999
Physical state	Crystalline solid; sold as 80% aqueous solution	IARC 1990; HSDB 1999
Color	80% aqueous solution is straw-colored or clear and colorless	NTP 1991; Hazleton UK 1992
Solubility	Soluble in water, methanol, ethanol; less than 1 mg/mL in DMSO, insoluble in ether, reaction with acetone	NTP 1991
Vapor pressure	80% aqueous solution: 1.0 mm Hg at 25°C	NTP 1991
pH	1 for aqueous solution of unspecified concentration; 2 for 80% aqueous solution	Hazleton UK 1991; Hazleton UK 1992
Melting point	154°C	Grasseli and Ritchey 1975
Boiling point	80% aqueous solution: 118°C	NTP 1991
Density (water = 1)	80% aqueous solution: 1.322 g/cm^3 at 17.8°C; 1.34 g/cm^3 at 20°C	NTP 1991; Hazleton UK 1992

OCCURRENCE AND USE

THPC is produced by the reaction of formaldehyde, phosphine, and hydrochloric acid (IARC 1990). THPC and its sulfur salt THPS are the predominant FR chemicals used for cotton apparel, especially children's sleepwear (NTP 1987). THPC was the most widely used FR for cotton fabrics in the 1950s. About 1,000-5,000 tons of THPC was used in the United States in 1987 (NTP 1987).

TOXICOKINETICS

Absorption

Dermal

No studies were identified that investigated the dermal absorption of THPC or other tetrakis(hydroxymethyl) phosphonium salts by humans. Dermal application of THPC to rats resulted in body weight loss and death (Aoyama 1975), indicating that THPC is absorbed by this route. Ulsamer et al. (1980), citing a 1953 report by the Wisconsin Alumni Research Foundation, stated that THPC can be absorbed through the skin in large amounts (1.5 gm/kg). It is not clear whether this is a derived amount or one based on animal data. The subcommittee could not locate a copy of the 1953 report.

Inhalation

No studies were identified that investigated the absorption of THPC following inhalation exposures.

Oral

Acute, subchronic, and chronic toxicity studies in rats and mice provide indirect evidence that THPC is absorbed through the gastrointestinal tract and becomes systemically bioavailable (see Hazard Identification section).

Distribution

No studies were identified that investigated the distribution of THPC or of other tetrakis(hydroxymethyl) phosphonium salts in humans or laboratory animals following dermal, inhalation, or oral exposure. Since the liver has been shown to be a target organ for THPC toxicity (see NTP 1987), it can be inferred that THPC is distributed systemically.

Metabolism

No studies were identified that investigated the metabolism of THPC or of other tetrakis(hydroxymethyl) phosphonium salts.

Excretion

No studies were identified that investigated the excretion of THPC or of other tetrakis(hydroxymethyl) phosphonium salts.

HAZARD IDENTIFICATION[2]

Dermal Exposure

Irritation

No skin reactions were observed in 100 volunteers (23 males, 77 females) aged 9-63 yr who were exposed to THPC-treated fabric for 72 hr (Osbourn 1971). Volunteers were treated topically (location not specified) with THPC-treated fabric patches, some of which were moistened with distilled water.

THPC was found to be non-irritating to the skin in 38 male volunteers who were dermally exposed to fabric patches containing Proban® 210, a THPC-based FR (Albright and Wilson 1982, as cited in IPCS 2000). The THPC content of the fabric patches was not reported. The fabric patches were applied to the forearms of the volunteers and covered for 48 hr. The test sites were then uncovered and examined for skin reactions 50 hr, 90 hr, 1 wk, and 2 wk after exposure.

Moderate to severe skin reactions were observed in male white rats and rabbits treated topically for 8 d with 0.75 mL of 15%, 20%, or 30% aqueous THPC (Aoyama 1975). In rats, skin redness was observed starting on d 4 for

---

[2]In this section, the subcommittee reviewed data on toxicity of tetrakis(hydroxymethyl) phosphonium salts, including the toxicity assessment prepared by the U.S. Consumer Product Safety Commission (Bittner 1999).

rats treated with 15% and 20% THPC, while redness was observed beginning on d 2 in rats treated with 30% THPC. Skin redness in the 30% dose group became very intense on d 6 followed by partial hair loss on d 7 and death on d 9. Rabbits treated topically for 6 d with 1 mL of 15% or 30% aqueous THPC developed skin redness on d 2-3, which became severe by no later than d 6. Skin necrosis developed on d 3-12. Total hair loss occurred in both dose groups by d 11-13, but hair began to regrow in both dose groups by d 18-19. Toxicity was comparatively more severe in rabbits treated with 30% THPC.

Ulsamer et al. (1980), citing the 1953 report by the Wisconsin Alumni Research Foundation, state that THPC is a mild skin irritant in the female rat and caused lethality, skin sloughing, and hyperemia after dermal application of >1.5 g/kg (species not given). The 1953 report was not located by the subcommittee.

Sensitization

There are reports of textile contact dermatitis occurring in children wearing nightdresses treated with THPC-based FRs (Martin-Scott 1966). A 2½-yr-old girl developed pruritis, enlarged regional glands, and blisters after wearing THPC-treated sleepware over a 4-mo period. A 6-yr-old girl developed an urticarial rash reaction after wearing THPC-treated nightwear for a second time. Both girls developed skin reactions after exposure to patches treated with Proban. Symptoms of sensitization subsided in both girls after the THPC-treated dresses were no longer worn. Negative skin reactions were observed in 100 other children given patch tests with no known exposure to THPC and in children with atypical infantile eczema (Martin-Scott 1966).

No skin sensitization reactions were observed upon re-challenge of 100 volunteers (23 males, 77 females) aged 9-63 yr exposed to THPC-treated fabric patches. Initially, the volunteers were exposed to the treated patches for 48 hr and the patches were then removed. Two wk after initial exposure, the volunteers were then re-challenged with THPC-treated fabric patches held in place for 72 hr. The patches were then removed and assessed for local skin reactions to the treatment.

THPC was not a sensitizer in albino guinea pigs tested using the Buehler method (Industrial Bio-test Labs 1975). Ten animals were treated topically on their shaved backs with nine consecutive applications of patches containing 0.5 mL of a 1% v/v dilution of commercial THPC (THPC concentration not reported). Each patch was applied for 5 hr. The animals were then rechallenged with THPC-treated patches 2 wk after the induction period and the application sites were examined for skin reactions at 24 and 48 hr after re-challenge.

Systemic Effects

Moderate to severe skin effects characterized by tissue changes were observed in male white rats and rabbits treated topically for 8 d with 0.75 mL of 15%, 20%, or 30% aqueous THPC (Aoyama 1975). Treatment was intended to be for 20 d in both species but was discontinued after 8 d in rats because of severe weight loss and after 6 d in rabbits due to severe skin reactions. Rabbits continued to be observed until d 20. In rats, histological examination of the skin showed atrophy, enhanced keratinization of the epidermis, and degeneration of the hair roots in all treated animals. In rabbits, histological examination of the skin showed severe subepidermal fibrosis without regeneration of epidermis papillae.

A high rate of deaths occurred in white mice treated repeatedly on their tails with aqueous extracts from fabric treated with a THPC-based FR (Afanas'eva and Evseenko 1971). Mice were treated with the extracts daily for 21 d. The extracts reportedly contained formaldehyde, hydrogen chloride, and organophosphorus compounds (not identified). The authors did not describe how the extracts were prepared, the sex of the animals used, or the number of animals tested. The authors reported that 50-70% of the treated animals died over the course of the experiment and exhibited weight loss and changes in the appearance of their fur. Tail-skin irritation was evident after 10-12 d of treatment and many of the tails fell off.

Immunological Effects

Afanas'eva and Evseenko (1971) reported that many (number not reported) of the mice treated with aqueous extracts from fabrics treated with a THPC-based FR developed leukopenia. The authors note that in addition to exposure to THPC, the extracts also contained formaldehyde, hydrogen chloride, and organophosphorus compounds and the exact composition of the extracts was not reported. Therefore, it is not possible to determine whether dermal exposure to THPC itself was the cause for the increased incidence of leukopenia in this study.

Neurological Effects

Mice treated with aqueous extracts from fabrics treated with a THPC-based FR became sluggish, had "reduced working capacity" for static work, and 20-40% lower cholinesterase activity levels (Afanas'eva and Evseenko 1971).

Developmental Effects

No studies were identified that investigated the toxic effects of THPC on reproduction or development following dermal exposure.

Cancer

Loewengart and Van Duuren (1977) found that THPC had tumor-promoting activity in a mouse skin carcinoma assay. Female ICR/Ha Swiss mice (20/group) were first topically treated once in a shaved area of the interscapular region with the tumor initiator 7,12-dimethylbenz[a]anthracene (DMBA) and were then topically treated at the same site with THPC (2 mg in 0.1 mL dimethyl sulfoxide [DMSO]), three times a wk for 400 d. Three of 20 animals developed skin papillomas. All three tumors progressed to squamous cell carcinomas over the course of the assay. In the same experiment, one skin papilloma developed among 20 mice that received topical applications of THPC (2 mg in 0.1 mL of DMSO) 3 times/wk but without pretreatment with a tumor initiator. The authors did not comment on the significance of this finding, but the report notes that none of the 20 animals receiving no treatment with any compounds developed skin tumors over the 400 d experimental period. No tumors were observed in animals treated initially with extracts from THPC-treated cloth and then topically treated with the tumor promoter phorbol myristate acetate (PMA) (2.5 µg in 0.1 mL acetone), 3 times/wk for 400 d. Similarly, no tumors were observed in animals treated initially with THPC extracts followed by dermal applications of acetone (0.1 mL), 3 times/wk for 400 d. In comparison, 19 of 20 animals treated initially with a single application of DMBA followed by treatment with PMA 3 times/wk for 400 d produced papillomas in 19 of 20 animals and squamous cell tumors developed in 9 of 20 animals. No tumors were reported in animals treated initially with DMBA followed by DMSO treatment 3 times/wk for 400 d. The initiating effect of DMSO was investigated for comparison. The authors reported that the use of DMSO as a solvent for THPC produced results that were unusually low in all respects as compared with acetone.

Van Duuren et al. (1978) found no evidence of carcinogenicity in female ICR/Ha Swiss mice treated topically with 2 mg THPC dissolved in 0.2 mL acetone/water, 3 times/wk for 496 d. All animals were killed at study termination. Skin, liver, and kidneys were examined microscopically for all animals that died prematurely and for 20% of the animals surviving to study completion. The incidence of tumors in THPC-treated mice was not significantly different from negative controls and animals treated with acetone.

Inhalation Exposure

No human or animal studies were identified that investigated the toxic effects of THPC following inhalation exposure.

Oral Exposure

Systemic Effects

Acute oral toxicity data for THPC are summarized in Table 18-2. The single-dose LD_{50} for THPC ranges between 161-282 mg/kg in rats and 280-600 mg/kg in mice. Administration of a single, oral dose of THPC to rats, once a day, for 14 consecutive days resulted in decreased body weight gain at dose levels ≥18.8 mg/kg and changes in physical appearance and death at dose levels of ≥75 mg/kg. Administration of a single dose of THPC to mice, once a day, for 14 consecutive days resulted in decreased body weight gain at dose levels ≥18.8 mg/kg and all animals died when treated with 300 mg/kg.

Ishizu (1975) reported dose-related increases in serum GOT and GPT activities and histological changes in the liver in both males and females of mice and rats given THPC in their drinking water for 1 mo. The author did not report how often THPC solutions were replaced. Therefore, there is some uncertainty regarding the actual concentrations of THPC that animals were exposed to in this study. Apparently, THPC undergoes oxidation in water, particularly at low concentrations (Hazleton UK 1992).

Thake et al. (1982) found that female F-344 rats gavaged with 80 mg THPC/kg-d for 45 or 90 d had decreased weight gain, lack of response to external stimuli, stiff gait, and paresis. Many of the rats also had abnormal neurobehavior and degeneration of the peripheral nerves (see Neurological Effects section). No treatment-related effects were reported among female rats similarly treated with 40 mg THPC/kg-d. These data were presented in an abstract and a follow-up study was not located in the searched literature.

Table 18-3 summarizes the results of subchronic (13-wk) and chronic (103-wk) toxicity studies for THPC in rats and mice conducted by NTP (1987). In the 13-wk study, all high-dose (60 mg/kg-d) male rats and five of 10 female rats died over the course of the study due to THPC treatment-related effects. Two of 10 males and one of 10 female rats died in the 15 mg/kg-d group, apparently due to gavage errors. Liver vacuolization was apparent in males at dose levels ≥7.5 mg/kg-d and in females ≥15 mg/kg-d. Paresis and axonal degeneration was apparent at 60 mg/kg-d. In the 13-wk study in mice, deaths, reduced body weights, paresis, and axonal degeneration occurred in both sexes

TABLE 18-2 Short-Term Oral Toxicity Studies on Tetrakis(hydroxymethyl) Phosphonium Chloride

Species, Sex, Number	Duration, Route	Dose (mg/kg-d)	Effect	Reference
Rats, NS, F, NS	1 dose, gavage	NS	"Oral toxicity" of 250 mg/kg	Anonymous 1953[a], as cited in Ulsamer et al. 1980
Rat, F344, F, NS	40 (N)	45 or 90 d	No effect	Thake et al. 1982[b]
	80 (L)	45 or 90 d	Decreased weight gain, neurotoxicity (behavioral, histological)	
Rat, F344, M/F, 5/sex/dose	1 dose, gavage (2-wk observation)	75-1,200	LD_{50} of 185 mg/kg for M LD_{50} of 161 mg/kg for F	NTP 1987
Mouse, B6C3F1, M/F, 5/sex/dose	1 dose, gavage (2-wk observation)	75-1,200	LD_{50} between 300-600 mg/kg for M LD_{50} of 280 mg/kg for F	NTP 1987
Rat, F344 rat, M/F, 5/sex/dose	14 d, gavage (1-wk observation)	9.4	No effect	NTP 1987
		18.8	Decreased body weight gain: 6% for M	
		37.5	Decreased body weight gain: 11% for M	
		75	Rough coats, arched backs, 2 of 5 M died	
		150	Yellow/tan or mottled red livers	
Mouse, B6C3F1, M/F, 5/sex/dose	14 d, gavage (1-wk observation)	18.8	Decreased body weight gain: 3% for M, 7% for F	NTP 1987
		37.5	Decreased body weight gain: 6% for M, 7% for F	
		75	Decreased body weight gain: 6% for M, 9% for F	
		150	3-6% body weight loss (18-20% decreased body weight)	
		300	All animals died	
Rat and Mouse	1 mo, drinking water	20-200 ppm[c]	Liver histological and enzyme changes; decreased body weight gain at $\geq$ 20 ppm, dose-related in both species	Ishizu 1975[b]

F, female; M, male; N, no-observed-adverse-effect level identified by reviewer; L, lowest-observed-adverse-effect level identified by reviewer.

[a]Unpublished, non-Good Laboratory Practice study.

[b]Published, non-Good Laboratory Practice study.

[c]Estimated dose for rats is 2.8-28 mg/kg-d (assume water intake of 0.049 L/d and body weight of 0.35 kg), and estimated dose for mice is 3.8-19 mg/kg-d (assume water intake of 0.0057 L/d and body weight of 0.03 kg).

TABLE 18-3 Subchronic and Chronic Gavage Studies on Tetrakis(hydroxymethyl) Phosphonium Chloride Conducted by NTP (1987)

Species, Strain, Sex, Number	Duration	Dose (mg/kg-d)	Effect
Rat, F 344, M/F, 10/sex/dose	5 d/wk, 13 wk	3.75 (N)	No effects at 3.75 mg/kg-d;
		7.5 (L)	Liver vacuolization[1] in 8 of 10 M at 7.5 mg/kg-d;
		15	Liver vacuolization in 9 of 10 M, 8 of 10 F; liver necrosis[2] of minimal severity in 9 of 10 M, 7 of 10 F;
		30	Liver vacuolization, necrosis in 10 of 10 M, F; decreased body weight gain in M;
		60	Most died; paresis, incoordination, axonal degeneration; liver vacuolization 10/10 M, F, liver necrosis in 7 of 10 M, 8 of 10 F
Mouse, B6C3F1, M/F, 10/sex/dose	5 d/wk, 13 wk	1.5	No effects
		4.5 (N)	No effects
		15 (L)	Liver vacuolization in 10/10 M, F
		45	Liver vacuolization in 10/10 M, F; necrosis in 10/10 M
		135	Most died, decreased weight gain; paresis, incoordination, axonal degeneration; liver vacuolization in 10/10 M, 9/10 F; liver necrosis in 8/10 M, 7/10 F
Rat, F344, M/F, 50/sex/dose	5 d/wk, 103 wk	3.75 (L)	Rough hair coats, diarrhea; liver vacuolization in 9/50 M (c = 0/50), 11/50 F (c = 3/50); liver cystic degeneration in 23/50 M (c = 12/50); spleen hematopoiesis in 9/50 F (c = 3/50)
		7.5	Rough hair coats, diarrhea; F had increased mortality after wk 70; liver vacuolization in 23/49 M (c = 0/50), 25/50 F (c = 3/50); liver cystic degeneration in 26/50 M (c = 12/50); spleen hematopoiesis in 15/50 F (c = 3/50)
Mouse, B6C3F1, M, 49 or 50/dose	5 d/wk, 103 wk	7.5 (L)	Rough hair coats; diarrhea; liver vacuolization in 39/49 M (c = 0/49)
		15	Rough hair coats; diarrhea; liver vacuolization in 44/50 M
Mouse, B6C3F1, F, 50/dose		15	Rough hair coats; diarrhea; liver vacuolization in 42/50 F (c = 0/49)
		30	Rough hair coats; diarrhea; liver vacuolization in 48/50 F; thyroid hyperplasia

[1]Liver vacuolization: hepatocellular periportal cytoplasmic vacuolization.
[2]Liver necrosis: periportal hepatocellular necrosis.
c, control group; F, female; M, male; N, no-observed-adverse-effect level identified by reviewer; L, lowest-observed-adverse-effect level identified by reviewer.

at 135 mg/kg-d. Hepatocyte periportal cytoplasmic vacuolization occurred in males and females at ≥15 mg/kg-d.

In the 2-yr study, high-dose female rats had lower survival than controls starting at wk 70 (21/50 vs. 37/50 for controls). No effects on survival were apparent in male rats or male or female mice. Body weights in both species were not affected by treatment with THPC. Clinical toxicity was apparent at all dose levels in both species characterized by rough hair coats and diarrhea. Dose-related increases in hepatocyte cytoplasmic vacuolization occurred in the treatment groups of both species. In rats, cystic degeneration of the liver and spleen hematopoiesis was observed in male and females, respectively. Increased incidence of thyroid follicular cell hyperplasia occurred in female mice at dose level of 30 mg THPC/kg-d, but were not considered treatment-related because it is a common degenerative lesion in aging rodents. No neuropathological findings were apparent in either species. LOAELs for liver lesions in this study were 3.75 mg/kg-d for rats and 7.5 mg/kg-d for mice. However, hepatocellular vacuolization can be a reversible lesion (Harada et al. 1999). Progression of the liver lesions in the NTP (1987) study to fibrosis or hyperplasia, which occurs with severe and persistent vacuolization, was not reported.

Immunological Effects

No studies were identified that investigated the immunological effects of THPC following oral exposure.

Neurological Effects

It was reported that female rats given 80 mg THPC/kg-d by gavage developed peripheral nerve (sciatic, tibial, muscle branches, and plantar) degeneration that was characterized as mild after 45 d of exposure and mild to severe after 90 d of exposure (Thake et al. 1982). These rats also had reduced spontaneous motor activity and reduced forelimb and hindlimb grip strength. No neurological effects were observed in females given 40 mg THPC/kg-d for 45 or 90 d. The THPC solvent and the number of animals tested were not specified.

Neurotoxic effects were seen in both rats and mice in the 13-wk studies conducted by NTP (1987). High-dose animals (60 mg/kg-d for rats and 135 mg/kg-d for mice) had paresis and incoordination of the rear limbs as well as rough hair coats, hunched backs, diarrhea, lethargy, and axonal degeneration. Axonal degeneration was seen in 2 of 10 high-dose female rats and in at least 19 of 20 high-dose mice and was characterized by swollen axon sheaths, missing or fragmented axons, and proliferation of neurolemma cells in the sciatic nerve, dorsal roots of the caudal spinal nerves, and spinal cord tracts.

Reproductive and Developmental Effects

Developmental toxicity studies for THPC are summarized in Table 18-4. In a range-finding study conducted by Hazleton UK (1991) in rabbits, maternal toxicity occurred at all dose levels characterized by red intestinal mucosa weight loss, and decreased food intake. No toxic effects were found in offspring from low-dose dams. One high-dose dam with severe weight aborted and died on d 19. Necropsy revealed ulceration of the stomach, pale jejunum and liver, clear fluid in the thoracic cavity, and red fluid in the uterus. Pre-implantation losses were not increased in any treatment group and post-implantation losses, litter size, and fetal weight were adversely affected in only high-dose dams. Treatment-related eye and/or limb malformations occurred in all high-dose litters. No malformations were observed in the mid-dose litters, but one low-dose fetus had arthrogryposis that was considered unrelated to treatment. The sex ratio of the offspring was not affected by treatment.

TABLE 18-4 Developmental Toxicity Studies of Tetrakis(hydroxymethyl) Phosphonium Chloride

Species, Strain, Sex, Number	Duration	Dose (mg/kg-d)	Effect	Reference (type)
Rabbit, New Zealand White, F, 5/dose	d 7-19 of gestation (killed on d 29)	6 (L)	Red intestinal mucosa;	Hazleton UK 1991
		18	Red intestinal mucosa; weight loss, decreased food intake;	
		60	Red intestinal mucosa; weight loss decreased food intake, 1 death; increased post-implantation loss, decreased litter size, decreased fetal weight; fetuses had eye/limb malformation	
Rabbit, New Zealand White, F, 16/dose	d 7-19 of gestation (killed on d 29)	2 (N)	No effects;	Hazleton UK 1992
		6 (L?)	Gastrointestinal irritation (1 animal); increased weight loss and fetal malformations were not clearly treatment-related;	
		18	One spontaneous abortion, increased weight loss and fetal malformation not clearly treatment-related	
		36	One spontaneous abortion; decreased weight gain; decreased food intake; gastrointestinal irritation; one total resorption; eye fetal malformations in one litter	

F, female; GLP, Good Laboratory Practive; L, lowest-observed-adverse-effect level identified by reviewer; L?, the evidence of an adverse effect occurring at this dose was equivocal; N, no-observed-adverse-effect level identified by reviewer.

In the full-scale study, maternal toxicity was apparent in dams treated with ≥6 mg/kg-d (Hazleton UK 1992). No maternal or fetal toxicity occurred at 2 mg/kg-d. There were no differences in uterine implantation rates, fetal weights, or sex ratio as compared with controls in any dose groups with the exception of one total resorption occurring in the 36 mg/kg-d dose group. The incidences of fetal malformations and external and visceral variations were not different in treated versus control offspring. Fewer of the offspring from treated animals had incomplete ossification of the 5th or 6th sternebra ($p < 0.01$) while a greater number of high-dose fetuses had supernumerary thoraco-lumbar ribs (83.1% vs. 62.6% for controls). The incidence of lowering of the pelvic girdle attachment position was significantly elevated in the offspring from high-dose dams (46.1% vs. 13.0% for controls; $p < 0.01$). The incidence of this malformation was also elevated in offspring from dams treated with 6 or 18 mg/kg-d (37.0% and 37.1%, respectively), but this was statistically significant only in the latter group ($p < 0.05$).

Cancer

No evidence of carcinogenicity was reported for rats and mice given THPC by gavage 5 d/wk for 103 wk (NTP 1987) (see Table 18-3). The incidence of mononuclear cell leukemia was increased in low-dose male rats (19/50, 25/50, and 16/50 for control, low, and high doses). However, this increase was not considered treatment-related because of a lack of dose-response and was statistically significant only in the life-table analysis. Animals were tested at or near the maximum tolerated dose (MTD) as evidenced by increased deaths in high-dose females and liver toxicity at all dose levels in both species.

Genotoxicity

Negative results were obtained when testing aqueous THPC in the *Salmonella* mutagenicity test using strains TA98 and TA100 with metabolic activation (Kawachi et al. 1980), or strains TA98, TA100, TA1535, and TA1537, with or without exogenous activation (MacGregor et al. 1980; Zeiger et al. 1987). The eluent (0.1 mL) of THPC-treated fabric samples incubated in physiological saline at 37°C for 18-24 hr (5 g fabric in 30 mL saline) also yielded negative results in *Salmonella* strains TA98, TA100, TA1537, and 1538, with or without exogenous metabolic activation (Huntingdon Research 1976).

Mutations were not induced in silkworms treated with THPC, but no experimental details were provided (Kawachi et al. 1980). THPC-induced mutations in mouse lymphoma cells without the use of exogenous metabolic activation

(Myhr et al. 1990). DMSO extracts of THPC-treated cotton fabrics were mutagenic to V79 hamster lung cells and also transformed BHK cells and BALB/3T3 cells, with and without metabolic activation (Ehrlich et al. 1980). Pure DMSO was not mutagenic when similarly tested. No difference in results were observed between dyed and undyed THPS-treated fabrics. Matthews et al. (1993) obtained an equivocal response in three transformation assay with A-31-1-13 BALB/c-3T3 cells (a positive response was obtained in one and a negative response in two). THPC was positive for genotoxicity in the *Bacillus subtilis rec* assay, with or without metabolic activation (Kawachi et al. 1980).

THPC induced chromosome aberrations in CHO cells without metabolic activation using either a conventional or delayed harvest protocol (18 hr instead of 10 hr), and with metabolic activation using the conventional harvest protocol (Loveday et al. 1989). Equivocal results were obtained for chromosome aberrations in hamster lung fibroblast cells in vitro without metabolic activation, and for rat bone marrow chromosome aberrations in vivo (Kawachi et al. 1980). THPC induced SCEs in cultured CHO cells with or without metabolic activation (Loveday et al. 1989). In vitro, THPC has been shown to form a stable adduct with the 2-amino group of guanosine (Van Duuren et al. 1978).

QUANTITATIVE TOXICITY ASSESSMENT

It is important to note that toxicity assessment values developed for THPC (such as RfDs) may not be a direct measure of the toxicity of THPC-treated fabric in a "real-world" exposure situation. This is because it is unknown whether any free THPC is present after polymerization onto the fabric. Information on the types, levels, and toxicities of THPC derivatives formed during the polymerization process is not available. An evaluation of THPC toxicity, however, is necessary as it provides a knowledge base for assessing the toxicological risks associated with this class of flame-retardant chemicals.

Noncancer

Dermal Assessment

There are no adequate dermal toxicity studies for deriving a dermal RfD for THPC. Long-term dermal studies conducted by Loewengart and Van Duuren (1977) and Van Duuren et al. (1978) only reported on the tumor findings, and therefore, could not be used for RfD development. Studies by Aoyama (1975) are of inadequate duration (6 d in rabbits and 8 d in rats).

Inhalation RfC

No studies were identified that investigated the noncancer effects of THPC inhalation, therefore an inhalation reference concentration (RfC) could not be determined.

Since THPC reacts (i.e., polymerized) directly with the fabric, it is expected that human inhalation exposure would be primarily to polymer derived from THPC attached to microscopic fibers of fabric. However, the subcommittee could not locate data that verify this hypothesis or toxicity data for THPC-treated fibers.

In the absence of relevant inhalation exposure data, the subcommittee chose to extrapolate inhalation RfCs from oral RfDs The subcommittee, however, recognizes that it is not an ideal approach and also recognizes that the estimated RfC levels might be considerably different than actual levels (if inhalation data were available). Extrapolating from one route of exposure (oral) to another (inhalation) requires specific knowledge about the uptake kinetics into the body by each exposure route, including potential binding to cellular sites. The subcommittee believes that its extrapolation of oral RfDs to inhalation RfCs is highly conservative; it assumes that all of the inhaled compound is deposited in the respiratory tract and completely absorbed into the blood. The NRC Committee on Toxicology (NRC 1985) has used this approach when inhalation exposure data were insufficient to derive inhalation exposure levels. The subcommittee believes that such an approach is justified for conservatively estimating the toxicological risk from exposure to THPC, and the RfC should be used as an interim or provisional level until relevant data become available for the derivation of an inhalation RfC.

For the purpose of estimating a hazard index for the inhalation of THPC, a provisional inhalation RfC of 0.0105 mg/m^3 was derived from the oral RfD for THPC using Equation 7 in Chapter 3.

Oral RfD

Of the available THPC studies, the most appropriate one to use for RfD derivation is the NTP (1987) 2-yr bioassay for THPC (see Table 18-3). Rats were dosed with 0, 3.75, or 7.5 mg/kg-d; male mice with 0, 7.5, or 15 mg/kg-d; and female mice with 0, 15, or 30 mg/kg-d (103 wk, 5 d/wk).

In the NTP (1987) study, the liver appears to be the critical target organ in both species. A dose-related increase in periportal hepatocyte cytoplasmic vacuolization occurred in both sexes of rats and mice, and was statistically significant at all dose levels tested. A clear dose-response was seen for the incidence of periportal hepatocyte cytoplasmic vacuolization in male and female rats. Similar results were observed in mice, but the increase in incidence was not linear. Liver toxicity occurred at the lowest dose tested in rats at 3.75 mg/kg-d which equates to 2.68 mg/kg-d when adjusted for discontinuous exposure (i.e., multiplied by 5/7).

The LOAEL of 2.68 mg/kg-d for liver toxicity in the rat was divided by the composite uncertainty factor (UF) of 1000 to yield an oral RfD of 0.003 mg/kg-d (see Table 18-5). The composite uncertainty factor was comprised of the following uncertainty factors: a factor of 3 (UF_L) was applied for LOAEL to NOAEL extrapolation. This uncertainty factor was reduced from the default of 10 to 3 because a minimal LOAEL was established due to the low frequency (18% in males, 22% in females) and low severity of the critical response (not life-threatening, possibly reversible). A factor of 3 (UF_D) was used for less than complete data (i.e., the toxicity database includes a chronic study in two species and a developmental toxicity study). A factor of 10 (UF_H) was used to account for intraspecies differences, and a factor of 10 (UF_A) was used for interspecies differences.

Cancer

IARC (1990) has determined that the tetrakis(hydroxymethyl) phosphonium salts, including THPC, are not classifiable as to their carcinogenicity to humans (Group 3), based on inadequate (i.e., lack of) evidence of carcinogenicity in animals and no data from human studies.

TABLE 18-5 Oral Reference Dose forTetrakis(hydroxymethyl) Phosphonium Chloride

Critical Effect	Species	Effect Level (mg/kg-d)	Uncertainty Factors		RfD (mg/kg-d)	Reference
Liver toxicity	Rat	LOAEL: 2.68	UF_A:	10	0.003	NTP (1987)
			UF_H:	10		
			UF_L:	3		
			UF_D:	3		
			Total:	900		

LOAEL, lowest-observed-adverse-effect level; RfD, reference dose; UF_A, interspecies variability; UF_H, intraspecies variability; UF_L, NOAEL (no-observed-adverse-effect level) for critical effect not determined; UF_D, inadequate or deficient toxicity database.

Dermal

Treatment-induced neoplasia was not found in the 496-d dermal carcinogenicity study conducted in female mice by Van Duuren et al. (1978). The results of an earlier dermal carcinogenicity study by the same group, in which a low incidence of skin papilloma was found, were compromised by solvent effects (Loewengart and Van Duuren 1977). The subcommittee concluded that data are inadequate to determine the carcinogenicity of THPC.

Inhalation

No studies were identified that investigated the carcinogenic effects of THPC following inhalation exposure. The subcommittee concluded that data are inadequate to assess the carcinogenicity of THPC via the inhalation route.

Oral

No evidence of carcinogenicity of THPC was found in rats or mice given THPC by gavage for 103 wk (NTP 1987). The subcommittee concluded that THPC is not likely to be carcinogenic via the oral route of exposure.

EXPOSURE ASSESSMENT AND RISK CHARACTERIZATION

This assessment assumes that exposure is to unreacted THPC and does not assess exposure to chemically altered forms of THPC that may form during the application of THPC to upholstery fabrics.

Noncancer

Dermal Exposure

Dermal exposure to THPC was estimated using the dermal exposure scenario described in Chapter 3. This exposure scenario assumes that an adult spends 1/4th of his or her time sitting on furniture upholstery treated with commercial THPC. It also assumes that 1/4th of the upper torso is in contact with the upholstery and that clothing presents no barrier.

The subcommittee concluded that THPC is an ionic substance and essentially not absorbed through the skin. In addition, THPC is likely to polymerize after application to the upholstery fabric; therefore, exposure to THPC is not likely to occur and it should not pose a toxic hazard by the dermal route of exposure when used as an FR in furniture upholstery.

Inhalation Exposure

Particles

Inhalation exposure estimates for THPC were calculated using the exposure scenario described in Chapter 3. This scenario assumes that a person spends 1/4th of his or her lifetime in a 30 m^3 room containing 30 m^2 of THPC-treated fabric and the room is assumed to have an air-change rate of 0.25/hr. It is also assumed that 50% of the THPC present in 25% of the surface area of the treated fabric is released over 15 yr and 1% of released particles are small enough to be inhaled.

Particle exposure was estimated using Equations 4 and 5 in Chapter 3. The subcommittee estimated an upholstery application rate (S_a) for THPC of 4.5 mg/cm^2. The release rate (μ_r) for THPC from upholstery fabric was estimated to be 2.3×10^{-7}/d (see Chapter 3, Equation 5) yielding a room airborne particle concentration (C_p) of 1.7 $\mu g/m^3$ and a short time-average exposure concentration of 0.43 $\mu g/m^3$. The time-averaged exposure concentration for particles was calculated using Equation 6 in Chapter 3.

Division of the time-average exposure concentration of 0.43 $\mu g/m^3$ by the provisional RfC for THPC of 0.0105 mg/m^3 gives a hazard index of 4.1×10^{-2}. These findings suggest that under this worst-case exposure scenario, inhalation of THPC particles from furniture upholstery is not likely to pose a noncancer toxicological risk to humans.

Vapors

In addition to the possibility of release of THPC in particles from worn upholstery fabric, the subcommittee considered the possibility of the release of THPC by evaporation. This approach is described in Chapter 3, and uses an exposure scenario similar to that described above for exposure to THPC particles.

The rate of flow of THPC vapor from the room is calculated using Equations 8-11 in Chapter 3. A saturated vapor concentration (C_v) of 10,300 mg/m^3 was estimated for THPC. The application density (S_a) for THPC in the treated upholstery was estimated as 4.5 mg/cm^2.

Using the parameters described, the equilibrium room-air concentration of THPC was estimated to be 8,700 mg/m^3. The short-term time-average exposure concentration for THPC was estimated as 2,175 mg/m^3 using Equation 12 in Chapter 3 and the equilibrium room-air concentration for THPC. It was also estimated that this air concentration could be maintained for approximately 16 hr. Division of the exposure estimate of 2,175 mg/m^3 by the provisional inhalation RfC of 0.0105 mg/m^3 yields a hazard index of 2.2×10^5.

These results indicate that if all of the THPC is released from the fabric into the air, THPC could be a toxic risk to persons entering the room. In reality, any FR that evaporated so rapidly would be useless in preventing upholstery flammability. Either THPC is much more strongly bound to the fabric than assumed in this scenario (so that the parameter γ in the analysis above is substantially less than unity), or the chemical is transformed during the application process. In either case, the emission rate would likely be controlled by some process other than diffusion through a boundary layer of air, as assumed here. The subcommittee believes that this exposure scenario provides no useful information about the potential toxicity of THPC vapors to humans associated with the emission of THPC vapors from treated furniture upholstery. Therefore, further investigation should be carried out to determine if exposure to THPC by this route poses a toxic risk to humans. These results suggest that the vapor inhalation scenario is unrealistic for THPC-treated furniture in a residential setting because evaporative loss of all THPC over 16 hr could not occur under normal conditions.

Since THPC is chemically cross-linked within the treated upholstery, the vapor pressure of THPC is assumed to be the vapor pressure for the polymerized form. In the absence of any published data, it is assumed that the polymerized form of THPC will have a vapor pressure approaching zero. Thus, the vapor inhalation noncancer risk from THPC-treated fabric may be assumed to be minimal if not zero.

Oral Exposure

The assessment of noncancer toxicological risk for oral exposure to THPC is based on the oral exposure scenario described in Chapter 3. This scenario assumes a child is exposed to THPC by sucking on 50 cm^2 of fabric treated with THPC, 1 hr/d for two yr. The subcommittee estimated an upholstery application rate (S_a) for THPC of 4.5 mg/cm^2. Oral exposure was calculated using Equation 15 in Chapter 3. The extraction rate (μ_w) for THPC was estimated to be 0.001 based on laundry data (Horrocks et al. 1992).

The worst-case, average oral daily dose for THPC was estimated as 0.00094 mg/kg-d. Division of the oral daily dose estimate of 0.00094 mg/kg by the oral

RfD for THPC of 0.003 mg/kg-d results in a hazard index of 0.313 and indicates that oral exposure to THPC is not likely to be a toxic risk under the given exposure scenario and conditions.

Cancer

Dermal

The evidence for the dermal carcinogenicity of THPC is equivocal. Van Duuren et al. (1978) found no evidence for the dermal carcinogenicity of THPC in mice. However, Loewengart and Van Duuren (1977) found equivocal evidence for the carcinogenicity of THPC when administered in combination with a known tumor promoter or initiator. The subcommittee concluded that data are inadequate to determine human carcinogenic potential by the dermal route.

Inhalation

No adequate data are available in humans or laboratory animals to assess the carcinogenicity of THPC vapors or particles containing THPC.

Oral

No evidence of carcinogenicity was found in rats or mice following chronic oral administration of THPC (see Systemic Effects section). The subcommittee concluded that THPC, used as an FR in upholstery fabric, is not likely to pose a cancer risk by the oral exposure route.

RECOMMENDATIONS FROM OTHER ORGANIZATIONS

Regulatory standards or guidelines have not been established for THPC or any of the other tetrakis(hydroxymethyl) phosphonium salts.

DATA GAPS AND RESEARCH NEEDS

Key information is needed on the types, amounts (including ratios), and toxicity of THPC derivatives present in THPC-treated cloth. It is important to note that THPC polymerizes within the fiber and fabric structure and may also react with other FR-formulation components present, so it might undergo other chemical changes that would alter its chemical properties and toxicity. It is also highly likely that oxidized forms of THPC will be present in or on the aged THPC FR-treated fabric. The subcommittee recommends that research be conducted to determine whether new chemical species are formed, and if so, to identify those chemical species.

Hazard indices for the inhalation of particles and oral exposure to THPC are less than one and therefore, these route of exposure are not anticipated to be a concern. Dermal exposure to THPC through contact with treated material is not expected to occur since THPC is chemically bound to the fabric.

REFERENCES

Afanas'eva, L.V., and N.S. Evseenko. 1971. Hygienic assessment of fireproof fabrics treated with an organophosphorus impregnating compound based on tetrahydroxymethylphosphonium chloride. Gig. Sanit. 36(3):102-103.

Albright and Wilson (Albright and Wilson, Ltd.). 1982. Study to Determine the Primary Irritancy of Untreated Fabric, Proban 210 Treated Fabric, Proban NX Treated Fabric, Laundered Proban NX Treated Fabric in Human Volunteers. J.R. Jackson, study director for Albright and Wilson Ltd. Study dated November 30, 1982. In: Product Summary: Proban® Tetrakis (hydroxyethyl)phosphonium compound with urea. Albright and Wilson Americas, Inc. Glen Allen, VA.

Albright and Wilson (Albright and Wilson, Ltd.). 1989. Composition and Chemistry of Tetrakishydroxymethylphosphonium Salts. TSCATS Fiche# OTS0519137, Doc# 89-890000200. Albright & Wilson, Ltd., West Midlands, UK.

Anonymous. 1953. Wisconsin Alumni Research Foundation Report to Oldbury Electrical Co.

Aoyama, M. 1975. Effect of anti-flame treating agents on the skin. Nagoya Med. J. 20(1):11-19.

Bittner, P. 1999. Toxicity Review for Tetrakis(hydroxymethyl) Phosphonium Salts Precondensate with Urea. Memorandum, dated March 6, 1999, from Patricia Bittner, Toxicologist, Division of Health Sciences, to Ronald Medford, Assistant Executive Director for Hazard Identification and Reduction, U.S. Consumer Product Safety Commission, Washington, DC.

Ehrlich, K., A. Hulett, and T. Turnham. 1980. Mammalian cell culture mutagenicity and carcinogenicity testing of dimethyl sulfoxide extracts of flame retardant-treated cotton fabrics. J. Toxicol. Environ. Health 6(2):259-271.

EPA (U.S. Environmental Protection Agency). 1986. Guidelines for Carcinogen Risk Assessment. Fed. Regist. 51(185):33992-34003.

EPA (U.S. Environmental Protection Agency). 1996. Proposed guidelines for carcinogen risk assessment. Fed. Regist. 61(Apr. 23):17960-18011.

EPA (U.S. Environmental Protection Agency). 1999. Carcinogen Risk Assessment Guidelines. Draft for SAB discussion. [Online]. Available: http://www.epa.gov/nceawww1/raf/SABamtg.pd

Grasseli, J.G., and W.M. Ritchey, eds. 1975. Pp. 120 in CRC Atlas of Spectral Data and Physical Constants for Organic Compounds, 2nd Ed., Vol. IV. Cleveland, OH: CRC Press.

Harada, T., A. Enomoto, G.A. Boorman, and R.R. Maronpot. 1999. Liver and gallbladder. Pp. 131 in Pathology of the mouse. R.R. Maronpot, G.A. Boorman, and B.W. Gaul, eds. Vienna, IL: Cache River Press.

Hazleton UK. 1991. THPC: Oral (Gavage) Range-finding Study in the Pregnant Rabbit. Report No. 6470-254/28. Hazelton UK, North Yorkshire, UK.

Hazleton UK. 1992. THPC: Oral (Gavage) Teratology Study in the Rabbit. Report No. 6700-254/29. Hazelton UK, North Yorkshire, UK.

Horrocks, A.R., J. Allen, S. Ojinnaka, and D. Price. 1992. Influence of laundering on durable flame retarded cotton fabrics. 1. Effect of oxidant concentration and detergent type. J. Fire Sci. 10(4):335-351.

HSDB (Hazardous Substances Data Bank). 1999. Hazardous Substances Data Bank. MEDLARS Online Information Retrieval System, National Library of Medicine. [Online]. Available: http://sis.nlm.nih.gov/cgi-bin/sis/htmlgen?HSDB Retrieved August 16, 1999.

Huntingdon Research (Huntingdon Research Center). 1976. Evaluation of Material Samples for Their Mutagenic Potential Utilizing the Ames Test Methodology. HRC Study #608675, dated August 27, 1976. In: Product Summary: Proban® Tetrakis (Hydroxyethyl)phosphonium Compound with Urea. Albright and Wilson Americas, Inc. Glen Allen, VA.

IARC (International Agency for Research on Cancer). 1990. Tetrakis(hydroxymethyl) phosphonium salts. Pp. 95-107 in IARC Monographs on the Evaluation of Carcinogenic Risk of Chemicals to Humans, Vol. 48, Some Flame Retartdants and Textile Chemicals, and Exposures in the Textile Manufacturing Industry. Lyon, France: IARC Press.

Industrial BIO-TEST Laboratories, Inc. 1975. Skin Sensitization Tests with Five Samples in Albino Guinea Pigs. IBT No. 601-06278. Doc# 878212099; Fiche# OTS0205956.

IPCS (International Programme on Chemical Safety). 2000. Environmental Health Criteria 218, Flame Retardants: Tris(2-Butoxyethyl) Phosphate, Tris(2-Ethylhexyl) Phosphate and Tetrakis (Hydroxymethyl) Phosphonium Salts. Geneva: World Health Organization.

Ishizu, S. 1975. Toxicity of organophosphorus fire retardants. Kobunshi 24:788-792.

Kawachi, T., T. Komatsu, T. Kada, M. Ishidate, M. Sasaki, T. Sugiyama, and Y. Tazima. 1980. Results of recent studies on the relevance of various short-term screening tests in Japan, In: The Predictive Value of Short-term Screening Tests in Carcinogenicity Evaluation. Appl. Methods Oncol. 3:253-267.

Loewengart, G., and B.L. Van Duuren. 1977. Evaluation of chemical flame retardants for carcinogenic potential. J. Toxicol. Environ. Health 2(3):539-546.

Loveday, K.S., M.H. Lugo, M.A. Resnick, B.E. Anderson, and E. Zeiger. 1989.

Chromosome aberration and sister chromatid exchange tests in Chinese hamster ovary cells in vitro: II. Results with 20 chemicals. Environ. Mol. Mutagen. 13(1):60-94.

MacGregor, J.T., M.J. Diamond, L.W. Mazzeno, Jr., and M. Friedman. 1980. Mutagenicity tests of fabric-finishing agents in Salmonella typhimurium: Fiber-reactive wool dyes and cotton flame retardants. Environ. Mutagen. 2(3):405-418.

Martin-Scott, I. 1966. Contact textile dermatitis (with special reference to fire proof fabrics). Br. J. Dermatol. 78(12):632-635.

Matthews, E.J., J.W. Spalding, and R.W. Tennant. 1993. Transformation of BABL/c-3T3 cells: V. Transformation responses of 168 chemicals compared with mutagenicity in Salmonella and carcinogenicity in rodent bioassays. Environ. Health. Perspect. Suppl. 101(Suppl. 2):347-482.

Myhr, B., D. McGregor, L. Bowers, C. Riach, A.G. Brown, I. Edwards, D. McBride, R. Martin, and W.J. Caspary. 1990. L5178Y mouse lymphoma cell mutation assay results with 41 compounds. Environ. Mol. Mutagen. 16(Suppl. 18):138-167.

NTP (National Toxicology Program). 1987. Toxicology and Carcinogenesis of Tetrakis(hydroxymethyl) Phosphonium Sulfate (THPS) (CAS No. 5566-30-8) and Tetrakis(hydroxymethyl) Phosphonium Chloride (THPC) (CAS No. 124-64-1) in F344/N Rats and B6C3F1 Mice (Gavage Studies). NTP Technical Report No. 296. NIH Publication No. 87-2552.

NTP (National Toxicology Program). 1991. Tetrakis(hydroxymethyl) Phosphonium Chloride. NTP Chemical Health and Safety Data. [Online]. Available: http://ntp-server.niehs.nih.gov/cgi/iH_Indexes/Chem_H&S/iH_Chem_H&S_Frames.html retrieved 8/99.

Osbourn, R.A. 1971. Report of Human Patch Tests on Substance: THPC® Safety Finishes, Cotton Knit Finish No. THPOH-NH_3. Hazelton Laboratories, Inc. Project No. 157-188. TSCATS Doc# 88-920001111; Fiche# OTS0537009.

Thake, D., P. Kurtz, and S. Carter. 1982. Neurotoxic evaluation of tetrakis hydroxymethyl phosphonium chloride in Fischer 344 rats. [Abstract No. 407]. Toxicologist 2(1):115.

Ulsamer, A.G., R.E. Osterberg, and J. McLaughlin, Jr. 1980. Flame-retardant chemicals in textiles. Clin. Toxicol. 17(1):101-131.

Van Duuren, B.L., G. Loewengart, I. Seidman, A.C. Smith, and S. Melchionne. 1978. Mouse skin carcinogenicity tests of the flame retardants tris(2,3-dibromopropyl)-phosphate, tetrakis(hydroxymethyl)phosphonium chloride, and polyvinyl bromide. Cancer Res. 38(10):3236-3240.

Zeiger, E., B. Anderson, S. Haworth, T. Lawlor, K. Mortelmans, and W. Speck. 1987. Salmonella mutagenicity tests: III. Results from the testing of 255 chemicals. Environ. Mutagen. 9 (Suppl. 9):1-109.

19

Chlorinated Paraffins

THIS chapter reviews the physical and chemical properties, toxicokinetics, toxicological, epidemiological, and exposure data on chlorinated paraffins.[1] The subcommittee used that information to characterize the health risk from exposure to chlorinated paraffins. The subcommittee also identified data gaps and recommended research relevant for determining the health risk from exposure to chlorinated paraffins.

PHYSICAL AND CHEMICAL PROPERTIES

The term "chlorinated paraffins" is commonly applied to chlorinated *n*-alkanes manufactured from straight-chain hydrocarbons (*n*-paraffins). Commercial chlorinated paraffins are mixtures that contain chlorinated paraffins of several carbon chain lengths with varying degrees of chlorination. Commercial chlorinated paraffins have carbon chain lengths between 10 and 38 carbon atoms and percent chlorination between 10% and 72%.

Chlorinated paraffins are named according to their average *n*-paraffin chain length and percent chlorination. For instance, a chlorinated paraffin that has an

---

[1]Category includes chlorinated α olefins which do not differ significantly from chlorinated paraffins with regard to structure, physical characteristics, or toxicity (EPA 1994).

average carbon chain length of 24 carbons and is 70% chlorine would be referred to as "C_{24}, 70% chlorine." The name may also include the range of carbon chains used in the manufacture of the mixture (e.g. C_{10-13}, 58% chlorine). Chlorinated paraffins with average carbon chain lengths of 10-13 carbons (C_{10-13}) are referred to as short-chain chlorinated paraffins. C_{14-20} chlorinated paraffins are categorized as medium-chain paraffins, while C_{20-30} chlorinated paraffins are referred to as long-chain paraffins. If chlorinated paraffins are used as flame retardants in residential furniture, long-chain chlorinated paraffins with 70% chlorination by weight are most likely to be used in a fabric backcoating application (Dr. Gary Stevens, University of Surrey, personal communication).

The physical and chemical properties of chlorinated paraffins vary depending on their carbon chain length and degree of chlorination (see Table 19-1). Chlorinated paraffins are insoluble in water or lower alcohols but can form emulsions or suspensions (EPA 1975). Chlorinated paraffins with low chlorine content (i.e., 35%) are usually mobile liquids. Chlorinated paraffins with higher degrees of chlorination (i.e., 40-60%) are viscous oils, while even higher chlorination of *n*-paraffins results in a waxy solid with a glassy sheen.

Commercial-grade chlorinated paraffins contain several contaminants. Alkenes (i.e., olefins) are unavoidably formed during dehydrohalogenation of chlorinated paraffins. Isoparaffins comprise about 1% of a chlorinated paraffin mixture; aromatic compounds are present at levels usually less than 100 ppm. Carbon tetrachloride, methylene chloride, chloroform, perchloroethylene, and metals have been detected in trace amounts (0.1-7.4 ppm) in chlorinated paraf-

TABLE 19-1 Physical and Chemical Properties of Representative Chlorinated Paraffins (adapted from IARC 1990)

Paraffin feedstock	Average chain length	Chlorine content (%)	Density (25°C, g/mL)	Viscosity (25°C, P)	Pour-point[a] (°C)	Heat stability (% HCl after 4 hr at 175°C)
C_{10}-C_{13} (short-chain)	C_{12}	60	1.36	35	-10	0.10
C_{13}-C_{17} (medium chain)	C_{15}	52	1.25	16	-10	0.10
C_{17}-C_{30} (long-chain)	C_{24}	39	1.12	7	-20	0.20
		42	1.17	30	0	0.20
		48	1.23	125	10	0.25
		70	1.65	Solid	NA	0.15

[a]Lowest temperature at which a substance flows under a specific condition.

fins (EPA 1975). Epoxidized soya bean oils, pentaerythritol, organometallic tin compounds, lead oxide, and cadmium compounds are added as stabilizers to chlorinated paraffins (IARC 1990).

OCCURRENCE AND USE

Approximately 70 million pounds of chlorinated paraffins were produced in the U.S. in 1998 (Chlorinated Paraffins Industry Association 1999). About half of all chlorinated paraffins consumed in the U.S. are used as extreme-pressure lubricant additives in the metal working industry (IARC 1990). Chlorinated paraffins have been used as flame retardants in commercial furniture, particularly in automobile upholstery. Chlorinated paraffins ($C_xH_{(2x-y+2)}Cl_y$) have been proposed as possible candidates for use as flame retardants in residential upholstered furniture in the U.S. (Fire Retardant Chemicals Association 1998). A recent survey of North American chlorinated paraffin industry did not identify textiles as a major area for use of chlorinated paraffins as flame retardants (Chlorinated Paraffins Industry Association 1999). Currently, chlorinated paraffins are not used as flame retardants in residential furniture in the U.S. (Fire Retardant Chemicals Association 1998) but are used as a flame-retardant backcoating for residential furniture upholstery in the United Kingdom.

For flame-retardant applications, chlorinated paraffins with approximately 70% chlorine are used. The carbon chain length of chlorinated paraffins used in flame retardants is dependent on the commercial application. C_{10-13}, 70% chlorine, is typically used as a flame retardant (FR) in rubber and soft plastics. C_{18-30}, 70% chlorine, is used in rigid plastics such as polyesters and polystyrene (IARC 1990). Long-chain, 70% chlorinated paraffins are used in upholstery backcoating in combination with antimony trioxide.

TOXICOKINETICS

Absorption

Short-chain Chlorinated Paraffins

No absorption data were located for any short-chain chlorinated paraffin following exposure by the dermal, oral, or inhalation routes.

Medium-chain Chlorinated Paraffins

C_{14-17}, 52% chlorine, was not absorbed through human skin in vitro at any detectable level after 56 hr of continuous contact (Scott 1989). About 0.70%

(Standard deviation [SD] = 0.15%) was recovered in excreta, expired air, and tissues of male Sprague-Dawley rats 96 hr after dermal application of 66 mg/cm^2 (≈2.0 g/kg body weight) of C_{18}, 50-53% chlorine (Yang et al. 1987). About 25% of an oral dose of 500 mg/kg C_{18}, 50% chlorine, was recovered after 24 hr in the excreta and 86% at 96 hr in Sprague-Dawley rats indicating that medium-chain chlorinated paraffin is absorbed to some extent through the rat GI tract.

Long-chain Chlorinated Paraffins

Less than 0.1% of a topically applied dose of C_{28}, 47% chlorine (66 mg/cm^2 or 2.0 g/kg body weight), was recovered after 96 hr in the excreta of Sprague-Dawley rats (Yang et al. 1987).

Distribution and Excretion

Short-chain Chlorinated Paraffins

Serrone et al. (1987)[2] reported that the highest levels of radioactivity in rats were found in the liver, kidney, adipose tissue, and the ovary following oral administration of C_{10-13}, 58% chlorine. Most of the dose was excreted in the feces.

Radiolabeled C_{12}, 17.4%, 55.9%, and 68.5% chlorination, were found to distribute to the liver, body fat, intestinal mucosa, bone marrow, salivary glands, and thymus within 24 hr of intravenous injection or oral gavage in C57B1 mice (Darnerud et al. 1982). Radioactivity continued to be detected in the liver, fat, adrenal cortex, and gonads after 4-12 d of exposure and retention in the liver and body fat increased with degree of chlorination. Intravenous injection of C_{12}, 17.4% and 55.9% chlorine, resulted in the retention of radioactivity in the central nervous system 30-60 d after injection. About 52% of C_{12}, 17.4% chlorination, was converted to CO_2 12 hr after administration and about 32% and 8% of the administered doses of C_{12}, 55.9% chlorination, and C_{12}, 68.5% chlorination, respectively, were converted to CO_2 over the same time period (Darnerud et al. 1982).

---

[2]Summary of studies conducted by International Research Development Corporation for the Chlorinated Paraffin Manufacturers Toxicology Testing Consortium (Aochem France, Caffaro Italy, Diamond Shamrock Chemical USA, Dover Chemical USA, Dynamit-Nobel AG Germany, Hercules Inc. USA, Hoechst AG Germany, Huls AG Germany, Imperial Chemical Industries plc UK, Keil Chemical USA, Neville Chemical USA, Rhone-Poulenc France, and Witco Chemical USA).

Medium-chain Chlorinated Paraffins

Birtley et al. (1980) reported that radio-labeled chlorinated paraffin was distributed to the body fat and liver of Wistar rats fed either 0.4 or 40 ppm of C_{14-17}, 52% chlorine, in their diet for 10 or 8 wk, respectively. Equilibrium was reached in the liver and body fat within 1 and 7 wk, respectively. No radioactivity was detected in the brain or adrenal glands. Radioactivity decreased to below background levels in the livers of the 0.4-ppm dose group within 3 wk after discontinuation of exposure. The half-life for radioactivity in the abdominal fat was estimated to be 8 wk. No attempts were made to chemically characterize the radioactivity in any tissue.

^{14}C-C_{16}, 34% chlorination, was readily absorbed and distributed to the intestinal mucosa, bone marrow, and exocrine glands when fed to C57B1 mice (Darnerud and Brandt 1982). When ^{14}C-C_{16}, 34% chlorination was given by intravenous administration, 33% of the dose was exhaled as CO_2 within 12 hr and 44% thereafter after administration.

Radiolabeled C_{16}, 69% chlorine (1.6 μmol/kg), was distributed to the bile, liver, kidney, and intestinal contents in C57B1 mice and quail within 24 hr after oral administration (Biessmann et al. 1983). Radioactivity was retained in the fat for >12 d and >30 d for quail and mice, respectively. In mice, radioactivity accumulated in the corpora lutea 1-4 d after exposure. In both species, 66 and 43 percent of radioactivity was eliminated in the feces following intravenous and oral administration, respectively, within 96 hr of administration. About 1% of an administered dose of C_{16}, 69% chlorination, was converted to CO_2 by C57B1 mice within 8 hr after gavage or intravenous injection. Urinary excretion was 3% in both cases. In quail, a combined 58% of the administered dose was eliminated in the urine and feces.

Radioactivity was detected in the liver, kidney, adipose tissue, and ovary in F-344 rats administered radio-labeled C_{14-17}, 52% chlorine, by oral gavage (Serrone et al. 1987). Most of the administered dose was excreted in the feces.

Poon et al. (1995) reports that radioactivity accumulated in both the liver and fat of rats fed 5-5,000 ppm (0.3-300.0 mg/kg-d, estimated dose levels) C_{14-17}, 52% chlorine, in their food for 90 d. Levels in the liver were approximately 20-60 times higher than in feed while radioactivity levels in fat were equal to those in the diet when measured at d 90 of the study.

Yang et al. (1987) found that about 3.3% of the radioactivity from a single oral dose of medium-chain paraffin (500 mg/kg) was distributed to the liver, intestines, and the fat 96 hr after dosing. About 0.12% of the radioactivity of a topically applied dose of 66 mg/cm^2 ($\approx$2.0 g/kg body weight) of [^{14}C]-labeled C_{18}, 50-53% chlorine, was present in intestines, liver, and fat, of male and female rats 96 hr after exposure.

Long-chain Chlorinated Paraffins

Radioactivity was detected in the liver, ovaries, blood, and adipose tissue in F-344 rats administered C_{20-30}, 43% chlorine, or C_{22-26}, 70% chlorine, by oral gavage (Serrone et al. 1987). Radioactivity levels were highest in the liver in both males and females. Radioactivity (i.e., the radio-labeled compound) was slowly eliminated from these animals, but no elimination rates or half-lives were determined.

Metabolism

No studies were found that attempted to identify the primary urinary metabolites of chlorinated paraffins formed in rodents or humans following exposure by any route. Ahlman et al. (1986) reported that injection of the radio-labeled short-chain chlorinated paraffin C_{16}, 65% chlorine, into the portal vein of Sprague-Dawley rats resulted in the excretion into the bile of *N*-acetylcysteine and glutathione conjugates. Less than 3% of the radioactivity that was excreted into the bile represented the unchanged parent compound.

Darnerud (1984) found that pretreatment of C57B1 mice with P450 inhibitors 30 min before administration of $[^{14}C]$-C_{12}, 69% chlorine, significantly decreased the rate and the amount of degradation of this compound to $^{14}CO_2$. Pretreatment of mice with the P450 inducer phenobarbital significantly increased the rate and the amount of $^{14}CO_2$ formed. Pretreatment with P450 inducers 3-methylcholanthrene or Arochlor 1254 did not increase $[^{14}C]$-C_{12}, 69% chlorine, degradation to $^{14}CO_2$. Administration of C_{14-17}, 52% chlorine, to C57B1 mice for 3 consecutive days enhanced the metabolism of C_{14-17}, 52% chlorine, to $^{14}CO_2$. Pretreatment with C_{10-13} did not significantly affect the rate of metabolism of C_{14-17}, 52% chlorine. Further studies showed that piperonyl butoxide-induced P450 enzymes may be important in the metabolism of highly chlorinated paraffins.

Enzyme Induction

Induction of enzyme activity in rats following administration of chlorinated paraffin has been reported by a number of authors. Epoxide hydrolase and glutathione transferase activity was induced in male Sprague-Dawley rats injected with C_{14-17}, 58% chlorination; C_{10-23}, 70% chlorination; or C_{23}, 70% chlorination; but not C_{22-26}, 42% chlorination, 1 g/kg-d for 5 d (Meijer et al. 1981). Hepatic uridine diphosphate (UDP)-glucuronosyltransferase and aminopyrine *N*-demethylase activities were increased in male and female Sprague-Dawley rats fed 5,000 ppm C_{14-17}, 52% chlorination, in feed for 13 wk (Poon et al. 1995). However, no significant alterations in arylhydrocarbon hydroxylase and 7-ethoxyresorufin O-deethylation activities were observed in either sex.

Nilsen et al. (1980, 1981) and Nilsen and Toftgard (1981) found that hepatic P450 protein content and liver weights were increased and deethylation activity was inhibited in rats 24 hr after intraperitoneal injection of short-chain chlorinated paraffin (Table 19-2). These effects were not observed for medium- or long-chain chlorinated paraffins.

Microsomal glucuronidation of thyroxin (T_4) was significantly increased in male rats treated with chlorinated paraffins by oral gavage (1 g/kg-d) for 14 d (Wyatt et al. 1993). The observed increase corresponded directly with statistically significant increases in plasma thyroid stimulating hormone (TSH) and enzyme activity characteristic of peroxisome proliferation. Mice receiving similar treatment were comparably more sensitive for these effects than rats. The authors concluded that increased T_4 glucuronidation levels could have caused the thyroid neoplasia observed in rodents exposed to C_{12}, 60% chlorination for 2 years (NTP 1986a).

Elcombe et al., Zeneca Central Toxicology Laboratory (1999, unpublished material) reported that oral administration of C_{10-13}, 56%chlorine; C_{10-13}, 58% chlorine; or C_{14-17}, 40% chlorine, for 14 d at 1 g/kg-d caused the induction of hepatic enzymes in male and female F-344 rats and B6C3F1 mice. Specifically, P450 4A1, 2B1, and 2B2 protein levels were elevated in treated animals. Ethoxycoumain-O-deethylation (ECOD), pentoxyresorufin-O-depentylation (PROD), lauryl acid hydroxylation (LAH), and liver β-oxidation activities were also increased following treatment . Increases in liver weight, hepatocellular hypertrophy, peroxisome proliferation, and proliferation of the smooth endoplasmic reticulum occurred in both rats and mice. In mice, PROD activities were not induced and only P450 4A1 protein levels were increased. No liver effects including enzyme induction were observed in male guinea pigs after oral administration of the aforementioned chlorinated paraffins for 14 consecutive days at 2 g/kg-d. Oral exposure to C_{20-30}, 43% chlorine, did not increase hepatic microsomal P450 content or hepatic enzyme activities in mice or rats. The effects of oral administration of C_{20-30}, 43% chlorine, was not assessed in male guinea pigs.

S.C. Hasmall et al., Zeneca Central Toxicology Laboratory (1999, unpublished material), provided further evidence that male guinea pigs respond differently to C_{10-13}, 58% chlorine, as compared with rats or mice. A statistically significant increase in hepatic *p*-nitrophenol-glucuronosyl transferase activity was observed in male guinea pigs given 1,000 mg/kg daily for 14 d. However,

TABLE 19-2 Effect of Carbon Chain Length and Percent Chlorination on Liver Weight, P450 Content, and O-Deethylation of 7-Ethoxyresorufin in Rats Treated with Chlorinated Paraffins by Intraperitoneal Injection

Substance	Liver Weight/Body Weight Ratio	nmol P450/mg Liver Protein	O-Deethylation of 7-Ethoxyresorufin
Short-chain chlorinated paraffins			
C_{10-13}, 49% chlorine	0.048 ± 0.002[a]	0.65 ± 0.07	80 ± 6
C_{10-13}, 59% chlorine	0.046 ± 0.002[a]	0.73 ± 0.09[a]	39 ± 5[a]
C_{10-13}, 71% chlorine	0.042 ± 0.001[a]	0.71 ± 0.05[a]	37 ± 4[a]
Medium-chain chlorinated paraffins			
C_{14-17}, 50% chlorine	0.038 ± 0.001[a]	0.62 ± 0.08	84 ± 6
Long-chain chlorinated paraffins			
C_{18-26}, 49% chlorine	0.034 ± 0.001	0.65 ± 0.08	86 ± 13
Controls	0.034 ± 0.001	0.55 ± 0.07	80 ± 8

[a] $p < 0.05$.
Source: Adapted from Nilsen et al. 1980, 1981.

neither liver β-oxidation nor T_4-glucuronosyl transferase activities were affected as compared with nonexposed controls. Similar results were reported for cultured guinea pig hepatocytes exposed to C_{10-13}, 58% chlorine (Williams et al., Zeneca Central Toxicology Laboratory, 1999, unpublished material).

HAZARD IDENTIFICATION[3]

Chlorinated paraffins are complex mixtures that are expected to differ with respect to their chemical content between "batches" or "runs" and between manufacturers. Chlorinated paraffins may differ in the number of carbons in the chain, chlorine content, and trace contaminants. Therefore, any toxicological risk assessment should be based on toxicological data generated for the specific commercial chlorinated paraffin to be used as an FR in residential furniture. Ideally, separate risk assessments should be conducted for each commercial chlorinated paraffin using toxicological data specific for that particular mixture.

[3]In this section, the subcommittee reviewed toxicity data on chlorinated paraffins, including the toxicity assessment prepared by the U.S. Consumer Product Safety Commission (Hatlelid 1999).

Dermal Exposure

Irritation

Birtley et al. (1980) assessed the dermal irritancy of seven classes of chlorinated paraffins in female Wistar rats (see Table 19-3). Topical application of 0.1 mL of chlorinated paraffin every other day for up to 6 d produced varying degrees of skin irritation in most cases. Chlorinated paraffins containing 10-13 and 14-17 carbons produced mild skin irritation independent of the degree of chlorination. Moderate skin irritation was inconsistently observed for C_{10-13}, 70% chlorination. Irritating chlorinated paraffins also produced mild erythema and desquamation responses by the third application. These responses improved with continued application. The authors noted that skin irritation may have been partly caused by chemical stabilizers. All chlorinated paraffins of longer chain length were characterized as nonirritating.

In studies summarized by EPA (1975), Abasov (1970) reported that KhP 470 produced no marked effect when applied to the skin.

Systemic Effects

No subchronic or chronic toxicity studies were identified for chlorinated paraffins following dermal exposure.

Available LD_{50} data for dermal application of chlorinated paraffins are summarized in Table 19-4. The dermal LD_{50} in rabbits for Chlorowax 500C (C_{12}, 59% chlorine) was reported to be greater than 10 g/kg body weight (Diamond Chemical Co. 1975, as cited in EPA 1975).

Birtley et al. (1980) reported no evidence of systemic toxicity in female Wistar rats topically treated with chlorinated paraffins (0.1 mL) every other day for up to 6 d with the chlorinated paraffins described in Table 19-3. However,

TABLE 19-3 Summary of Chlorinated Paraffins Assessed for Dermal Toxicity by Birtley et al. (1980) in Female Wistar Rats

Number of Carbon Atoms	Extent of Chlorination of *n*-Paraffin (by weight)		
in *n*-Paraffin Chain	41-50%	51-60%	61-70%
10-13	X	X	X
14-17		X	
20-30	X	X	X

the report does not specify which chlorinated paraffins were tested for systemic toxicity.

Injection of male Sprague-Dawley rats with short- or medium-chain chlorinated paraffins produced various clinical and hepatic effects, mainly peroxisome proliferation (Nilsen et al. 1980, 1981; Nilsen and Toftgard 1981; Meijer et al. 1981). These studies are summarized in the Metabolism section.

Neurological Effects

Intravenous injection of mice with $C_{10\text{-}13}$, 49% chlorination (300 mg/kg) produced a statistically significant decrease in motor capacity as compared with vehicle control mice (Eriksson and Kihlström 1985). Motor capacity was not significantly decreased in mice injected with this chlorinated paraffin at 30, 97.5, 165, or 232.5 mg/kg. Decreases in motor activity apparently did not occur in mice injected with $C_{10\text{-}13}$, 70% chlorine, at 30, 97.5, 165, 232.5, or 300 mg/kg. The authors noted that at higher dose levels, both chlorinated paraffins caused an unwarranted cessation of movement (one forepaw in the air during walking). No other studies were found regarding the effects of chlorinated paraffins on the nervous system.

Other Systemic Effects

No studies were identified that investigated the immunological, reproductive, developmental, or carcinogenic effects of chlorinated paraffins following dermal exposure.

Inhalation Exposure

There were no clinical signs of toxicity in rats exposed to Chlorowax 500C at exposure concentration of 3.3 mg/L for one hr (Diamond Shamrock Chemical Company 1975, as reviewed by EPA 1975). No other studies were located that investigated the systemic toxicity of chlorinated paraffins following inhalation exposure.

No studies were identified that investigated the immunological, neurological, reproductive, developmental, or carcinogenic effects of chlorinated paraffins following inhalation exposure.

TABLE 19-4 Acute and Short-term Toxicity Studies of Chlorinated Paraffins

Duration, Route	Effects	NOAEL/ LOAEL	Reference
Short-chain chlorinated paraffins			
C_{10-13}, 56% Cl_2			
14 d, gavage[a]	Significant increase in relative liver weights. Peroxisome proliferation, hepatocyte hypertrophy. Increase in total microsome P450 content. Increased hepatic P450 4A1, 2B1, and 2B2 protein levels. Induction of ECOD, PROD, LAH, and liver β-oxidation activities.	ND	Elcombe et al., Zeneca Central Toxicology Laboratory, 1999, unpublished material
14 d, gavage[a]	Significant increase in relative liver weights, peroxisome proliferation, and hepatocyte hypertrophy. Total P450 microsomal content increased. Increased hepatic P450 4A1 protein levels. Induction of ECOD, LAH, and liver β-oxidation activities.	ND	Elcombe et al., Zeneca Central Toxicology Laboratory, 1999, unpublished material
14 d, gavage[a]	Increase in relative liver weight. No changes in liver morphology. No enzyme activity induction (ECOD, PROD, EROD, LAH, β-oxidation).	ND	Elcombe et al., Zeneca Central Toxicology Laboratory, 1999, unpublished material
14 d, gavage[a]	Statistically significant decreases in body weight gain, relative liver weight. Increase in p-nitrophenol-glucuronosyl transferase activity at 1,000 mg/kg-d. Neither liver β-oxidation nor T_4-glucuronosyl transferase activities were affected at either dose levels as compared with nonexposed controls.	ND	S.C. Hasmall et al., Zeneca Central Toxicology Laboratory, 1999, unpublished material

15 d, gavage[a]	Relative liver weights, liver β-oxidation activity, and T_4- and p-nitrophenol glucuronidation activity levels increased in males and females. Increases in relative kidney weights in males. T_4 levels decreased in both dose groups, both sexes. Centrilobular hypertrophy.	ND	Wyatt et al., Zeneca Central Toxicology Laboratory, 1999, unpublished material
14 d, gavage[a]	Significant increase in absolute liver weight. Peroxisome proliferation. T_4 levels were significantly decreased and TSH levels significantly increased in animals exposed at 1,000 mg/kg-d.	Liver weight LOAEL: 10 mg/kg-d Peroxisome proliferation NOAEL: 500 mg/kg-d; LOAEL: 1,000 mg/kg-d	Wyatt et al. 1993
14 d, gavage[a]	Significant increase in absolute liver weight at 100 mg/kg-d. Peroxisome proliferation. T_4 levels significantly decreased and TSH levels significantly increased in animals exposed at 1,000 mg/kg-d.	Liver weight NOAEL: 50 mg/kg-d; LOAEL: 100 mg/kg-d Peroxisome proliferation NOAEL: 100 mg/kg-d; LOAEL: 250 mg/kg-d	Wyatt et al. 1993
C_{10-13}, 58% Cl_2			
14 d, gavage[a]	Significant increase in relative liver weights. Peroxisome proliferation, hepatocyte hypertrophy. Total P450 microsomal content and hepatic P450 4A1, 2B1, and 2B2 protein levels increased. Induction of ECOD, PROD, LAH, and liver β-oxidation activities.	ND	Elcombe et al., Zeneca Central Toxicology Laboratory, 1999, unpublished material
14 d, gavage[a]	Significant increase in relative liver weights, peroxisome proliferation and hepatocyte hypertrophy. Total P450 microsomal content and hepatic P450 4A1 protein levels increased. Induction of ECOD, LAH, and β-oxidation activities.	ND	Elcombe et al., Zeneca Central Toxicology Laboratory, 1999, unpublished material

TABLE 19-4 *(Continued)*

Duration, route	Effects	NOAEL/ LOAEL	Reference
14 d, gavage[a]	Slight increase in relative liver weight. No changes in liver morphology. No induction of total P450 microsomal content or ECOD, PROD, EROD, LAH, or liver β-oxidation activities.	ND	Elcombe et al., Zeneca Central Toxicology Laboratory, 1999, unpublished material
14 d, gavage[a]	No increase in absolute liver weight starting at 100 mg/kg-d. Peroxisome proliferation. T_4 levels were significantly decreased and TSH levels significantly increased in animals exposed at 1,000 mg/kg-d.	Liver weight NOAEL: 50 mg/kg-d; LOAEL: 100 mg/kg-d Peroxisome proliferation NOAEL: 100 mg/kg-d; LOAEL: 250 mg/kg-d	Wyatt et al. 1993
14 d, gavage[a]	Significant increase in absolute liver weight at ≥500 mg/kg-d. Peroxisome proliferation. T_4 levels were significantly decreased and TSH levels significantly increased in animals exposed at 1,000 mg/kg-d.	Liver weight NOAEL: 100 mg/kg-d; LOAEL: 250 mg/kg-d Peroxisome proliferation NOAEL: 100 mg/kg-d; LOAEL: 250 mg/kg-d	Wyatt et al. 1993
14 d, gavage[a]	Animals inactive after dosing. No deaths or toxicty observed.	Toxicity NOAEL: 13,600 mg/kg-d	NTP 1986a
14 d, gavage[a]	Animals inactive after dosing. No deaths or toxicty observed.	Toxicity NOAEL: 27,200 mg/kg-d	NTP 1986a

16 d, gavage[b]	One of five males and two of five females of the 7,500 mg/kg-d group died. High-dose rats had diarrhea, body-weight gains decreased. Livers enlarged in three of five animals of all dose groups except 469-mg/kg-d female rats.	Liver effects LOAEL: 469 mg/kg-d	NTP 1986a
16 d, gavage[b]	Deaths at ≥ 3,750. Final mean body weights of survivors were not different from controls; diarrhea occurred in all dosed mice. Livers were enlarged in dosed mice that survived the study.	Liver effects LOAEL: 938 mg/kg-d	NTP 1986a
14 d, diet	Increased liver weight accompanied by hepatocellular hypertrophy.	Liver effects NOAEL: 30 mg/kg-d; LOAEL: 100 mg/kg-d	IRDC 1983, as reviewed by Serrone et al. 1987 (438-002)[c]
14 d, gavage[a]	Increased liver weight accompanied by hepatocellular hypertrophy.	Liver effects NOAEL: 30 mg/kg-d; LOAEL: 100 mg/kg-d	IRDC 1981, as reviewed by Serrone et al. 1987 (438-006)[c]
Medium-chain chlorinated paraffins			
C_{14-17}, 40% Cl_2			
14 d, gavage[a]	Significant increase in relative liver weights. Peroxisome proliferation, hepatocyte hypertrophy. Increased total P450 microsomal content and hepatic P450 4A1, 2B1, and 2B2 protein levels. Induction of ECOD, PROD, LAH, and liver β-oxidation activities.	ND	Elcombe et al., Zeneca Central Toxicology Laboratory, 1999, unpublished material
14 d, gavage[a]	Significant increase in relative liver weights, peroxisome proliferation, and hepatocyte hypertrophy. Total P450 microsomal content and P450 4A1 protein levels increased. Induction of ECOD, LAH, and liver β-oxidation activities. No induction PROD or EROD activities.	ND	Elcombe et al., Zeneca Central Toxicology Laboratory, 1999, unpublished material

TABLE 19-4 *(Continued)*

Duration, route	Effects	NOAEL/ LOAEL	Reference
14 d, gavage[a]	No increase in relative liver weight. No changes in liver morphology observed. No induction of total P450 microsomal content or ECOD, PROD, EROD, LAH, or liver β-oxidation activities.	ND	Elcombe et al., Zeneca Central Toxicology Laboratory, 1999, unpublished material
15 d, gavage[a]	Relative liver weights, liver β-oxidation activity, and T_4- and p-nitrophenol glucuronidation activity levels increased in all dosed animals. Increases in relative kidney weights in males of both dose groups. T_4 levels were decreased in both dose groups. Centrilobular hypertrophy.	ND	Wyatt et al., Zeneca Central Toxicology Laboratory, 1999, unpublished material
14 d, gavage[a]	Significant increase in absolute liver weight, but no consistent dose-response. Peroxisome proliferation. T_4 levels significantly decreased and TSH levels significantly increased in animals exposed at 1,000 mg/kg-d.	Liver weight LOAEL: 10 mg/kg-d Peroxisome proliferation NOAEL: 250 mg/kg-d; LOAEL: 500 mg/kg-d	Wyatt et al. 1993
14 d, gavage[a]	Significant increase in absolute liver weight at 1,000 mg/kg-d. Peroxisome proliferation. T_4 levels significantly decreased and TSH levels significantly increased in animals exposed at 1,000 mg/kg-d.	Liver weight NOAEL: 500 mg/kg-d; LOAEL: 1,000 mg/kg-d Peroxisome proliferation NOAEL: 250 mg/kg-d; LOAEL: 500 mg/kg-d	Wyatt et al. 1993

C_{14-17}, 52% Cl_2			
14 d, diet	Increased liver weight accompanied by mild, diffuse hepatocellular hypertrophy..	Liver effects NOAEL: 500 ppm; LOAEL: 1,500 ppm for F, 5,000 ppm for M	IRDC 1981, as reviewed by Serrone et al. 1987 (438-003)[c]
14 d, gavage[a]	No compound-related effects.	Toxicity NOAEL: 30 mg/kg-d	IRDC 1981, as reviewed by Serrone et al. 1987 (438-005)[c]
Long-chain chlorinated paraffins			
C_{23}, 43% Cl_2			
14 d, gavage[d]	No significant increase in relative liver weights or changes in liver morphology. No change in total P450 microsomal content or hepatic P450 4A1, 2B1, and 2B2 protein levels. No increase in ECOD, EROD, PROD, LAH, and liver β-oxidation activities.	ND	Elcombe et al., Zeneca Central Toxicology Laboratory, 1999, unpublished material
14 d, gavage[d]	No significant increase in relative liver weights or changes in liver morphology. No change in total P450 microsomal content or hepatic P450 4A1, 2B1, and 2B2 protein levels. No increase in activity levels for ECOD, EROD, PROD, LAH, and liver β-oxidation activities.	ND	Elcombe et al., Zeneca Central Toxicology Laboratory, 1999, unpublished material
Gavage	Animals inactive after dosing. No deaths or toxicty observed.	Toxicity NOAEL: 23,400 mg/kg-d	NTP 1986b
16 d, gavage[b]	No deaths occurred and no clinical signs of toxicity were observed in treated animals. Body weights of dosed animals were not different from those of controls.	Toxicity NOAEL: 3,750 mg/kg-d	NTP 1986b

TABLE 19-4 *(Continued)*

Duration, route	Effects	NOAEL/ LOAEL	Reference
16 d, gavage[b]	No deaths occurred and no clinical signs of toxicity were observed in treated animals. Body weights of dosed animals were not different from those of controls.	Toxicity NOAEL: 7,500 mg/kg-d	NTP 1986b
$C_{22\text{-}26}$, 70% Cl_2			
14 d, diet	No compound-related effects were found in the 14-d study in either male or female rats.	Toxicity NOAEL: 15,000 pp	IRDC 1981, as reviewed by Serrone et al. 1987 (438-004)[c]

ECOD, ethoxycoumain-o-deethylase; EROD, ethoxyresorufin-o-deethylase; F, female; LAH, lauryl hydroxylation; M, male; PROD, pentoxyresofu-o-depentylation.

[a]Test compound administered as a single dose on d 1 in corn oil unless noted otherwise.

[b]Animals were exposed at $C_{20\text{-}30}$, 43% chlorination.

[c]Study results summarized in Serrone et al. (1987).

[d]Repeat dose studies. Test compounds were administered daily for 16 d. Highest doses for mice were administered as split doses, 5 hr apart, twice a day.

Oral Exposure

Systemic Effects

Gosselin et al. (1976) report that chlorinated paraffins are "practically nontoxic" in humans with a probable oral lethal dose above 15 g/kg, or greater than 2.2 pounds, for a 70-kg person.

Table 19-4 provides a summary of oral acute and short-term toxicity data for chlorinated paraffins. Based on these data, it appears that chlorinated paraffins are not very toxic with gross toxic effects (i.e., death, clinical signs of toxicity) occurring at doses greater than 469 mg/kg-d. Increases in liver weights occur at doses greater than 10-30 mg/kg-d.

Additional oral acute toxicity studies in rats conducted by Birtley et al. (1980) of $C_{10-13,}$ $C_{14-17,}$ or C_{20-30} containing either 41-50%, 51-60%, or 61-70% chlorine found that only C_{10-13} caused clinical toxicity and at dose levels of 4 g/kg or greater. All animals exhibiting signs of toxicity usually recovered within 7 d of treatment. EPA (1975) reviewed the LD_{50} data for a variety of commercial chlorinated paraffins administered by various routes of administration. No compound-related mortality was reported in these studies. Doses tested ranged from 5 to 60 g/kg. Some of these studies only reported the amounts of chlorinated paraffins used in the tests; these amounts ranged between 0.464 and 21.4 mL/kg chlorinated paraffin.

Table 19-5 provides a summary of the subchronic and chronic toxicity data available for short-, medium-, and long-chain chlorinated paraffins.

Short-chain Chlorinated Paraffins

Subchronic and chronic (90-d, 6-mo, 12-mo, and 2-yr) oral toxicity studies have been carried out for short-chain chlorinated paraffins by the U.S. National Toxicology Program (NTP 1986a). Subchronic studies have been carried out by IRDC (as reviewed by Serrone et al. 1987) and Wyatt et al., Zeneca Central Toxicology Laboratory (1999, unpublished material). Major toxic effects associated with subchronic or chronic oral exposure to short-chain chlorinated paraffins include a statistically significant increase in relative liver weights in rats (NTP 1986a; Serrone et al. 1987; Wyatt et al. 1993), and mice (NTP 1986a). Microscopic examination revealed hepatocyte hypertrophy in rats (NTP 1986a; Serrone et al. 1987; Wyatt et al. 1993) and mice (NTP 1986a). Increased relative liver weights with associated pathological changes occurred at dose levels greater than 10 mg/kg-d in rats and 250 mg/kg-d in mice. Increases in thyroid (NTP 1986a) and kidney weights (NTP 1986a; Serrone et al. 1987;

TABLE 19-5 Subchronic and Chronic Toxicity Studies of Chlorinated Paraffins

Species, Strain, Sex	Dose	Duration, Route	Effects	NOAEL/LOAEL	Reference
Short-chain chlorinated paraffins					
C_{10-13}, 58% Cl_2					
Rat, F344, M/F	313, 625, 1,250, 2,500, 5,000 mg/kg-d	90 d, gavage	Gross: No compound-related deaths or clinical signs of toxicity observed. Body weights decreased in males of 2,500 and 5,000 mg/kg-d groups. Organ Weight Changes: Relative liver weights significantly ($p < 0.01$) increased in males and females at 313 mg/kg-d or higher dose groups. Microscopic: No treatment-related lesions noted in livers from males or females. Kidney nephrosis in males and females of 5,000 mg/kg-d dose group. Rats in other dose groups were not examined microscopically.	Liver pathology LOAEL: 313 mg/kg-d Body weight: NOAEL: 1,250 mg/kg-d LOAEL: 2,500 mg/kg-d	NTP 1986a[a]
Mouse, B6C3F1, M/F	125, 250, 500, 1,000, 2,000 mg/kg-d	90 d, gavage	Gross: 26 deaths occurred due to gavage-related trauma. No clinical signs of toxicity apparent, but body weights of males in 1,000 and 2,000 mg/kg-d dose groups lower than controls. Organ Weight Changes: Relative liver weights were significantly ($p < 0.01$) increased in males and females in 500 mg/kg-d or higher dose groups. Microscopic: Hepatocyte hypertrophy and focal hepatic necrosis in both males and females.	Liver pathology: NOAEL: 250 mg/kg-d LOAEL: 500 mg/kg-d Body weight: NOAEL: 500 mg/kg-d LOAEL: 1,000 mg/kg-d	NTP 1986a[a]
Rat, F344, M/F	0, 312, 625 mg/kg-d	6 mo, gavage	Gross: Male body weights were significantly lower ($p < 0.01$) than controls in both dose groups Organ Weight Changes: Relative liver and kidney weights were significantly ($p < 0.01$) increased in male and female rats dosed at levels 312 mg/kg-d or higher. Microscopic: Hepatocyte hypertrophy and renal	Liver/kidney weights: NOAEL: ND LOAEL: 312 mg/kg-d	NTP 1986a[a]

			nephropathy observed with increased frequency in all exposed animals as compared with controls. No chemical-related changes in relative adrenal glands, thymus, heart, brain, or spleen observed.		
Rat, F344, M/F	0, 312, 625 mg/kg-d	12 mo, gavage	Gross: Male body weights were significantly lower ($p < 0.01$) than controls in both dose groups. Organ Weight Changes: Relative liver and kidney weights were significantly ($p < 0.01$) increased in male and female rats dosed at levels 312 mg/kg-d or higher. Microscopic: Hepatocyte hypertrophy and renal nephropathy observed with increased frequency in all exposed animals as compared with controls. No chemical-related changes in relative adrenal glands, thymus, heart, brain, or spleen observed.	Liver/kidney weights: NOAEL: ND LOAEL: 312 mg/kg-d	NTP 1986a[a]
Rat, F344, M/F	0, 312, 625 mg/kg-d	2 yr, gavage	Gross: Survival significantly ($P < 0.001$) lower among dosed males beginning at wk 89; survival of low-dose females significantly lower after wk 92. Decreased activity among exposed rats beginning wk 90; some high-dose females had distended or firm abdomens. At 24 mo, pale skin and eyes were observed in both exposed males and females. Exposed males judged to be emaciated with a high incidence of stained and wet fur in the pelvis/perianal area. High-dose males had sunken or small eyes and abnormal breathing. Mean body weights were reduced compared with controls beginning at wk 17 for males and wk 42 for females. Microscopic: Liver necrosis/hypertrophy observed in male and females beginning at 312 mg/kg-d. Nephropathy was more severe in dosed male rats and the incidence was increased in dosed females. Renal tubular cell hyperplasia was increased in high-dose males. Increased incidence of inflammation and hyperkeratosis of the forestomach in males beginning at 312 mg/kg-d.	Liver/kidney effects: NOAEL: ND LOAEL: 312 mg/kg-d	NTP 1986a[a]

TABLE 19-5 *(Continued)*

Species, Strain, Sex	Dose	Duration, Route	Effects	NOAEL/LOAEL	Reference
Mouse, B6C3F1, M/F	0, 125, 250 mg/kg-d	2 yr, gavage	Gross: Survival significantly reduced among high-dose females after wk 100. Survival comparable to controls for low-dose females and exposed males. Exposed females had decreased body weights (6-12%) compared with controls after wk 36. Microscopic: Incidence of kidney nephrosis was increased among high-dose females. No other significant non-neoplastic pathological changes were reported.	Kidney effects: Females: NOAEL: 125 mg/kg-d LOAEL: 312 mg/kg-d	NTP 1986a[a]
Rat, F344, M/F	0, 312, 625 mg/kg-d	90 d, gavage	Gross: Male rats in the high-dose group had significantly decreased body weights as compared with controls. Statistically significant increases in relative liver and kidney weights in both males and females of both dose groups. Microscopic: Hepatocyte and follicular cell hypertrophy in males and females in all dose groups. Protein accumulation in kidneys of males in the high-dose group. Small increase in α2u-globulin protein in the proximal convoluted tubules of males in the 625 mg/kg-d dose group, but accumulation was not restricted to hyaline droplets. Biochemical: Free plasma T_4 levels were significantly decreased in male and female rats. Increased activity levels of hepatic β-oxidation of fatty acids and glucuronidation of T_4 and p-nitrophenol in both sexes in both dose groups.	Liver/kidney weight: NOAEL: ND LOAEL: 312 mg/kg-d Liver/thyroid pathology: NOAEL: ND LOAEL: 312 mg/kg-d	Wyatt et al., Zeneca Central Toxicology Laboratory, 1999, unpublished material
Rat, F344, M/F	10, 100, 625 mg/kg-d	90 d, diet	Gross: No overt toxicity observed. Body weight gains slightly reduced, changes in water consumption, slight skin atonia at 625 mg/kg-d. Organ Weight Changes (relative): 10: None in males or females	Liver weight: NOAEL: 10 mg/kg-d LOAEL: 100 mg/kg-d Liver pathology: NOAEL: 10 mg/kg-d	IRDC 1984, as reviewed by Serrone et al. 1987 (438-

			100: Increase in liver weight in males and females 625: Increase in liver, kidneys, and thyroid in males and females Microscopic (dose, mg/kg-d): 10: No effects in males or females 100: Hepatocellular hypertrophy of liver in males and females, chronic nephritis of the kidney and hypertrophy and hyperplasia of the thyroid in males 625: Hepatocellular hypertrophy of liver and hypertrophy and hyperplasia of the thyroid in males and females, chronic nephritis of the kidney in males, renal tubule pigmentation of kidneys in females	LOAEL: 100 mg/kg-d	022/035)
Rat, F344, M/F	10, 100, 625 mg/kg-d	90 d, gavage	Gross: No overt toxicity observed. Body weight gains slightly reduced, changes in water consumption, slight skin atonia at 625 mg/kg-d. Organ Weight Changes (dose, mg/kg-d) (relative): 10: None in males or females 100: Increase in liver weight in males and females 625: Increase in liver, kidneys, and thyroid weights in males and females Microscopic: 10: No effects in males or females 100: Hepatocellular hypertrophy of liver in males and females, chronic nephritis of the kidney and hypertrophy and hyperplasia of the thyroid in males 625: Hepatocellular hypertrophy of liver and hypertrophy and hyperplasia of the thyroid in males only, chronic nephritis of the kidney in males, renal tubule pigmentation of kidneys in females	Liver weight: NOAEL: 10 mg/kg-d LOAEL: 100 mg/kg-d Liver pathology: NOAEL: 10 mg/kg-d LOAEL: 100 mg/kg-d	IRDC 1984, as reviewed by Serrone et al. 1987 (438-022/029)

TABLE 19-5 *(Continued)*

Species, Strain, Sex	Dose	Duration, Route	Effects	NOAEL/LOAEL	Reference
Medium-chain chlorinated paraffins					
C_{14-17}, 52% Cl_2					
Rat, Sprague-Dawley M/F	0, 5, 50, 500, 5,000 ppm (0.3, 3, 30, 300 mg/kg-d)[b]	90 d, diet	Gross: No gross lesions were noted at necropsy other than enlarged livers in the high-dose animals; one female in the 5,000 ppm group had an enlarged lung with pulmonary congestion. Organ Weight Changes (relative): 30: Increase in liver weight in females, none in males 300: Increase in liver and kidney weights in male and females Microscopic: 3: Changes in thyroid in females 30: Changes in liver and thyroid in males and females; changes in inner medulla tubules of kidney in females 300: Changes in liver and thyroid in males and females; changes in kidney proximal tubules in males; changes in inner medulla tubules of kidney in females Biochemical: 3: Serum cholesterol increased in females 30: Increased urinary ascorbic acid excretion, decreased hepatic vitamin A level in males and females, serum cholesterol increased in females 300: Elevated hepatic UDP-glucuronosyltransferase activity, increased urinary ascorbic acid excretion, decreeased hepatic vitamin A level in males and	Liver weight: NOAEL: 3 mg/kg-d LOAEL: 30 mg/kg-d Liver pathology: NOAEL: 0.3 mg/kg-d LOAEL: 3 mg/kg-d Liver biochemistry: NOAEL: 0.3 mg/kg-d LOAEL: 3 mg/kg-d	Poon et al. 1995

			females; females: increased aminopyrine N-demethylase activity, N-acetylglucosaminidase activity, serum cholesterol		
Dog, beagle, M/F	0, 10, 30, 100 mg/kg-d	90 d, diet	Gross: No deaths or clinical signs of toxicity noted. No effect was seen on weight gain, hematological parameters, or urinalysis results. Organ Weight Changes (relative): Significant increase ($p < 0.05$) in relative liver weight in male dogs in 100 mg/kg-d dose group Microscopic: Cloudy, pale hepatocytes and proliferation of hepatocyte smooth endoplasmic reticulm in dogs dosed with 30 or 100 mg/kg-d Biochemical: Statistically significant ($p < 0.05$) increases in serum alkaline phosphatase activity in male dogs in 100 mg/kg-d group.	Liver weight: NOAEL: 30 mg/kg-d LOAEL: 100 mg/kg-d Liver pathology: NOAEL: 10 mg/kg-d LOAEL: 30 mg/kg-d Liver biochemistry: NOAEL: 30 mg/kg-d LOAEL: 100 mg/kg-d	Birtley et al. 1980
Rat, Wistar, M/F	0, 250, 500, 2,500, 5,000 ppm (15, 30, 150, 300 mg/kg-d)[b]	90 d, diet	Gross: Reduced weight gain as a result of reduced food consumption was observed in male rats at wk 1 onward in 2,500 and 5,000 ppm dose groups and wk 4 onward in 500 ppm dose group. No deaths or clinical abnormalities observed in test animals. Authors report that there was a tendency towards congestion of the kidney with increasing dietary concentration Organ Weight Changes (relative): 15: None 30: Increase in liver weight in females 150: Increase in liver weight in males and females 300: Increase in liver and kidney weights in males and females Microscopic: 15: None 30: Liver: proliferation of the smooth endoplasmic reticulm	Liver weight: NOAEL: 15 mg/kg-d LOAEL: 30 mg/kg-d Liver pathology: NOAEL: 15 mg/kg-d LOAEL: 30 mg/kg-d Liver biochemistry: NOAEL: ND LOAEL: 15 mg/kg-d Other effects: Inadequate dose-response information given for kidney congestion and reported biochemical effects. More information is needed to derive a NOAEL or LOAEL for these effects.	Birtley et al. 1980

TABLE 19-5 *(Continued)*

Species, Strain, Sex	Dose	Duration, Route	Effects	NOAEL/LOAEL	Reference
			150: Liver: proliferation of the smooth endoplasmic reticulm 300: Liver: proliferation of the smooth endoplasmic reticulm Biochemical: Authors report "a marked decline" in alkaline phosphatase activity in all dose groups; serum glutamic pyruvate transaminase activity also tended to decline except for a slight increase observed in males in 5,000 ppm dose group.		
Rat, F344, M/F	10, 100, 625 mg/kg-d	90 d, diet	Organ Weight Changes (relative): 10: Increase in liver weight in males (not considered biologically significant), none in females 100: Increase in liver and kidney weights in males and females 625: Increase in liver and kidney weights in males and females; increase in thyroid and adrenal gland weights in males Microscopic (dose, mg/kg-d): 10: No changes detected in males or females. 100: No changes detected in liver and kidney of male and females 625: Hepatocellular hypertrophy of liver in males and females; thyroid hyperplasia of the thyroid in males only, chronic nephritis of the kidney in males, renal tubule pigmentation of kidneys in females; no changes detected in adrenal glands in male rats	Liver weight NOAEL: 10 mg/kg-d LOAEL: 100 mg/kg-d Liver pathology: NOAEL: 100 mg/kg-d LOAEL: 625 mg/kg-d	IRDC 1981, as reviewed by Serrone et al. 1987 (438-023/026)

C_{14-17}, 40% chlorine					
Rat, F344, M/F	0, 312, 625 mg/kg-d	90 d, gavage	Gross: Statistically significant increases in relative liver and kidney weights in males and females of both dose groups. Microscopic: Hepatocyte and follicular cell hypertrophy in males and females in all dose groups. Protein accumulation in kidneys of males in high-dose group. A small, but statistically significant increase in α2u-globulin protein in proximal convoluted tubules was observed in male rats in high-dose group, but this was not restricted to hyaline droplets. Biochemical: Free plasma T_3, T_4, and TSH levels were not significantly affected in either sex at either dose level. Increased activity levels of hepatic β-oxidation of fatty acids and glucuronidation of T_4 and *p*-nitrophenol in both sexes in both dose groups.	Liver/kidney weight: NOAEL: ND LOAEL: 312 mg/kg-d Liver/thyroid pathology: NOAEL: ND LOAEL: 312 mg/kg-d	Wyatt et al. 1993; Zeneca Central Toxicology Laboratory, 1999, unpublished material
Long-chain chlorinated paraffins					
C_{20-30}, 43% Cl_2					
Rat, F344, M/F	100, 900, 3,750 mg/kg-d	90 d, gavage	Gross: No significant adverse effects on body weight gain, water/food consumption, haematology, or clinical biochemistry measurements. Macroscopic: 100: No increases in absolute liver or kidney weights in males. Increased liver weight in females. 900: No increases in absolute liver or kidney weights in males. Increased liver weight in females. 3,750: No increases in absolute liver or kidney weights in males. Increased liver weight in females. Microscopic: 100: Males; no effects. Multifocal granulomatous hepatitis, inflammation, and necrosis in livers from females. 900: Males; no effects. Multifocal granulomatous hepatitis, inflammation, and necrosis in livers from females.	Liver weight: NOAEL: ND for females; 3,750 for males Liver pathology: NOAEL: ND for females; 3,750 for males	IRDC 1981, as reviewed by Serrone et al. 1987 (438-005)

TABLE 19-5 *(Continued)*

Species, Strain, Sex	Dose	Duration, Route	Effects	NOAEL/LOAEL	Reference
			3,750: Nephrosis of the kidney in males. Multifocal granulomatous hepatitis, inflammation, and necrosis in livers and mineralization of the kidney in females.		
Rat, F344, M/F	0, 235, 469, 938, 1,875, 3,750 mg/kg-d	90 d, gavage	Gross: No significant decreases in body weight were observed in exposed male or females as compared with controls. No deaths or clinical signs of toxicity were observed in either males or females orally exposed to chlorinated paraffin Microscopic: A dose-related increase in the incidence of granulomatous inflammation of the liver was observed in females.	Liver: granulomatous inflammation: Male: LOAEL: ND NOAEL: 3,750 mg/kg-d Female: LOAEL: 235 mg/kg-d NOAEL: ND	NTP 1986a[c]
Mouse, B6C3F, M/F	0, 469, 938, 1,875, 3,750, 7,500 mg/kg-d	90 d, gavage	Gross: No significant decreases in body weight was observed in exposed male or females as compared with controls. No deaths or clinical signs of toxicity were observed in either males or females orally exposed to chlorinated paraffin Microscopic: No gross lesions or changes tissue histology were detected that could be attributed to oral exposure to chlorinated paraffin	No adverse effects detected at 469 mg/kg-d	NTP 1986a[c]
Rat, F344, M/F	Male: 0, 1,875, 3,750 mg/kg-d Female: 0, 100, 300, 900 mg/kg-d	6 mo, gavage	Gross: No compound-related clinical signs of toxicity. Some clinical chemistry values altered at 1,875 mg/kg-d in male and 100 mg/kg-d in females. Macroscopic: Dose-related and statistically significant increases in relative liver weights observed in treated females at 900 mg/kg-d. Microscopic: Dose-related increase in the incidence of	Liver weight: Females: NOAEL: 300 mg/kg-d LOAEL: 900 mg/kg-d Liver: granulomatous inflammation: Females:	NTP 1986a[c]

			granulomatous inflammation of the liver occurred in treated females. Granulomatous inflammation of the liver confirmed by clinical chemistry and hematology studies.	NOAEL: ND LOAEL: 100 mg/kg-d	
Rat, F344, M/F	Male: 0, 1,875, 3,750 mg/kg-d Female: 0, 100, 300, 900 mg/kg-d	12 mo, gavage	Gross: No compound-related clinical signs of toxicity. Some clinical chemistry values altered at 1,875 mg/kg-d in male and 100 mg/kg-d in females. Significant alterations were observed at dose levels of 3,750 mg/kg-d in dosed males and 300 mg/kg-d in dosed females at 12 mo. Macroscopic: Dose-related and statistically significant increases in relative liver weights observed in treated males and females. Microscopic: Dose-related increase in the incidence of granulomatous inflammation of the liver occurred in treated males and females. Granulomatous inflammation of the liver confirmed by clinical chemistry and hematology studies.	Liver weight: Females: NOAEL: 100 mg/kg-d LOAEL: 300 mg/kg-d Males: NOAEL: 1,875 mg/kg-d LOAEL: 3,750 mg/kg-d Liver: granulomatous inflammation: Females: NOAEL: ND LOAEL: 100 mg/kg-d Males: NOAEL: ND LOAEL: 1,875 mg/kg-d	NTP 1986a[c]
Rat, F344, M/F	Male: 0, 1,875, 3,750 mg/kg-d Female: 0, 100, 300, 900 mg/kg-d	2 yr, gavage	Gross: Mean body weights not affected. No clinical signs of toxicity observed among dosed animals. No significant decreases in survival as compared with controls Microscopic ($p < 0.05$): Males: Lymphocytic and granulomatous inflammation of liver; pigmentation of the liver; spleen congestion; granulomatous inflammation of pancreatic and mesenteric lymph nodes; mesenteric and pancreatic lymphoid hyperplasia beginning at 1,875 mg/kg-d. Hepatocellular hyperplasia and granulomatous inflammation of pancreatic and mesenteric lymph nodes beginning at 3,750 mg/kg-d.	Liver: granulomatous inflammation: Females: NOAEL: ND LOAEL: 100 mg/kg-d Males: NOAEL: ND LOAEL: 1,875 mg/kg-d	NTP 1986a[c]

TABLE 19-5 *(Continued)*

Species, Strain, Sex	Dose	Duration, Route	Effects	NOAEL/LOAEL	Reference
			Females: Lymphocytic and granulomatous inflammation of liver; pigmentation of the liver/kidney; nephropathy; hepatocellular hypertrophy; spleen congestion; granulomatous inflammation of pancreatic and mesenteric lymph nodes; pancreatic lymphoid hyperplasia beginning at 100 mg/kg-d. Lymphoid hyperplasia of the pancreatic lymph node beginning at 300 mg/kg-d; skin hyperkeratosis at 900 mg/kg-d.		
Mouse, B6C3F1, M/F	0, 2,500, 5,000 mg/kg-d	2 yr, gavage	Gross: Mean body weights comparable with controls throughout the study. Survival was altered in females from treated and controls after wk 65 due to utero-ovarian infection. Microscopic: No significant increases in the incidence of nonneoplastic lesions are reported for male or females.	No adverse effects detected at 2,500 mg/kg-d	NTP 1986a[c]
C_{22-26}, 70% Cl_2					
Rat, F344, M/F	100, 900, 3,750 mg/kg-d	90 d, diet	Gross: Slight but statistically significant decreases in body-weight gain. Food consumption slightly increased in male rats at all doses. Macroscopic: 100: No increases in relative or absolute organ weights 900: No increases in relative or absolute organ weights 3,750: Increase in relative and absolute liver weight in males; increase in relative liver weight in females Microscopic: 100: No liver, kidney, or thyroid findings in males or females	Liver weight: NOAEL: 900 mg/kg-d LOAEL: 3,750 mg/kg-d Liver pathology: NOAEL: 900 mg/kg-d LOAEL: 3,750 mg/kg-d Liver biochemistry: NOAEL: 900 mg/kg-d LOAEL: 3,750 mg/kg-d	IRDC 1981, as reviewed by Serrone et al. 1987 (438-024/027)

900: No liver, kidney, or thyroid findings in males or females

3,750: Hepatocellular hypertrophy and cytoplasmic vacuolation (fat accumulation) of the liver in both males and females. Chronic nephritis in males. No effects were seen in the thyroid in either sex.

Biochemical: Increases in alanine aminotransferase activity observed in male and female rats at 3,750 mg/kg-d. Increased asparatate aminotransferase activity in females at 3,750 mg/kg-d.

---

[a]Authors note compound was $C_{10\text{-}12}$, 60% chlorination.

[b]Chlorinated paraffins incorporated in feed. Daily dose estimated using standard default for rat daily food consumption of 0.015 kg and standard default for rat average body weight of 0.25 kg.

[c]Authors note that compound was $C_{22}\text{-}_{26}$, 43% chlorination.

Wyatt et al. 1993) were also observed in male and female rats exposed to short-chain chlorinated paraffins. Changes in organ weight were accompanied by pathological changes in the corresponding organs.

Medium-chain Chlorinated Paraffins

Poon et al. (1995), IRDC (as reviewed by Serrone et al. 1987), and Birtley et al. (1980) evaluated the toxicity of the medium-chain chlorinated paraffin C_{14-17}, 52% chlorine, in 90-d subchronic toxicity studies. Statistically significant increases in relative liver weight were observed in both rats (Birtley et al. 1980; Serrone et al. 1987; Poon et al. 1995) and Beagle dogs (Birtley et al. 1980). Histopathological examination showed proliferation of the endoplasmic reticulum of liver hepatocytes in both rats and dogs (Birtley et al. 1980) and hepatocellular hypertrophy (Seronne et al. 1987). Poon et al. (1995) found minimal to mild single-cell necrosis at 300 mg/kg-d (estimated) and other hepatic effects starting at 30 mg/kg-d (estimated). Poon et al. (1995) and Serrone et al. (1987) reported that significant increases in relative kidney and thyroid weights were also observed in rats along with pathological changes in these tissues. Effects on biochemical parameters occurred at dose levels of 3 mg/kg-d (estimated) or greater.

Significant increases in liver and kidney weight were also observed in rats gavaged with C_{14-17}, 40% chlorine, for up to 90 d at 312 or 625 mg/kg-d. Increases in liver weight were accompanied by hepatocyte hypertrophy and an increase in liver β-oxidation activity and T_4 glucuronidation. Free plasma T_3, T_4, and TSH levels were not affected at any dose level in either sex. A statistically significant increase in α2u-globulin protein present in the proximal convoluted tubules was also observed in male rats at 625 mg/kg-d. This effect was not due to hyaline droplets formation. Thyroid follicular-cell hypertrophy was identified in all dosed males and females.

Long-chain Chlorinated Paraffins

Multifocal granulomatous hepatitis has been consistently observed in F-344 rats exposed to C_{20-30}, 43% chlorine (Serrone et al. 1987), or C_{23}, 43% chlorine (NTP 1986b), for 90 d or longer. These changes were accompanied by an increase in liver weight and necrosis in female rats administered daily doses of C_{20-30}, 43% chlorine, by gavage for 90 d at 100, 900, or 3,750 mg/kg-d (Serrone et al. 1987). These effects were not observed in similarly-treated male rats. Kidney toxicity occurred in high-dose males and females characterized by nephrosis and mineralization. Treatment did not produce any clinical signs of toxicity, any effects on body weight, or on other measures of toxicity (see Table 19-5).

NTP (1986b) found a dose-related increase in the incidence of granulomatous inflammation of the liver in female rats given C_{23}, 43% chlorine, by gavage for 90 d at 235, 469, 938, 1,875, or 3,750 mg/kg-d. These lesions were not observed in similarly-treated male rats. Treatment for 90 d did not produce any clinical signs of toxicity or any effects on body weight. There were no gross lesions or adverse changes in tissue histology observed in male or female B6C3F1 mice exposed to C_{23}, 43% chlorine, for 90 d at dose levels of 469, 938, 1,875, 3,750, or 7,500 mg/kg-d (NTP 1986b). No deaths or clinical signs related to chemical exposure were observed in males or females in any of the dose groups. No changes in body weight were observed in treated males or females.

In the two year study, the incidence of hepatocellular hyperplasia was increased in male rats exposed daily by gavage with 3,750 mg/kg-d and in females gavaged with 300 or 900 mg/kg-d (see Table 19-5). The incidence of granulomatous inflammation and lymphoid hyperplasia of the pancreatic and mesenteric lymph nodes was also increased among treated rats. The incidence of nephropathy of the kidney was significantly elevated in female rats dosed with 300 or 900 mg/kg-d. Brown staining around the mouth occurred in low- and high-dose males after wk 43. Dosed females showed a high incidence of distended abdomens during the latter part of the study. No significant difference in survival was calculated for treated rats as compared with controls after survival was adjusted for gavage-related deaths. No significant increases in the incidence of nonneoplastic lesions were reported for male or female B6C3F1 mice administered C_{23}, 43% chlorine, for up to 2 yr at dose levels of 2,500 or 5,000 mg/kg-d (NTP 1986b). Mean body weights in the high-dose groups were comparable with controls throughout study. Survival was low in all female groups including controls after wk 65 due to a high incidence of utero-ovarian infection.

Statistically significant increases in absolute liver weight and serum abnormalities were detected in F-344 rats subchronically exposed to C_{22-26}, 70% chlorine, at 3,750 mg/kg-d for 90 d (Serrone et al. 1987). Pathological examination revealed hepatocellular hypertrophy and cytoplasmic vacuolation in males and females and chronic nephritis in males exposed to 3,750 mg/kg-d. The NOAEL for absolute and relative organ weight increases and pathological findings for both sexes was 900 mg/kg-d.

Immunological

No studies were identified that investigated the immunological effects of chlorinated paraffins following oral exposure.

Neurological

Administration of a single peroral dose of 1.4 μmol/kg of polychlorohexadecane (chain length and percent chlorination not reported) produced a significant decrease (65%) in the V_{max} value for the presynaptic, sodium-dependent uptake of choline in the brain of 10-d-old mice (Eriksson and Nordberg 1986). This suggests that chlorinated paraffin may have an effect on the cholinergic system, but no comment was made about the significance of this effect on behavior or other neurological parameters. It was also found that the binding affinity of [^{3}H]quinuclidinyl benzilate (QNB), a muscarinic cholinergic receptor antagonist, to mouse cerebral cortex P2 fractions from 10-d-old mice was not affected when measured 7 d after exposure.

Reproductive and Developmental Effects

The teratogenic effects of chlorinated paraffins in rodents are summarized in Table 19-6. Serrone et al. (1987) notes that pregnant rabbits were more sensitive to C_{10-13}, 58% chlorine, C_{14-17}, 52% chlorine, or C_{22-26}, 70% chlorine, than pregnant rats based on maternal toxicity, but that none of the four chlorinated paraffins tested had a teratogenic effect on either rat or rabbit fetuses.

There was an increased number of post-implantation losses, resorptions, fetal deaths, and adactyly and/or shortened digits in pregnant dams exposed to C_{10-13}, 58% chlorine, at the highest dose level. However, eight of 25 pregnant dams from the high-dose group died following administration. Signs of maternal toxicity were observed in both mid- and high-dose groups. Serrone et al. (1987) reports that the study authors concluded that the digital malformations could not be interpreted as a direct teratogenic effect and was more likely to be a secondary response due to maternal toxicity.

No studies were identified that addressed the effects on reproduction.

Cancer

The National Toxicology Program (NTP) has conducted oral carcinogenicity bioassays in both rats and mice administered C_{10-12}, 60% chlorine, and C_{22-26}, 43% chlorine, (NTP 1986a, 1986b). Results of these studies are summarized in Table 19-7.

Evidence of carcinogenicity was observed in F-344/N rats given C_{10-12}, 60% chlorine, by gavage 5 d/wk for 104 wk. Dose-related and statistically significant increases in the incidences of hepatocellular carcinomas and of hepatic carcinomas plus adenomas combined were observed in both males and females.

Statistically significant increases in the incidence of mononuclear-cell leukemia occurred in male rats as compared with controls. The incidence of tubular-cell adenomas and adenocarcinomas of the kidney were elevated in male rats, but not significantly. Statistically significant increases in the incidence of thyroid follicular-cell carcinomas plus adenomas combined were observed in female rats. A dose-dependent and statistically significant increase in hepatocellular adenomas and hepatocellular adenomas and carcinomas was observed in B6C3F1 mice given C_{10-12}, 60% chlorine, by gavage for 2 yr (NTP 1986a). Dose-dependent and statistically significant increases in the incidences of alveolar and bronchiolar carcinomas, and thyroid follicular-cell adenomas and carcinomas occurred in male and female mice, respectively.

No tumor types were found to be significantly increased in male F-344/N rats given C_{22-26}, 43% chlorine, by gavage for 5 d/wk for 103 wk. However, the incidence of pheochromocytomas of the adrenal medulla was significantly increased in female rats. A dose-related increase in the incidence of malignant lymphoma occurred in male mice given C_{22-26}, 43% chlorine, at 2,500 or 5,000 mg/kg-d, 5 d/wk for 103 wk (NTP 1986b). The incidence of hepatocellular adenomas and carcinomas combined was elevated in female mice, but not significantly. Survival among females was low among treated and control animals due to a high incidence of utero-ovarian infection and may have decreased the potential of the study to detect treatment-related carcinogenicity (NTP 1986b).

The European Union (EU) is currently evaluating the health risks posed by chlorinated paraffins. At technical meetings on October 1-3, 1996, and February 19-21, 1997, it was agreed by specialized experts representing the Member States that C_{10-13} chlorinated paraffins are not genotoxic. However, no agreement could be reached regarding the significance of the tumors reported by NTP (1986a) or their relevance to man. In a June 1997 meeting of the Commission Group of Specialized Experts in the fields of Carcinogenicity, Mutagenicity, and Reprotoxicity, it was concluded that no significance could be placed on the slight excess in lung, pancreas, stomach, Hadrian gland tumors, or leukemias reported in NTP (1986a). The Specialized Experts decided that only the observed tumors of the liver, kidney, and thyroid should be considered significant. It was also agreed that the liver and thyroid tumors could be attributed to peroxisomal proliferation which could cause a hormonal imbalance and that humans would be much less sensitive to peroxisome proliferation than rats and mice. No plausible mechanism for the observed kidney tumors was agreed upon, but it was thought that either α2u-globulin accumulation or chronic nephropathy might be contributing factors. In 1998, the EU Scientific Committee for Toxicity, Ecotoxicity, and the Environment generally agreed with the conclusions of the Specialized Experts with the exception of ruling out the significance of the increased incidence of lung tumors in male mice. However,

TABLE 19-6 Teratogenicity Studies on Chlorinated Paraffins

Species, Strain	Doses (mg/kg-d)	Duration, Route	Effects	Reference
Short-chain chlorinated paraffins				
C_{10-13}, 58% Cl_2				
Rat, Charles River	100, 500, 2,000	d 6 -19 of gestation, gavage	Dams: Eight of 25 pregnant dams from the high-dose group died following administration. Signs of maternal toxicity were observed in both mid- and high-dose groups. Fetuses: Increased number of post-implantation losses, increased number of early and late resorptions, and a decrease in viable fetuses per dam in the high-dose group. Adactyly and/or shortened digits were observed in the high-dose group. No effects were observed at the lowest dose level. Authors interpretations: Digital malformations observed at a dose that produced significant mortality in treated dams cannot be interpreted as a direct teratogenic effect and are more likely to be a secondary response due to maternal toxicity.	IRDC 1982, as reviewed by Serrone et al. 1987 (438-016)
Rabbit, Dutch Belted	10, 30, 100	d 6 -27 of gestation, gavage	Embryotoxicity was seen in both the mid-dose and the high-dose groups. No effect was observed on the occurrence of malformations at any dose level.	IRDC 1982/83, as reviewed by Serrone et al. 1987 (438-031; 037)
Medium-chain chlorinated paraffins				
C_{14-17}, 52% Cl_2				
Rat, Charles River	500, 2,000, 5,000	d 6 -19 of gestation, gavage	Dams: Maternal toxicity was observed in the high-dose group. Fetuses: There were no biologically or meaningful relevant statistically significant differences in Caesarean section observations or in the incidence of fetus malformations in litters belonging to the treated groups as compared with corresponding controls.	IRDC 1981/83/84, as reviewed by Serrone et al. 1987 (438-017; 034; 047)

Rabbit, Dutch Belted	10, 30, 100	d 6 -27 of gestation, gavage	Dams: No maternal toxicity was observed but body weight losses were observed in the high-dose group dams. Fetuses: There were no biologically meaningful differences in the numbers of litters/fetuses with malformations or in the developmental and genetic variations when the treated groups were compared with controls.	IRDC 1982/83, as reviewed by Serrone et al. 1987 (438-020; 032; 036)
Long-chain chlorinated paraffins				
C_{20-30}, 43% Cl_2				
Rat, Charles River	500, 2,000, 5,000	d 6 -19 of gestation, gavage	Dams: Maternal toxicity observed in the high-dose group. One female died on d 18 of gestation. Fetuses: No differences in the incidence of fetal malformations were observed when treated groups were compared to controls.	IRDC 1981/83, as reviewed by Serrone et al. 1987 (438-015; 033)
Rabbit, Dutch Belted	500, 2,000, 5,000	d 6 -27 of gestation, gavage	Two of 12 dams at the high dose and one of the 13 dams at the mid-dose aborted. In the high-dose group, there was a slight increase in mean post-implantation loss and a slight decrease in the mean number of viable fetuses when compared to controls. The data did not indicate a teratogenic response.	IRDC 1981, as reviewed by Serrone et al. 1987 (438-018; 030)
C_{22-26}, 70% Cl_2				
Rat, CD	500, 2,000, 5,000	d 6 -19 of gestation, gavage	Dams: No maternal toxicity was observed in any of the pregnant rats that were given chlorinated paraffin. Fetuses: There were no biologically meaningful differences in the incidence of developmental variations and malformations in the treated groups as compared with controls.	IRDC 1983, as reviewed by Serrone et al. 1987 (438-045; 046)
Rabbit, Dutch Belted	100, 300, 1,000	d 6 -27 of gestation, gavage	Does: No biologically meaningful differences were observed in maternal appearance, behavior, or body-weight gain. Fetuses: There were no adverse treatment-related differences in Caesarean section observations or in the incidence of fetus malformations in the liters of the treated groups as compared with controls. There was no difference between treated and control fetuses in the occurrence of genetic and developmental variation.	IRDC 1981, as reviewed by Serrone et al. 1987 (438-018; 030)

TABLE 19-7 Carcinogenicity Studies on Chlorinated Paraffins

Species, Strain, Sex, Number	Dose (mg/kg-d)	Duration, Route	Results	Reference
Short-chain chlorinated paraffins				
C_{12}, 60% Cl_2				
Rat, F344, M/F, 50/sex/dose	0, 312, 625	104 wk, gavage	Tumor incidence: Hepatocellular carcinoma / Adenoma+carcinomas / Test for trend Males: 0 — 0/50 — 0/50 312 — 10/50 — 13/50 625 — 16/48 — 16/48 — $p < 0.001$ Females: 0 — 0/50 — 0/50 312 — 5/50 — 6/50 625 — 7/50 — 7/50 — $p = 0.005$ Other tumor incidences significantly elevated: Males: mononuclear cell leukemia Females: thyroid follicular cell adenomas/carcinomas	NTP 1986a
Mouse B6C3F1, M/F, 50/sex/dose	0, 125, 250	103 wk, gavage	Tumor incidence: Hepatocellular adenomas / Adenoma+carcinomas / Test for trend Males: 0 — 11/50 — 20/50 125 — 20/50 — 34/50 250 — 29/50 — 38/50 — $p < 0.001$ Females: 0 — 0/50 — 3/50 125 — 18/50 — 22/50 215 — 22/50 — 28/50 — $p < 0.001$ Other tumor incidences significantly elevated: Males: alveolar/bronchiolar carcinomas Females: thyroid follicular cell adenomas/carcinomas	NTP 1986a

Long-chain chlorinated paraffins				
C_{23}, 43% Cl_2				
Rat, F344, M/F, 50/sex/dose	Males: 0, 1,875, 3,750 Females: 0, 100, 300, 900	103 wk, gavage	Tumor Incidence: The incidence of phaeochromocytomas of the adrenal medulla was significantly increased in females: control, 1/50; low-dose, 4/50; mid-dose, 6/50; high-dose, 7/50 ($p = 0.046$, tumor test for trend).	NTP 1986b
Mouse, B6C3F$_1$, M/F, 50/sex/dose	0, 2,500, 5,000	103 wk, gavage	Tumor incidence: The incidence of malignant lymphomas was significantly increased in males: they occurred in 6/50 controls and in 12/50 low-dose and 16/50 high-dose males ($p = 0.009$, life-table test for trend; $p = 0.011$ tumor test for trend). Tumor trend not significant for hepatocellular carcinomas and adenomas combined (4/50 in controls, 3/49 in low-dose, and 10/50 in high-dose animals). Authors note that toxicity and premature death in females may have decreased the potential of the study to detect carcinogenic effects.	NTP 1986b

it was felt that this reconsideration does not alter the outcome of the risk characterization for man. The Committee also noted that one cannot totally discount the possibility that liver and thyroid tumors could arise in man following exposure to short-chain chlorinated paraffins, but there would be very large differences in carcinogenic sensitivity between animals and humans in the induction of these tumors. The Commission also in agreed with the use of a NOAEL of 100 mg/kg-d for kidney carcinogenicity in male mice (EUSCTEE 1998).

The International Agency for Research on Cancer (IARC) last evaluated the carcinogenicity of chlorinated paraffins in 1989-1990 (IARC 1990). At that time, based on the available data IARC concluded that there was sufficient evidence for the carcinogenicity of the chlorinated paraffin $C_{12,}$ 60% chlorine, in experimental animals and that it was possibly carcinogenic to humans (Group 2B). IARC (1990) concluded that there was limited evidence for the carcinogenicity of C_{23}, 43% chlorine, in experimental animals.

As of September 1999, chlorinated paraffins are not listed in EPA's IRIS (Integrated Risk Information System) database. However, based on the available carcinogenicity of these compounds, EPA decreed in 1994 (EPA 1994) that there was sufficient evidence for adding short-chain chlorinated paraffins to the list of chemicals subject to reporting under section 313 of the Emergency Planning and Community Right-to-Know Act (EPCRA). EPA concluded that long-chain chlorinated paraffins should not be classified as potential carcinogens based on the animal data (EPA 1994).

Genotoxicity

The chlorinated paraffins—C_{10-13}, 58% chlorine; C_{14-17}, 52% chlorine; C_{20-30}, 43% chlorine; and C_{22-26}, 70% chlorine—did not cause chromosomal or chromatid aberrations in rat bone marrow cells from male F-344 rats administered by gavage at 5 g/kg-d for 5 consecutive days (Serrone et al. 1987).

Administration of C_{10-13}, 58% chlorine, by gavage to male Charles River rats at dose levels of 250, 750, or 2,000 mg/kg-d for 5 consecutive days did not result in an increased frequency of dominant lethal mutations in offspring following mating with 20 potentially mated females (see Serrone et al. 1987).

The chlorinated paraffins—C_{10-13}, 50% chlorine (Birtley et al. 1980); C_{10-12}, 60% chlorine (NTP 1986a); C_{14-17}, 52% chlorine (Birtley et al. 1980); C_{20-30}, 42% chlorine (Birtley et al. 1980); C_{23-26}, 43% chlorine (NTP 1986b); and C_{10-23}, 70% chlorine (Meijer et al. 1981)—have been tested in the Ames assay, with or without exogeneous metabolic activation at or near toxic doses. All were found to be negative for mutagenicity.

Hepatocytes taken from male rats exposed to near toxic doses of C_{10-12}, 60%

chlorine, of up to 2 g/kg-d were negative for unscheduled DNA synthesis 2 or 12 hr after dosing (Ashby et al. 1990). Moderate hepatocyte proliferation was observed 12, 24, and 36 hr after exposure to dose levels of 1,000 or 2,000 mg/kg-d of C_{10-12}, 60% chlorine. Quantitatively greater responses were measured for hepatocytes taken from rats exposed to positive control compounds. Chlorinated paraffins—C_{10-13}, 50% chlorine; C_{20-30}, 42% chlorine; C_{14-17}, 52% chlorine—all with or without a stabilizer, did not induce cell transformation in baby hamster kidney cells in vitro at concentrations of 0.25, 2.5, 25, and 2,500 µg/mL.

C_{10-12}, 60% chlorine, inhibited intercellular communication in a rat liver epithelium-derived cell line (IAR 20) in vitro, when assessed 1 and 12 hr after exposure (Warngard et al. 1996).

QUANTITATIVE TOXICITY ASSESSMENT

Noncancer

Dermal Assessment

No dermal toxicity studies of adequate duration (i.e., subchronic or chronic) were found in the literature for deriving a dermal RfD for any chlorinated paraffins.

Inhalation RfC

The subcommittee did not locate any inhalation toxicology studies of adequate duration (i.e., subchronic or chronic) for deriving an inhalation RfC for any of the chlorinated paraffins.

Oral RfD

Chlorinated paraffins, if used as FRs in residential furniture, will most likely be applied as a latex backcoating, possibly in combination with antimony trioxide and other FRs. According to chlorinated paraffin manufacturers, the chlorinated paraffins that are most likely to be used in latex back coating applications are the long-chain chlorinated paraffins (C_{20-30}) with high chlorine content (i.e., 70%). This is because of the high flame-retardancy of these chemicals as compared with the medium- or short-chain chlorinated paraffins.

Currently, no chronic toxicity data are available for a long-chain chlorinated paraffin with high chlorine content (i.e., 70%). A subchronic toxicity study is available for C_{22-26}, 70% chlorine (Serrone et al. 1987). The subcommittee concluded that this study provides sufficient data for calculating an oral RfD for long-chain, highly chlorinated paraffins (the details of this study were reviewed in the Hazard Identification section under Systemic Effects of Oral Exposure). The critical effect observed in this study was liver toxicity (increases in relative liver weight, hepatocellular hypertrophy, cytoplasmic fat vacuolation, increases in serum alanine aminotransferase and aspartate aminotransferase activities). These effects occurred at a dose level of 3,750 mg/kg-d. The authors also report a slight increase in the incidence of kidney nephritis in males at the 3,750 mg/kg-d dose level. No toxic effects were observed in either male or female rats at the dose level of 900 mg/kg-d or lower. Therefore, in this study a LOAEL of 3,750 mg/kg-d and a NOAEL of 900 mg/kg-d was identified.

Based on the NOAEL for liver and kidney toxicity for C_{22-26}, 70% chlorine, in male and female rats, the estimated RfD for long-chain chlorinated paraffins (C_{20-30}), 70% chlorine, is 0.3 mg/kg-d (see Table 19-8). To derive the oral RfD, the NOAEL was divided by a composite uncertainty factor (UF) of 3,000 which consists of a factor of 3 to extrapolate from animals to humans (factor UF_A). A factor of 3 instead of 10 was used because the toxicokinetics and dynamics of long-chain chlorinated paraffins are anticipated to be similar in rodents and humans. A factor of 10 was used to account for the possible sensitivity of children and elderly to long-chain chlorinated paraffins (factor UF_H). A factor of 10 was applied to account for less-than-lifetime toxicity data (factor UF_S) and another factor of 10 was applied because the toxicity database for the long-chain chlorinated paraffin C_{22-26}, 70% chlorine, is incomplete (UF_D).

The subcommittee has moderate to high confidence that the RfD calculated for C_{22-26}, 70% chlorine, is sufficiently protective of human health for several reasons. A NOAEL was available for C_{22-26}, 70% chlorine. Dose levels at which toxicity occurred for C_{22-26}, 70% chlorine, reported by Serrone et al. (1987) are comparable across studies. In a 2-year study (NTP 1986b), liver and kidney effects occurred in female rats at a dose level of 300 mg/kg-d, but did not occur at 100 mg/kg-d. In addition, the application of two 10-fold uncertainty factors (less-than-lifetime data, insufficient toxicity database) in the oral RfD derivation should be sufficiently protective if a significant critical effect occurs at dose level of less than 900 mg/kg-d. The use of a three-fold uncertainty factor to account for possible differences in sensitivity between rodents and humans is also sufficiently conservative because current mechanistic data suggest that it is not likely that humans are more sensitive than rodents to chlorinated paraffins.

TABLE 19-8 Oral Reference Dose for C_{22-26}, 70% Chlorine

Critical Effect	Species	Effect Level (mg/kg-d)	Uncertainty Factors		RfD (mg/kg-d)	Reference
Liver and kidney toxicity	Rat	NOAEL: 900	UF_A:	3	0.3	IRDC 1981, as reviewed by Serrone et al. 1987
			UF_H:	10		
			UF_S:	10		
			UF_D:	10		
			Total:	3,000		

NOAEL, no-oberved-adverse-effect level; RfD, reference dose; UF_A, extrapolation from animals to humans; UF_H, intraspecies variation; UF_S, extrapolation from a study of less-than-lifetime in duration; UF_D, inadequate or deficient toxicity database.

Cancer

Short-chain Chlorinated Paraffins

There is adequate evidence for the carcinogenicity of the short-chain chlorinated paraffin C_{12}, 60% chlorine, in rodents (NTP 1986a). Chronic oral administration of C_{12}, 60% chlorine, induced dose-dependent increases in the incidence of hepatocellular carcinomas and adenomas in both F-344 and B6C3F1 mice. The combined incidence of thyroid follicular-cell carcinomas and adenomas was significantly elevated in female mice and rats. The incidence of alveolar and bronchiolar carcinomas was significantly elevated in male mice, and the incidences of tubular-cell adenomas and adenocarcinomas of the kidney and mononuclear-cell leukemia were significantly elevated in male rats.

Tests for the genotoxicity of C_{12}, 60% chlorine, have been consistently negative suggesting that it may be causing cancer in rodents by nongenotoxic mechanisms. For example, liver peroxisome proliferation may be responsible for the liver, thyroid, and kidney tumors observed in the NTP (1986a) bioassay (Bentley et al. 1993; Cattley et al. 1998). Peroxisomal proliferation is associated with hepatocarcinogenesis in rats and mice (Ashby et al. 1990; Bentley et al. 1993; Cattley et al. 1998). Morphological analysis of the livers from rats and mice exposed to C_{10-13}, 58% chlorine, has consistently shown a pattern of peroxisomal proliferation, hepatocyte hypertrophy, and the induction of hepatic P450 4A1, fatty acetyl CoA oxidase, and T_4 glucuronidation product levels (Ashby et al. 1990; Wyatt et al. 1993). It is believed that the increase in the incidence of thyroid follicular cell carcinomas and adenomas in females may be the result of increased T_4 glucuronidation in conjunction with peroxisomal proliferation which decreases plasma T_4 levels resulting in the stimulation of pituitary TSH production. In turn, increased TSH production stimulates thyroid hypertrophy and hyperplasia and the eventual formation of follicular cell carcinoma (Wyatt et al. 1993). Hypertrophy and hyperplasia of the thyroid have been reported to occur in F-344 rats exposed to short-chain chlorinated paraffins for 90 d (Serrone et al. 1987).

At this time, there is no adequate explanation for the formation of kidney tumors observed in male rats. It has been suggested that an unidentified sex-specific protein may stimulate sustained DNA synthesis in the kidney following chronic administration of short-chain chlorinated paraffins (Wyatt et al. 1993). Increased protein accumulation has been observed in dosed males accompanied by nephropathy, manifested by increased incidences of regenerating tubules. Immunocytochemical staining for α2u-globulin in the tubules shows that this protein is present but is not the predominant protein that accumulates, and no hyaline droplets have been observed in male rats treated with high doses of short- or medium-chain chlorinated paraffins for 90 d (Wyatt et al. 1993). No hypothesis has been presented for the possible mechanisms associated with the increased incidence of alveolar and bronchiolar carcinomas in male mice or mononuclear cell leukemia in male rats. The subcommittee does acknowledge that these tumor types, along with kidney tumors in male rats, did not occur across species, and therefore less weight should be given to these findings in the evaluation of the carcinogenicity of C_{10-13}, 58% chlorine.

Medium-chain Chlorinated Paraffins

Currently, there are no human or animal data available for evaluating the carcinogenicity of medium-chain chlorinated paraffins.

Long-chain Chlorinated Paraffins

Currently, there are no epidemiological or cancer bioassay data for long-chain chlorinated paraffins (C_{22-26}) paraffins with 70% chlorination, which are the type of chlorinated paraffin that are most likely to be used as FRs in residential furniture. Meijer et al. (1981) tested a long-chain chlorinated paraffin of similar chemistry (C_{10-23}, 70% chlorine) for mutagenicity in three strains of *S. typhimurium* and found that this compound was negative for mutagenicity at all concentrations tested in two of the three strains. A positive response was observed in one strain at the highest concentration tested, but this may have been a chance occurrence since no dose-response for mutagenicity was observed. The authors also point out that no toxic effects were seen suggesting that this compound may not penetrate into bacterial cells. No other genotoxicity data are known to be available for long-chain chlorinated paraffins with 70% chlorination.

The subcommittee believes that the best study available for evaluating the carcinogenicity of long-chain chlorinated paraffins with 70% chlorination is the NTP (1986b) rodent carcinogenicity bioassay for the long-chain chlorinated paraffin $C_{22\text{-}26}$, 43% chlorine (see Table 19-7). In this bioassay, no tumor types were consistently elevated across species in F-344 rats and B6C3F1 mice administered C_{23}, 43% chlorine, for 103 wk. In mice, there was a dose-related increase in the incidence of malignant lymphoma among males as compared with controls. The incidence of hepatocellular adenomas and carcinomas combined was elevated in female mice in the high-dose group, but the incidence of this tumor category was not significant across dose levels as determined by incidental tumor tests for trend. However, early mortality due to infection was common among females in both the treated and control groups and may have prevented the identification of increased incidences in late-forming tumor-types. In rats, the incidence of pheochromocytomas of the adrenal medulla was increased in females, and no tumor types were found to be significantly increased in F-344 males exposed to chlorinated paraffin as compared with controls. $C_{22\text{-}26}$, 43% chlorine, was not genotoxic in the Ames assay with or without exogenous metabolic activation (NTP 1986b) and did not induce chromosomal aberrations in F-344 rats administered toxic doses (Serrone et al. 1987).

Based on the animal and genotoxicity data for $C_{22\text{-}26}$, 43% chlorine, the subcommittee concluded that there is limited evidence for its carcinogenicity in rodents. This conclusion is in agreement with that of IARC (1990) and EPA (1994).

Derivation of a Cancer Potency Factor

The subcommittee concluded that the derivation of cancer potency factor (i.e., $0.1/LED_{10}$) for long-chain chlorinated paraffins is not warranted based on the lack of cancer data for long-chain chlorinated paraffins, 70% chlorine, and the limited evidence for the carcinogenicity of C_{23}, 43% chlorine. The subcommittee does acknowledge that there are adequate data for the carcinogenicity for $C_{10\text{-}12}$, 60% chlorine, in rodents, but these chlorinated paraffins are not likely to be used as FRs in residential furniture. Therefore, the subcommittee concluded that the derivation of a cancer potency estimate for short-chain chlorinated paraffins was not necessary.

EXPOSURE ASSESSMENT AND RISK CHARACTERIZATION

Noncancer

Dermal Exposure

Dermal exposure to chlorinated paraffins was estimated using the dermal exposure scenario described in Chapter 3. This exposure scenario assumes that an adult spends 1/4th of his or her time sitting on furniture upholstery back-coated with chlorinated paraffins and also assumes 1/4th of the upper torso is in contact with the upholstery and clothing presents no barrier. Exposure to other chemicals present in the backcoating was not included in this assessment.

First Iteration

As a first estimate of exposure, it was assumed that skin, clothing, and the upholstery did not impede dermal exposure to chlorinated paraffins present in the backcoating. It was also assumed that there would be sufficient water present from sweat to facilitate dissolution of chlorinated paraffins from the back-coating and absorption through the skin. In this scenario, only the dissolution rate of chlorinated paraffins from the backcoating is assumed to be the limiting factor in absorption by the body. It is assumed that all of the chlorinated paraffins that dissolves is immediately absorbed into the body by the sitting person.

Dermal exposure was estimated using Equation 1 in Chapter 3. For this calculation, the subcommittee estimated an upholstery application rate (S_a) for chlorinated paraffins of 3 mg/cm^2. The extraction rate (μ_w) for chlorinated paraffins was estimated to be 0.025 based on extraction data for hexabromo-cyclododecane in polyester fiber (McIntyre et al. 1995). The release rate from the fiber for estimating extraction was 0.04/d at 28°C calculated using the equation $2d/2\pi R$ (d = film thickness, R = fiber radius) with a correction from fiber to film of a factor of 0.63.

Using these assumptions, an estimated absorbed daily dose of 0.59 mg/kg was calculated for chlorinated paraffins. In the absence of a dermal RfD, the subcommittee believes it is appropriate to use the oral RfD for $C_{22\text{-}26}$, 70% chlorine, as the best estimate of the internal dose from dermal exposure. A hazard index of 1.97 was calculated for this first iteration by dividing the estimated daily dermal dose of 0.59 mg/kg-d by the oral RfD for chlorinated paraffins of 0.3 mg/kg-d. This hazard index of 1.97 indicates that dermal exposures to long-chain chlorinated paraffins at the worst-case levels might be a health concern. The subcommittee recommends that the dermal absorption of these substances from treated fabric be investigated.

Alternative Iteration

The estimated dermal daily dose for chlorinated paraffins can be calculated using an estimate of the dermal penetration rate for chlorinated paraffins (Chapter 3, Equations 2 and 3). Instead of assuming that all dissolved chlorinated paraffins immediately penetrates the skin and enters systemic circulation, it is assumed that the skin slows the absorption of chlorinated paraffins to a specific amount of chemical absorbed per unit of time. This estimate can be measured experimentally and is referred to as the skin permeability coefficient K_p. However, the dermal penetration constant for chlorinated paraffins has not been measured experimentally. However, K_p can be estimated from a correlation between the octanol-water partition coefficient (K_{ow}) and molecular weight (mass/unit amount of substance) using Equation 2 in Chapter 3 yielding an alternate K_p of 5.77×10^{-5} cm/d.

The water solubility of long-chain chlorinated paraffin, 70% chlorination, was not available. Therefore the alternative exposure estimate could not be calculated. However, it was determined that the calculated dose rate for chlorinated paraffins would only be a concern in this scenario if the water solubility of long-chain chlorinated paraffins exceeded 650 g/liter—which is not possible.

Inhalation Exposure

Particles

Inhalation exposure estimates for chlorinated paraffins were calculated using the exposure scenario described in Chapter 3. This scenario assumes that a person spends 1/4th of his or her lifetime in a 30-m^3 room containing 30 m^2 of chlorinated paraffins-treated fabric and the room is assumed to have a air-change rate of 0.25/hr. It is also assumed that 50% of the chlorinated paraffins present in 25% of the surface area of the treated fabric is released over 15 yr and 1% of released particles are small enough to be inhaled.

Particle exposure was estimated using Equations 4 and 5 in Chapter 3. The subcommittee estimated an upholstery application rate (S_a) for chlorinated paraffins of 3 mg/cm^2. The release rate (μ_r) for chlorinated paraffins from upholstery fabric was estimated to be 2.3×10^{-7}/d (see Chapter 3, Equation 5) yielding a room airborne particle concentration (C_p) of 1.1 μg/m³ and a short time-average exposure concentration of 0.28 μg/m³. The time-averaged exposure concentration for particles was calculated using Equation 6 in Chapter 3.

In the absence of relevant inhalation exposure data, the subcommittee chose to estimate inhalation RfCs from oral RfDs The subcommittee, however, recognizes that it is not an ideal approach and also recognizes that the estimated RfC levels might be considerably different than actual levels (if inhalation data were available). Extrapolating from one route of exposure (oral) to another (inhalation) requires specific knowledge about the uptake kinetics into the body by each exposure route, including potential binding to cellular sites. The subcommittee believes that its extrapolation of oral RfDs to inhalation RfCs is highly conservative; it assumes that all of the inhaled compound is deposited in the respiratory tract and completely absorbed into the blood. The NRC Committee on Toxicology (NRC 1985) has used this approach when inhalation exposure data were insufficient to derive inhalation exposure levels. The subcommittee believes that such an approach is justified for conservatively estimating the toxicological risk from exposure to this FR. The RfC should be used as interim or provisional levels until relevant data become available for the derivation of inhalation RfC.

To calculate a hazard index for the inhalation route, a provisional inhalation RfC of 1.05 mg/m³ was derived using the oral RfD for $C_{22\text{-}26}$, 70% chlorine (see the following section for the derivation of the oral RfD) and Equation 7 in Chapter 3.

Division of the time-average exposure concentration of 0.28 μg/m³ by the provisional RfC for chlorinated paraffins of 1.05 mg/m³ yields a hazard index of 2.7×10^{-4}. This suggests that under the subcommittee's worst-case exposure assumptions, Chlorinated paraffins would not be considered to be a toxic risk by the inhalation route of exposure.

Vapors

Volatility data for chlorinated paraffins were not located. Therefore, the subcommittee did not calculate worst-case exposure estimates for this exposure.

Oral Exposure

The assessment of noncancer toxicological risk for oral exposure to chlorinated paraffins is based on the oral exposure scenario described in Chapter 3.

This scenario assumes a child is exposed to chlorinated paraffins by sucking on 50 cm^2 of fabric backcoated with chlorinated paraffins, 1 hr/d for two yr. The subcommittee estimated an upholstery application rate (S_a) for chlorinated paraffins of 3 mg/cm^2. Oral exposure was calculated using Equation 15 in Chapter 3. The extraction rate (μ_w) for chlorinated paraffins was estimated to be 0.025 based on extraction data for hexabromocyclododecane in polyester fiber (McIntyre et al. 1995). The release rate from the fiber for estimating extraction was 0.04/d at 28°C calculated using the equation $2d/2\pi R$ (d = film thickness, R = fiber radius) with a correction from fiber to film of a factor of 0.63.

The worst-case average oral daily dose for chlorinated paraffins was estimated to be 0.16 mg/kg-d. Division of the dose estimate by the oral RfD for chlorinated paraffins of 0.3 mg/kg-d gives a hazard index of 0.053. This suggests that under the subcommittee's worst-case exposure assumptions, chlorinated paraffins do not pose a noncancer toxicological risk when incorporated into residential furniture upholstery at the estimated application levels.

Cancer

Dermal Exposure

There are inadequate data for assessing the carcinogenicity of chlorinated paraffins when exposure occurs by the dermal route of exposure.

Inhalation Exposure

There are inadequate data for assessing the carcinogenicity of chlorinated paraffins when exposure occurs by inhalation.

Oral Exposure

EPA has concluded that long-chain chlorinated paraffins should not be classified as potential carcinogens (EPA 1994). This conclusion is based on limited evidence for the carcinogenicity of the long-chain chlorinated paraffin, C_{22-26}, 43% chlorine, in rodents and the fact that cancer data are not available for long-chain chlorinated paraffin, 70% chlorine. It is the opinion of the subcommittee that long-chain chlorinated paraffins are not likely to be human carcinogens and derivation of a cancer potency factor for this class of chlorinated paraffins is not warranted (see Cancer in Quantitative Toxicity Assessment for conclusions regarding short- and medium-chain chlorinated paraffins).

RECOMMENDATIONS FROM OTHER ORGANIZATIONS

The subcommittee is not aware of any exposure limits recommended by the regulatory agencies or other organizations.

DATA GAPS AND RESEARCH NEEDS

Chronic toxicity data are not available for $C_{22\text{-}26}$, 70% chlorine, for any route of exposure. Properly conducted reproductive/developmental studies on chlorinated paraffins are not available. Human dermal absorption data for chlorinated paraffins are not available either. Information on the leachability of long-chain chlorinated paraffins from latex or other types of backcoating following exposure to simulated human secretions (saliva, sweat) is also needed. The volatility of $C_{22\text{-}26}$, 70% chlorine, is also not reported in the literature.

Based on a dermal hazard index greater than one, the dermal absorption of chlorinated paraffins from treated fabric should be investigated.

REFERENCES

Abasov, D.M. 1970. Toxicology of new chloroparaffin KhP 470. Tr. Azerb. Nauch.-Issled. Inst. Gig. Tr. Profzabol. 5:180-183. As cited in EPA (1975).

Ahlman, M., A. Bergman, P.O. Darnerud, B. Egestad, and J. Sjovall. 1986. Chlorinated paraffins: Formation of sulphur-containing metabolites of polychlorohexadecane in rats. Xenobiotica 16(3):225-232.

Ashby, J., P.A. Lefevre, and C.R. Elcombe. 1990. Cell replication and unscheduled DNA synthesis (UDS) activity of low molecular weight chlorinated paraffins in the rat liver in vivo. Mutagenesis 5(5):515-518.

Bentley, P., I. Calder, C. Elcombe, P. Grasso, D. Stringer, and H.J. Wiegand. 1993. Hepatic peroxisome proliferation in rodents and its significance for humans. Food Chem. Toxicol. 31(11):857-907.

Biessmann, A., P.O. Darnerud, and I. Brandt. 1983. Chlorinated paraffins: Disposition of a highly chlorinated polychlorohexadecane in mice and quail. Arch. Toxicol. 53(1):79-86.

Birtley, R.D., D.M. Conning, J.W. Daniel, D.M. Ferguson, E. Longtstaff, and A.A. Swan. 1980. The toxicological effects of chlorinated paraffins in mammals. Toxicol. Appl. Pharmacol. 54(3):514-525.

Cattley, R.C., J. DeLuca, C. Elcombe, P. Fenner-Crisp, B.G. Lake, D.S. Marsman, T.A. Pastoor, J.A. Popp, D.E. Robinson, B. Schwetz, J. Tugwood, and W. Wahli. 1998. Do peroxisome proliferating compounds pose a hepatocarcinogenic hazard to humans? Regul. Toxicol. Pharmacol. 27(1 pt. 2):47-60.

Chlorinated Paraffins Industry Association. 1999. Letter to Darryl Arfsten (NRC), from Robert J. Fensterheim, Executive Director, CPIA, regarding chlorinated paraffins for use in furniture flame retardants. September 29, 1999. Chlorinated Paraffins Industry Association, Washington, DC.

Darnerud, P.O. 1984. Chlorinated paraffins: Effect of some microsomal enzyme inducers and inhibitors on the degradation of 1-14C-chlorododecanes to $14CO_2$ in mice. Acta Pharmacol. Toxicol. 55(2):110-115.

Darnerud, P.O., and I. Brandt. 1982. Studies on the distribution and metabolism of a ^{14}C-labeled chlorinated alkane in mice. Environ. Pollut. (Series A) 27:45-56.

Darnerud, P.O., A. Biessmann, and I. Brandt. 1982. Metabolic fate of chlorinated paraffins: Degree of chlorination of [1-14C]-chlorododecanes in relation to degradation and excretion in mice. Arch. Toxicol. 50(3-4):217-226.

Diamond Shamrock Chemical Company. 1975. Personal Communication, Cleveland, OH.

EPA (U.S. Environmental Protection Agency). 1975. Investigation of Selected Potential Environmental Contaminants: Chlorinated Paraffins. Final Report. Office of Toxic Substances, U.S. Environmental Protection Agency, Washington DC. EPA-560/2-75-007.

EPA (U.S. Environmental Protection Agency). 1994. Addition of Certain Chemicals; Toxic Chemical Release Reporting; Community Right-to-Know; Final Rule. Fed. Regist. 59(Nov. 30):61462

Eriksson, P., and J.E. Kihlstrom. 1985. Disturbance of motor performance and thermoregulation in mice given two commercial chlorinated paraffins. Bull. Environ. Contam. Toxicol. 34(2):205-209.

Eriksson, P., and A. Nordberg. 1986. The effects of DDT, DDOH-palmitic acid, and a chlorinated paraffin on muscarinic receptors and the sodium-dependent choline uptake in the central nervous system of immature mice. Toxicol. Appl. Pharmacol. 85(2):121-127.

EUSCTEE (European Union Scientific Committee for Toxicity, Ecotoxicity, and the Environment). 1998. Opinion on the results of the Risk Assessment of: Alkanes, C_{10-13}, chloro {SCCP} carried out in the framework of Council Regulation (EEC) 793/93 on the evaluation and control of the risks of existing substances - Opinion expressed at the 6th CSTEE plenary meeting, Brussels, 27 November 1998. [Online]. Available: http://europa.eu.int/comm/dg24/health/sc/sct/out23_en.html

Fire Retardant Chemicals Association. 1998. Textile Flame Retardant Applications by Product Classes for 1997 Within and Outside of the United States: Halogenated Oelfins and Paraffins. Fire Retardants Chemicals Association. Lancaster, PA.

Gosselin, R.E., H.C. Hodge, R.P. Smith, and M.N. Gleason. 1976. Clinical toxicology of Commercial Products. 4th Edition. Baltimore: Williams and Wilkins. Cited in the Hazardous Substance Data Bank file for Chlorinated Paraffins. File updated March 3, 1998.

Hatlelid, K. 1999. Toxicity Review for Chlorinated Paraffins. Memorandum, dated March 25, 1999, from Kristina Hatlelid, Toxicologist, Division of Health Sciences, to Ronald L. Medford, Assistant Executive Director for Hazard Identification and Reduction. U.S. Consumer Product Safety Commission, Washington, DC.

IARC (International Agency for Research on Cancer). 1990. Chlorinated Paraffins. In: IARC Monographs on the Evaluation of Carcinogenic Risks to Humans. Volume 48. Some Flame Retardants and Textile Chemicals, and Exposure in the Textile Manufacturing Industry. Lyon, France: IARC Press.

McIntyre, J.E., I. Holme, and O.K. Sunmonu. 1995. The desorption of model compounds from poly(ethylene terephthalate) fibre. Colourage 41(13)77-81.

Meijer, J., M. Rundgren, A. Astrom, J.W. DePierre, A. Sundvall, and U. Rannug. 1981. Effects of chlorinated paraffins on some drug-metabolizing enzymes in rat liver and in the Ames test. Adv. Exp. Med. Biol. 136(Pt. A):821-828.

Nilsen, O.G., and R. Toftgard. 1981. Effects of polychlorinated terphenyls and paraffins on rat liver microsomal cytochrome P-450 and *in vitro* metabolic activities. Arch. Toxicol. 47(1):1-11.

Nilsen, O.G., R. Toftgard, and H. Glaumann. 1980. Changes in rat liver morphology and metabolic activities after exposure to chlorinated paraffins. Pp. 525-528 in Mechanisms of Toxicity and Hazard Evaluation: Proceedings of the Second International Congress on Toxicology, Brussels, Belgium, July 6-11, 1980. B. Holmstedt, R. Lauwerys, M. Mercker, and M. Roberfrold, eds. New York: Elsevier/North Holland Biomedical Press.

Nilsen, O.G., R. Toftgard, and H. Glaumann. 1981. Effects of chlorinated paraffins on rat liver microsomal activities and morphology. Importance of the length and the degree of chlorination of the carbon chain. Arch. Toxicol. 49(1):1-13.

NRC (National Research Council). 1985. Emergency and Continuous Exposure Limits for Selected Airborne Contaminants, Volume 5. Committee on Toxicology. Board on Toxicology and Environmental Health Hazards, National Research Council. Washington, DC: National Academy Press.

NTP (National Toxicology Program). 1986a. NTP Technical Report on the Toxicology and Carcinogenesis Studies of Chlorinated Paraffins (C_{12}, 60% chlorine) in F344/N rats and B6C3F1 Mice (Gavage Studies). NTP TR 308. NIH Publication No. 86-2564. U.S. Department of Health and Human Services, National Institutes of Health.

NTP (National Toxicology Program). 1986b. NTP Technical Report on the Toxicology and Carcinogenesis Studies of Chlorinated Paraffins (C_{23}, 43% chlorine) in F344/N rats and B6C3F1 Mice (Gavage Studies). NTP TR 305. NIH Publication No. 86-2561. U.S. Department of Health and Human Services, National Institutes of Health.

Poon, R., P. Lecavalier, P. Chan, C. Viau, H. Håkansson, I. Chu, and V.E. Valli. 1995. Subchronic toxicity of a medium-chain chlorinated paraffin in rat. J. Appl. Toxicol. 15(6):455-463.

Scott, R.C. 1989. In vitro absorption of some chlorinated paraffins through human skin. Arch. Toxicol. 63(5):425-426.

Serrone, D.M., R.D. Birtley, W. Weigand, and R. Millischer. 1987. Toxicology of chlorinated paraffins. Food. Chem. Toxicol. 25(7):553-562.

Warngard, L., Y. Bager, Y. Kato, K. Kenne, and U.G. Ahlborg. 1996. Mechanistical studies of the inhibition of intercellular communication by organochlorine compounds. Arch. Toxicol. Suppl. 18:149-159.

Wyatt, I., C.T. Coutts, and C.R. Elcombe. 1993. The effect of chlorinated paraffins on hepatic enzymes and thyroid hormones. Toxicology 77(1-2):81-90.

Yang, J.J., T.A. Roy, W. Neil, A.J. Krueger, and C.R. Mackerer. 1987. Percutaneous and oral absorption of chlorinated paraffins in the rat. Toxicol. Ind. Health 3(3):405-412.

Appendix A

Biographical Sketches

Donald E. Gardner is president of Inhalation Toxicology Associates, Inc. He received his Ph.D. in environmental health from the University of Cincinnati. Dr. Gardner's research interests include environmental and occupational toxicology, immunotoxicology, pulmonary toxicology, and host defense mechanisms. He has worked in both the federal and private sectors, holding senior positions with the U.S. Public Health Service, the U.S. EPA, and industry. He was awarded the Society of Toxicology's 1992 Inhalation Toxicology Specialty Section Lifetime Achievement Award for Inhalation Toxicology, and he served as president of the society's inhalation specialty section, the metal specialty section, and the immunotoxicology section. He was also president of the North Carolina chapter of the Society of Toxicology. He is currently the chair of NRC's Subcommittee on Guidelines for Spacecraft Maximum Allowable Concentrations for Space Station Contaminants. Dr. Gardner is editor-in-chief of the journal *Inhalation Toxicology*, and is on the editorial board of *Toxic Substance Journal* and *Environmental and Nutritional Interactions*.

Joseph F. Borzelleca is emeritus professor of pharmacology and toxicology in the Department of Pharmacology and Toxicology at Virginia Commonwealth University, Medical College of Virginia. He received his Ph.D. in pharmacology from Thomas Jefferson University Medical College. His research interests include the toxicology of food additives, pesticides, water contaminants, and biodisposition of chemicals including drugs. Dr. Borzelleca is past president of the Society of Toxicology and past chair of the NRC Committee on Toxicology. He has served on several other NRC committees, including the Committee on Food Additives Survey Data, the Committee on Toxicology's Subcommittee on Disinfectants, and the Committee on Safe Drinking Water. He has also served on the editorial board of several scientific journals and on numerous national and international panels.

David W. Gaylor is associate director for Risk Assessment Policy and Research and acting associate director for Research at the National Center for Toxicological Research. He received his Ph.D. in statistics from North Carolina State University. Dr. Gaylor is actively engaged in statistical design and the analysis of toxicologic experiments and quantitative risk assessments. He is a former member of the Environmental Health Committee of EPA's Science Advisory Board and currently serves as a member of the NRC Committee on Toxicology. He is on the editorial boards of *Risk Analysis, Human and Ecological Risk Assessment,* and *Regulatory Toxicology and Pharmacology.*

Sidney Green is an associate professor in the Department of Pharmacology at Howard University. He received his Ph.D. in pharmacology in 1972 from Howard University, Washington, D.C. Dr. Green was director of the Division of Toxicological Research at the U.S. Food and Drug Administration in the Center for Food Safety and Applied Nutrition. He was also director of the Toxic Effects Branch in the Office of Toxic Substances at the Environmental Protection Agency. Dr. Green has expertise in genetic toxicology and systemic toxicology. He is currently a member of the National Research Council Committee on Toxicology, and serves on the Subcommittee on Military Smokes and Obscurants. Dr. Green is past president of the American College of Toxicology. He is on the editorial boards of the *Journal of Toxicology*, *Human and Ecological Risk Assessment*, and *Human and Experimental Toxicology*.

Richard Horrocks is dean of the Faculty of Technology and director of research at Bolton Institute in the United Kingdom. He received his Ph.D. in chemistry from the University of East Anglia. His research focuses on industrial fiber durability and degradation, textile-fiber flammability and flame retardancy. Dr. Horrocks is a member of the Association of Professors and Heads in Textiles, United Kingdom, and the Fiber Society, U.S.A. He is also a member of several British standards committees. He has served on several committees of the Textile Institute (Professional and Educational Affairs Committee, Accreditation Committee, Examination Board) and British Apparel and Textile Confederation (Technical Committee, Textiles). He is on the editorial board of *Fire and Materials* and serves on the Committee of the United Kingdom Fire Chemistry Discussion Group of the Society of Chemical Industry.

Michael A. Jayjock is a senior research fellow and manager of Risk Assessment at the Rohm and Haas Company. He received his Ph.D. from Drexel University in Environmental Engineering and Science. His principal research interest includes the development of predictive and cost-efficient exposure models. He has served on the Integrated Human Exposure Committee of the EPA Science Advisory Board, Canadian National Research Council, Chemical Manufacturers Association, International Society of Exposure Assessment, and American Industrial Hygiene Association. He has also served on the NRC Committee on Advances in Assessing Human Exposure to Airborne Pollutants. Dr. Jayjock is currently a member of the Exposure Assessment Strategies and Risk Assessment Committees of the American Industrial Hygiene Association and chair of the Technical Issues Panel on Exposure of the Generic Research Program of the Chemical Manufacturer's Association. He is on the editorial board of the *American Industrial Hygiene Association Journal.*

Sam Kacew is a professor in the Department of Pharmacology, University of Ottawa. He received his Ph.D. in pharmacology from Ottawa University. Dr. Kacew is a fellow of the Academy of Toxicological Sciences and recipient of the Velyien E. Henderson Award from the Society of Toxicology of Canada (1983), and the Achievement Award from the Society of Toxicology (1986). Dr. Kacew is a member of the Expert Advisory Committee, Canadian Network. Dr. Kacew is on the grants committee of the National Institutes of Health and has served on several national and international panels. He is editor-in-chief of the *Journal of Toxicology and Environmental Health* and an associate editor of *Toxicology and Applied Pharmacology*, and is on the editorial boards of several journals.

James N. McDougal is a senior scientist at Geo-Centers, Inc. and associate professor of pharmacology and toxicology at Wright State University. He was previously employed as director of the Toxicology Division of Air Force Armstrong Laboratory. Dr. McDougal received his Ph.D. in pharmacology from the University of Arizona. His research focuses on dermal absorption, pharmacokinetic modeling, and risk assessment. He has served on several scientific panels for EPA, ATSDR, the National Toxicology Program, and the Air Force. He is the recipient of two "Paper of the Year" awards from scientific societies.

Richard K. Miller is a professor of obstetrics and gynecology and environmental medicine at the University of Rochester. He is also Associate Chair of Obstetrics and Gynecology, and Director, Division of Research, School of Medicine and Dentistry. He received his Ph.D. in pharmacology and toxicology from Dartmouth Medical School and was a Fulbright (Distinguished Professor)

Fellow in 1988. He directs an NIEHS-funded New York Teratogen Information Service. His research interests, include vertical transmission of HIV-1 and reproductive and developmental toxicity of anti-HIV therapies, metal, and vitamins. Dr. Miller is past president of the Teratology Society. He served as chair of the Reproductive and Developmental Toxicology Program Review Committee, member of the Board of Scientific Counselors for the National Toxicology Program, and as chair of the Independent Scientific Advisory Committee on Retinoids. Dr. Miller has served on several NRC committees. Dr. Miller is currently the editor-in-chief of *Trophoblast Research* and serves on the editorial board of *Placenta*, *Proceedings of the American Society of Experimental Biology and Medicine*, and *Reproductive Toxicology*.

Robert Snyder is professor and chair of the department of pharmacology and toxicology at Rutgers University College of Pharmacy, and is associate director of the Environmental and Occupational Health Sciences Institute. He received his Ph.D. in biochemistry from the State University of New York at Syracuse. His research interests include solvent toxicology, chemically induced bone marrow depression, liver toxicity, and chemical carcinogenesis. Dr. Snyder was the recipient of the 1987 Rutgers University Board of Trustees Award for Excellence in Research, winner of the 1990-1991 Humboldt Research Award for Senior U.S. Scientists, and recipient of the 1996 Society of Toxicology Education Award. He is a councilor of the Academy of Toxicological Sciences and of the American College of Toxicology. Dr. Snyder served as president of the mechanisms section of the Society of Toxicology and a former president and vice president of the Middle Atlantic Society of Toxicology. He also served as chair of NRC's Committee on Alkyl Benzene Derivatives, chair of the COT's Panel on Psychochemicals, and as a member of several COT subcommittees. He currently serves as a member of the EPA's National Advisory Committee on Acute Exposure Guideline Levels. He is also currently the chair of the NRC's Subcommittee on Chronic Reference Doses for Selected Chemical Warfare Agents.

Gary C. Stevens is director of the Polymer Research Centre at the University of Surrey, United Kingdom. He is also managing director of the University of Surrey Environmental Body. Dr. Stevens received his Ph.D. in physics from the University of London. Dr. Steven's research includes work on the fire performance of materials including the mechanisms and role of flame retardants in resisting fire and reducing fire hazards and risks. Dr. Stevens has led two multidisciplinary research projects within the Polymer Research Centre on the risks and benefits of flame retardants and the effectiveness of the UK furniture fire regulations.

Robert G. Tardiff is president of Sapphire Group, Inc., a consulting group that focuses on hazard assessment, chemical interaction, risk assessment, and risk communication. He received his Ph.D. in toxicology and pharmacology from the University of Chicago. Previously, he was chief of the Toxicological Assessment Branch of the U.S. EPA and executive director of the NRC's Board on Toxicology and Environmental Health Hazards. Dr. Tardiff served as president of the Society for Risk Analysis and is an editor on a number of toxicology and environmental health journals.

Mary Vore is a professor of toxicology and Director of the Graduate Center for Toxicology at the University of Kentucky. She received her Ph.D. in pharmacology from the Vanderbilt School of Medicine. Dr. Vore's research focuses on the hepatic toxicity of chemicals. She has served on the NIH Pharmacology Study Section and the National Advisory Environmental Health Science Council, the NIEHS External Review Working Group, and as president of the Society of Toxicology's (SOT) Mechanisms Specialty Section. She is a member of the NRC Committee on Toxicology. She has also served on SOT's Board of Publications and is currently on the editorial board of *Drug Metabolism and Disposition*. She has previously served on editorial boards of *Toxicology and Applied Pharmacology*, *Environmental Health Perspectives*, and *Toxicology*.

Appendix B

Flame-Retardant Composition in Fabrics:

Their Durability and Permanence

FLAME retardancy may be conferred on textile fabrics by use of inherently flame resistant fibers, use of chemical after-treatments or both (Horrocks 1986, 1996, In press). Inherently flame retardancy may arise from a chemical structure which is thermally stable in the first instant or transforms to one (e.g., the polyaramids or other aromatic structures), incorporation of flame-retardant additives during the production of man-made fibers (e.g., FR viscose), or by the synthesis of conventional fiber-forming polymers which include flame retardant comonomers (e.g., FR polyester). Chemical after-treatments include surface or topical treatments, coatings and functional finishes which become a part of the final fiber structure. Table B-1 summarizes a selection of the current types available with selected examples.

The most durable to laundering and service are the inherently FR fibers in the first instance although additive leaching may be a problem. For chemically after-treated textiles the durability depends on the strength with which the formulation adheres or bonds to the fiber surface (including internal voids) and/or molecules. Some functional finishes are as durable as the fiber structure itself. Others, however, have a durability level, which is limited to specific end-use requirements. This is typically the case for furnishing fabrics which are

TABLE B-1 Durably-finished and Inherently Flame Retardant Fibers in Common Use

Fiber	Flame Retardant Structural Components	Mode of Introduction
Natural:		
COTTON	Organophosphorus and nitrogen-containing monomeric or reactive species e.g.. Proban CC (Albright and Wilson), Pyrovatex CP (Ciba), Aflammit P and KWB (Thor), Flacavon WP (Schill & Seilacher)	F
WOOL	Antimony-organo-halogen systems e.g., Flacavon F12/97(Schill & Seilacher), Myflam, (B F Goodrich, formerly Mydrin)	F
	Zirconium hexafluoride complexes, e.g. Zirpro (IWS); Pyrovatex CP (Ciba), Aflammit ZR (Thor)	F
Regenerated: VISCOSE	Organophosphorus and nitrogen/sulphur-containing species e.g., Sandoflam 5060 Clariant, formerly Sandoz) in FR Viscose(Lenzing); polysilicic acid and complexes e.g., Visil AP (Sateri)	A A
Inherent Synthetic:		C/A
POLYESTER	Organophosphorus species: Phosphinic acidic comonomer e.g., Trevira CS, (Trevira GmmbH, formerly Hoechst); phosphorus-containing additive, Fidion FR (Montefiber).	
ACRYLIC (modacrylic)	Halogenated comonomer (35-50% w/w) plus antimony compounds e.g., Velicren (Montefiber); Kanecaron (Kaneka Corp.)	C
POLY-PROPYLENE	Halo-organic compounds usually as brominated derivatives, e.g., Sandoflam 5072 (Clariant, formerly Sandoz)	A
POLYHALO-ALKENES	Polyvinyl chloride, e.g., Clevyl (Rhone-Poulenc) Polyvinylidene chloride, e.g., Saran (Saran Corp.)	H
High Heat and Flame Resistant (Aromatic):		Ar
POLY-ARAMIDS	Poly(m-phenylene isophthalamide), e.g., Nomex (DuPont), Conex (Teijin); poly(p-phenylene terephthalamide), e.g., Kevlar (DuPont), Twaron (Enka)	
POLY (ARAMID-ARIMID)	e.g., Kermel (Rhone-Poulenc)	Ar
POLYBEN-ZIMIDAZOLE	e.g., PBI (Hoechst-Celanese)	Ar
Key:		
F	Chemical finish	
A	Additive introduced during fiber production	
C	Copolymeric modifications	
H	Homopolymer	
Ar	Aromatic homo- or copolymer	

assumed to be either dry-cleaned or cleaned by localized sponging under aqueous conditions.

FLAME RETARDANT APPLICATION METHODS

Successful flame-retardant finishes are those which combine acceptable levels of flame retardancy at an affordable cost and are applicable to textile fabrics using conventional textile finishing and coating equipment. Figure B-1 attempts to present an overall summary of four basic processes shown schematically and as they would be used on open-width textile fabrics. Each relates to one or more of the examples of flame retardant finishes/treatments for cellulosic textiles identified above in Table B-2, wool finishes in Table B-3, and synthetic textile finishes in Table B-4. It is of interest to note that alternative application methods to padding may be used in processes i-iii such as foam application; padding perhaps represents the most commonly used technique. Each process, i-iv, relates to finish type as follows:

Process (i):

This simple pad/dry technique is applicable with most nondurable and water-soluble finishes such as the ammonium phosphates and similar finishes.

Process (ii):

This sequence is typical of those used to apply crease-resistant and other heat-curable textile finishes. In the case of flame-retardant finishes it finds best use for application of the phosphonamide systems such as Pyrovatex (Ciba), Afflamit (Thor) and the now obsolete Antiblaze TFR1 (Albright and Wilson, pers. commun., 1998) which are applied with resin components like the methylolated melamines. Because the process requires the presence of acidic catalysts (e.g. phosphoric acid), the wash-off stage will include an initial alkaline neutralization stage.

This same sequence without the washing-off stage may be used to apply semidurable finishes where a curing stage allows a degree of interaction to occur between the finish and the cellulose fiber; typical examples are ammonium phosphates which during curing at about 160°C give rise to phosphorylation of the cellulose. Thus the finish develops a degree of resistance to water soak and gentle laundering treatments.

Process (iii):

This is best exemplified by the THPC-based Proban process, which requires an ammonia gas curing process in order to polymerize the applied finish into the internal fiber voids. In this way the Proban CC condensate of tetrakis (hydroxy methyl) phosphonium chloride (THPC) and urea after padding and drying on to the fabric, is passed through a patented ammonia reactor which cross-links the condensate to give an insoluble polymeric finish. In order to increase the stability and hence durability of the finish, a subsequent oxidative "fixation" stage is required before finally washing off and drying.

Process (iv):

Back-coating describes a family of application methods where the flame-retardant formulation is applied in a bonding resin to the reverse surface of an otherwise flammable fabric. In this way the aesthetic quality of the face of the fabric is maintained while the flame-retardant property is present on the back or reverse face. Flame retardants must have an element of transferability from the back into the whole fabric and so they almost always are based on the so-called vapor-phase active antimony-bromine (or other halogen) formulations as typified by Myflam (B F Goodrich, formerly Mydrin) and Flacavon (Schill & Seilacher) products which comprise brominated species such as decabromodiphenyl oxide or hexabromocyclododecane and antimony III oxide (see Tables B-2 and B-4). Application methods include doctor blade or knife coating methods and the formulation is as a paste or foam. These processes and finishes are used on fabrics where aesthetics of the front face are of paramount importance such as furnishing fabrics and drapes.

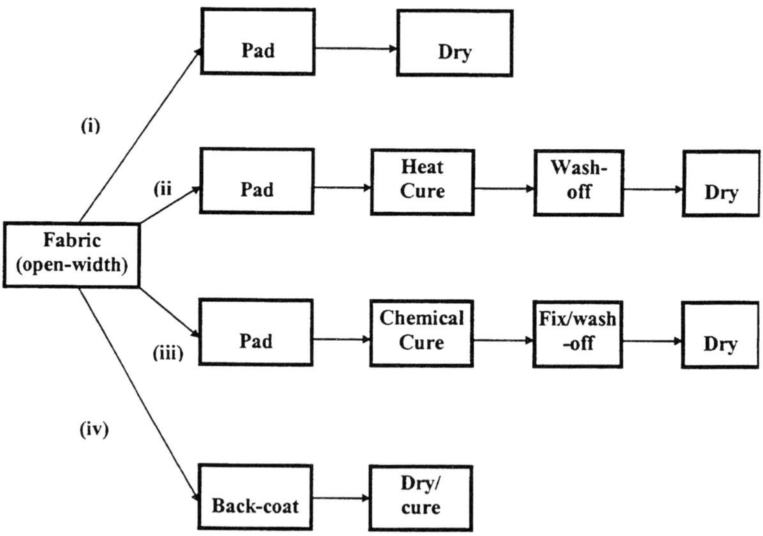

FIGURE B-1 Schematic representation of the most common methods of application of flame retardants to textile fabrics.

TABLE B-2 Commonly Available Flame-Retardant Finishes for Cotton

Type	Durability	Structure/formula
Salts :		
(i) Ammonium polyphosphate	Non- or semi durable (dependent on n)	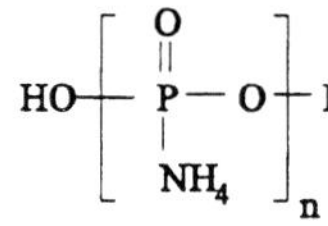
(ii) Diammonium phosphate	Non-durable	$(NH_4)_2HPO_4$
Organophosphorus :		
(i) Cellulose reactive methylolated phosphonamides	Durable to more than 50 launderings	$(CH_3O)_2\overset{O}{\overset{\Vert}{P}}.CH_2.CH_2.CO.N(H)(CH_2OH)$ e.g Pyrovatex CP (Ciba) Antiblaze TFR 1 (Albright & Wilson) Aflammit KWB (Thor)
(ii) Polymeric tetrakis (hydroxy methylol) phosphonium salt condensates	Durable to more than 50 launderings	THPC - urea - NH_3 condensate e.g. Proban CC (Albright and Wilson), Aflammit P (Thor)
(Back) Coatings :		
(i) Chlorinated paraffin waxes	Semi-durable	$C_nH_{(2n-m+2)} \cdot Cl_m$ e.g. Flacavon FK (Schill & Seilacher)
(ii) Antimony/halogen (aliphatic or aromatic bromine - containing species)	Semi- to fully durable	Sb_2O_3 (or Sb_2O_5) + Decabromodiphenyloxide or Hexabromocyclododecane + Acrylic resin e.g. Myflam (Mydrin) Flacavon F12 (Schill & Seilacher)

APPLICATION LEVELS AND FABRIC WEIGHTS

It is impossible to list accurately the application levels because of the diversity of formulations and fabric structures available. However, in order to pass the test requirements defined in BS5852:1979 for domestic furnishing fabrics in the UK for cigarette and simulated match sources (this latter is similar to the conditions in the proposed CPSC test method), worst possible levels of selected generic formulations may be presented. However, these are not the same as those necessary to pass the more stringent requirements of commercial furnishings in the UK but a safety factor of 1.5 could confidently be applied to the levels listed below if need be.

TABLE B-3 Non- and Semi-Durable[a] Flame-retardant Finishes for Wool and Wool Blends

Trade Name	Chemical Constitution	Durability
Albright & Wilson		
Antiblaze FSD	Ammonium polyphosphate (APP) + ammonium bromide	Dry cleaning
Antiblaze RD1	Ammonium salt of phosphonic acid	-
Antiblaze LR3	APP powder (30% w/w P) for use in coatings	Dry cleaning
Antiblaze LR4	APP powder (27% w/w P) for use in coatings	Dry cleaning
Ciba		
Flovan BU	Inorganic (ammonium?) halide (bromide?)	Dry cleaning
Flovan CGN	Ammonium acid phosphonate	-
Thor		
Flammentin ASN	Ammonium phosphate (APP or DAP?)	Dry cleaning
Flammentin HM	Ammonium salts (phosphates?); useful up to 30% wool in blends	Dry cleaning
Flammentin KRE	Organic phosphorus-nitrogen compound	-
Flammentin MCFC	Cross-linking silicone + P & N-containing compounds	40°C water soak, dry cleaning
Schill & Seilacher		
Flacavon RNEU	Organic P- and N-containing compound	Dry cleaning

[a]Semi-durable relates to durability to simple water soak and dry cleaning protocols as defined in appropriate standards.

Table B-5 lists these maximum levels assuming a range of fabric weights between 150 and 300 g (6-12 oz) and that lighter fabrics generally require higher treatment levels than heavier fabrics. Wholly thermoplastic fiber (e.g. polyester, polyamide)-containing fabrics require higher resin levels because of the need to provide a charring scaffold sufficient to maintain a barrier in spite of the melting fiber characteristics (see below). The list is restricted to those chemical species, which are most commonly used in the UK. However, it may be possible to use the behavior of these well-used chemical systems to model the behavior of little-used FRs given their basic chemical and physical properties (e.g. structure, solubility, volatility). Furthermore, the list does not include those FR chemicals which are used mainly in the plastics sector like alumina trihydrate, zinc hydroxide, zinc borate, etc., which find little or no use in the UK furnishing sector. These are excluded because they are effective only at

TABLE B-4 Durable[a] Finishes for Synthetic Fiber-Containing Textiles

Trade Name	Chemical Constitution/Comments
Albright & Wilson	
Antiblaze N/NT	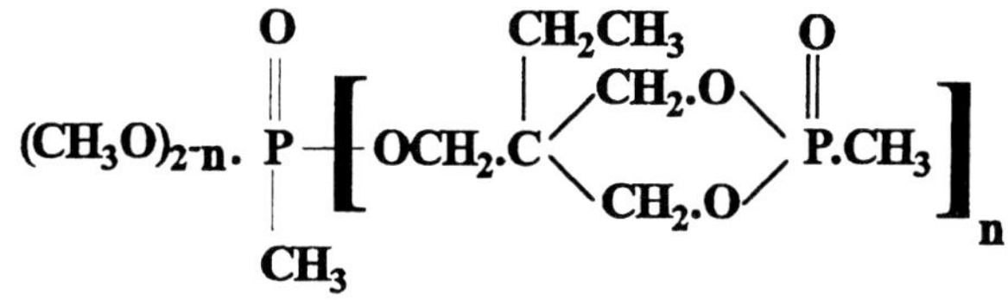 N=1; cyclic oligomeric phosphonate; pad-dry (110-135°C)-cure(185-200 °C) Primary use: polyester Secondary uses: polyamide, polypropylene
Thor	
Aflammit PE	As above for Antiblaze CU, polyester
Aflammit NY	Organic nitrogen and sulphur compound (probably a thiourea derivative) and a reactive cross-linking compound; polyamide. Cure at 150-170°C for 45-60s.
Schill & Seilacher	
Flacavon AM	Nitrogen and sulphur-containing compound (thiourea derivative?); polyamide; 100-110°C dry only; durable to dry cleaning
Flacavon AZ	Organic phosphorus compound (as for Antiblaze CU?); polyester
Flacavon H12/10	Organic phosphorus and nitrogen-containing compound (+ binder)
Flacavon H14/587	Antimony oxide + bromine compound (+ binder); all fibers especially polyester-cotton blends
Apex Chemicals (US)	
Apex Flameproof 334	Organohalogen compounds: polyester
Apex Flameproof 1510	Organohalogen compounds: polyester, polyamide
Emco Services (US)	
Flame Out PE-60	Organohalogen compounds: polyester
Flame Out PE-19	Cyclic phosphorus compound: polyester
Flame Out N-15	Organic nitrogen compound: polyamide
Glo-Tex International (US)	
Guardex PFR-DPH	Organohalogen-phosphorus compound: polyester, polyamide, polypropylene
Guardex FR-MEHN	Organophosphorus compound: polyamide
Sybron Chemicals (US)	
Flame Gard PE conc	Organophosphorus compound: polyester
Flame Gard 908	Organic nitrogen compound: polyamide

[a]Durable relates to levels associated with normal dry cleaning and laundering protocols as defined in appropriate standards.

TABLE B-5 Typical Maximum (Rounded) Application Levels on Furnishing Fabrics

Fabric/Fibers	Formulation/ Finish	Effective Element, %w/w Fabric	Total add-on of FR Agent, %	Add-on on 300 gsm (11 oz) Fabric
Cotton Viscose Cellulosic blends with wool, synthetics, etc.	Semi-durable P salt, e.g. ammonium polyphosphate	3% P	10% (25% as back-coating)	30 (75 as back-coating)
Cotton Viscose Cellulosic blends with wool, synthetics, etc.	THPC	3%P	15%	45
Cotton Viscose Cellulosic blends with wool, synthetics, etc.	Phosphonic acid derivative, e.g. Pyrovatex	2%P	12% (excl. resin componen t)	36
Wool and wool-rich blends	Zirconium salt	2.5% Zr	6%	18
All fabrics	Sb-Br/DBDPO	12% Br, 7% Sb	25%	75 (50 gsm DBDPO; 25 gsm Sb_2O_3)
All fabrics	Sb-Br/HEXA	12% Br, 7% Sb	25%	75 (53 gsm HEXA; 22 gsm Sb_2O_3)
All fabrics	Chloroparaffins	7% Cl	10%	30
Polyester and blends	Cyclic phosphonates	2%P (as 50% replacement for Sb/Br system)	12%	37
All fabrics	Aromatic phosphate plasti-cizer for back-coatings		30%	45

[a]All fabric/FR formulation combinations must be char-forming to pass BS5852, Sources 0 & 1, hence formulations active on 100% fusible synthetic fiber-containing fabrics (e.g. polyester, polyamide, polypropylene) are included at higher levels sufficient for acceptable char levels).

Note: The right hand column figures are "worst possible" figures based on the highest commercially used add-ons normally applied to the lightest most flammable or a synthetic fibre-containing fabric but here applied to a heavy 300 gsm fabric. In practice the heavier fabrics require the lowest add-ons and so a realistic add-on might be less than one half of the concentrations in column 5. In addition, the right hand column figures are area density dependent and so would be reduced on lighter fabrics proportionately although increased to compensate for the lighter structure. The former reduction would probably be greater than the latter and so a net decrease will follow in concentration per unit area of fabric as fabric weight decreases.

very high levels (with adverse affects on aesthetics), are ineffective on some fiber types and/or they cannot withstand the water-soak durability requirement.

It is important for the purpose of this study to emphasise that while fabric add-ons determine the overall FR performance, surface chemical concentrations are both add-on and area density dependent. Thus the same add-on on a heavy fabric compared with a lighter structure gives a much higher concentration of potentially extractable chemical per unit fabric area. Thus its toxicological risk factor will be greater.

THE SPECIAL CASE OF BACK-COATING

Antimony-halogen flame retardants are currently the most successful within the back-coated textile areas based on cost and effectiveness. Unlike the fiber-reactive, durable phosphorus- and nitrogen-containing flame retardants used for cellulosic fibers, they can only be applied topically in a resin binder, usually as a back-coating. For textiles, most antimony-halogen systems comprise antimony III oxide and bromine-containing organic molecules such as decabromodiphenyl oxide (DBDPO) or hexabromocyclododecane.

Depending on the nature of the resin binder, often an acrylic copolymer or ethylene-vinyl acetate copolymer, these coated systems may have some char-forming character. This enables them to be used successfully on synthetic fiber—containing furnishing fabrics, for example, which must have a means of counteracting the effects of fiber thermoplasticity if they are to pass composite tests such as BS5852, ISO 1891/2, EN 1021, etc. Thus the percentage resin component may be as high as 60-70% (w/w) of the total add-on, which itself may be as high as 50% (w/w) on fabric, if the char formed from it has to support the melting fibers in the fabric to which it is attached. Examples would be 100% polyester and polyamide fabrics and the highest add-ons would be on lightweight (<200-g) fabrics.

In addition to the char-forming character and possible flame retardancy of vinyl chloride-containing copolymers, resins must be sufficiently hydrophobic to enhance the durability of flame-retardant additives that have low solubilities, and they must have T_g values low enough to maximize softness and handle.

DURABILITY AND LEACHING BEHAVIOUR

Durability and leachability are two different but complementary terms. The former is considered with respect to the flame retardant and its ability to maintain an acceptable level of flame-retardant behavior during the lifetime of the textile.

Thus, durability is determined by

- launderability, aftercare and defined cleansing requirements,
- weatherability, and
- exposure to light, heat, and atmospheric agents, usually together, present in indoor environments.

There are considerable data regarding specific FR durability. Because most textile performance standards require a specified cleansing treatment prior to FR testing, formulations often select themselves for specific applications. The most durable FR treatments are those where there is reaction with or polymerization within the fiber structure. Less durable treatments are often surface treatments or coatings, which require the presence of binding agents or resins.

Leachability, however, refers to the removal of flame retardants, their degradation products and associated materials (e.g. resin components and plasticisers) with respect to a solvent medium. The fate and effects of such leached materials are more obviously important in determining their subsequent toxicological and ecotoxicological properties. It must be remembered that while quantitative data is not generally available, it is known that leached materials are often different from the applied flame-retardant chemistry. For example, leached FR-related materials may be

- soluble surface-located forms of the reacted FR system,
- hydrolysis products which may closely resemble the applied flame retardant prior to its fixation or curing,or
- degradation products.

In addition, these products may be temperature dependent in terms of rate of formation (e.g. hydrolysis, degradation) and relative rates of release. Thus elevated temperature extracted material types and concentrations using normal leaching or laundering test conditions (which are rarely less than 40°C) may not extrapolate easily to lower temperatures associated with ambient room or external body temperatures. Thus, while data exists with regard to the ease of removal of FRs to a variety of agencies, the chemical characteristics of removed materials are little known.

When durability data are available, rates of removal of some FR-related materials may be estimated, given a number of assumptions. Ecotoxicological risks may then be calculated by correlating the estimated rates of removal with known data, i.e., no-observed-adverse-effect levels (NOAELs), lowest-adverse-effect-levels (LOAELs), oral reference doses (RfDs), and inhalation reference concentrations (RfCs).

Case 1

Removal of THPC-related Materials from FR Cotton

Consider a heavy 300-g (11oz) THPC-treated (e.g. Proban) cotton upholstery fabric containing about 15% (w/w) finish (see Table B-5) equivalent to a phosphorus content of 3% (w/w). Note that in practice, to pass the UK furnishing fabric regulations, a level of 2% phosphorus would be more appropriate for such a heavy fabric.

Assumptions:

i. Removal in aqueous environments is essentially a hydrolytic process, which regenerates the THP moiety as TPHOH.
ii. Laundering data for loss of phosphorus may be used to calculate losses at body temperature (40°C) in aqueous contact.
iii. An area for infant sucking contact is $100 cm^2$.

Experimental data (Horrocks et al. 1992) for a number of 75 °C wash cycles with a variety of hot water and detergent systems, shows that 50 cycles removes about 0.5% phosphorus from fabrics. If each total cycle, including rinses, is of about 1 hr duration, and if a reduction in the temperature by 8°C is assumed to halve the rate of removal (Arrhenius Law), then

50 hr at 75 °C ≡ 1600 hr at 35 °C
≡ 0.5% phosphorus loss from fabric
≡ 2.5% loss of THP finish (see Table B-5).

For a 300-g (11 oz) fabric, 2.5% THP finish loss is equivalent to release of 7.5 g THPOH during 1600 hr of exposure to an aqueous solution at 35 °C. Thus,

7.5 g in 1600 hr ≡ 4.7 mg THPOH removed in 1 hr at 35 °C
≡ $0.047 mg/100 cm^2/hr$ at 35 °C.

If it is assumed that external body contact temperatures are lower than 35 °C, e.g. by a further 8 °C increment to give 27 °C, then the extraction rate will be halved again to $0.023 mg/100 cm^2/hr$ at 27 °C. This corresponds to a fractional loss rate of 0.0013/d.

Case 2

Removal of Phosphonic Acid Derivative-Related Materials from FR Cotton

We can consider a phosphonic acid derivative-finished (e.g. Pyrovatex) 300-g fabric in a similar manner as above except that the initial level of application is about 12% FR (see Table B-5) to give a phosphorus level of 2% (w/w) (although 1.5% would be closer to the commercial situation for such a heavy fabric). Laundering experiments show that the same conditions remove about 0.25% (w/w) phosphorus. This corresponds to 1.5% solubilized phosphonic acid derivative. Using the above arguments for the THPC-treated fabric, this relates to a loss rate at 35°C of 0.028 mg/100 cm^2/hr and at 27°C, 0.014 mg/100 cm^2/hr. This latter corresponds to a fractional loss rate of 0.0009/d.

Case 3

Ambient Atmospheric Hydrolysis and Solubilization of Cured Phosphonic Acid Derivatives Applied to Cellulosics

The long-term ambient hydrolysis of these compounds has been estimated to be at the rates: 10% over 2 years, 20% over 4 years and 40% over 8 years (Ciba-Geigy circa 1994, unpublished material). This could be assumed to be approximately linear at 5% (w/w) loss per annum with respect to the original concentration applied.

If we accept a cured formulation which both is typical and contains one of the highest levels of condensed formaldehyde, then given this figure, we can calculate

i. The annual release of soluble Pyrovatex precursor within the fabric for a range of typical fabric weights and application rates, and
ii. The rate of formaldehyde evolution associated with this release.
iii. A typical formula for a cured Pyrovatex-type of molecule applied with a trimethylolmelamine resin and bridged to the cellulose molecule is

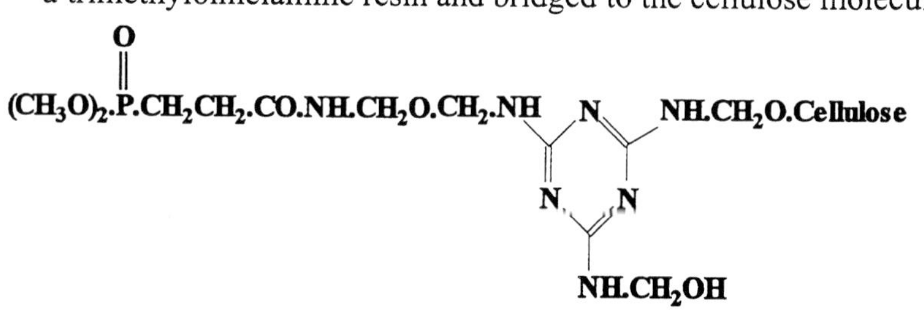

Cured substituent MW = 396.
Pyrovatex component MW =214.
Formaldehyde released is 1 mole per hydrolytic scission (MW = 30).

Rate of Solubilization

A typical 280-g (10 oz) fabric with a Pyrovatex or equivalent content giving 2% (w/w) phosphorus and solids content of 12% (w/w) (ignoring the associated bridging resin) equates to 33.6 g solids per square meter. At 5% (w/w) hydrolysis per year, 1.7 g is solubilized which (if not removed during laundering), yields 8.5, 17 and 25.5 g solubilized material over 5, 10, and 15 year periods respectively. For a 100-cm^2 model exposure area, this corresponds to masses of extractibles of 0.08, 0.17, and 0.255 g respectively. These amounts could represent maximum daily intakes for a single contact for a child sucking this area. Higher areas of contact (e.g. 2,200 cm^2 for a "body contact" exposure) could yield proportionately higher levels of extractable, although complete removal by one exposure would be difficult to imagine unless the body in contact is wet.

Rate of Evolution of Formaldehyde

Assuming that 396 g of hydrolyzed material gives rise to 30 g of formaldehyde (HCHO) (ignoring evolution from any concurrent hydrolysis of associated resin component), then each year the 280-g fabric will generate (30/396) × 1.7 = 0.129 g of HCHO per square meter. This equates to a fabric content rising to 516 ppm per square meter assuming no loss by volatilization. Volatilization should be quite efficient and could give rise to an average release of 1.4 ppm per day HCHO per square meter. If a typical suite of furniture comprises 30 square meters of fabric, then it may release (30 × 0.129)/365 = 0.011 g of HCHO/d.

If this is released into an average bedroom or living room (4 × 4 × 3 = 48 ≅ 50 m^3, then a maximum air concentration of 0.16 ppm HCHO will occur each day.

REFERENCES

Albright & Wilson Ltd., UK. 1998. Personal communication to A R Horrocks.

Ciba-Geigy. circa 1994. Pyrovatex CPnew, Durability of the Flame Retardance and Care of the Goods, Technical manual. (Unpublished material). Pfersee, Germany.

Horrocks, A.R. 1986. Flame retardant finishing of textiles. J. Soc. Dyers Colour. 16:62-101.

Horrocks, A.R. 1996. Developments in flame retardants for heat and fire resistant textiles—the role of char formation and intumescence. Polym. Degrad. Stab. 54(2/3): 143-154.

Horrocks, A. R. In press. Flame Retardant Finishes and Finishing in Textile Finishing. D. Heywood, ed. Society of Dyers and Colourists, Bradford, England.

Horrocks, A.R., J. Allen, S. Ojinnaka, and D. Price. 1992. Influence of laundering on durable flame retarded cotton fabrics. 1. Effect of oxidant concentration and detergent type. J. Fire Sci. 10(4):335-351.